Environmental Toxicology

Environmental Topics

A series of books edited by J. Rose

Volume 1

Environmental Health

The Impact of Pollutants

Edited by J. Rose

Volume 2

Environmental Concepts, Policies and Strategies

Edited by J. Rose

Volume 3

Water and the Environment

Edited by J. Rose

Volume 4

Acid Rain

Current Situation and Remedies

Edited by J. Rose

Volume 5

Human Stress and the Environment

Edited by J. Rose

Volume 6

Water Quality for Freshwater Fish

Further Advisory Criteria

Edited by Gywneth Howells

Volume 7

Environmental Toxicology: Current Developments

Edited by J. Rose

This book is part of a series. The publisher will accept continuation orders which may be cancelled at any time and which provide for automatic billing and shipping of each title in the series upon publication. Please write for details.

Environmental Toxicology

Current Developments

Edited by

J. Rose

University of Central Lancashire, Preston (UK)

GORDON AND BREACH SCIENCE PUBLISHERS
Australia • Canada • China • France • Germany • India • Japan • Luxembourg
Malaysia • The Netherlands • Russia • Singapore • Switzerland • Thailand

Printed in India.

Amsteldijk 166
1st Floor
1079 LH Amsterdam
The Netherlands

British Library Cataloguing in Publication Data

Environmental toxicology : current developments.
(Environmental topics ; v. 7)
1. Environmental toxicology
I. Rose, J. (John), 1917–
571.9'5

ISBN: 90-5699-140-X

Printed and Bound at Ajanta Offset, New Delhi, India

To my nephew

Dr Eliezer Rozenbaum, as a mark of my admiration for his valuable work for his patients.

Contents

Preface

This collection of 18 chapters from experts in various areas of environmental toxicology addresses a number of topics of considerable relevance to our environment and well-being. It is impossible to deal with all aspects of the subject in the confines of one volume, hence this book focuses on a number of select and important themes in modern toxicology. The topics range from arsenic to pesticides and volcanic gases.

The first chapter of the book, "General Principles of Toxicology", an introduction to the subject, endeavours to give a balanced view of old and new knowledge of toxicology that disregards various fashionable trends that appear and disappear on the tide of new discoveries. The authors analyse the important principles of toxicology, viz. dose-response which states that there is a toxic dose and a safe dose for every chemical. It is stressed that exposure includes both dose and time. Some challenging questions are raised by the authors which may not please some toxicologists. However, out of controversy grows progress!

Another challenging topic is dealt with in Chapter 2 entitled "Environmental Pollution, Neurotoxicity and Criminal Violence". The author assumes that neurotoxic metals absorbed in the brain can affect its functions, when pollution interacts with poverty, poor diet, alcohol, drug abuse and social stress. Professor Masters tests this hypothesis against available data and finds extensive support for his assumption.

The next chapter by Drs McGregor and Partensky addresses the vital subject of carcinogens and mutagens. It is possible that two-thirds of cancer deaths in the USA can be linked to diet, smoking, obesity and lack of exercise; a similar pattern exists in other Western industrialised countries. The authors also consider other factors, such as viruses and digenean trematodes. While occupational carcinogens have a relatively low community aspect, the risk of developing cancer may be quite large among relatively small groups of occupationally exposed people. The whole matter is subjected to an up-to-date rigorous analysis.

Chapters 4 and 5 are concerned with geochemistry. The first by Professor F. R. Siegel addresses metal toxins and development planning in the context of earth materials. A large number of metals are studied, ranging from arsenic to zinc, in the context of pollution and health. The matter of element mobility is analysed in depth, since this may lead to a more effective environmental clean-up and remediation. Co-operation between geochemists and medical/toxicological researchers would lead to meaningful research in this area. This chapter is followed by that on the effect of toxic volcanic gases on the European environment, especially those in the eruptions in Iceland and Italy. Some interesting historical facts are presented by the author who suggests that volcanic gases have the potential to intensify the environmental impact of pollution episodes which are primarily of an anthropogenic origin.

The problem of pesticides is examined in the next two chapters. The first by Pimentel *et al.* addresses the problem of the economic and environmental costs of pesticide use. They stress that most benefits of pesticides are based on direct crop returns and do not include the indirect environmental and economic costs associated with their use. These effects are analysed by the authors of the chapter. In the USA alone the environmental and social costs of pesticide use amount to a staggering $8.3 billion p.a.! Hence it is essential to examine this matter in detail in order to develop and implement a balanced policy of using these chemicals.

A somewhat different approach is adopted by Lappé in the second chapter of this group dealing with the effects of exposure to pesticides. The author analyses these secondary effects in the context of the regulatory climate. It is emphasised that programmes ought to be developed to assess health risks for pesticides around themes of common toxicity.

Chapter 8 considers an important class of environmental pollutants, viz. chlorinated fatty acids. The authors deal with the physiological effects of such compounds, many of them related to reproduction processes. In view of the complexity of this subject, the authors advocate a great deal of research work into the origin, effects and properties of these pollutants.

A particular case of pollution due to arsenic is considered in Chapter 9. The major concern of the chapter relates to chronic exposure to potentially carcinogenic species. The author, Dr Rossman, reviews the toxicology and metabolism of arsenic compounds, and considers in some detail their molecular and genotoxic effects. It is found that human cells are much more sensitive to arsenites than those of rodents. The analysis of these differences may provide a clue why arsenic compounds are carcinogenic to humans and not to rodents.

Chapter 10 continues with the theme of pollution and health. In particular, the author considers urban air pollution and health. Epidemiological techniques for investigating this matter are described. The author describes the difficulties involved in such studies and the various factors involved. Research programmes in this area at the University of Edinburgh (UK) are examined in order to illustrate the general issues under discussion.

The effect of toxic substances on the development of diseases in aquatic organisms is reviewed in Chapter 11. The author shows that although there is evidence to support the presence of harmful substances in the aquatic environment, the link to disease is unclear. Harmful chemicals, especially heavy metals, may accumulate in aquatic organisms and lead to immunosuppression, reduction in metabolism and damage to gills and epithelia in fish. The author expresses scepticism over generalised statements concerning the role of polluted seas and their effect on disease. He advocates more definitive experiments in order to establish the truth.

A similar suggestion is delineated in Chapter 12 dealing with toxicology in the working environment. The review looks at the range of substances used in occupational settings and the potential hazards to workers. The need for more relevant research is stressed in the area of toxicology of substances, coupled with an international adoption of precautionary policies and standards in order to construct wide toxicity margins. The role of such measures is advocated because of large gaps in knowledge and suitable techniques.

Chapter 13 is a part of four chapters (13–16) concerned with various toxins found in nature. D'Mello and Macdonald deal with the subject of fungal toxins and their role as disease elicitors. Fungi synthesise many secondary metabolites that are known to be toxic to plants, animals and humans. These mycotoxins have been implicated in many animal and human disorders, particularly in relation to their carcinogenic, neurotoxic, hepatotoxic and immunosuppressive effects. This problem needs strict regulations and research. It appears that attempts to exploit disease-resistant plant genotypes represent a promising strategy for the reduction in mycotoxin contamination of primary food commodities.

Hanley and MacDonald continue with the theme of toxins by discussing natural plant toxins, their benefits and risk (Chapter 14). In the authors' view, natural plant toxins comprise a vast number of compounds. For some members of the class, there are beneficial and also harmful effects. In particular, diet related disease is a clear indication that altering exposure to natural plant toxins could have a significant effect on health and fitness.

In Chapter 15 the authors continue with the theme of plant-associated toxins in the human food supply. They believe that low level exposure of humans to such toxins present in the diet is of major concern. As in some previous chapters by other authors, it is stressed that levels of actual exposure and the possible health effects are usually unknown. Hence research is essential to obtain the necessary information on bioavailability, pharmacology and toxic bioactivities of low level plant-associated poisonous chemicals.

The last chapter (16) in the series dealing with plant and other toxins addresses the case of toxic cyanobacteria blooms, which often lead to fatal animal poisoning. The chapter reviews the problem of toxic cyanobacteria, citing many examples. The author discusses several management strategies and then the role of modelling in future bloom management.

Chapter 17 is a general paper concerned with the use of stable isotope ratios in freshwater and marine biomagnification studies. In fact, trophic interactions make it very difficult to accurately determine an organism's food web position and, thus, to understand the biomagnification of persistent contaminants through freshwater and marine food webs. The chapter reviews the recent use of stable isotope ratios to characterise food web interactions, especially in regard to mercury and organochlorines. This technique appears to be of importance in obtaining valuable information on the contribution of different food sources to a predator's contaminant burden.

The last chapter (18) addresses the problem of risk assessment. The authors advocate the use of a tiered approach in order to avoid wasted effort. This involves a series of appraisals in which the intensity and detail are increased till a definitive assessment can be made. In their view, properly validated models can play an important part in risk assessment. An outline of a risk assessment scheme is described in regard to losses of pollutants from agricultural land to water, weather, soil and land use.

It is clear that the 18 chapters of this multi-authored book cover a large range of up-to-date and valuable topics relating to health. Extensive references provided at the end of each chapter are an invaluable source of advances in the field of toxicity in the environment. Hence, the book will be of interest and importance to toxicology experts, public health personnel, medical researchers and practitioners, biologists, botanists, food scientists, technologists in various industries, agriculturists and environmentalists of various hues and shapes. The book could also serve as a textbook for undergraduate and postgraduate students in relevant disciplines. It is hoped that the work will induce its readers to pursue the themes presented in the various chapters and thus make a signal contribution to plant, animal and human health.

Finally, I would like to thank the various contributors for their hard and valuable work, the many referees for their patience and assistance and the Publishers, Gordon and Breach, for their support and goodwill.

J. Rose

List of Contributors

ADDISCOTT, T. M.	IARC, Rothamsted, Harpenden, Hertfordshire AL5 2JQ, UK
AUSTIN, B.	Department of Biological Sciences, Heriot-Watt University, Riccarton, Edinburgh EH14 4AS, UK
BASHORE, T.	College of Agriculture and Life Sciences, Cornell University, Ithaca, New York 14853-0901, USA
BEVERLAND, I. J.	Department of Public Health Sciences, University of Edinburgh, Medical School, Teviot Place, Edinburgh EH8 9AG, UK
BJÖRN, H.	Department of Ecology, Lund University, Ecology Building, S-223 62 Lund, Sweden
COLEGATE, S. M.	Commonwealth Scientific and Industrial Research Organisation, Division of Animal Health, Australian Animal Health Laboratory, Private Bag 24, Geelong, Victoria, Australia 3220
D'MELLO, J. P. F.	Department of Crop Science and Technology, The Scottish Agricultural College, West Mains Road, Edinburgh EH9 3JG, UK
DOSHI, A.	Department of Government, Dartmouth College, Hanover, New Hampshire 03755, USA
DOULL, J.	Department of Pharmacology, Toxicology and Therapeutics, University of Kansas Medical Center, Kansas City, Kansas 66160, USA
EDGAR, J. A.	Commonwealth Scientific and Industrial Research Organisation, Division of Animal Health, Australian Animal Health Laboratory, Private Bag 24, Geelong, Victoria, Australia 3220

GRATTAN, J. P.	Institute of Earth Studies, University of Wales, Aberystwyth SY23 3DB, UK
GREINER, A.	College of Agriculture and Life Sciences, Cornell University, Ithaca, New York 14853-0901, USA
HANLEY, A. B.	CSL Food Science Laboratory, Norwich Research Park, Colney, Norwich NR4 7UQ, UK
HONE, B.	Department of Government, Dartmouth College, Hanover, New Hampshire 03755, USA
HOWARD, A.	Aquatic Environments Research Centre, University of Reading, Reading, Berkshire, UK
KIDD, K. A.	Freshwater Institute, 501 University Cresent, Winnipeg, Manitoba, Canada, R3T 2N6
LAPPÉ, M.	Center for Ethics and Toxics, 39175 S. Highway One, P.O. Box 673, Gualala, CA 95445, USA
MASTERS, R. D.	Department of Government, Dartmouth College, Hanover, New Hampshire 03755, USA
MACDONALD, A. M. C.	Department of Crop Science and Technology, The Scottish Agricultural College, West Mains Road, Edinburgh EH9 3JG, UK
MACDONALD, S. J.	CSL Food Science Laboratory, Norwich Research Park, Colney, Norwich NR4 7UQ, UK
MCGREGOR, D.	International Agency for Research on Cancer, 150 cours Albert Thomas, 69372 Lyon 08, France
PARTENSKY, C.	International Agency for Research on Cancer, 150 cours Albert Thomas, 69372 Lyon 08, France
PIMENTEL, D.	College of Agriculture and Life Sciences, Cornell University, Ithaca, New York 14853-0901, USA
ROSSMAN, T. G.	The Nelson Institute of Environmental Medicine and the Kaplan Comprehensive Center, New York University Medical Center, 550 First Avenue, New York, NY 10016, USA
ROZMAN, K. K.	Department of Pharmacology, Toxicology and Therapeutics, University of Kansas Medical Center, Kansas City, Kansas 66160, USA
SIEGEL, F. R.	Department of Geology, George Washington University, Washington D.C. 20052, USA

SMITH, P. — IARC, Rothamsted, Harpenden, Hertfordshire AL5 2JQ, UK

STEGELMEIER, B. L. — Commonwealth Scientific and Industrial Research Organisation, Division of Animal Health, Australian Animal Health Laboratory, Private Bag 24, Geelong, Victoria, Australia 3220

WATTERSON, A. — Centre for Occupational and Environmental Health, De Montfort University, Scraptoft Campus, Scraptoft, Leicester LE7 9SU, UK

1. GENERAL PRINCIPLES OF TOXICOLOGY

KARL K. ROZMAN [*,†] and JOHN DOULL [*]

WHAT IS TOXICITY?

Toxicology traditionally has been defined as the science of the study of qualitative and, more importantly, quantitative aspects of injurious effects of chemicals and physical agents in a subject or in a population of subjects. Paracelsus had already recognized nearly five hundred years ago that there is no such thing as non-poisonous and that the dose alone makes a poison not to be poisonous. Even endogenous body constituents and food stuffs can be deleterious to an organism if present in excessive quantities over prolonged periods of time. Thus, in addition to the dose, time is the second important variable that the science of toxicology deals with.

What then is toxicity? It is the accumulation of injury over short or long periods of time, which renders an organism incapable of functioning within the limits of adaptation. Therefore, a more appropriate definition of the scope of toxicology would be that it is the science that elucidates the causality chain of interactions and their time course (exposure) between biological entities (subjects) of different intrinsic susceptibility and chemical and physical agents of different intrinsic potency. Thus, modern toxicology determines in a broader sense exposure-responses consisting of dose- and time-responses thereby establishing practical thresholds which define the safety of chemicals.

EXPOSURE

This is a qualitative notion, which in itself has no toxicological connotation. As a liminal condition, entry of a single molecule into an organism represents exposure. The major portals of entry into higher organisms are per os, via the lungs and through the skin. However, any surface area is a potential site of exposure, e.g. the eyes. Artificial portals of entry can be created, e.g. by intravenous, intramuscular or subcutaneous injections. The duration and/or frequency of exposure to chemicals is one critical, quantitative aspect of toxicology. It deals with the time course of absorption, distribution, biotransformation and excretion of chemicals. Its quantitative treatment is termed pharmacokinetics or more properly toxicokinetics. The above mentioned factors are critical determinants of the residency time of a toxic agent in an organism, which in turn represents one of the two time scales of toxicology to be discussed.

EFFECTS

Similar to the notion of exposure, effect is a qualitative designation. It follows from the definition of toxicity that any biological manifestation beyond the limits of

* Department of Pharmacology, Toxicology and Therapeutics, University of Kansas Medical Center, Kansas City, Kansas 66160 (USA).
† Section of Environmental Toxicology, GSF-Institut fur Toxikologie, Neuherberg, 85758 (Germany).

adaptation for a sufficiently long period of time is an adverse effect. The manifestation of an adverse or toxic effect is the second time-dependent function of toxicology because it is determined by the reversibility/irreversibility of the injury. For example, the high affinity binding of carbon monoxide to hemoglobin leads to rapid asphixiation once a critical carrier capacity reduction for oxygen has been reached. However, moderately elevated blood pressure will not result in frank injury until after years of its persistence. It needs to be pointed out that this second time scale (pharmacodynamic or more correctly toxicodynamic time scale) is seldom, if ever, identical with the toxicokinetic time scale. This would be the case only if a toxic interaction in an organism would be instantaneously and entirely reversible with the disappearance of the causative agent.

The time course of manifestation of toxic effects and their reversibility has been widely studied. Traditionally, toxicologists conducted acute, subchronic and chronic experiments. Acute studies usually involve single doses with an observation period of 14 days. Repeated (usually daily) doses are administered to experimental animals for up to 90 days in a typical subchronic study. Any repeated dose administration of chemicals for periods longer than 3 months has been considered a chronic study. The carcinogenicity bioassay was initially designed to last for the life-time of the experimental animals. Later it was limited to 2 years. As some readers of this chapter may recognize, the terms "acute", "subchronic" and "chronic" are qualitative or at best, semi-quantitative epithets. However, from the theoretical point of view there are more important quantitative criteria that define a toxic effect as will be shown in subsequent subheadings.

DOSE-RESPONSE

$(\partial E/\partial c)_t$ is a mathematical formulation of what Paracelsus first recognized as a clear quantitative causality link between exposure and toxic effect about 500 years ago. A modern toxicological interpretation of his writings is that there is nothing that is non-toxic. It is the dose (the amount of a chemical) that determines the extent of an adverse effect or the lack thereof. Haber[1] formulated an, as yet, incomplete quantitative relationship between dose and time to effect. Here it suffices to state that there are three critical factors in toxicology. These are the dose, the toxicokinetic time scale and the toxicodynamic time scale. Since these factors are linked to each other the only meaningful way to study any one of them is by keeping the two others constant. Therefore, a strictly defined dose-response study requires that it be conducted at a constant toxicokinetic time scale (=steady state) and at constant time to effect. Many studies conducted in the past satisfy these criteria, but the majority probably does not. If a compound has a half-life of several hours and the time to manifestation of effect is in the order of minutes then a single dose experiment is largely valid. However, if both occur on the time scale of hours then the experiment does not fulfill these criteria since the toxicokinetic time scale is not constant, viz. a significant proportion of the toxicant will be eliminated during the time to effect. Another example of a valid dose-response study is a carcinogenicity bioassay administering daily doses of a compound with a half-life of a day or two for two years and then sacrificing all animals. In this instance the two time scales are kept constant.

In general, there are three types of dose-responses as depicted in Fig. 1. Reactive dose-responses *per se* are not considered toxic effects because they represent a reaction of the organism to the entry of a noxious agent with the aim of countering

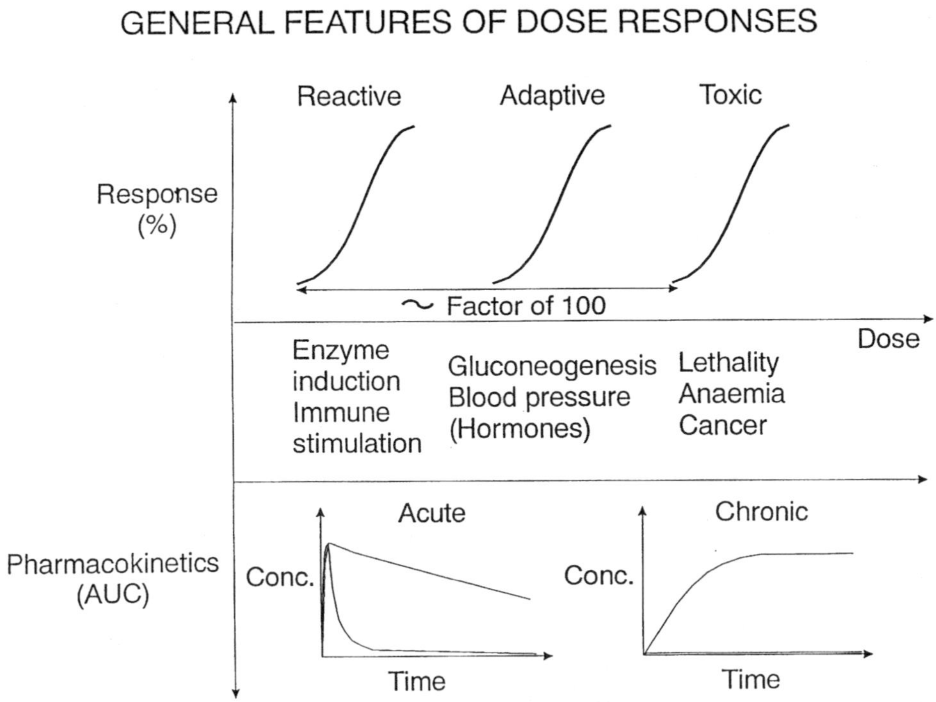

Figure 1 The upper panel depicts schematically the three different types of dose-responses that toxicology deals with. The lower panel illustrates the importance of the toxicokinetic time scale for the manifestation of toxicity.

its adverse effect by enhancing its elimination through enzyme induction or neutralizing its effect due to production of antibodies. In rare instances these—*per se*—beneficial effects can become detrimental (toxic), e.g. through production of more toxic metabolites or sustained stimulation of the immune system leading to autoimmune disease.

Adaptive dose-responses are also not *a priori* detrimental to an organism as they represent normal physiological responses to maintain homeostasis. For example, a chemical may produce changes in LH/FSH blood levels. If this compound has a short half-life and exposure is limited to a single or a few doses there may be no adverse consequences as transient changes in LH/FSH levels are also part of normal cycling. However, if the compound in question has a long half-life, it will result in persistent changes in LH/FSH levels resulting in ovulation arrest, which must be viewed as a toxic effect. Similarly, a chemical with a short half-life affecting gluconeogenesis may be harmless if given as a single dose, but could cause profound disturbance in glucose homeostasis if administered repeatedly leading to an affective steady state concentration.

Increasing the dose (and/or time) will eventually lead to toxic responses with all compounds. One compound can give rise to several or many dose-responses in any of the three dose-ranges but particularly in the toxic dose-range. As a result one may obtain overlapping dose-responses, which in practice can lead to truncated dose-responses. For example, if hemorrhage and anemia are both caused by a chemical in a comparable dose range then any given experiment in a small animal population will yield incomplete (or truncated) dose-responses for both effects.

Different chemicals in turn will give rise to different dose-responses which may or may not be parallel. Parallel dose-responses of various chemicals resulting in the same end point of toxicity usually imply the same mechanism of action. The position of dose-responses to each other provides a measure of the relative potency of the compounds in question. Chemicals exerting their effect by the same mechanism of action are either agonists or antagonists. These interactions occur according to the Mass Action Law. If the dose-responses of two or more chemicals are not parallel then their relative potency cannot be characterized by a single number. Non-parallel dose-responses are indicative of different modes of action. No interaction is likely to occur between such compounds. In rare instances non-parallel dose-responses can result in synergistic (more than additive) effects. The cause of synergism is thought to be an additional interaction at the same target cell or molecule. The role of synergism in toxicology has been exaggerated. There are few synergistic interactions and they usually have a toxicokinetic origin.

The issue of U-shaped or hormetic dose-responses has of late received considerable attention. This is the experimental finding that low doses of chemicals have the opposite affect of high doses. There is a significant amount of data to support this notion[2]. For example the notorious environmental pollutant TCDD (dioxin) stimulates the immune system of rats at low doses but severely suppresses it at high doses[3]. It also reduces mammary tumor rates from nearly 80% in old female rats to almost zero, but causes liver cancer at high doses[4]. There are numerous examples for such dose-responses also in the field of radiation biology.

A universal scale to express dose-responses is their molecular presentation. Since all toxicological effects represent interactions between molecules it is best to present dose-responses on this scale. This requires a simple conversion of the dose given in units of weight/kg to number of molecules/kg (=weight divided by molecular weight multiplied by Avogadro's number). This form of presentation has the advantage of one molecule (zero on the logarithmic scale) being the universal point of references for all chemicals and relationships between dose-responses can be readily seen on this absolute scale rather than on an often (unintentionally) manipulated relative scale (Fig. 2). This scale is also quite suitable to establish a molecular perspective of controversial issues in toxicology. Figure 3 illustrates this for a number of issues related to one of the most controversial environmental chemicals, i.e. 2,3,7,8-tetrachlorodibenzo-*p*-dioxin (TCDD).

TIME-RESPONSES

As alluded to earlier, there are two time scales in toxicology which sometimes can be similar but usually are quite different. Therefore, two time integrals are needed to correctly describe most toxic effects that occur over extended periods of time. When both time scales are very short then considerations regarding time as a variable can be neglected as it has been the case in the early days of toxicology and again since the advent of quantitative risk assessment. But time is always a variable as we all know from the experience of aging, which is the result of accumulating injury from very low potency toxic insults (food stuffs) and occasionally, more potent toxicants to the point where it becomes incompatible with life.

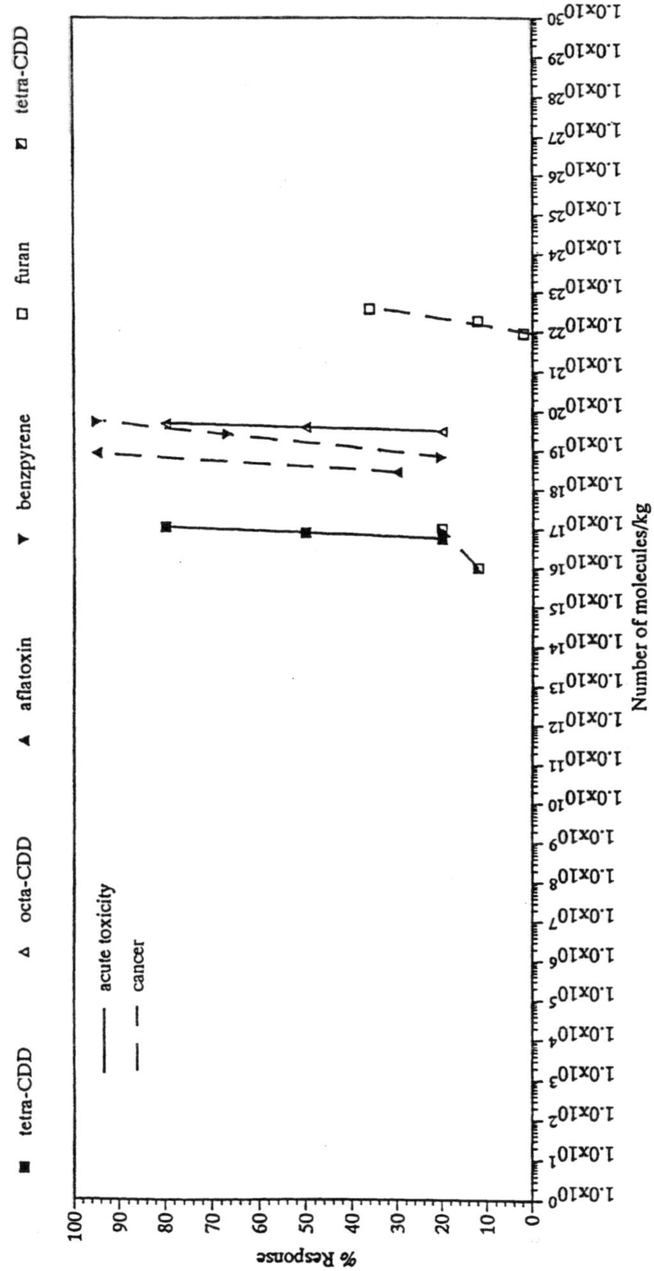

Figure 2 Molecular presentation of different dose-responses from various compounds.

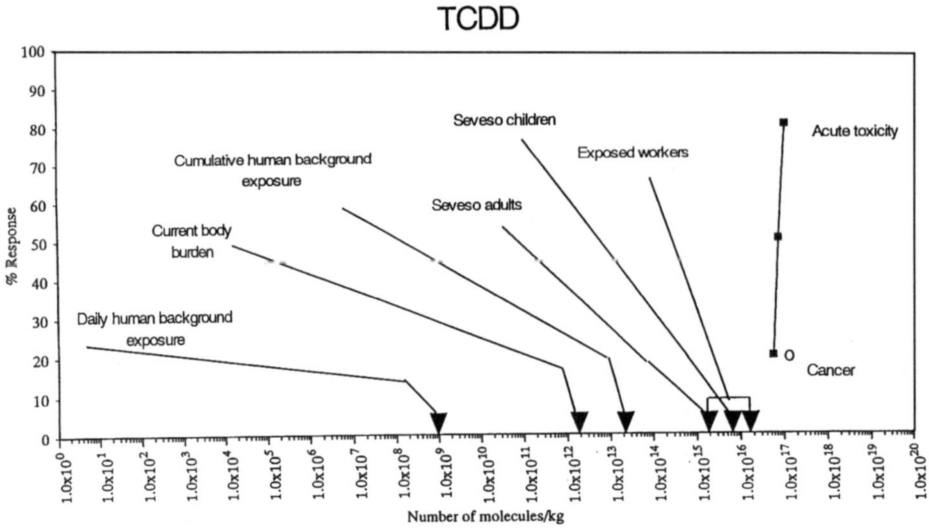

Figure 3 Molecular presentation of different issues in the risk assessment of TCDD. In comparison EPA's 10^{-6} risk is about 10^7 molecules/kg. Acute toxicity and cancer data are derived from rat studies.

TOXICOKINETIC TIME SCALE

$(\partial c/\partial t)_E$ describes the residency time of a substance in an organism, which is known as the discipline of toxicokinetics. Many books and papers have been published on this subject[5]. The time-dependence of a dose in an organism is well-understood. Therefore, it is not necessary to dwell on it here after having put it in the conceptual frame of toxicology. Although not stated anymore, the fact that toxicokinetic equations are strictly valid at constant or no effect levels is implicitly part of toxicokinetics.

TOXICODYNAMIC TIME SCALE

$(\partial E/\partial t)_c$ describes the time-dependence of an effect at constant dose and at toxicokinetic steady state. This deals with the development of an effect and its reversibility. This is the least understood part of toxicology although it has been studied extensively. The reason for this discrepancy may be a lack of conceptual frame work under which most of the experiments were conducted. Most of the chronic experiments were conducted by keeping the toxicokinetic and toxicodynamic time scales constant, which is appropriate to establish a dose response. However, in order to study the toxicodynamic time scale it is necessary to keep the toxicokinetic time scale and the (cumulative) dose constant. To our knowledge only Druckrey[6] conducted systematic long-term studies under conditions of constant cumulative dose. Sometimes he obtained the relationship of $c \times t =$ constant, which is the same relationship when the time to effect (toxicodynamic time) or the toxicokinetic time scale (no biotransformation or saturation of biotransformation) were kept constant (Haber's Law of Inhalation Toxicity). Sometimes Druckrey and others obtained dose/time relationships, which can be described as $c \times t^x =$ constant. This is clearly due to a lack of our understanding of the complexity of the contribution of the two time integrals to toxicity and not to a lack of validity of the herein outlined principles of toxicology. The carcinogenicity bioassays would have been conceptually sound except for the introduction of the maximum tolerated dose (MTD). As is well known, the MTD was introduced on grounds other than science. As we have seen earlier, to study a dose-response it is necessary to keep both the toxicokinetic and toxicodynamic time scales constant. This was indeed done by conducting the bioassays (for the most part) under steady state conditions and terminating the studies at 104 weeks. However, the MTD truncated most dose-responses because the steep part of the dose-response curve would have occurred for most chemicals above the MTD. Therefore, the vast majority of carcinogenicity dose-responses are restricted to the initial shallow and linear portion of the full dose-response. This unfortunate development gave rise to the scientifically totally meritless linear risk assessments of the past 20 years. That in turn fueled the multifaceted chemical-induced cancer hysteria, when in fact cancer formation obeys the laws of toxicology like any and all other end points of toxicity.

THREE-DIMENSIONAL DOSE-TIME-RESPONSE

Dose is a "pure" variable in the sense that the dose-response depends only on the Mass Action Law, but time is a complex variable as can be gleaned from earlier discussions on time-responses.

Haber[1], for acute exposure to inhalation toxicants and Druckrey and Küpfmüller[7], for carcinogens, showed that under certain circumstances the principle that dose × time = constant in terms of response exists. Rozman *et al.*[8] later confirmed that this relationship was also valid for TCDD with various end points. The argument has been advanced that there are too many exceptions to this principle to be useful in toxicology[9]. This misunderstanding arose by not recognizing that Haber–Druckrey's laws are special cases of the general law of toxicology, in which two kinds of time-dependencies exist corresponding to toxicokinetic and toxicodynamic time functions. There has been a revival of discussions lately regarding the role of time in toxicology. Rozman *et al.*[10] and Rozman[11] made various recent attempts to provide a theoretical frame-work for the relationship between dose and time in toxicology, among them a three-dimensional presentation of the dose-time-response curve indicating that time causes a curvature of the $c \times t$ integral (Fig. 4). By no means have all issues been resolved. Nevertheless, it is now widely recognized that the dose alone will not provide all answers to risk assessments of toxic agents. Therefore, Paracelsus' famous Latin paraphrasing "Dosis facit venenum" may need revision into "Dosis tempusque facit venenum".

FUNDAMENTAL PRINCIPLES OF TOXICOLOGY

There are several critical issues which make the interpretation of toxicological data often controversial. Among them the most important ones are species-to-species extrapolation and high-to-low dose extrapolation. The critical issue here is the existence or lack of existence of a threshold dose below which no adverse effect will occur. The other critical issue is the evaluation of the toxicity of mixtures particularly in light of potential synergistic interactions.

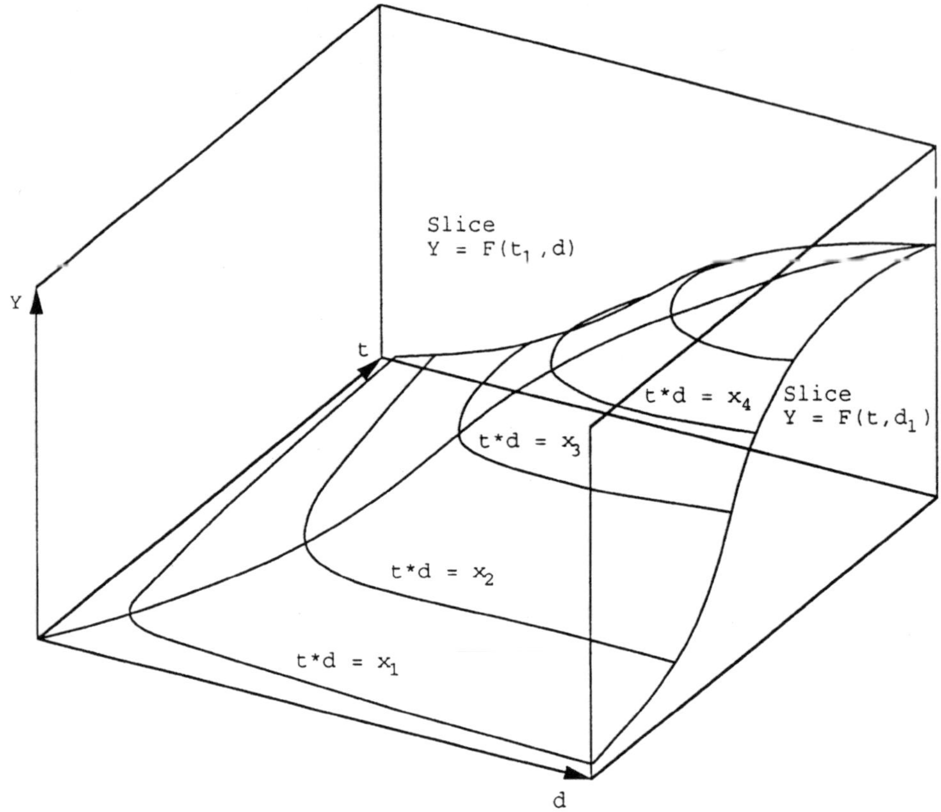

Figure 4 Three-dimensional plotting of the dose-time-response surface.

1. *Species-to-Species Extrapolation*

It is quite clear that the toxicity of chemicals cannot be studied in humans for ethical reasons using *a priori* designed toxicological experiments. Therefore, most toxicological information has been generated in surrogates, viz. animals. In general, animals are good but imperfect surrogates for humans. Dose- and time-dependent functions may be different in animals resulting in a shift of the dose-time-response in animals in comparison to the human dose-time-response, either towards more or less susceptibility to a given compound. Therefore, testing for toxicity always involves multiple species to identify the most sensitive species. In the absence of sufficient information the prudent assumption is that humans are more sensitive to the toxic action of chemicals than the most sensitive animal model. As knowledge about the mode of action of a chemical increases this assumption may turn out to be incorrect. In fact humans, for the most part, are less sensitive to the toxic effects of chemicals than are commonly used laboratory rodents. There are a number of reasons for this, all of them related to biological factors that allow the longer natural life-span of humans compared to rodents. In rare instances, humans can be more sensitive to the toxicity of a given chemical e.g. when it comes to a metabolic activation in humans which does not occur or occurs to a lesser extent in rodents or other laboratory animals. Therefore, the notion of most sensitive species can be given up only if enough mechanistic information is available. Otherwise safety factors are introduced [no observable adverse effect level (NOAEL) in animals divided by judgmental factor] to assure that a projected safe value for a chemical will do no harm to even the most sensitive humans.

2. *High-to-Low Dose Extrapolation*

This issue is at the core of the interpretation of toxicological data. Since toxicological studies are always conducted in a limited number of animals and since effects have a certain biological distribution the question is what would be the shape of the dose-response curve if a given study was conducted in a larger number of animals. Expressed differently, what is the best way of extrapolating from the region of experimental data points to low doses, which are inaccessible to experimentation. Traditionally the probit approach was and still is used for most end points of toxicity. This approach is based on the observation that most if not all effects are normally distributed. An exception was made for cancer by introducing various mathematical no-threshold linear extrapolations. These models were based on the assumption that a single molecule has a finite likelihood to hit a critical target molecule (DNA) in the organism which via mutation and clonal expansion may result in cancer. In fact, logically it is not possible to claim the existence of a theoretical threshold in an infinitely large population or in a finite population of eternal life. Although not stated any more, the critical assumption for all these linear models was that for low doses time as a variable is not important. (Indeed for subthreshold doses time may not be important within the life-span of a given species.) However, time as a variable is almost always important and particularly so for low doses. In fact, time is the variable that allows a theoretical justification of the threshold concept. That there is a relationship between dose and time has been known for some time as discussed earlier, although this relationship has not been fully explored. Nevertheless, it should be obvious that there is a natural limit to human life, say 100 or 120 years. Thus, toxicologists, unlike risk assessors, have no intention or obligation to protect dead bodies from the potential toxicity of chemicals, because that is exactly what linear no-threshold extrapolations do. Therefore, in the context of a three-dimensional dose-time-response continuum, there is no basis for no threshold extrapolations. In fact, the natural life-span of any species contains implicitly the notion of practical threshold for any end point of toxicity including cancer.

3. *Mixtures*

The no-threshold extrapolation came about as a consequence of Racheal Carson-induced cancer/chemophobia. A recent book on endocrine disruptors is likely to serve the same role for chemical-induced synergistic interactions. Because of the universal validity of the Mass Action Law, if there is interaction between chemicals it usually is additive. In rare instances interactive dose-responses can take on the appearance of synergism (=over additivity). This is not a widespread phenomenon in toxicology. It should be understood that this rare phenomenon usually has a toxicokinetic explanation or sometimes it is due to interaction of a chemical at two different binding sites of the same biological target (molecule). In the vast majority of cases when humans encounter complex mixtures there is no interaction at all and chemicals act independently. When there is interaction, it occurs in the form of agonism or antagonism according to well understood principles of mass action at a biological receptor. Any exception to this rule requires careful scrutiny and understanding of the mechanism of what would constitute a synergistic interaction.

CONCLUSIONS

Toxicology examines the dose- and time-dependence of interactions between exogenous and endogenous chemicals and physical agents. The first and most important principle of toxicology is that of dose-response (mass action) which states that there is a toxic dose and a safe dose for every chemical. Just as there are no inherently "safe chemicals" (safe under all conditions of exposure) there are also no chemicals which cannot be used safely by simply reducing the exposure. Toxicologist recognize that even if we were dealing with an infinite population or a finite population of eternal life, one molecule could cause an effect. In the real world, however, where the total population is less than six billion and the average life-span is less than 100 years, there are practical and pragmatic thresholds at much higher levels for all chemicals including mutagens and carcinogens. The recognition that all toxic effects are dose-related also means that the use of labels such as toxic chemical, liver poison, irritant, carcinogen, teratogen etc., has little descriptive or regulatory value unless we include information about the dose or exposure. It needs to be pointed out, that labels are not bright lines providing yes/no answers to safety questions or regulatory issues. We also need to remember that exposure includes both dose and time. When Paracelsus made his dose-response observation nearly 500 years ago, he did not specifically mention time as a variable although it was implicit in his observations. Time was formally introduced into the dose-response equation in 1924 by Fritz Haber[1]. Druckrey subsequently proved that time is a variable together with dose in producing cancer and Rozman[10,11] generalized the dose-time-response in a three-dimensional continuum. Indeed toxicology has now become a scientific discipline with a core theory. Thus attempts at falsification of the theory should eventually prove or disprove it rather than hypothesis testing.

References

1. F. Haber, *Fünf Vorträge aus den Jahren 1920–1923* (Verlag von Julius Springer, Berlin, 1924) pp. 77–92.
2. Multi-authored, "Low dose linearity: the role or the exception" **In**: *Belle Newsletter* (E. J. Calabrese, Ed.) **Vol. 6**, No. 1, pp. 1–27 (1997).
3. F. Fan, D. Wierda and K. K. Rozman, "Effects of 2,3,7,8-tetrachlorodibenzo-*p*-dioxin on humoral and cell-mediated immunity in Sprague-Dawley rats" *Toxicology* **106**, 221–228 (1996).
4. R. J. Kociba, D. G. Keyes, J. E. Beyer, R. M. Carreon, C. E. Wade, D. A. Dittenber, R. P. Kaluius, L. E. Franson, C. N . Park, S. D . Bernard, R. A. Hummel and C. G. Humiston, "Results of a two-year chronic toxicity and oncogenicity study of 2,3,7,8-tetrachlorodibenzo-*p*-dioxin in rats" *Toxicol. Appl. Pharmacol.* **46**, 279–303 (1978).
5. M. Gibaldi and D. Perrier, "Pharmacokinetics" **In**: *Drugs and Therapeutical Sciences* (Series of Textbooks and Monographs, J. Swarbrick, Ed.) 2nd ed., Marcel Dekker, New York/Basel (1982) pp. 1–494.
6. H. Druckrey, "Quantitative aspect in chemical carcinogenicity" **In**: *Potential Carcinogenic Hazard from Drugs. Evaluation of Risk.* (R. Truhaut, Ed.) UICC Monograph Series, **Vol. 7** (Springer-Verlag, Berlin, 1967) pp. 60–78.
7. H. Druckrey and K. Küpfmüller, "Quantitative Analyse der Krebsentstehung" *Z. Naturforschg.* **36**, 254–266 (1948).

8. K. K. Rozman, W. L. Roth, B. U. Stahl, H. Greim and J. Doull, “Relative potency of chlorinated dibenzo-*p*-dioxins (CDDs) in acute, subchronic and chronic (carcinogenicity) toxicity studies: implications for risk assessment of chemical mixtures” *Toxicology* **77**, 39–50 (1993).
9. M. O. Amdur, “Air pollutants” **In**: *Toxicology* (L. J. Casarett and J. Doull, Eds.) (Macmillian, New York/Toronto/London, 1975) pp. 527–554.
10. K. K. Rozman, L. Kerecsen, M. K. Viluksela, D. Osterle, E. Deml, M. Viluksela, B. U. Stahl, H. Greim and J. Doull, “A toxicologist’s view of cancer risk assessment” *Drug. Metab. Rev.* **28**, 29–52 (1996).
11. K. K. Rozman, Quantitative definition of toxicity: a mathematical description of life and death with dose and time as variables. *Medical Hypotheses* (In press).

2. ENVIRONMENTAL POLLUTION, NEUROTOXICITY, AND CRIMINAL VIOLENCE

ROGER D. MASTERS *, BRIAN HONE † and ANIL DOSHI ‡

"the Opinion of this mischievous Effect from Lead is at least above Sixty Years old; and you will observe with Concern how long a useful Truth may be known and exist, before it is generally receiv'd and practis'd on."

Benjamin Franklin[1]

"Regarding violence in our society as purely a sociologic matter, or one of law enforcement, has led to unmitigated failure. It is time to test further whether violence can be amenable to medical/public health interventions."

Dr. C. Everett Koop and Dr. George Lundberg[2]

I. INTRODUCTION

This chapter will explore the hypothesis that uptake of neurotoxic metals may be among the many factors contributing to the unusually high and widely varying rates of violent crime in the United States. The hypothesis rests on findings that loss of impulse control and increased aggressive behavior can be related to abnormalities of brain chemistry caused by a complex interaction of insufficiencies of essential vitamins and minerals, toxic uptake, alcoholism, and social stress. After reviewing evidence at the level of individual neurochemistry, ecological data will be presented to show that, controlling for standard socio-economic and demographic variables, environmental releases of lead and manganese predict geographical differences in rates of violent crime. Although this approach to criminal violence might seem at first unduly reductionist, analysis of the complex interactions between brain biochemistry, environment, and behavior explains otherwise puzzling variations in crime rates and suggests potentially effective approaches to crime prevention.

To provide an adequate account of violence in contemporary industrial societies, it will be useful to take an epidemiological approach to geographical and historical variations in crime rates that are not well explained—and often not examined—in conventional analyses[3]. This is particularly appropriate in the United States, where violence is, in the words of two leading physicians, "a public health emergency, largely unresponsive to methods thus far used in its control"[2]. Despite a recent levelling of rates of violent crime, the United States still faces what the Center for Disease Control has called an "epidemic" of violence. Although prevailing theories of violent crime properly implicate a host of social, economic, and psychological variables[4,5], these factors alone do not adequately explain why American counties have rates of violent crime that vary from less than 100 to over 3000 per 100,000 population (Map 1).

* Department of Government, Dartmouth College and Chair, Executive Committee, Gruter Institute for Law and Behavioral Research.
† Wing. Net.
‡ Dartmouth College.

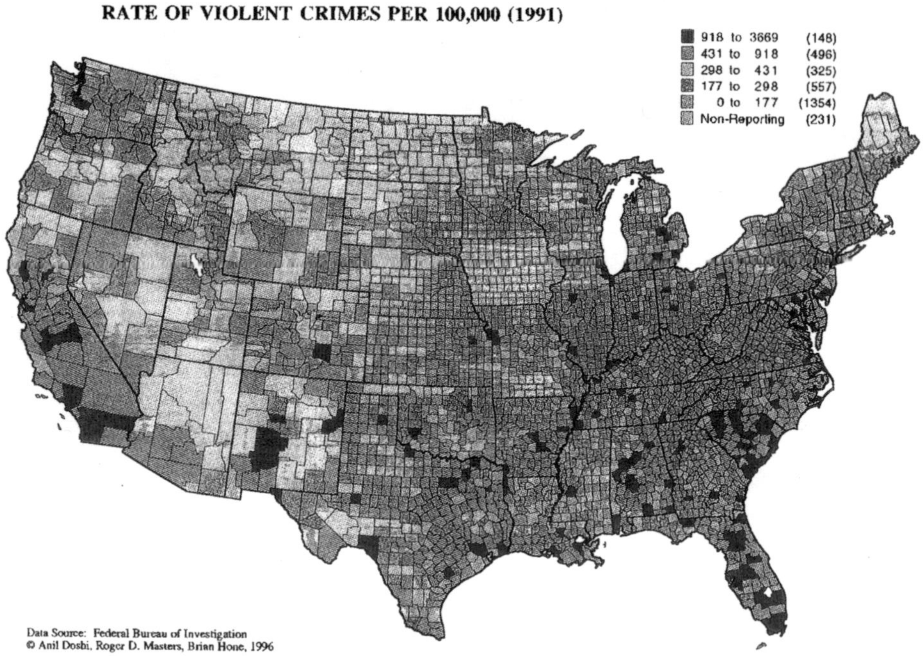

Map. 1

Although much is known about the behavioral effects of abnormal brain biochemistry, this information has hitherto not been brought together to account for these variations in rates of criminal violence. In part this is because most discussions of environmental toxins focus on such health risks as cancer. There is, however, increasing awareness that neurotoxins such as lead and manganese also have subclinical effects on brain biochemistry leading to learning disabilities, poor impulse control, and an increased risk of aggressiveness. An analysis of such relationships between environmental pollution and violence must rest on a complex, multi-causal analysis of human behavior. In this view, neurotoxicity is only one cause among many, at most functioning as a catalyst which, in addition to poverty, social stress, alcohol or drug abuse, individual character, and other social factors, increases the likelihood that an individual will commit a violent crime (below, Section II).

As in other epidemiological studies, five distinct relationships need to be established to confirm the hypothesized contribution of environmental neurotoxicity to high rates of violence: (1) correlation, (2) prediction, (3) function, (4) transmission, and (5) ecological verification.

(1) **Correlation** *associating toxic uptake with criminal violence at the individual level*. To be credible, the first condition of the neurotoxicity hypothesis must be evidence that individuals who engage in criminal violence are more likely to have absorbed a toxic chemical than comparable controls. It has long been known that serious behavioral and cognitive deficits are caused by exposure to lead, especially during infancy and childhood. Subclinical lead poisoning has been correlated with learning disabilities, attention deficit disorder, and other psychological conditions sometimes associated with deviant behavior. Similar effects have been found for manganese. In seven populations of criminal offenders whose bodily uptake of toxic

metals has been studied, lead, manganese, cadmium, or other toxic metals have been significantly elevated in the violent offenders compared to nonviolent criminals or controls (below, Section IIIA).

(2) **Prediction** *showing that young children with higher levels of toxic uptake are more likely, later in life, to engage in aggressive or violent behavior.* Correlational studies need to be confirmed by long-term, prospective research showing that toxicity allows prediction of future violent behavior. Such research only seems to have been conducted with regard to lead. In two studies of this neurotoxin, using different methods, lead uptake at age 7 was significantly predictive of juvenile delinquency or increased aggression in teen-age years and early adulthood (below, Section IIIB)

(3) **Functional effects** *of the neurotoxin that could account for loss of impulse control and increased violence.* Lead has both negative consequences for neuroanatomical development and functional effects degrading catecholamines and other basic neurotransmitters. Of particular importance is the deleterious effect of lead on glia, the brain cells that play an essential role in inhibition and detoxification. Manganese has the effect of lowering levels of serotonin and dopamine, neurotransmitters associated with impulse control and planning. Other neurotoxins may contribute to violent behavior, since cadmium, aluminum, and other metals have also been found to have deleterious effects on the brain. These biochemical factors interact in complex ways: for example, because lead degrades the detoxification capacity of the brain, exposure to lead pollution will enhance the effects of alcohol, drugs, or other toxins. Combinations of several toxic elements are probably synergistic rather than additive, with the extent of brain dysfunction also depending on diet, allergies, social status, stress, and individual experience (below, Section IV).

(4) **Transmission** *by known pathways must deliver neurotoxic elements to individuals in quantities associated with violent behavior.* Despite the prohibition of leaded gasoline and paint, environmental pollution of lead and other neurotoxins remains a serious problem. In four American cities, high traffic corridors contain soil residues of leaded gasoline that cause unhealthy levels of lead absorption in children. In the state of Massachusetts, controlling for other socio-economic factors, individuals absorb significantly higher amounts of lead in towns where industrial factories are located than in other localities. In addition to contemporary effects of industrial releases of neurotoxins, residues of particulate matter remain in soils for long periods, contaminating dirt with which children play. Other pathways include lead and manganese in aging public water systems or pipes within residential units and peeling leaded paint. In addition to such environmental exposure, dietary sources of manganese may be important. Infants absorb manganese in high levels from baby formula, and some crops absorb manganese from soil and fertilizers. Susceptibility to neuronal uptake of toxic metals is greatly increased for individuals with a diet low in calcium, zinc, and other essential vitamins. Lead, manganese, and other toxic elements thus probably have a disproportionate effect on the poor, since the combination of dietary insufficiency and environmental pollution has effects not observed when only one of the two is present (below, Section V).

(5) **Ecological measures** *of environmental pollution, controlling for other variables, should correlate with higher rates of violent crime.* Although the foregoing data are consistent with the possibility that neurotoxic metals could cause loss of impulse control, antisocial behavior, and violence (as well as learning disabilities, memory deficits, or physical disability), the role of environmental pollution needs to be more directly tested. Geographical data from counties in the United States provide a valuable test since ecological differences in environmental pollution should predict otherwise unexplained variations in rates of criminal violence. For this purpose, a dataset of all

counties in the United States was constructed, integrating the U.S. Environmental Protection Agency's Toxic Release Inventory (TRI), for lead and for manganese, crime reports from the Federal Bureau of Investigation, rates of alcoholism from the Department of Health and Human Services, and socio-economic and demographic data from the Census Bureau. Controlling for such conventional factors as income, population density, and ethnic composition, environmental pollution had an independent effect on rates of violent crime (measured as total homicide, sexual abuse, aggressive assault, and robbery). When all counties are dichotomized into presence or absence of industrial lead pollution (Map 2) and presence or absence of industrial manganese releases (Map 3), and higher and lower than average rate of alcoholism (Map 4), counties with all three factors of neurotoxicity have rates of violent crime over 3 times that of the national average (Fig. 1, see below, Section VI).

Neurotoxicity is obviously only one of many factors contributing to violence, but it may be especially important in explaining why rates of crime have differed so widely by geographical region and by ethnic group. Local rates of pollution are largely independent of such variables as unemployment rates, high school dropouts, and police per capita. Both multiple regression analysis and a structural co-variate model indicate that, controlling for socio-economic and demographic factors, environmental pathways of neurotoxic metals significantly contribute to rates of violent crime.

This exploration of relationships between brain biochemistry, diet, neurotoxic metals and violent behavior has obvious relevance to public policy. Crime prevention and improved educational performance may be greatly enhanced by parent-training in breast-feeding and proper diet. Vitamin and mineral supplementation, which some studies suggest may even increase IQ, could be particularly important in improved school performance and cognitive development. If releases of neurotoxic metals are

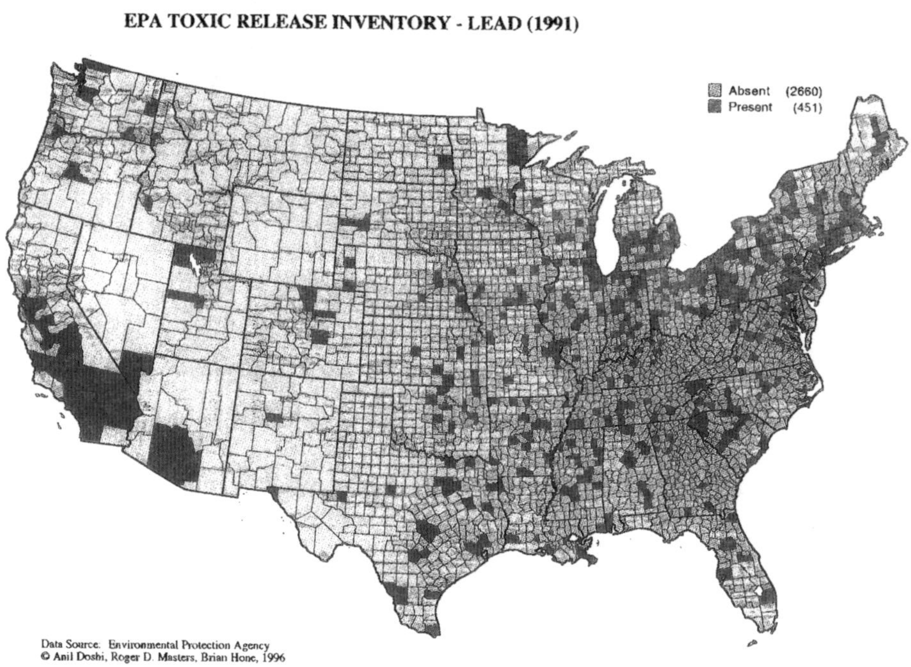

Map. 2

EPA TOXIC RELEASE INVENTORY - MANGANESE (1991)

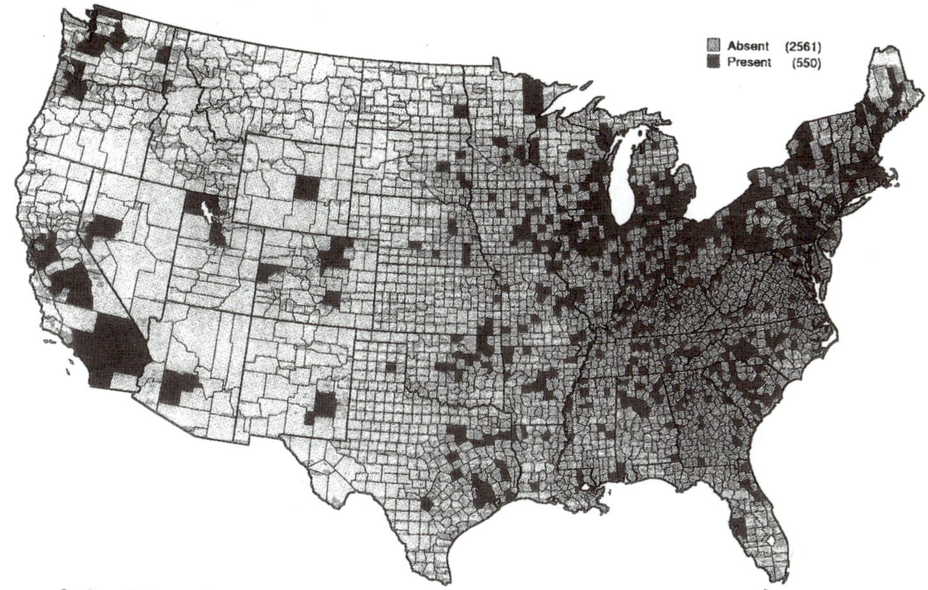

Data Source: Environmental Protection Agency
© Anil Doshi, Roger D. Masters, Brian Hone, 1996

Map. 3

RATE OF ALCOHOL RELATED DEATHS

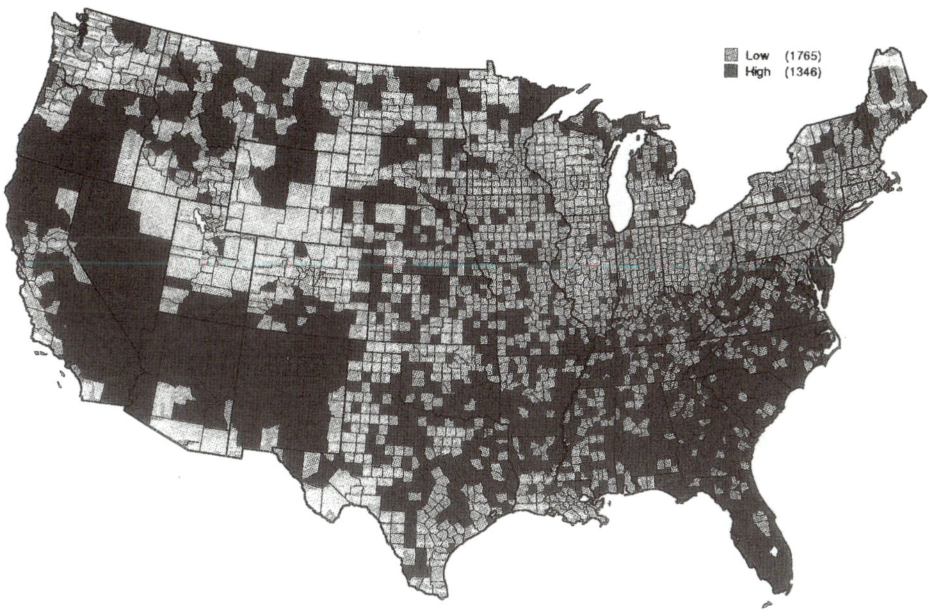

Data Source: US Health and Human Services, National Institute of Alcohol Abuse and Alcoholism
© Anil Doshi, Roger D. Masters, Brian Hone, 1996

Map. 4

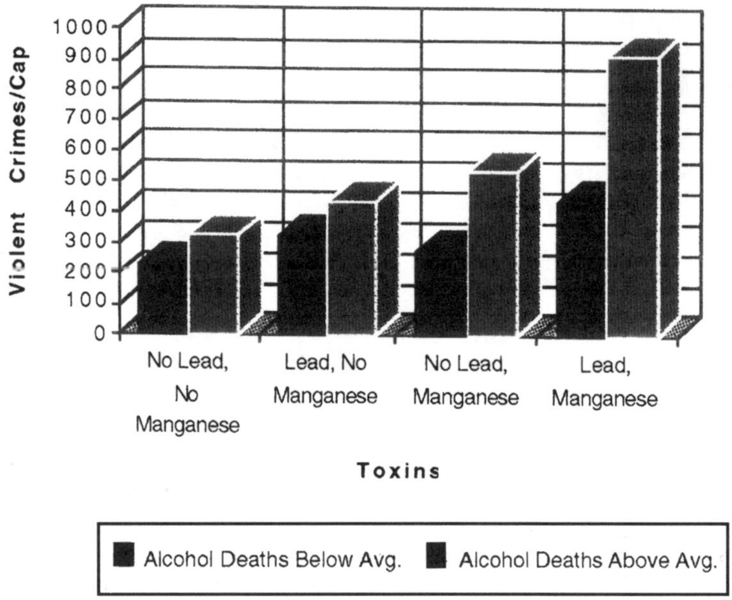

Figure 1 Association of manganese and lead pollution and rates of alcoholism with violent crime (50 state sample, 2899 Counties). *Notes:* Three-way ANOVA, data from all 2899 counties in U.S. reporting 1991 crime statistics to FBI. Rates of Death from Alcoholism dichotomized at national average (47.2/ 10,000). Significance of main effects: Presence or Absence of Manganese Pollution (TRI): t ratio = 11.32, $p < .0001$, $F_{1,2898} = 128.25$; Lead Pollution (TRI): t ratio = 9.66, $p < 0.0001$, $F_{1\ 2898} = 93.22$; Alcoholism: t ratio = 11.99, $p < 0.0001$, $F_{1,2898} = 143.64$. Significance of interactions: Manganese and alcoholism: t ratio = 6.86, $p < 0.0001$, $F_{1,2898} = 47.04$; Lead and Alcoholism: t ratio = 3.00, $p < .0027$, $F_{1,2898} = 9.01$; Lead and Manganese: t ratio = 3.91, $p < 0.0001$, $F_{1,2898} = 15.30$; Lead, Manganese and Alcoholism: $p < 0.0169$, t ratio = 2.39, $F_{1,2898} = 5.72$.

associated with rates of crime, reducing environmental pollution takes on higher priority. Such a finding could also aid the criminal justice system by improving predictions of recidivism, which are currently little better than chance, and by pointing to the use of vitamin and mineral normalization to improve rehabilitation (below, Section VII). Before major policy changes are proposed, however, it is essential to confirm the hypothesized relationships with further cross-cultural and experimental studies. This caution is especially necessary because the suggestion that violent crime might be associated with abnormalities in brain biochemistry has been highly controversial.[6]

II. NEUROTOXICITY, BRAIN BIOCHEMISTRY, AND BEHAVIOR

The study of possible links between environmental pollution and crime requires a focus on biochemical toxins that change brain structure and function, often in complex interactions with diet, alcohol or drug use, stress, and other ecological or cultural factors. This approach assumes that neurotoxicity is only one cause among many, at most functioning as a catalyst which, in addition to poverty, individual character, and other social factors, sometimes increases the risk of violent crime[7]. To clarify the relationship between these multiple levels of analysis, the pathways linking toxic exposure to behavior will be discussed in general before describing the specific effects attributed to lead and manganese. This is particularly important because discussions of environmental toxins have often focused on such health risks as cancer rather than on brain biochemistry and behavior.

Exposure to toxic elements, poor diet, and substance abuse have interacting effects on brain function and social behavior: "Changed brain chemistry can alter behaviour, and changed behaviour can alter brain chemistry: the interaction is two way. It therefore follows that behaviour, cognition, social interactions, and other expressions of brain function are subject not only to the social environment but also to certain aspects of the chemical environment. The relevant chemical factors include (a) neurotoxic pollutants in general, of which lead is evidently now the most serious in its impact, (b) certain common nutrient deficiencies, particularly of zinc, and (c) neurotoxins of voluntary abuse, of which ethanol is still probably producing the most widespread social damage"[8].

Among the biochemical factors that influence brain function, metal ions such calcium, potassium, and zinc play an important role[9]. As in other human cells, neuronal cell surfaces must have an equilibrium of positive ionic charges ("cations") and negative ionic charges ("anions"). Because neurons form an electrochemical system of communication, deficits in calcium and zinc can lead to neuronal uptake of toxic cations like lead and manganese; the resulting abnormalities of bioinorganic chemistry can be harmful when concentrations of key elements are either too low or too high[10].

Toxic elements can have both direct and indirect effects on behavior. On the one hand, neurotoxicity may result in abnormal neuronal development or destruction of neurons and brain structures; on the other, neurotoxins can interfere with normal brain biochemistry, degrading the level of necessary neurotransmitters and regulatory substances. In many cases, these effects depend on ratios or interactions between potentially toxic metals rather than absolute values of any single element[11,12]. Of particular importance may be synergistic interactions between elements whose toxicity is greatly multiplied when they are combined.

- **Direct or "frank" toxicity.** In some cases, uptake of neurotoxic elements leads to clinically diagnosed conditions. For example, prolonged exposure to manganese dust from mines or factories leads to the condition known as "industrial manganism," with symptoms like those of Parkinson's Disease[13]. Extreme concentrations of manganese have also been associated violence in environments with mining operations or industrial exposure, as on Groote Eylandt off Australia[14,15]. Exposure to copper during neonatal development has been associated with abnormal structures of the hippocampus, a key brain structure for learning. And, of course, the lasting effects of fetal alcohol syndrome and lead poisoning on normal brain development are well documented[16].
- **Indirect or subclinical toxicity.** Changes in brain biochemistry that are at first not obviously associated with environmental pollution can also have significant effects on behavior. Neurotoxic elements can lower levels of basic neurotransmitters, disturbing normal brain function by reducing inhibition and thereby contributing an additional "risk factor" for violence. For example, neuronal uptake of manganese has the effect of reducing neuronal levels of serotonin while increasing serotonin concentrations elsewhere in the body (Fig. 2).

The effects of abnormal levels of neurotoxic metals on neurotransmitter function and behavior depend on many factors. Dietary deficits in calcium, zinc, and essential vitamins or minerals can result in greater absorption of lead, manganese, and other toxic metals from water supplies or food and uptake of such neurotoxic elements in

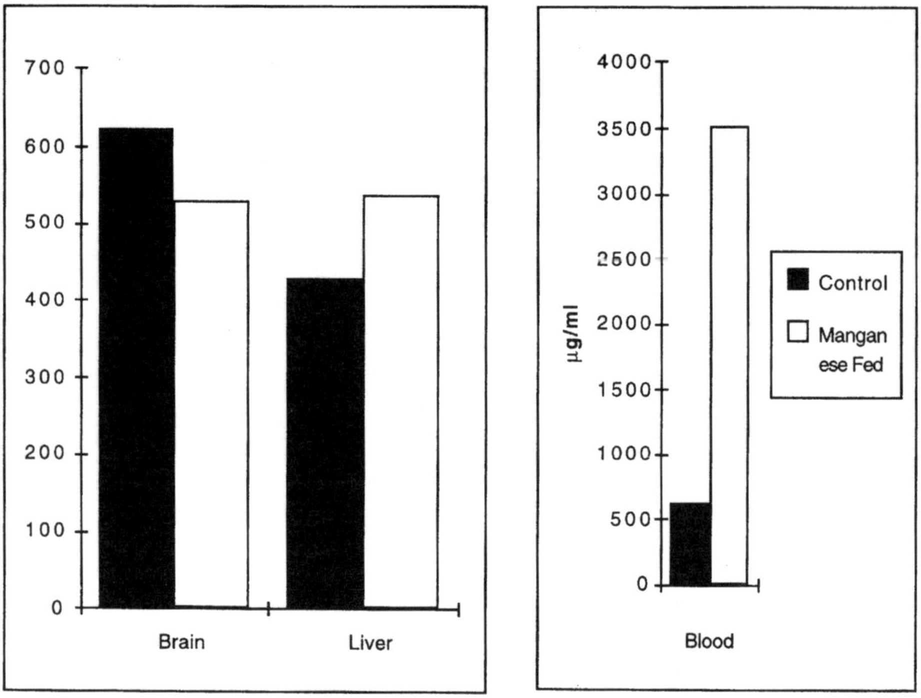

Figure 2 Effects of mangnese feeding on serotonin levels. Serotonin levels in brain and liver: mg/g tissue, wet weight; in blood: mg/ml. Significance: effects on brain and blood, $p<0.05$; on liver, n.s. *Source:* M. Kimura, N. Yagi and Y. Itokawa, "Effect of subacute manganese feeding on serotonin metabolism in the rat," *J. Toxicology and Environmental Health* **4**, 701–707 (1978).

brain cells. For example, laboratory animals whose diet included excess manganese did not absorb it when calcium levels were normal, whereas manganese uptake became significant when their diet was deficient in calcium (Fig. 3)[17]. As will be noted below, this effect could be associated with social class both because diets are likely to be poorly balanced and because toxic substances are more likely to be present in environments marked by industrial pollution, aging water systems, and housing with lead paint.

Because individuals differ in the level and activity of neurotransmitters like serotonin[18–20], the effects of neurotoxic elements that degrade neurotransmitter function are also likely to interact with personality. Equally important is social status. Serotonin levels are usually higher among dominant individuals than subordinates, a difference that occurs as a consequence of higher status in humans as well as nonhuman primates[21]. Conversely, chronic stress (which is often associated with marginal social status or personal insecurity) increases cortisol levels in the brain and thereby lowers functional levels of catecholamines such as serotonin and dopamine[22,23].

Additional biochemical imbalances have been associated with hypoglycemia, a biochemical deficiency in glucose, particularly when combined with alcohol or drugs[24,25]. Alcohol or drug consumption possibly represent crude efforts at self-medication triggered by these imbalances. Criminal violence could thus be traced to the interactive effects of personality, dietary imbalance, poverty and social disintegration, social marginality, substance abuse, and toxic elements in the environment[26].

The traditional concept of causality adopted by most social scientists is thus not appropriate for the analysis of the ways brain biochemistry and behavior are

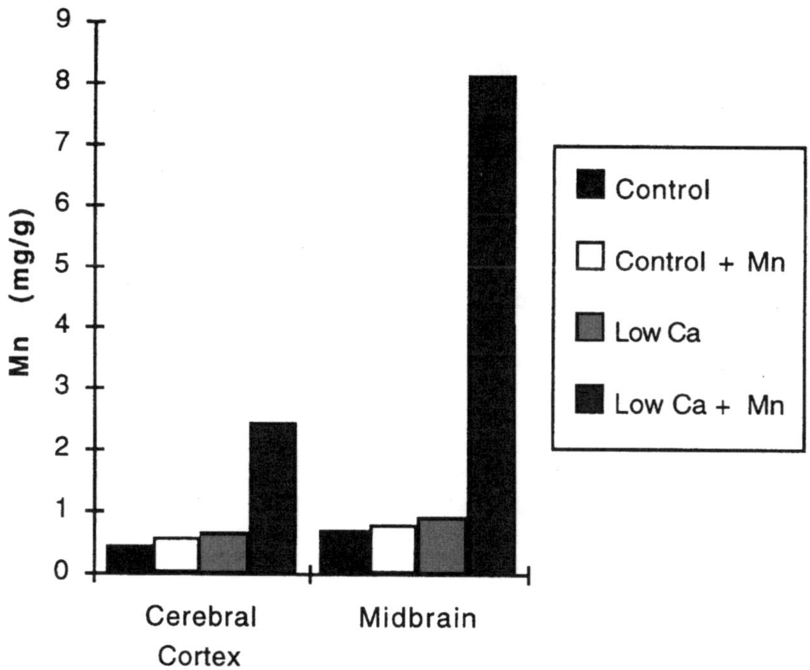

Figure 3 Calcium deficiency and manganese uptake. Source: V. A. Murphy, J. M. Rosenberg, Q. R. Smith, and S. I. Rapoport. "Elevation of brain manganese in calcium-deficient rats," *Neuro Toxicology*, **12**, 255–264 (1991). *Note:* Control: normal diet (7 g Ca/kg and 0.055 g Mn/kg food with 1.92 g/l sodium acetate trihydrate in water); Control + Mn: manganese supplement only (normal diet plus 1.79 g/l Mn (II) acetate tetrahydrate) in water, 0.4 g Mn/l); Low Ca: calcium deficiency (0.1 g Ca/kg and 1055 g Mn/kg food with Na acetate in water); Low Ca + Mn: manganese supplement and calcium deficiency combined. In Lo Ca + Mn group, levels of manganese were within range of values in figure: cerebral cortex = 2.45 mg/g; Caudate nucleus = 5.48 mg/g; Hippocampus = 3.64 mg/g; Hypothalamus = 6.32 mg/g; Thalamus = 4.44 mg/g; Midbrain Colliculi = 8.10 mg/g; Cerebellum = 4.27 mg/g; Pons-Medulla = 6.27 mg/g; Spinal Cord = 3.20 mg/g.

influenced by the interactions of poverty and social stress, lifestyle, substance abuse, personality, and environmental toxins. Instead, it is necessary to conceive of multiple risk factors, some originating in the social environment, some from the physical environment, and some from the genetics and individual life history of individuals. While complex, this approach avoids a puzzle in the typical explanations of violence as the product of social factors. Sociological approaches to crime lack a theoretical framework that explains why factors like violence on TV or availability of handguns and drugs trigger violent behavior in only a small proportion of the population as well as why this proportion should vary from one place to another. Part of the missing linkage may concern the way physical as well as social environments and lifestyles effect brain chemistry and behavior.

III. NEUROTOXICITY AND VIOLENCE AT THE INDIVIDUAL LEVEL

If neurotoxicity is to be linked to criminal violence, the first requisite is evidence that individuals who have absorbed toxic elements from the environment are significantly more likely to engage in aggressive or violent behavior. As in the study of epidemiology in public health, such evidence takes two forms:

(1) **correlational studies**, in which violent offenders are compared to nonviolent criminals and law-abiding controls at a specific time; and

(2) **prospective studies**, in which individuals are followed from childhood to maturity, with measures of toxic uptake in childhood used to predict later behavior.

Studies using both approaches have found that individuals who have absorbed toxic metals such as lead and manganese are significantly "at risk" to engage in violent or aggressive behavior.

Various methods have been used to assess the uptake of neurotoxic elements by an individual. Although one frequent technique for assessing neurotoxicity is the analysis of head hair[27,28], this method is sometimes questioned on grounds of contamination by shampoos or hair treatments. Other measures, such as blood, teeth, bone, and saliva, also have their limitations, however, and all have been used in studies cited here correlating toxic uptake with behavioral abnormalities. While the bodily tissue or fluid used as a measure may influence results in some cases, the specific assay used in any one study cannot be used to deny the general finding that uptake of toxic elements has deleterious behavioral consequences for humans.

Section A. Correlational Studies

A review of the literature indicates seven different samples of prison inmates whose levels of toxic metals were studied. In all seven groups, either lead and cadmium or manganese were significantly higher in head hair of violent offenders than in non-violent inmates or controls (Table I)[12,29–32]. In addition, silicon was significantly elevated in two samples of violent offenders, and, in one group, mercury was abnormally high. Equally interesting is the fact that lithium, which has been found to detoxify manganese, was abnormally low in two of the seven samples. Other correlational studies are consistent with the hypothesis that violent offenders are more likely to have abnormal brain biochemistry than non-violent criminals or law abiding citizens[33,34].

Section B. Prospective Studies

Claims that neurotoxicity has behavioral effects, like supposed discoveries of the cause of any new disease, are properly subject to the objection that correlational studies are in themselves not conclusive[35]. To meet this objection, prospective research provides critical evidence that neurotoxicity in childhood predicts aggressive behavior and crime among juveniles and young adults. Because such prospective studies measure levels of neurotoxicity years before violent behavior is observed, they provide even stronger evidence that individual absorption of toxic elements is a major risk factor for violence. In the largest and longest of these studies, a longitudinal biosocial study of 1000 black residents of Philadelphia from birth to 22[8], both lead intoxication and anemia at age 7 were significant predictors of the number of juvenile offenses, seriousness of juvenile offenses and number of adult offenses for males (Table II)[36]. Another study found lead absorption in bones in childhood to predict aggressive behavior later in life[37].

Other approaches to the prediction of violent behavior also implicate abnormal brain biochemistry as a potential factor. Among these is the study of recidivism among criminals found guilty of impulsive homicide and arson. In this subset of violent offenders, a combination of hypoglycemia, low levels of serotonin, and

Table I Element levels in hair of violent offenders: *Ratio of Violent/Control*

Element	Pihl and Ervin[5]	Marlow *et al.*[6]	Gottschalk *et al. #1*[8]	Gottschalk *et al. #2*[8]	Gottschalk *et al. #3*[8]	Schauss *et al.*[7]	Schrauzer *et al.*[13]
Lead	2.31***	2.38*	1.66	NS	NS	Sig.**	NS
Cadmium	1.50***	2.06*	0.68	NS	NS	Sig.**	NS
Manganese	1.04	1.50	7.33***	3.39***	2.15**	0.78	2.28**
Chromium	NS	1.14	NS	NS	NS	NS	NS
Sodium	1.33	1.41	NS	NS	NS	NS	NS
Mercury	NS	1.41*	NS	NS	NS	NS	NS
Silicon	NS	1.26*	NS	NS	NS	NS	NS
Copper	0.95	2.07*	NS	NS	NS	Sig.**	NS
Cobalt	0.91	1.39	NS	NS	NS	NS	NS
Lithium	NS	0.58*	NS	NS	NS	NS	0.28***
Calcium	0.87	NS	NS	NS	NS	NS	1.14
Magnesium	0.73	NS	NS	NS	NS	NS	0.64
Potassium	1.33	NS	NS	NS	NS	NS	NS
Iron	0.95	NS	NS	NS	NS	NS	NS
Zinc	0.97	NS	NS	NS	NS	NS	NS

$^*p < 0.05$, $^{**}p < 0.01$, $^{***}p < 0.0005$, NS = Not Significant.

Note: Comparable data on learning disabled children reveals significantly higher levels of lead (ratio 5.75*), cadmium (ratio 1.59*), maganese (ratio 1.43*), chromium (ratio 2.78*) and sodium (ratio 1.78*), and lower levels of cobalt (ratio 0.70*) and lithium (ratio 0.55*). *Source*: Bryce-Smith D. "Lead Induced Disorder of Mentation in Children." *Nutrition and Health*. 1983; **1**: 179–194.

Table II Factors predicting juvenile delinquency, males in Philadelphia biosocial study ($n = 487$)

Independent variables	Ages	Number of juvenile offenses, ages 7–17	Seriousness of junvenile offenses, ages 7–17
Pregnancy and delivery conditions	Birth	NS	NS
Mother's education	Birth	NS	NS
Father's education	Birth	NS	NS
Family income	Birth	NS	NS
Time father unemployed	Birth	0.179***	0.115**
Hand preference	1	NS	NS
Stanford-Binet	4	NS	NS
Hand preference	4	NS	NS
Eye preference	4	NS	0.059**
Foot preference	4	NS	NS
Neurological abnormalities	7	NS	NS
Abnormal movements	7	NS	NS
Abnormal vision	7	NS	NS
Lead Intoxication	7	0.149***	0.143**
Anemia	7	NS	0.044*
Intellectual status	7	NS	NS
Speech	7	0.045*	NS
Foster parents	7	NS	NS
Father absence	Birth–7	NS	NS
Household moves	Birth–7	0.092*	0.112
Persons supported	7	NS	0.09**
Family income	7	NS	NS
r^2		0.157	0.128

*$p < 0.05$, **$p < 0.01$, ***$p < 0.0005$, NS = Not Significant. *Source:* Denno D. W. Gender, Crime, and the Criminal Law Defenses. *Journal of Criminal Law & Criminology*. 1994; **85**: 80–180 (Table II).

alcoholism were among factors that effectively identify repeat offenders at the time of first sentencing[38,39].

IV. NEUROTOXICITY, LOSS OF IMPULSE CONTROL AND VIOLENCE

To be plausible, the neurotoxicity hypothesis requires the identification of precise biochemical mechanisms underlying correlations between individual concentrations of toxic metals and violent behavior. Correlational or prospective data do not in themselves explain how or why individuals who have absorbed a toxic chemical might be more likely to engage in violent behavior or crime than others. Empirical evidence that neurotoxins have functional effects on brain structure and behavior, undermining impulse control and triggering violence, is therefore of great importance. Such effects, however, are different for each toxin.

Section A. Lead Toxicity, Behavior, and Violence

Behavioral and cognitive deficits caused by lead, noted in antiquity by Hippocrates and two centuries ago by Benjamin Franklin, have been the subject of widespread scientific analysis[40–44]. Such subtle effects have often been ignored in the light of more obvious health defects produced by very large doses of lead poisoning, such as those from industrial pollution or peeling paint in aging buildings. Less attention has been paid to the effects of lead upon the brain, despite strong evidence that lead absorption lowers IQ (Fig. 4) and otherwise disturbs cognitive development and social behavior. Because exposure of infants to lead has effects lasting through puberty and beyond, the long-term developmental damage from lead can interact with the short term behavioral consequences of neurotoxic metals.

Although legislation prohibiting lead in gasoline additives and paint reduced lead levels in blood by 78% between 1976–1978 and 1988–1991[45], toxic release of lead and lead compounds remains a serious problem due to aging water systems and industrial pollution[46]. Exposure to subclinical doses of this neurotoxic metal can be a major hazard for four principal reasons: (1) children absorb up to 50% of lead they ingest, compared to 8% for adults;[47] (2) prolonged exposure to even very low doses of lead can cause neuronal damage during early development, resulting in lasting cognitive and behavioral deficits;[7,48] (3) current lead levels have direct effects on neurotransmitter function, influencing cognition and reducing impulse control[2,17,22–26,33], and (4) highest levels of lead uptake are reported for the demographic groups most likely to commit violent crimes[49,50]. Because the consequences of exposure depend on lifestyle factors such as alcoholic consumption, smoking, and diet, moreover, averages for a population will tend to mask the risks arising from combinations of co-factors[51].

Prolonged exposure of low levels of lead can influence dopaminergic, cholinergic, and glutamatergic neurotransmitter functioning, producing learning deficits that include impairment of passive avoidance learning—i.e., the capacity to be deterred

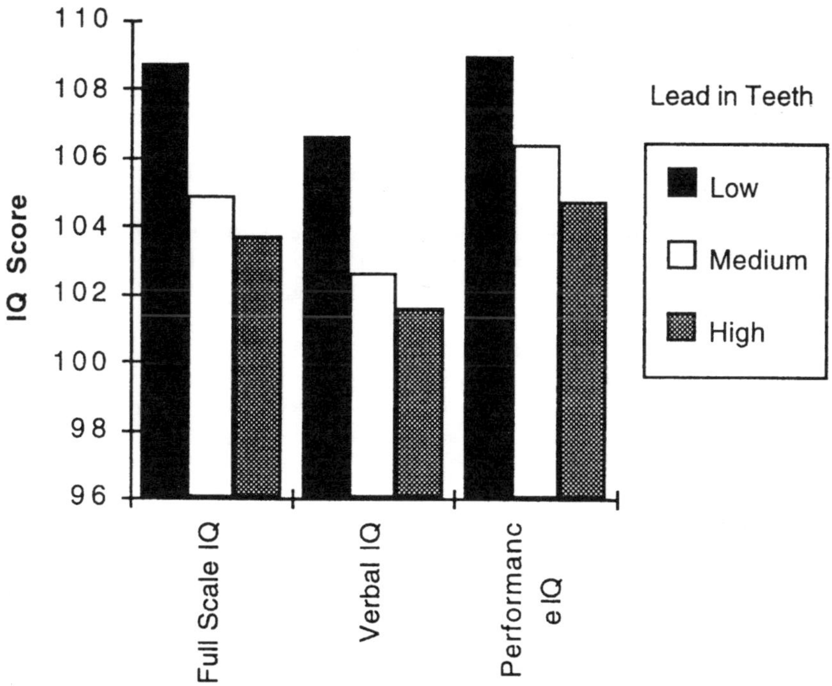

Figure 4 Effects of lead absorption on IQ. *Source:* D. Bryce-Smith, "Environments chemical influences on behaviour and mentation," John Jeyes Lecture—1984, *Chemical Society Review*, **15**, 93 123 (1985).

by future punishment[5,13]. Attention deficit disorder (ADD) and attention deficit hyperactivity disorder (ADHD), conditions frequently associated with juvenile delinquency, have also been traced to prolonged exposure to lead[5,13,15,52,53]. Because lead interferes with the detoxification functions of glial cells[54], moreover, it can enhance the effects of other toxic substances. Apart from its effects on such health concerns as hypertension[55], therefore, it should not be surprising that lead uptake at age seven is predictive of both juvenile and adult crime (Table II)[8].

Successful medical treatment of learning and behavioral deficits due to early exposure to lead provides even more convincing evidence of the effects of subclinical toxicity: "the involvement of lead as a cause of some types of mental retardation is clearly demonstrated by the observation made by a number of different workers that reducing the burden of lead in the child with chelating agents, usually penicillamine, brings about a substantial IQ increase (typically of 7 IQ points) in about two children out of three. Similar detoxification has been applied to children exhibiting hyperactive behaviour problems"[56]. Particularly given the contributions of attention deficit disorder and school failure to violence, it should hardly be surprising that head hair analyses of criminals often show abnormally high levels of lead (Table I).

Section B. Manganese Toxicity and Violent Behavior

Manganese is a cation whose valence can change among several states (Mn^{2+}, Mn^{3+}, Mn^{4+}); whatever its toxicity elsewhere in the body, within the brain manganese can lower the levels of essential neurotransmitters[57]. Once present in abnormal levels, manganese can therefore produce subtle behavioral effects in the absence of clinical signs of disease. Because other toxic metals may have similar effects, disturbances in behavioral inhibition could follow from interactions between heavy metals rather from amounts of manganese as an isolated factor. Of particular relevance, for example, is the finding that occupational exposure to lead also has the effect of significantly increasing blood levels of manganese[58].

Manganese has been associated with subclinical behavioral disturbances due to its effects on catecholamine function[59–62]. Although industrial manganism can be traced to direct exposure to large quantities of manganese dioxide or manganese nitrate,[63–65] chronic exposure to low levels of manganese is probably more relevant to loss of impulse control and outbursts of violent behavior, especially under stress. Exposure to manganese lowers levels of serotonin in the brain, while paradoxically increasing serotonin levels in blood and body tissue (Fig. 2)[66]; manganese also degrades dopamine and reduces levels of essential minerals in brain cells[67–69]. While the mechanism is not clearly understood, some of these effects may be due to altered levels of monoamine oxidase (MAO A)[29,32,38]. However produced, low levels of serotonin in the brain are associated with mood disturbances, poor impulse control, and increases in aggressive behavior—effects that have increasingly been treated with Prozac and other psychotropic medications which enhance serotoninergic function[70,71].

Vitamin and mineral deficiencies play a central role in lead and manganese uptake. In laboratory studies, for example, mere environmental exposure to manganese does not lead to toxicity, whereas animals with deficits in calcium intake show significant manganese retention[72]. Other aspects of diet also influence the absorption of manganese. In particular, laboratory animals raised on infant formula have cellular retention of manganese that is five times greater than controls raised on human mother's milk (Fig. 5)[73]. Full-term human infants raised on infant formula had manganese retention of 2.8 μg/kg, over five times the level (0.43 μg/kg) in similar infants who were breast fed[74]. The combination of calcium insufficiency and

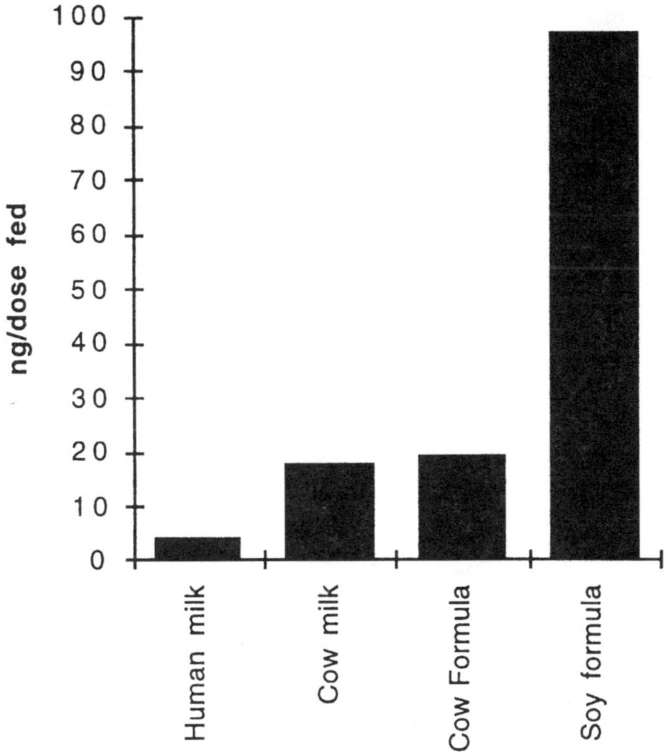

Figure 5 Manganese retention from milks and formulas. *Source:* Carl L. Keen, Janet G. Bell and Bo Lönnerdal, "The effect of age on manganese uptake and retention from milk and infant formulas in Rats," *J. Nutrition*, **116**, 395–402 (1986). *Note:* although the *rate* of absorbing manganese in brain tissue is actually higher in human mother's milk (1.4% of dose) than either cow milk formula (0.9% of dose) or soy formula (1.0% of dose), the total *amount* of manganese in mother's milk (0.01 mg/ml), is 4 times lower than in cow milk (0.04 mg/l), 5 times lower than cow milk formula (0.05 mg/ml), and 30 times lower than soy formula (0.30). "As the concentration of manganese in most infant formulas exceeds that found in human milk, it is evident that the *amount* absorbed and retained from cow milk and the infant formulas can far exceed that from human milk." (p. 400).

manganese toxicity could therefore be described to the general public as "reverse Prozac." Similar interactions between diet, lifestyle, and toxic uptake have been reported for lead (see Section IIIA above).

Section C. Other Neurotoxic Metals and Brain Biochemistry

Because a number of metals may disturb normal brain function[75–77], there are other, often unsuspected correlations in addition to those already noted for lead and manganese. Analysis of individuals with learning disabilities or records of criminal violence sometimes reveal imbalances in heavy metals, some of which—such as chromium, cadmium and sodium—seem to be associated with violent behavior, while other metals such as lithium and cobalt seem to reduce aggressive or inappropriate impulses (Table II)[78]. Dietary deficiencies in calcium and vitamin D may also play the key role in the uptake of these elements and minerals.

Conversely lithium lowers the levels of manganese and other toxic metals in the brain—and is lower in hair samples of violent offenders than controls (Table II)[31,34]; at the population level, moreover, homicide levels in Texas were found to be inversely related to levels of lithium in the water supply[26]. Not surprisingly, lithium has been used successfully to prevent violence in mental patients as well as prison inmates[79].

Other interactions between social behavior, brain biochemistry and violence are also important. As many observers have noted, alcohol and drug use is highly correlated with loss of impulse control and with violence[80]. For some offenders, this may be associated with hypoglycemia and low levels of serotonin[40]. The culture of juvenile gangs provides another factor, creating needs for alcohol and drugs as a condition of belonging to a meaningful peer group[81].

Evidence for an interaction between diet, neurotoxicity, and violence might also be derived from long-term studies of policies that have provided nutritional supplements to poor and disadvantaged populations. Programs like AFDC or WIC that have the potential to improve the diet and parenting-skills of poor, unwed mothers, may be especially important because they can reduce prenatal and infant deficiencies in essential vitamins and minerals. Use of infant formula instead of mother's milk, which as noted leads to increased uptake of manganese, has been significantly associated with lower IQ scores for premature infants (Fig. 6)[82];

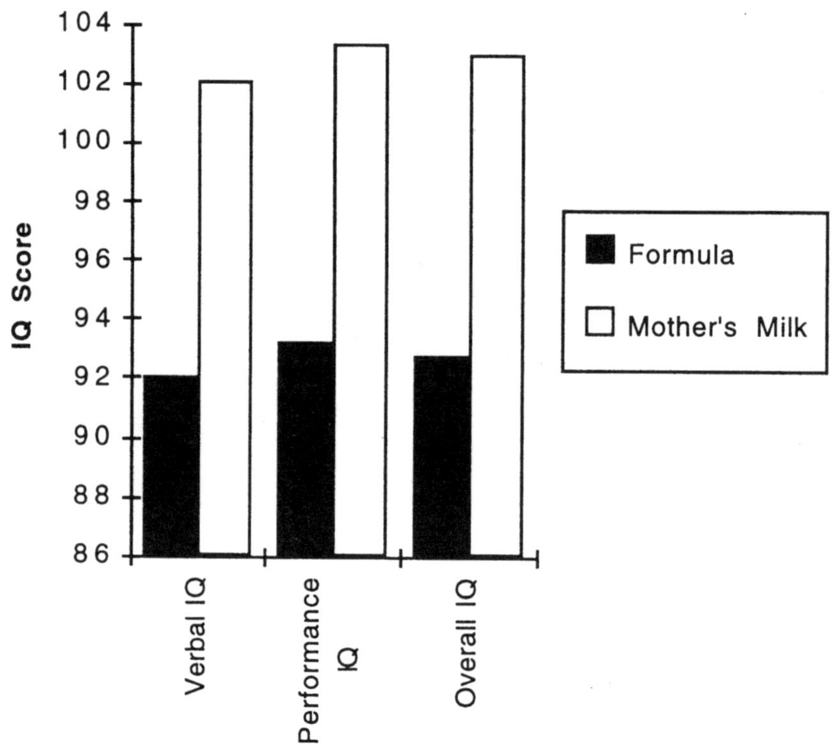

Figure 6 Breast milk and subsequent IQ (age 7.5–8 years). *Source:* A. Lucas, R. Morley, T. J. Cole, G. Lister, and C. Leeson-Payne, "Breast milk and subsequent intelligence quotient in children born preterm," *Lancet*, **339**, 261–264 (1992). *Note:* Preterm infants babies under 1850 g at birth divided into mothers who chose formula ($n=90$) and those who chose to nurse with their own milk ($n=210$). Of latter group in above chart, 17 were unable to provide milk; although included in the nursing group, their infants had IQ scores at age 7.5–8 comparable to infants raised on formula (overall IQ = 94.8, compared to 92.8 for formula and 103.7 for successful breast-fed infants). Controlling for social class, mother's education, female sex, and days of ventilation, mother's milk had a net contribution of 8.3 points to higher IQ.

similar though somewhat smaller differences were found for full-term infants[83]. More direct measures of manganese toxicity have also been linked to learning disabilities[84]. Many studies confirm that higher rates of school completion and employment as well as lower rates of criminal violence are among the multiple benefits flowing from concerted early intervention programs that include dietary improvement[85].

Section D. Alcoholism, Brain Biochemistry and Synergy

Ethanol in alcoholic beverages is a toxin that has serious developmental and functional effects[86]. In addition, ethanol greatly increases the deleterious effects of other toxic metals. In laboratory animals, the effects of a toxic element (cadmium) were "significantly altered" when combined with ethanol, especially because "animals exposed to ethanol absorb much more cadmium than their unexposed counterparts"[87]. In humans, ethanol interacts with the physiological effects of lead on blood pressure and other physiological processes[88,89] as well as with manganese[90]. As a result, the combination of alcohol consumption and poor diets, often found in marginal young males, puts them at particular risk. Given the interaction between alcohol and various imbalances in brain biochemistry, there should be little surprise that consumption of alcohol has often been associated with a high proportion of violent crimes[2,4,91,92]; consistent with individual data, the geographical distribution of alcoholism resembles that of violent crime (cf. Maps 1 and 4). But not everyone who consumes alcoholic beverages becomes violent. That multiplicative interaction—or synergy—between neurotoxic elements and ethanol could be an important part of the explanation is supported by the county data on rates of violent crime (see Fig. 1 and Section VI below).

Although heroin, cocaine, and other psychoactive drugs have often been implicated in violent crime, somewhat less is known about the precise causal effects of the brain biochemistry involved. In any event, drug users often also consume alcohol, making a differential assessment of behavioral effects difficult. In some cases, moreover, the role of hard drugs has been associated with conflict over the "turf" associated with illegal markets rather than the disinhibiting effects of the drug itself[93]. Because we have not located reliable geographical statistics for drug use comparable to those for alcohol, data analysis will focus on the latter as a general measure of the role of substance abuse in violence.

Anthropologists have also noticed a correlation between violence and hypoglycemia, the tendency for lower than average uptake of glucose[94]. Associated with abnormally high levels of insulin, this condition is frequently associated with alcoholism, though it can be independent of it[34].

V. ENVIRONMENTAL PATHWAYS OF TOXIC ELEMENTS

In exploring the neurotoxicity hypothesis, it is not sufficient to show that abnormal brain chemistry could be a contributory "risk factor" for violent behavior. From an epidemiological perspective, it is also necessary to show that environmental pathways have the effect of transmitting toxic elements to individuals in quantities associated with violent behavior. Among the multiple pathways of exposure to lead, manganese, and other neurotoxic metals, some have now been studied in sufficient detail to demonstrate the linkage between environmental sources and individual absorption (especially with regard to lead and manganese).

Despite the belief that direct exposure leaded gasoline and house-paint have been the primary pathways for lead absorption, other sources now seem equally if not more important [95]. Perhaps the most comprehensive data concern residues of lead in the soil along highly travelled automobile corridors in urban areas. Although legislation prohibiting lead in gasoline additives and paint reduced lead levels in blood by 78% between 1976–1978 and 1988–1991[96], toxic elements can remain in soils for prolonged periods, especially in the absence of extensive forest cover facilitating absorption in plant material[97]. Lead-contaminated soils have been demonstrated to cluster in urban neighborhoods along heavily traveled urban roadways in Baltimore[98], Minneapolis[99–102], and New Orleans[103,104]; these studies show, moreover, that the contact of children's hands with contaminated soils in playgrounds and backyards is directly responsible for their absorption of lead[105,106].

Industrial plants are a second source of pollution with toxic metals. In a survey of all towns in Massachusetts, controlling for other variables, individual measures of lead uptake were correlated with the presence of old industrial factories, suggesting that both long-lasting residues and current releases of neurotoxic metals in industry can be effectively transmitted to children. In this study, the average age of housing stock—associated with the presence of peeling lead paint—was also correlated with high lead levels among children[107].

Like lead, manganese is another toxic metal often used in industrial processing. In addition, public water supplies may also be sources of these neurotoxic metals, since aging cast iron pipes contain both lead and manganese that can leach into residues. Early in this century, cast iron was the most typical material for pipes and conduits; #20 Grey cast iron normally contains 0.4–0.6% manganese along with 2.25% silicone, 3.5% carbon, 0.4% Phosphorous, and 0.1% sulfur[108]. Aging or rusting pipes could therefore leach toxins like lead and manganese into water supplies, particularly in the decaying inner cities now inhabited by the poor. For example, the superintendent of the Edgartown, Massachusetts water department recently described renovations in the town's water system: "When we dismantled the standpipe, we found two and a half feet (*sic*) of a substance with the consistency of pudding"[109]. Other sources of toxicity occur in water systems within buildings. Even in newer, multi-story construction, water supplies in the upper floors may be particularly high in lead: in a recently constructed science building at Dartmouth College, where the first three floors had lead levels around the EPA water standard of 15 mg/dL, the fourth floor faucet had levels between 100 and 200 mg/dL[110].

A further source of manganese toxicity arises from diet and lifestyle. Because laboratory studies indicate that cellular uptake of manganese from formula based on cows' milk is five times greater than from mother's milk and 20 times greater for soy formula (Fig. 5)[111]. The practice of bottle-feeding infants greatly increases the infant's exposure to toxicity[112]. Bottle-feeding will have even worse effects if water polluted with manganese or lead is used to mix the formula. For those with diets deficient in calcium, zinc, vitamin D, and other necessary vitamins and minerals, rates of bottle feeding are likely to be higher and its effects on the infant more damaging.

Consumption of alcohol and psychoactive drugs also delivers toxic substances to the brain that tend to disinhibit behavior. These effects can have highly complex interactions with the uptake of neurotoxic elements like lead and manganese. On the one hand, alcohol and cocaine provide short-term bursts of neurotransmitters like serotonin and dopamine, whose activity is reduced by lead and manganese; hence the desire to use alcohol and drugs may be increased by neurotoxic exposure as well as by social stress. On the other hand, drugs like cocaine are often processed with chemicals like potassium permanganate and, over the long run, tend to reduce endogenous production of catecholamines. Hence environment–lifestyle interactions may generate positive feedback loops in which the presence of neurotoxins trigger synergistic reactions of the sort noted previously.

As the foregoing suggests, all individuals are not equally vulnerable to ecological pathways of toxins. The human body, and especially the brain, has evolved defense mechanisms against neurotoxic metals. For example, under normal circumstances, virtually all ingested manganese is excreted, which is hardly surprising insofar as manganese is naturally the 12th most abundant element in soil. Because neuronal uptake occurs for individuals with vitamin and mineral deficiencies, however, factors in the social environment have the effect of facilitating the transmission of neurotoxicity.

America's poor often suffer from dietary deficiencies, including deficits in calcium and vitamin D (the rate limiting factor for cellular binding of calcium). For example, the 1976–1980 National Health and Nutrition Examination Survey, based on a national probability sample, found that black teenage males consume on average only two-thirds as much calcium (887 mg/day) as do whites (1332 mg/day); a follow-up survey revealed that calcium intake among Hispanics (1195 for Mexican Americans, 1255 for Puerto Ricans, and 1185 for Cubans) was also below the white average.[113] The calcium needs of pregnant or breast feeding women or adolescents (1200 mg/day) are higher than average—and even higher for pregnant teenagers (1600 mg/day). This presents a specific problem for pregnant minority women: for example, non-Hispanic black women between the ages of 18 and 39 get only 467 mg/day of calcium as compared to 642 mg/day among white women of the same age[114]. Given the increased uptake of neurotoxic metals associated with calcium deficiencies in laboratory studies, calcium deficits among the poor may have particularly deleterious effects during infant development and childhood.

Finally, poor mothers typically do not breast-feed their infants. "By 1986–87, 73 percent of infants born to mothers with more than 12 years of education were breastfed compared with 49 percent of infants born to mothers with 12 years of education and 31 percent of mothers with less than 12 years of education"; this difference interacts with ethnicity since "white infants were nearly three times as likely to be breastfed as were black infants"[115]. The effects of manganese toxicity associated with infant formula are thus greatest for the poor, for ethnic minorities, and for those with little education.

Factors in the social environment are likely to interact with environmental pollution from industrial sources and the urban infrastructure. While it may be difficult to isolate how much toxicity is transmitted by each of the interacting pathways, an ecological approach to behavioral deregulation and criminal violence seems reasonable. If the neurotoxicity hypothesis is correct, it should be directly confirmed by geographic differences in rates of crime that are associated with the known pathways of toxic elements. Using multivariate analysis, moreover, it should be possible to assess how various socio-economic and demographic factors compare with neurotoxicity as risk factors for violence.

VI. THE ECOLOGY OF VIOLENCE: ENVIRONMENTAL POLLUTION AND RATES OF VIOLENT CRIME

County-level reports of toxic releases and other ecological factors can be used to study violence, much as they have been proposed as a way to measure the impact of ambient pollution on human health[116]. To test the hypothesis that neurotoxicity is among the co-factors related to violent crime, we used ecological data on the distribution of environmental pollutants and alcoholism to predict the rates of criminal violence. The Environmental Protection Agency's Toxic Release Inventory (TRI) for lead and manganese was correlated with 1991 FBI crime reports from all counties ($n = 3141$) in the US; counties with no reported incidence of either violent or property crimes were dropped from the analysis ($n = 242$, of which 231 are in the continental US shown on Map 1).

Although county-level reports in toxic release inventory measure current levels of pollution, they are likely to indicate the existence of industrial plants that have left lasting residues of toxicity in soil and on buildings. The report of environmental releases of lead or manganese can thus be viewed as a proxy variable for industrial activities resulting in environmental pollution with any neurotoxic substance. To be sure, our data concern current residence rather than an individual's childhood environment, and hence do not provide prospective measures like those cited above. Given barriers to geographic mobility into neighborhoods of different social classes, however, it is not implausible that the county of current residence is similar to environments during infancy, especially for poor and minority populations; if so, for many of those most at risk, present exposure to toxic metals should also be a rough measure of environments during development.

The distribution of manganese and lead pollution is highly skewed: over 80% of counties have no reported release of either manganese and its compounds, or lead and its compounds. The EPA's recorded releases of these toxic metals are not predicted by other demographic or socio-economic variables normally associated with crime: a multiple regression model using 19 factors (population density, ethnic composition, ethnic poverty, unemployment, income, police per capita, public sewers, public water supplies, rate of infant death, age of housing stock, welfare, education, alcoholism, and other toxic release) predicts less than 5% of the variance of reported levels of lead ($r^2 = 0.031$) or manganese ($r^2 = 0.045$).

Because environmental releases of these toxic metals have non-linear effects on behavior, EPA reports were dichotomized as the presence or absence of reported lead or manganese toxicity. To assess whether toxicity might exacerbate the effects of alcoholism, rates of death from all causes associated with alcoholism were dichotomized at the national mean (47.2 per 100,000). The three way ANOVA (Fig. 1) shows that all three variables (death from alcoholism as well as lead and manganese pollution) are highly significant predictors of rates of violent crime (lead: $p < 0.0001$, t ratio $= 9.66$, $F_{1,2898} = 93.22$; manganese, $p < 0.0001$, t ratio $= 11.32$, $F_{1,2898} = 128.25$; alcoholism, $p < 0.0001$, t ratio $= 11.99$, $F_{1,2898} = 143.64$), with significant two-way interactions between manganese and alcoholism ($p < 0.0001$, t ratio $= 6.86$, $F_{1,2898} = 47.04$), lead and alcoholism ($p < 0.0027$, t ratio $= 3.00$, $F_{1,2898} = 9.01$), lead and manganese ($p < 0.0001$, t ratio $= 3.91$, $F_{1,2898} = 15.30$), and a significant three-way interaction between alcoholism, lead, and manganese ($p < 0.0169$, t ratio $= 2.39$, $F_{1,2898} = 5.72$).

In counties with no reported releases of lead or manganese and below average deaths from alcoholism, rates of violent crime are below average (216 per 100,000 compared to the national mean of 298). In contrast, the 52 counties with toxic releases from both metals and above average rates of alcoholism have almost four times as much violent crime (920 per 100,000). Although epidemiologists are increasingly aware of the importance of such synergistic interactions, most conventional models of violent crime have looked at individual variables rather than complex effects of ecological and lifestyle factors on brain chemistry and behavior[2,3,9].

The correlations between environmental pollution and crime interact significantly with population density. For example, in counties with no reported pollution from either lead or manganese and below average alcoholism, population density makes little difference in crime rates (which are 170 per 100,000 in the 677 counties with below average density, and 265 in the 565 counties with above average density). In contrast, where lead and manganese pollution are accompanied by high rates of alcoholism, densely populated counties ($n=48$) report 970 violent crimes per 100,000, or three times the national average, while the four low density counties with similar neurotoxicity have only 138 violent crimes per 100,000. Since the stress involved in urban living has neurochemical correlates that exacerbate the effects of neurotoxicity[28,40,45], toxic pollution and rates of alcoholism seem to be hitherto unsuspected risk factors that contribute to geographic differences in violence.

Multivariate Analysis To ascertain how these effects might relate to other risk factors associated with violence, a multiple regression was computed, controlling for 10 socio-economic and demographic variables in addition to toxic variables and their interactions (Table IIIa). Consistent with a multi-causal model, twelve of these variables had significant effects, predicting over a third of the variance (adjusted $r^2=0.369$). Controlling for other factors (population density, median grade of education, police per capita, percent blacks and percent hispanics in poverty, per capita income, unemployment rate, percent households on public water supplies, rates of infant death, and percent housing built before 1950), lead and manganese toxicity and rates of death from alcoholism are significant predictors of violent crime, with significant two-way and three-way interactions among the three measures of toxicity.

Some findings from the multiple regression model are contrary to conventional wisdom. Unemployment is not associated with overall rates of violence, and the higher the median grade of school completed, the *higher* the crime rate (suggesting that educational failure *per se* is not a source of violent behavior). Conversely, sociological theories of crime do not predict that the percent of households on public water supplies could be a significant risk factor of crime. Since aging pipes in public water systems often contain lead and manganese[117–120], this finding provides additional support for the neurotoxicity hypothesis. Finally, age of housing stock—often assumed to measure infants' exposure to pica from leaded paint—is *negatively* associated with violence, suggesting that water supplies and environmental pollution are more likely than paint to serve as pathways for lead[121].

To insure that the findings were not an artifact of inaccurate statistics, these results were checked by using the same factors to predict rates of aggravated assault (a category of violent crime not confounded by reporting errors attributed to sexual assault, yet more frequent and hence more reliable in rural areas than homicide), property crime (to measure non-violent deviance), per capita welfare loads (an index of social disintegration), and median grade of education completed (Table IIIb–e). Findings are consistent with the hypothesis that environmental pollution is a risk factor in violence, in the disintegration of stable families, and—to a lesser extent—in educational failure.

These results are also consistent with the evidence that exposure to environmental pollutants does not effect all people in the same way. Since deficiencies in calcium, zinc or other vitamins and minerals are associated with uptake of lead and manganese, population groups with poor diets and low levels of breast feeding should have greater vulnerability to toxic chemicals[5,12,14,17,38,57,122–126]. Comparison of the regression equations predicting violent and property crime (Table IIIa–c) is consistent

Table III Multiple regression model for 50 states

Factors	IIIa: All Violent Crimes	IIIb: Aggravated Assault	IIIc: Property Crimes	IIId: Welfare Recip/Cap	IIIe: Median Grade Completed
Population Density (log)	82.42**** (20.24)	38.77**** (12.26)	490.91**** (22.39)	0.00574**** (24.56)	−0.08903**** (−8.23)
Per-capita Income	−0.007**** (−2.74)	−0.0059*** (−2.81)	NS	NS	0.00017**** (25.23)
Unemployment	NS	2.27* (1.71)	16.05* (1.74)	0.00021** (2.10)	−0.03665**** (−8.06)
%Black Poverty	40.06** (2.33)	27.62** (2.07)	NS	NS	−0.13322*** (−2.88)
%Hispanic Poverty	62.11*** (2.79)	41.23** (2.39)	541.5**** (4.52)	NS	NS
Police/Capita	153423**** (16.56)	89318**** (12.42)	821385**** (16.47)	4.97007**** (9.34)	NS
Infant Death Rate	(1.813)*** (2.78)	1.173** (2.32)	NS	0.000065* (1.74)	NS
%Housing pre-1950	−526.75**** (−13.43)	−447.34**** (−14.69)	−3722**** (−17.63)	0.007379*** (3.28)	0.93843**** (9.02)
Public Water/Cap	225.34**** (4.07)	197.4**** (4.59)	2618**** (8.78)	NS	1.00734**** (6.82)
Median Grade Completed	24.68*** (3.50)	16.82*** (3.07)	586.7**** (15.46)	0.001109** (2.74)	N/A
Lead Present/ Absent	40.80**** (4.67)	18.49* (2.73)	134.5*** (2.86)	0.003491**** (6.95)	NS
Manganese Present/Abs.	58.71**** (6.68)	39.26**** (5.75)	155.3**** (3.28)	0.004482**** (8.88)	0.05672** (2.40)
Alcohol Death Rate	101.62**** (11.55)	62.46**** (9.14)	188.4*** (3.97)	0.001186** (2.35)	−0.10258**** (−4.35)
Alcohol * Lead	21.48** (2.54)	NS	NS	0.001626*** (3.34)	0.05316** (2.34)
Alcohol * Manganese	55.40**** (6.54)	36.79**** (5.59)	117.6** (2.58)	0.001578*** (3.24)	0.04051* (1.78)
Lead * Manganese	34.89**** (4.11)	15.57** (2.36)	NS	0.003126**** (6.41)	NS
Alcohol * Lead * Manganese	19.21** (2.27)	NS	NS	0.001764*** (3.63)	NS
adj r^2	0.369	0.259	0.467	0.369	0.415
F	97.45	58.60	145.12	97.17	125.02
DF	17,2783	17,2783	17,2783	17,2783	16,2784
p	0.0000	0.0000	0.0000	0.0000	0.0000

**** $p<0.0001$, *** $p<0.005$, ** $p<0.05$, * $p<0.10$.
Note: Cell entries are unstandardized coefficients (t ratio in parentheses).

with this explanation: while population density, police per capita, and age of housing have similar effects on both types of criminal behavior, lower per capita income and black poverty—which presumably are associated with higher risks of dietary insufficiency—are not correlated with property crime as they are for total violent crime or aggravated assault.

As would be predicted by the neurotoxicity hypothesis, percent of households on public water supplies—a probable pathway of toxicity—is correlated with increased crime of all kinds, but not with percent of population on welfare. Moreover, neither ethnic poverty nor alcoholism is significantly associated with percent of population on welfare, suggesting that these factors associated with crime do not have a direct impact on familial and social disintegration.

Structural Co-Variate Model Although multiple regression and analysis of variance are widely used by social scientists to assess the effects of diverse variables on phenomena such as crime, these statistical methods have been criticized. The statistics reported in Table IIIa–e are mathematically based on the debatable assumption that the predictive factors are "independent;" in multiple regression models where the predictive variables are interdependent, results are highly unreliable. In a survey of geographic analyses of homicide rates, Land, McCall, and Cohen demonstrated the importance of this objection by showing how apparently parallel statistical studies implicated quite diverse risk factors to explain variations between cities, metropolitan areas, or states[127].

To circumvent this analytical problem, Land, McCall and Cohen introduced a more complex methodology. First, they computed a principal components analysis of the set of predictive variables, thereby identifying underlying dimensions or factors that might account for interrelated variables (such as *per capita* income and rates of unemployment); each geographical entity was then given a "factor score" for each of these underlying factors or dimensions—and it is these scores that are used to predict crime rates. Using this method, called a "structural co-variate model," Land, McCall, and Cohen found stable findings for homicide rates in cities and states, with three "structural indexes/covariables—population structure, resource deprivation/affluence, and percentage of the male population divorced" showing "statistically significant relationships to the homicide rate . . . across all time periods and levels of analysis"[128]. In this study, "population structure" represents the average population density and population size of the geographic units; "resource deprivation/affluence" reflects income, poverty, economic inequality, blacks as a percentage of population and percentage children under 18 not living with both parents. Many of the variables used in our study were not included by Land, McCall, and Cohen, and their results also showed that cities in the "south" had, controlling for all factors, consistently higher homicide rates than elsewhere in the US.

To insure that our findings are not vitiated by the statistical questions raised by Land, McCall, and Cohen, our set of 21 predictor variables was factor analyzed (Table IVa). Given the larger variability introduced by studying counties, it was not surprising that we found seven distinct factors: wealth and inequality, comparable to their "resource deprivation/affluence" (reflecting per capita income, median grade of education completed, percent white poverty, and—more weakly—unemployment); urban-rural, comparable to their "population structure" (reflecting population density, welfare load per capita, and welfare expenditures per capita); old housing-white population (reflecting % housing pre-1950, % white, and negatively % hispanic); public goods (reflecting sewers per capita, households on public water supplies per capita and police per capita), rates of alcoholism (reflecting two measures of alcohol disease and death), black population (reflecting % black population and black poverty), and industrial pollution (reflecting the toxic release inventories for lead and manganese). It is worth stressing that this approach confirms that the geographical distribution of our measures of neurotoxicity are independent of conventional socio-economic and demographic variation.

Table IV Principal components analysis and structural covariate model

IVa Principal components analysis			IVb Structural co-variate model		
	21 variables		Estimate	*t*-ratio	Std. error probability
Wealth			0.0155	5.85	0.00266 < 0.0001
PC income	0.77				
Med. grade	0.62				
White poor	−0.89				
Unemployment	−0.51				
(Hispanic poor)*	(−0.43)				
Eigenvalue		3.6			
% Variance		17.30%			
Old housing/White			−567.19	−21.11	26.8969 < 0.0001
Housing pre 1950	0.82				
White %	0.62				
Hispanic %	−0.51				
Eigenvalue		2.8			
% Variance		13.30%			
Public goods			496.648	16.7	29.738 < 0.0001
Sewers/Capita	0.89				
Public H2o/Cap	0.83				
Police/Cap	0.74				
Eigenvalue		2.3.			
% Variance		11.20%			
Rural			−48.071	−18.56	2.59 < 0.0001
Pop. Dens./Sq. Mi.	−0.61				
AFDC $/Capita	−0.77				
AFDC #/Capita	−0.76				
Eigenvalue		1.6			
% Variance		7.50%			
Alcoholism			−3.62	−9.32	0.388 < 0.0001
Deaths—Alcoholism	−0.83				
All alcohol related	−0.76				
Eigenvalue		1.4			
% Variance		6.60%			
Black population			150.183	5.53	27.173 < 0.0001
% Black	0.73				
Black poverty	0.51				
(Infant death)*	0.46				
Eigenvalue		110.00%			
% Variance		5.10%			
Industrial pollution**			−31.698	−5.34	5.936 < 0.0001
Lead TRI	0.77				
Manganese TRI	0.71				
Eigenvalue		1			
% Variance		5.00%			
Intercept in co-variate model			6.461	0.16	40.787 ns
Adjusted r-square in co-variate model					0.411

* Variables poorly predicted by factor structure (commonality < 0.25)

** Entered in structural co-variate model as a dichotomous variable counties either with or without pollution from either lead or manganese. (Negative sign means less crime where no pollution.) A parallel model with continuous variation on this co-variate showed little difference, with seven factors all highly significant and explaining comparable variance (adjusted *r*-square = 0.407).

Following the procedure suggested by McCall, Land, and Cohen, we then computed the multiple regression using this structural co-variate model (Table IVb). As will be seen, the principal components or factors reflecting environmental pollution and alcoholism—the direct measures of neurotoxicity—remain highly significant. Moreover, this approach probably understates the effects of environmental pollution. On the one hand, the toxicity of soils in the highly traveled automobile corridors studied by Mielke and his collaborators is a characteristic of urban, high density counties (and hence is including in the rural–urban co-variate); on the other, aging public water supplies are included in the public goods co-variate. That these pathways of toxicity are more important than aging house-paint is confirmed by the negative sign for the co-variant for % white and housing pre-1950.

VII. CONCLUSIONS AND POLICY IMPLICATIONS

In these findings, urbanism, ethnicity and toxicity emerge as important correlates of violent crime. If so, the traditional approaches to crime in the United States need to be reconsidered from an ecological point of view. The environment seems to increase the probability that some individuals lose capacities for impulse control, leading to increased rates of crime and reinforcing patterns of school dropout, family disintegration, and unemployment.

Apart from studies cited above, the principal geographic explanation of violence in the literature is Nisbett and Cohen's analysis of the Southern "culture of honor"[129,130]. Although this hypothesis is not inconsistent with the present findings, the data presented by Nisbett and Cohen focus on differential rates of violence in small cities and towns as well as broad regional patterns of culture. Moreover, their attribution of an upland herding "culture of violence" to "Scotch–Irish" immigration could well introduce co-variation with rates of alcoholism, a trait that may have a genetic component[131]. In any event, regional cultural attitudes, state legislation, and differences in homicide rates between cities under 200,000 in "hills and high plains" and in "moist plains"[117] do not explain many of the county-level variations we have observed and analyzed.

The neurotoxicity hypothesis may also help to explain historical trends that have otherwise puzzled many experts. Evidence that rates of violent crime in New York and other cities have fallen about 10% (*New York Times*, 23 July 1995, Section 4, pp. 1, 4; *New York Times*, 13 August 1995, pp. 1, 18) may be linked to changes in diet and neurotoxicity. Lead consumption in the U.S. which was over 0.201×10^6 mt per year in 1977–1978, declined to 0.119×10^6 mt per year in 1982, and to pre-1943 levels of 0.028×10^6 mt per year in 1988–1989, with corresponding declines in deposition in air and even in forest soils[97]. At the same time, calcium consumption, which had risen to historically high levels during World War II, declined from 1945 through the 1960s before beginning a modest increase. Because there is a lag of 15–20 years between the effects of lead uptake on neonatal or infant development and the onset of criminal behavior in juvenile or early adult years, increased rates of violence between 1955 and 1975 may have reflected the combination of greater emissions of lead and reduced calcium intake. Further research is therefore needed to assess whether the recent declines in violence might be in part related to the maturation of the first generations of youth to benefit from the prohibition of leaded gasoline and other initiatives to reduce environmental pollution.

Cross-national data are also be relevant. It is sometimes claimed, for example, that the neurotoxicity hypothesis is contradicted by low rates of violence in Canada, where MMT—a manganese based gasoline additive—has been used since leaded gasoline became illegal. In fact, the statistics may actually suggest the contrary. Although homicide and robbery rates in Canada are lower than in the United States (perhaps in part due to gun control and other public policies), when all four types of crime classified as violent by the FBI (homicide, aggravated assault, sexual assault, and robbery) are combined, rates are actually higher in Canada than the U.S. (Fig. 7). This effect is clearly related to rates of aggravated assault, the most frequent kind of interpersonal violence in both countries. More important for present purposes, since the introduction of MMT, the rate of all violent crimes has increased at a faster rate in Canada than in the U.S. Since factors such as guns and ethnicity that are often associated with high crime rates in the U.S. are absent in Canada, the increasing difference between the two countries contradicts the assertion that the Canadian experience with MMT is evidence against a relationship between manganese and violent behavior.

If confirmed, the neurotoxicity hypothesis would have obvious implications for public policy. Crime prevention and improved educational performance may be greatly enhanced by parent-training in proper diet. Studies indicate that breast feeding increases IQ in both preterm, low birthweight infants[132] and full-term infants[133]. Vitamin supplements may be useful in improving some aspects of school performance and cognitive development, at least for those with poor diets[5,134–138].

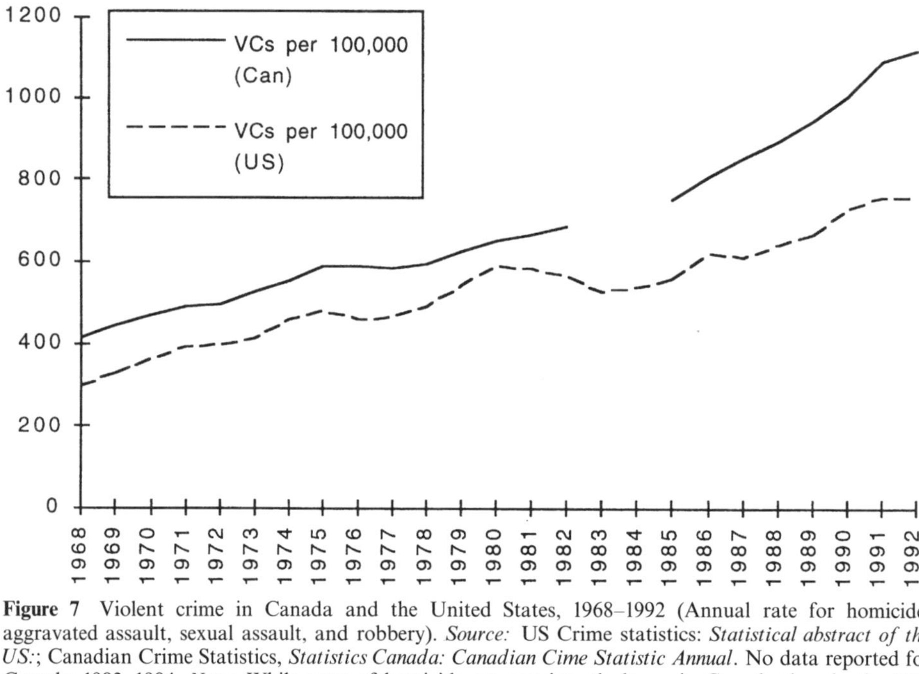

Figure 7 Violent crime in Canada and the United States, 1968–1992 (Annual rate for homicide, aggravated assault, sexual assault, and robbery). *Source:* US Crime statistics: *Statistical abstract of the US:*; Canadian Crime Statistics, *Statistics Canada: Canadian Cime Statistic Annual*. No data reported for Canada, 1983–1984. *Note:* While rates of homicide are consistently lower in Canada than in the U.S. (presumably as a result of such factors as gun control legislation and urbanization), the phenomena of violent crime are most closely correlated with and better measured by aggravated assault, whose rates closely parallel the overall rates shown here. Cf. F. Zimring, "Firearms, violence, and public policy," *Scientific American* **265**, 48–57 (1991); A. A. Leenaars, D. Lester, "Effects of gun control on homicide in Canada," *Psychological Reports*, **75**, 81–82 (1994). While cross-cultural differences in reporting categories and criteria might explain the higher rates in Canada, they would not explain the differing trends since 1985.

Pre-school programs, which in some cases have been found to reduce rates of crime[139,140], need to be evaluated in terms of their effects on diet rather than by conventional educational assessments; indeed, the short-term nature of the gains in IQ reported in follow-up studies of Head Start may reflect the essential role of nutritional supplementation in remedial education[141,142]. If releases of neurotoxic metals are associated with rates of crime, as this study suggests, reducing environmental pollution takes on higher priority. In our criminal justice system, because existing means of predicting recidivism are little better than chance, the assay of toxic metals in head hair might provide an inexpensive marker for potential violence in probation decisions[35]. With adequate identification of the precise biochemical imbalances associated with violent crime, vitamin and mineral normalization could contribute to improved rehabilitation[13,35,40].

One issue of immediate importance is posed by MMT, the manganese-based gasoline additive used in Canada. The introduction of MMT into the United States has long been prohibited under an EPA finding that it violated the Clean Air Act[143]. The recent reversal of this decision by a Federal Appellate Court (*Wall Street Journal*, October 25, 1995) increases the importance of the present findings[144]. While the manufacturers have claimed that "no adverse health effects have been observed" from the use of MMT in Canada, traditional studies use standards of occupational exposure that do not consider interactions between manganese, lead, alcoholism, and dietary insufficiencies, and focus on health rather than on behavioral outcomes such as learning disabilities and crime that are sensitive to subclinical levels of toxicity[145]. In animal models, ingestion of MMT alters catecholamine function[146]. And in one study comparing Canadian blue collar workers and garage mechanics, higher atmospheric exposure to manganese was associated with higher levels of the metal in garage mechanics head hair[147]. This study is, however, far from conclusive: the contribution of atmospheric manganese was estimated as 1% of total absorption, and the same investigators found these garage mechanics to be exposed to significantly higher levels of manganese in their household tap water[148]. At most, therefore, it can be said that more study is needed, including consideration of occupational effects of MMT vapor and skin contact.

Given the extraordinary level of violence that persists in urban America and the failure of traditional policies to meet it, further confirmation of the hypothesized relationships between dietary deficiencies, neurotoxic uptake, and violence is urgently needed. Of particular importance will be direct studies correlating environmental exposure, diet, neurotoxic uptake, and behavior. Equally important will be experimental studies of the effects of vitamin and mineral normalization for violent offenders. Although changes in the criminal justice system should not be implemented without further evidence, the neurotoxicity hypothesis may require important changes in the way Americans think about crime and its prevention.

APPENDIX ON METHODOLOGY

Our analysis is based on a dataset constructed from multiple sources; all data are by county, a reasonable geographic unit for measuring the effects of industrial pollution and geographic variability in crime rates[14]. Counties without data for FBI crime reports were dropped from the analysis. Variables, with definition and source, are as follows:

Violent Crime. Data for 1991: the sum of (1) Murder and nonnegligent manslaughter ("the willful (nonnegligent) killing of one human being by another");

(2) Forcible rape ("carnal knowledge of a female forcibly and against her will. Assaults or attempts to commit rape by force are also included; however, statutory rape (without force) and other sex offenses are excluded"); (3) Robbery ("taking or attempting to take anything of value from the care, custody, or control of a person or persons by force or threat of force or violence and/or by putting the victim in fear"); and (4) Aggravated assault ("unlawful attack by one person upon another for the purpose of inflicting severe or aggravated bodily injury"). Source: Federal Bureau of Investigation[149].

Property Crime. Data for 1991: the sum of (1) burglary ("the unlawful entry of a structure to commit a felony or theft"). (2) larceny-theft ("the unlawful taking, carrying, leading, or riding away of property from the possession or constructive possession of another [including] shoplifting, pocket-picking, purse snatching, thefts from motor vehicles, thefts of motor vehicle parts and accessories, bicycle thefts, etc. in which no use of force, violence or fraud occurs. This crime category does not include embezzlement, 'con' games, forgery, worthless checks, and motor vehicle theft). (3) Motor vehicle theft ("the theft or attempted theft of a motor vehicle"). Source: Federal Bureau of Investigation, Uniform Crime Reporting (UCR) Program[80].

Toxic Release Inventory. Total in pounds per year for 1991, computed separately for lead, lead compounds, manganese, and manganese compounds (sum of midpoint of range for "point air release, land release, chemicals transferred to an off-site facility, chemicals transferred to publicly owned treatment works, underground injection releases, discharges to water"). Source: U.S. Environmental Protection Agency, Toxic Release Inventory, 1987–1993[150].

Population. Total population in county as recorded in the FBI database. Source: Federal Bureau of Investigation, Uniform Crime Reporting (UCR) Program[80].

Population Density. Definition. "Persons per square mile is the average number of inhabitants per square mile of land area." Source: U.S. Bureau of Census, Population Division, July 1, 1992[80].

Houses with Water Supply from a Public System or Private Company. Households served by a common source supplying water to five or more units ... supplied by a city, county, water district, water company, etc. or a well which supplies water to five or more housing units[80].

Houses on Public Sewer Systems. Households served by "a public sewer ... operated by a government body or by a private organization. Source: U.S. Bureau of the Census, Census of Government, Public Employment"[80].

Housing Built before 1950. Percentage of households living in units constructed before 1950. (Comparable data also analyzed for housing units built before 1939.) Source: U.S. Bureau of the Census, Census of Government, Public Employment[80].

Police Officers. Police protection employees of local government with arrest power, as "full time equivalent." Source: U.S. Bureau of the Census, Census of Government, Public Employment[80].

School Drop-outs. Educational Attainment—Persons 25 years and over completing less than 9th grade. Source: U.S. Bureau of the Census, Census of Population and Housing[80].

High School Drop-outs. Educational Attainment—Persons 25 years and over completing 9th to 12th grade, no diploma. Source: U.S. Bureau of the Census, Census of Population and Housing[80].

Educational Achievement — BA or Higher Education. "Persons 25 or older—Bachelor's, graduate, or professional degree." Source: U.S. Bureau of the Census, Census of Population and Housing[80].

Per capita Money Income. Aggregate money income ("wages and salaries, non-farm self-employment; farm self-employment; Social Security; public assistance; and all other regularly received income such as veteran's payments, pensions, unemployment compensation, and alimony") divided by total resident population. Source: U.S. Bureau of the Census, Census of Population and Housing[80].

Unemployment Rate. Number of unemployed as a percent of the civilian labor force ("all civilians 16 year old and over classified as employed or unemployed"). Source: U.S. Bureau of Labor Statistics[80].

White Population. U.S. Census Bureau classification based on guidelines in Federal Statistical Directive #15, OMB ("Includes persons who indicated their race as 'White' or reported entries such as Canadian, German, Italian, Lebanese, Near Easterner, Arab or Polish"). Source: U.S. Bureau of the Census, Population Division, July 1, 1992[80].

Black population. U.S. Census Bureau classification ("Includes persons who indicated their race as 'Black or Negro' or reported entries such as African American, Afro-American, Black Puerto Rican, Jamaican, Nigerian, West Indian, or Haitian"). Source: U.S. Bureau of the Census, Population Division, July 1, 1992[80].

Hispanic Population. U.S. Census Bureau classification ("Includes persons who classified themselves in one of the specific Hispanic origin categories listed on the [self-identification] questionnaire—'Mexican,' 'Puerto Rican,' or 'Cuban'—as well as those who indicated that they were of 'other Spanish/Hispanic' origin. ... Persons of Hispanic origin may be of any race"). Source: U.S. Bureau of the Census, Population Division, July 1, 1992[80].

Persons below Poverty Level (by racial or ethnic category). U.S. Census Bureau definition of poverty, computed separately for whites, blacks, and hispanics ("poverty thresholds are computed on a national basis only... Poverty status is derived on a sample basis"). Source: U.S. Bureau of the Census, Census of Population and Housing[80].

Aid to Families with Dependent Children (AFDC)—Number Served. Total recipients, month of February ("Federal grants to help defray State costs of providing financial assistance to needy children who are under age 18 (or under age 21 and attending school); living in the home of a parent or other relative; and deprived of parental support or care because of the death, continued absence from the home, or physical or mental incapacity of a parent—or if a State elects, the unemployment of a father"). Source: U.S. Social Security Administration[80].

Aid to Families with Dependent Children (AFDC)—Payments. Total payments, in dollars, for February. Source: U.S. Social Security Administration[80].

Women, Children and Youth Program (ACYF)—Number Served. Total number of children enrolled in WIC program[151].

Woman, Children and Youth Program (ACYF)—Payments. Average monthly expenditures, ACYF program[82].

Infant Deaths per 1000 Live Births. "infant death rates represent the number of deaths of infants under 1 year of age per 1,000 live births. They exclude fetal deaths."

Source: U.S. National Center for Health Statistics, Vital Statistics of the United States[80].

All Alcohol Related Causes of Death. "Any death certificate with explicit mention of alcohol, diseases often caused by alcohol or accident often caused by alcohol. If, for example, 42% of auto deaths involve alcohol, 42% of traffic deaths are included." Source: U.S. Department of Health and Human Services, National Institute on Alcohol Abuse and Alcoholism[152].

Causes of Death with Explicit Mention of Alcohol. "Alcoholic psychoses, alcohol dependent syndrome; nondependent abuse of alcohol; alcoholic polyneuropathy, cariomyopathy, gastritis, fatty liver, hepatitis (acute), cirrhosis of the liver, liver damage unspecified; excess blood alcohol content, accidental poisoning by ethyl alcohol, not elsewhere specified." Source: U.S. Department of Health and Human Services, National Institute on Alcohol Abuse and Alcoholism[80].

Parameters for analyses of variance (ANOVA) and multiple regression analyses are noted in Tables and Figures. Earlier work with samples of counties in eight states[153] and sixteen states[154] showed similar patterns of correlation between toxic metals, alcoholism, and violent crime. While more complex statistical manipulations are possible (e.g. path analysis), inferential statistics must in any event be supplemented by direct observation of individual diet and neurotoxic uptake in polluted versus non-polluted environments. Such tests of the neurotoxicity hypothesis are urgently needed to assess causal factors in learning disabilities and health risks as well as probabilities of engaging in violent behavior[155].

References

1. Letter from Benjamin Franklin to Benjamin Vaughan "on the bad Effects of Lead taken inwardly," (31 July 1786) **In**: (Lemay, ed.) *Complete Writings of Benjamin Franklin* (Library of America, New York, 1987) pp. 1163–1166. I thank T. Brader from bringing this passage to my attention.
2. C. E. Koop and G. D. Lundberg, "Violence in America: A Public Health Emergency" *J. Am. Med. Assn.* **267**, 3075–3076 (1992).
3. For example, the reasons for recently reported declines in homicide rates in many American cities, like those for prior historical changes, have been described as "mysterious" (*New York Times*, 23 July 1995, Section 4, pp. 1, 4) and "unclear" (*New York Times*, 13 August 1995, pp. 1, 18).
4. F. J. Earls, "Violence and Today's Youth" *Future of Children* **4**, 4–23 (1994).
5. J. Wilson and J. Petersilia (Eds.) *Crime* (Institute for Contemporary Studies, San Francisco, 1995).
6. W. W. Gibbs, "Seeking the Criminal Element" *Scientific American* **272**, 100–107 (1995).
7. As a leading scientific journalist puts it: "...the nature–nurture dichotomy is itself an illusion. As many scholars are now realizing, everything we associate with 'nurture' is at some level a product of our biology—and every aspect of our biology, from brain development to food preference, has been shaped by an environment." G. Cowley, "It's Time to Rethink Nature and Nurture" *Newsweek* 52–53 (27 March, 1995).
8. D. Bryce-Smith, "Environmental Chemical Influences on Behaviour and Mentation" *Chem. Soc. Rev.* **15**, 93–123 (1986).
9. S. J. Lippard and J. M. Berg, *Principles of Bioinorganic Chemistry* (University Science Books, Mill Valley, CA, 1994).
10. "If the concentration of a given essential metal ion is too low, processes that need to use that ion will be adversely affected, and the organism can suffer from metal-ion deficiency. Once the concentration of a given metal ion is above a lower threshold, there will be enough of that ion to fulfill its biological functions. The concentration cannot be increased indefinitely without adverse consequences, however. Above an upper threshold, the effects of metal-ion toxicity will arise. ... Some metal ions have no known or presumed biological function... metal ions not utilized in biological systems can be quite toxic, often because they tend to bind nonspecifically, but with high affinity, to certain types of sites. Because of this tight binding, which is often a consequence of kinetic inertness, these metals may bind to sites where they inhibit some normal process in such a manner that they are not easily removed

and excreted." (S. J. Lippard and J. M. Berg, *Principles of Bioinorganic Chemistry* (University Science Books, Mill Valley, CA, 1994) pp. 139–140.)
11. J. Schubert, E. J. Riley and S. A. Tyler, "Combined Effects in Toxicology—A Rapid Systematic Testing Procedure: Cadmium, Mercury, and Lead" *J. Toxic. and Env. Health* **4**, 763–776 (1978).
12. P. E. Cromwell, B. R. Abadie, J. T. Stephens and M. Kyler, "Hair Mineral Analysis: Biochemical Imbalances and Violent Criminal Behavior" *Psychological Reports* **64**, 259–266 (1989).
13. L. T. Fairhall and P. A. Neal, *Industrial Manganese Poisoning* National Institute of Health Bulletin, No. 182. (U.S. Government Printing Office: Washington, DC, 1943).
14. L. Gottschalk, T. Rebello, M. S. Buchsbaum, H. G. Tucker and E. L. Hodges, "Abnormalities in Trace Elements as Indicators of Aberrant Behavior" *Comprehensive Psychiatry* **32**, 229–237, at 235 (1991).
15. For a review of the evidence of manganese toxicity, see Brief of *Amici Curiae* Violence Research Foundation and Citizens for Health, *Ethyl Corp. v. Browner*, U.S. Court of Appeals for District of Columbia Circuit, Case No. 94-1505 (November, 1994).
16. D. Bryce-Smith, "Lead Induced Disorder of Mentation in Children" *Nutrition and Health* **1**, 179–194 (1983).
17. V. A. Murphy, J. M. Rosenberg, Q. R. Smith and S. I. Rapoport, "Elevation of Brain Manganese in Calcium-Deficient Rats" *Neuro Toxicology* **12**, 255–264 (1991).
18. C. R. Cloninger, "A Unified Biosocial Theory of Personality and its Role in the Development of Anxiety States" *Psychiatric Developments* **3**, 167–226 (1986).
19. C. R. Cloninger, "A Systematic Method of Clinical Description and Classification of Personality Variants" *Archives of General Psychiatry* **44**, 573–588 (1987).
20. C. R. Cloninger, D. M. Svrakic, and T. R. Przybeck, "A Psychobiological Model of Temperament and Character" *Archives of General Psychiatry* **50**, 957–990 (1993).
21. R. Masters, M. T. McGuire (eds.) *The Neurotransmitter Revolution* (S. Ill. Univ. Press: Carbondale, IL, 1993), esp. ch. 9–10.
22. R. Sapolsky, *Stress* (MIT: Cambridge, 1992).
23. R. Sapolsky, "Why Stress is Bad for Your Brain" *Science* **273**, 749–750 (1996).
24. J. Virkkunen, M. Eggert, R. Rawlings and M. Linnoila, "A Prospective Follow-up Study of Alcoholic Violent Offenders and Fire Setters" *Archives of General Psychiatry* **53**, 523–529 (1996).
25. M. Linnoila *et al.*, "Serotonin and Violent Behavior" **In:** *The Neurotransmitter Revolution* (R. Masters and M. T. McGuire, eds.) (Southern Illinois University Press, Carbondale, Illinois, 1993) pp. 61–95.
26. G. N. Schrauzer and K. P. Shrestha, "Lithium in Drinking Water and the Incidences of Crimes, Suicides, and Arrests Related to Drug Addictions" *Biological Trace Element Research* **25**, 105–113 (1990).
27. T. H. Maugh, "Hair: A diagnostic tool to complement blood serum and urine" *Science* **202**, 1271–1273 (1978).
28. M. Laker, "On Determining Trace Element Levels in Man: the Uses of Blood and Hair" *Lancet* 259–262 (31 July, 1982).
29. R. O. Pihl and F. Ervin, "Lead and Cadmium Levels in Violent Criminals" *Psych. Rep.* **66**, 839–844 (1990).
30. M. Marlowe, H. G. Schneider and L. B. Bliss, "Hair: A Mineral Analysis in Emotionally Disturbed and Violence Prone Children" *Biosocial Med. Research* **13**, 169–179 (1991).
31. A. G. Schauss, "Comparative hair-mineral analysis results in a random selected behaviorally normal 15–59 year old population and violent criminal offenders" *Int. J. Biosocial Research* **1**, 21–41 (1981).
32. L. Gottschalk, T. Rebello, M. S. Buchsbaum, H. G. Tucker and E. L. Hodges, "Abnormalities in Trace Elements as Indicators of Aberrant Behavior" *Comprehensive Psychiatry* **32**, 229–237 (1991).
33. M. Marlowe, "Hair Mineral Analysis in Emotionally Disturbed and Violence Prone Youth" (In Preparation, Appalachian State University, Boone, NC).
34. G. N. Schrauzer, K. P. Shrestha and M. F. Flores-Arce, "Lithium in Scalp Hair of Adults, Students, and Violent Criminals" *Biological Trace Element Research* **34**, 161–176 (1992).
35. W. W. Gibbs, "Seeking the Criminal Element" *Scientific American* **272**, 100–107 (1995).
36. D. W. Denno, "Gender, Crime, and the Criminal Law Defenses" *J. Criminal Law & Criminology* **85**, 80–180 (1994).
37. H. L. Needleman, J. A. Riess, M. J. Tobin and G. E. Biesecker, "Greenhouse JB. Bone Lead Levels and Delinquent Behavior" *JAMA* **275**, 363–369 (1996).
38. J. Virkkunen, M. Eggert, R. Rawlings and M. Linnoila, "A Prospective Follow-up Study of Alcoholic Violent Offenders and Fire Setters" *Archives of General Psychiatry* **53**, 523–529 (1996).
39. M. Linnoila *et al.*, "Serotonin and Violent Behavior" **In**: *The Neurotransmitter Revolulion* (R. Masters and M. T. McGuire, eds.) (Southern Illinois University Press, Carbondale, Illinois, 1993) pp. 61–95.
40. D. Hunter, *Diseases of Occupation* (Little Brown, Boston,1972).

41. M. Rutter and R. R. Jones (eds.), *Lead versus Health* (John Wiley, New York, 1983).
42. R. Rabin, "Warnings Unheeded: A History of Child Lead Poisoning" *Am. J. Publ. Health* **79**, 668–674 (1989).
43. H. L. Needleman (ed.), *Human Lead Exposure* (CRC Press, Boca Raton, FL, 1989).
44. D. C. Rice, "Behavioral Deficit (Delayed Matching Sample) in Monkeys Exposed from Birth to Low Levels of Lead" *Toxicol. Applied Pharm.* **75**, 337–345 (1994).
45. J. Pirkle *et al.*, "The Decline in Blood Lead Levels in the United States" *JAMA* **272**, 284–291 (1994).
46. D. A. Cory-Sletcha, "Relationships between Lead Induced Learning Impairments and Changes in Dopaminergic, Cholinergic, and Gutamatergic Neurotransmitter System Functioning" *Ann. Rev. Pharm. Toxic.* **35**, 391–395 (1995).
47. P. B. Hammond, "Metabolism of Lead" **In**: *Lead Absorption in Children* (J. J. Chisholm and D. M. O'Hara, eds.) (Urban and Schwartzenberg: Baltimore, MD, 1988) pp. 11–20.
48. D. Bryce-Smith, "Lead Induced Disorder of Mentation in Children" *Nutrition and Health* **1**, 179–194 (1983).
49. D. Brody *et al.*, "Blood Lead Levels in the U.S. Population" *JAMA* **272**, 277–283 (1994).
50. In one laboratory, analysis of lead in head hair of large samples over the last decades reveals lead levels between 30% and 100% higher among blacks than whites (Dr. R. Smith, Doctor's Data Lab., W. Chicago, IL, pers. com.)
51. M. Berode, V. Wietlisbach, M. Richenbach and M. P. Guillemin, "Lifestyle and environmental factors as determinants of blood lead levels in a Swiss population" *Environ. Res.* **55**, 1–17 (1991).
52. H. Needleman *et al.*, "The Long-term Effects of Exposure to Low Doses of Lead in Childhood" *N.E. Journ. Medicine.* **322**, 83–88 (1990).
53. H. Needleman and B. Gatsonis, "Meta-analysis of 24 Studies of Learning Disabilities due to Lead Poisoning" *JAMA* **265**, 673–678 (1991).
54. E. Tiffany-Castiglioni, M. E. Legare, L. A. Schneider, W.H. Hanneman, E. Zenger and S. J. Hong, "Astroglia and Lead Neurotoxicity" **In**: *The Role of Glia in Neurotoxicity* (M. Aschner and H. K. Kimelberg, eds.) (CRC Press Boca Raton, FL, 1996) 175–200.
55. H. Hu, A. Aro, M. Payton, S. Korrick, D. Sparrow, S. T. Weiss and A. Rotnitzky, "The Relationship of Bone and Blood Lead to Hypertension" *JAMA* **275**, 1171–1176 (1996).
56. D. Bryce-Smith, "Lead Induced Disorder of Mentation in Children" *Nutrition and Health* **1**, 179–194 (1983).
57. Conventional accounts of the role of manganese in human physiology and disease emphasize metabolic processes, with little attention to the brain except for "industrial manganism" See F. H. Nielson, "Manganese" *Modern Nutrition in Health and Disease*, 8th ed., pp. 275–277 (1994). Because there is little consensus on the normal function of manganese in humans, nutritional experts have not even agreed on an established minimum daily requirement of manganese (Dr. Denis Bier and Dr. Curtis Hunt, Remarks at Conference on "*Law Medicine–and the Juvenile Justice System*," School of Criminology and Criminal Law, Florida State University, Tallahassee, FL, November 29–December 1, 1994). From this perspective, the evidence for direct correlations between dietary imbalance and violence seems confusing and unreliable: "Diet and Criminal Behavior," Letter of Dr. Kenneth Moritsugu, Assistant Surgeon General, Federal Bureau of Prisons to Congressman Thomas J. Bliley, Jr., May 27, 1994.
58. R. Truckenbroct, L. Wirter and K. H. Schaller, "Effect of occupational lead exposure on various elements in the human blood" [in German] *Zentralol. Bakeriol. Microbiol. Hyg.* **179**, 1187–1197 (1984).
59. J. Donaldson, F. S. Labella and H. Gesser, "Enhanced autooxidation of dopamine as a possible basis of manganese neurotoxicity" *Neurotoxicology* **2**, 53–64 (1981).
60. J. Donaldson and A. Barbeau, "Metal ions in Neurology and Psychiatry. Manganese Neurotoxicity" **In**: *Neurology and Neurobiology* (S. Gabay, J. Harris and B.T. Ho, eds.) (Alan R. Liss, NY, 1985) **Vol. 15**, pp. 259–285.
61. J. Donaldson, "The Physiopathologic Significance of Manganese in Brain: Its Relation to Schizophrenia and Neurodegenerative Disorders" *Neurotoxicology* **8**, 451–462 (1987).
62. J. Emord, Brief of *Amici Curiae* Violence Research Foundation and Citizens for Health, *Ethyl Corp. v. Browner*, U.S. Court of Appeals for District of Columbia Circuit, Case No. 94–1505 (November 1994).
63. L. T. Fairhall and P. A. Neal, *Industrial Manganese Poisoning*. National Institute of Health Bulletin, No. 182. (Washington, DC, U.S. Government Printing Office, 1943).
64. C. Kilburn, "Manganese, Malformation and Motor Disorders: Findings in a Manganese exposed population" *Neurotoxicol.* **30**, 421–430 (1987).
65. J. Cawte and M. T. Florence, "A Manganic Milieu in North Australia" *Int. J. Biosocial Med. Res.* **11**, 43–56 (1989).
66. These effects are due to a multiplicity of complex biochemical interactions. Manganese acts as an oxidant of dopamine, possibly when divalent manganese salts are oxidized to the trivalent state; this

process seems to be most likely in regions of the brain with an appropriate neurochemical mileu (for a review, see E. H. Hodges and F. Crinella, "*Effects of Nutritional Supplemention on Violent Behavior of Incarcerated Youthful Offenders*." Dept. Pediatrics, Univ. Cal. Irvine, 1994). Although there is some question about the precise mechanism through which manganese influences serotonin, it is known that abnormal levels of monoamine oxidase A (MAO A) can lead to unusually low or high levels of serotonin, either of which has been demonstrated to lead to higher levels of aggressive behavior (Masters and McGuire, 1993; Cases, 1995). According to some, manganese increases the activity of MAO A, thereby depleting serotonin and dopamine (M. N. Subhash and T. S. Padmashree, 1990. Regional distribution of dopamine B-hydroxylase and monamine oxidase in the brains of rats exposed to manganese, *Federation of Chemical Toxicology*, **28**, 567–570), perhaps through Manganese Superoxide Dismutase and the inhibition of free radical activity due to increased levels of malondialdehyde (Hodges and Crinella, 1994). It may be, however, that the effects are mediated by the destruction of serotonergic receptors rather than—or in addition to—reductions in levels of serotonin. In addition, a trivalent cation like manganese may interfere with normal brain function by other mechanisms not directly associated with the catecholamines. For example, since cations such as cadmium, nickel, cobalt and magnesium function as blockers of neuronal calcium channels (J. E. Richmond, E. Sher and I. M. Cooke, "Characterization of the Ca^{2+} Current in Dissociated Crustacean Peptidergic Neuronal Somata" *J. Neurophysiology* **73**, 2357–2368 (1995); B. R. Christie, L. S. Eliot, K.-I. Ito, H. Miyakawa, and D. Johnston, "Different Ca^{2+} Channels in Soma and Dendrites of Hippocarnpal Pyramidal Neurons Mediate Spike-Induced Ca^{2+} Influx" *J. Neurophysiology* **73**, 2553–2557 (1995)), ratios of manganese to calcium could have direct effects on the response patterns of neurons in such key neuroanatomical structures as the hippocampus.

67. M. Kimura, N. Yagi and Y. Itokawa, "Effect of Subacute Manganese Feeding on Serotonin Metabolism in the Rat" *J. Toxicol. Environ. Health* **4**, 701–707 (1978).
68. E. H. Hodges and F. Crinella, *Effects of Nutritional Supplementation on Violent Behavior of Incarcerated Youthful Offenders* (Dept. Pediatrics, Univ. Cal. Irvine, 1994).
69. F. C. Wedler, "Manganese" **In**: *The Role of Glia in Neurotoxicity* (M. Aschner and H. K. Kimelberg, eds.) (CRC Press, Boca Raton, FL, 1996) pp. 155–174.
70. R. Masters, M. T. McGuire (Eds.), *The Neurotransmitter Revolution*. (S. Ill. Univ. Press: Carbondale, IL, 1993).
71. P. D. Kramer, *Listening to Prozac* (Viking, New York, 1993).
72. V. A. Murphy, J. M. Rosenberg, Q. R. Smith and S. I. Rapoport, "Elevation of Brain Manganese in Calcium-Deficient Rats" *Neuro toxicology* **12**, 255–264 (1991).
73. C. L. Keen, J. G. Bell and B. Lönnerdal, "The Effect of Age on Manganese Uptake and Retention from Milk and Infant Formulas in Rats" *J. Nutrition* **116**, 395–402 (1986).
74. K. Dorner *et al.*, "Longitudinal manganese and copper balances in young infants and preterm infants fed on breast-milk and adapted cow's milk formulas" *British J. Nutr.* **61**, 559–572 (1989).
75. M. R. Werbach, "Aggressive Behavior" **In**: *Nutritional Influences on Mental Illness: A Sourcebook of Clinical Research*. (Third Line Press, Tarzana, CA, n.d.) pp. 6–15.
76. Doctor's Data, Inc., "Nutritional and Metabolic Findings for Behavior-Disordered Children and Teenagers" (Doctor's Data, Inc., West Chicago, IL, 1994).
77. Doctor's Data, Inc., "A Summary of Literature Regarding Elements in Human Hair" (Doctor's Data, Inc., West Chicago, IL, 1986).
78. E.g., in hair assays of Patrick Purdey, murderer of five Stockton California children in 1989, and George Hennand, another mass murderer from Texas, levels of cations calcium, sodium, magnesium, potassium were all more than one standard deviation above normal. (Hair assays courtesy of Everett L. Hodges, President, Violence Research Foundation, Tustin, CA.)
79. K. Tardiff, "Mentally Abnormal Offenders: Evaluation and Management of Violence" *Clinical Forensic Psychiatry* **15**, 553–567 (1992).
80. A. Raine, *The Psychopathology of Crime* (Academic Press, San Diego, 1993).
81. L. E. Cohen and R. Machalek, *Behavioral Strategy: A Neglected Element in Criminological Theory and Crime Policy* (Unpublished manuscript, 1993).
82. A. Lucas, R. Morley, T. J. Cole, G. Lister and C. Leeson-Payne, "Breast-milk and subsequent intelligence quotient in children born preterm" *Lancet* **339**, 261–284 (1992).
83. W. J. Rogan and B. C. Gladen, "Breast-feeding and cognitive development" *Early Human Development* **31**, 181–193 (1993).
84. P. J. Collipp, S. Y. Chen and S. Maitinsky, "Manganese in Infant Formulas and Learning Disability" *Ann. Nutr. Metabl.* **27**, 488–494 (1983).
85. A. Zervigon-Hakes and M. Graham, *Florida's Children: Their Future is In Our Hands* (Florida State University Center for Prevention and Early Intervention Policy, Tallahassee, Florida, 1994).
86. A. K. Snyder, "Responses of Glia to Alcohol" **In**: *The Role of Glia in Neurotoxicity* (M. Aschner and H. K. Kimelberg, eds.) (CRC Press: Boca Raton, FL, 1996) pp. 111–136.

87. G. Sharma, R. Sandhir, R. Nath and K. Gill, "Effect of Ethanol on Cadmium Uptake and Metabolism of Zinc and Copper in Rats Exposed to Cadmium" *J. Nutrit.* **121**, 87–91 (1991).
88. C. Cezard, C. Demarquilly, M. Boniface and J. M. Haguencoer, "Influence of the Degree of Exposure to Lead on Relations between Alcohol Consumption and the Biological Indices of Lead Exposure" *Brit. J. Indust. Med.* **49**, 645–647 (1992).
89. H. W. Hense, B. Filipiak, L. Novak and M. Stoeppler, "Nonoccupational Determinants of Blood Lead Concentrations in a General Population" *Int. J. Epidemiology* **21**, 753–762 (1992).
90. M. Ledig, G. Tholey, L. Megias-Megias, P. Kopp and F. Wedler; "Combined effects of ethanol and manganese on cultured neurons and glia" *Neurochem. Res.* **16**, 591–596 (1991).
91. R. O. Pihl and J. B. Peterson, "Attention-deficit hyperactivity disorder, childhood conduct disorder, and alcoholism: Is there an association?" *Alcohol Health and Rese. World* **15**, 25–31 (1991).
92. A. Raine, *The Psychopathology of Crime* (Academic Press, San Diego, 1993).
93. D. Boyum and M. A. R. Kleiman, "Alcohol and Other Drugs" **In**: *Crime* (J. Q. Wilson and J. Petersilia, eds.) (Institute for Contemporary Studies, San Francisco, 1995) pp. 295–326.
94. For a review, see H. Caton, "A New Approach to the Revolutionary Crowd" *Australian Journal of Politics and History* **40**, 187–202 (1994).
95. H. W. Mielke, "Lead Dust-Contaminated Communities and Minority Health: A New Paradigm" **In**: *The National Minority Health Conference* (B. L. Johnson, R. C. Williams and C. M. Harris, eds.) (Princeton Scientific Publishing Co., Princeton, NJ, 1992) pp. 101–112.
96. J. Pirkle *et al.*, "The Decline in Blood Lead Levels in the United States" *JAMA* **272**, 284–291 (1994).
97. E. K. Miller and A. J. Friedland, "Lead Migration in Forest Soils: Response to Changing Atmospheric Inputs" *Environmental Science and Technology* **28**, 662–669 (1994).
98. H. W. Mielke, J. C. Anderson, K. J. Berry, P. W. Mielke, R. L. Chaney and M. Leech, "Lead Concentrations in Inner-City Soils As a Factor in the Child Lead Problem" *Amer. J. Public Health* **73**, 1366–1369 (1983).
99. H. W. Mielke, B. Blake, S. Burroughs and N. Hassinger, "Urban Lead Levels in Minneapolis: The Case of the Hmong Children" *Environmental Research* **34**, 64–76 (1984).
100. H. W. Mielke, S. Burroughs, R. Wade, T. Yarrow and P. W. Mielke, "Urban Lead in Minnesota: Soil Transect Results of Four Cities" *J. Minn. Acad. of Sc.* **50**, 19–24 (1984).
101. H. W. Mielke and J. L. Adams, *Environmental Lead Risk in the Twin Cities* (University of Minnesota, Center for Urban and Regional Affairs, Minneapolis, Minn., 1989).
102. H. W. Mielke, J. L. Adams, P. L. Reagan and P. W. Mielke, Jr., "Soil-dust Lead and Childhood Lead Exposure as a Function of City Size and Community Traffic Flow: The Case for Lead Abatement in Minnesota" **In**: *Lead in Soil: Issues and Guidelines Supplement to Environmental Geochemistry and Health* (B. E. Davies and B. G. Wixson, eds.) Vol. 9, pp. 253–271.
103. H. W. Mielke, "Lead in New Orleans Soils: New Images of an Urban Environment" *Environmental Geochemistry and Health* **16**, 123–128 (1994).
104. H. W. Mielke, "Lead Dust Contaminated U.S.A. Communities: Comparison of Louisiana and Minnesota" *Applied Geochemistry* **Sup. 2**, 257–261 (1993).
105. H. W. Mielke, J. E. Adams, B. Huff, J. Pepersack, P. L. Reagan, D. Stoppel, and P. W. Mielke, Jr., "Dust Control as a Means of Reducing Inner-City Childhood Pb Exposure" *Trace Substances in Environmental Health* **15**, 121–128 (1992).
106. L. Viverette, H. W. Mielke, M. Brisco, A. Dixon, J. Schaefer, and K. Pierre, "Environmental Health in Minority and Other Underserved Populations: Benign Methods for Identifying Lead Hazards at Day Care Centres of New Orleans" *Environmental Geochemistry and Health* **18**, 41–45 (1996).
107. A. J. Bailey, J. D. Sargent, D. C. Goodman, J. Freeman and M. J. Brown, "Poisoned Landscapes: The Epidemiology of Environmental Lead Exposure in Massachusetts Children 1990–1991" *Social Science Medicine* **39**, 757–766 (1994).
108. *ASME Metals Properties Handbook* (McGraw Hill, New York, 1954) p. 6.
109. M. Lovewell, *Vineyard Gazette*, 11 July 1995, p. 6.
110. L. Johnson, *Sources of Residential Lead in Rural New England* (Thesis, Dept. Chem., Dartmouth College, 1993).
111. C. L. Keen, J. G. Bell and B. Lönnerdal, "The Effect of Age on Manganese Uptake and Retention from Milk and Infant Formulas in Rats" *J. Nutrition* **116**, 395–402 (1986).
112. Collipp, *et al.*, *op. cit.*
113. *Nutrition Monitoring in the U.S., Chartbook I: Selected Findings from the National Nutrition Monitoring and Related Research-Program* (B. Ervin and D. Reed, eds.) (Public Health Service, Hyattsville, MD, 1993) Figure 56.
114. *Nutrition Monitoring in the U.S., Chartbook I: Selected Findings from the National Nutrition Monitoring and Related Research Program* (B. Ervin and D. Reed, eds.) (Public Health Service, Hyattsville, MD, 1993) Figure 57.

These ethnic differences have persisted despite recent increases in calcium intake. Hence the 1988–91 National Heath and Nutrition Examination Survey (K. Alaimo, M. A. McDowell, R. R. Briefel, A. M. Bischof, C. R. Caughman, C. M. Loria and C. L. Johnson, *Dietary Intake of Vitamins, Minerals, and Fiber of Persons Ages 2 Months and Over in the United States* U.S. Department of Health and Human Services, Center for Disease Control, Washington, 1994.) revealed that black males 16–19 years old had only 78% of the calcium intake (1076 mg/day) of comparable white males (1373 mg/day). Since there is usually a delay of at least 15 years between infant environment and the onset of criminal careers, it would be valuable to explore the possibility that changes in diet and environmental toxicity might be related to the recently reported declines in rates of violent crime reported in some cities (see note 10 above).

115. *Nutrition Monitoring in the U.S., Chartbook I: Selected Findings from the National Nutrition Monitoring and Related Research Program* (B. Ervin and D. Reed, eds.) (Public Health Service, Hyattsville, MD, 1993) p. 19.
116. D. R. Wernette, L. A. Nieves, "Breathing Polluted Air" *EPA Journal* 16–17 (March/April, 1992).
117. *Inorganic Lead Exposure: Metabolism and Intoxications* (N. Castellino, P. Castellino and N. Sannolo, eds.) (Lewis Publishers, Boca Raton, FL, 1995).
118. Safe Drinking Water Amendments Act of 1995, Report of the committee on environment and public works, United States Senate on S. 1316; 104th Congress, 1st session, Senate report 104-169. Nov. 7, 1995
119. *Valley News* [Lebanon, NH] page 1, (May 4, 1996).
120. *N.Y. Times*, p. B7 (May 9, 1996).
121. The negative correlation between older housing and violent crime may be explained by a study in Seattle which found higher levels of lead in *recently* built homes: A. R. Sharret, A. P. Carter, R. M. Orheim and M. Feinleib, "Daily Intake of lead, cadmium, copper and zinc from drinking water: the Seattle study of trace metal exposure" *Environ. Res.* **28**, 456–475 (1982). Cf. *Lead Poisoning*, Hearings before the subcommittee on health and environment of the committee on energy and commerce, House of Representatives, 102nd Congress, 1st session. Including H.R. 2840, a bill to amend the public health service act to reduce human exposure to lead in residences, schools for young children, and day care centers, including exposure to lead in drinking water. April 25 and July 26, 1991. Serial No. 102-28. Washington, U.S. Government Printing Office.
122. *Nutrition Monitoring in the U.S., Chartbook I: Selected Findings from the National Nutrition Monitoring and Related Research Program* (B. Ervin and D. Reed, eds.) (Public Health Service, Hyattsville, MD, 1993).
123. National Center for Health Statistics, *Advance Data Number 258*, (Hyattsville, MD, Public Health Service, 1994), Table 1. Data are based on two national probability samples (National Health and Nutrition Surveys I and III, 1976–1980 and 1988–1991). Although intake of calcium has risen somewhat more among blacks than whites over this period, in 1988–1991, black males 16–19 years old still had only 78% the calcium intake (1076 mg/day) of comparable white males (1373 mg/day). RDA for adolescents is 1200 mg/day.
124. L. G. Borrud, P. C. Pillow, P. K. Allen, R. S. McPherson, M. Z. Nichaman and G. R. Newell, "Food Group Contributions to Nutrient Intake in Whites, Blacks, and Mexican Americans in Texas" *J. Am. Diet Assn.* **89**, 1061–1069 (1989).
125. J. J. DiIulio, "The Question of Black Crime" *Public Interest* **117**, 3–32 (1994).
126. T. Sahi, "Genetics and epidemiology of adult-type hypolactasia" *Scand. J. Gastroenterol.* **Suppl. 202**, 7–20 (1994).
127. K. C. Land, P. L. McCall and L. E. Cohen, "Structural Covariates of Homicide Rates: Are There Any Invariances across Time and Social Space?" *Am. J. Soc.* **95**, 922–963 (1990).
128. K. C. Land, P. L. McCall and L. E. Cohen, "Structural Covariates of Homicide Rates: Are There Any Invariances across Time and Social Space?" *Am. J. Soc.* **95**, 947 (1990).

In comparing these results to those reported here, it should be noted that Land, McCall and Cohen analyze either cities ($n = 526$–896), metropolitan areas ($n = 182$–257), or entire states ($n = 50$); the resulting sample sizes are far smaller than when using county data—and in the urban analyses, they eliminate a number of cities with abnormally high homicide rates as statistical "outliers." Because many of these cities are highly polluted (e.g., Greenville, South Carolina; Gary, Indiana; E. St Louis, MO)—and their approach masks the full range of population densities sampled across counties, it should be not expected that the results of the findings of Land, McCall and Cohen concerning homicide will coincide in every respect with our study of all forms of violent crime.

129. R. E. Nisbett, "Violence and U.S. Regional Culture" *Amer. Psychologist* **48**, 441–449 (1993).
130. R. E. Nisbett and D. Cohen, *Culture of Honor* (Westview, Boulder, CO, 1996).
131. K. Blum, J. G. Cull, E. R. Braverman and D. E. Comings, "Reward Deficiency Syndrome" *Amer. Scientist* **84**, 132–145 (1996).

132. A. Lucas, R. Morley, T. J. Cole, G. Lister and C. Leeson-Payne, "Breast-milk and subsequent intelligence quotient in children born preterm" *Lancet* **339**, 261–284 (1992).
133. W. J. Rogan and B. C. Gladen, "Breast-feeding and cognitive development" *Early Human Development* **31**, 181–193 (1993).
134. D. Benton and G. Roberts, "Effect of Vitamin and Mineral Supplementation on Intelligence of a Sample of Schoolchildren" *Lancet* **i**, 140–143 (1988).
135. I. K. Crombie, J. Todman, G. McNeill, C. D. V. Florey, I. Menzies and R. A. Kennedy, "Effect of Vitamin and Mineral Supplementation on Verbal and Non-verbal Reasoning of Schoolchildren" *Lancet* **335**, 744–747 (1990).
136. D. Benton and J. P. Buts, "Vitamin/mineral supplementation and intelligence," *Lancet* **335**, 1158–1160 (1990).
137. D. Benton and R. Cook, "Vitamin and Mineral Supplements Improve the Intelligence Scores and Concentration of Six-year-old Children" *Pers. Ind. Diff.* **12**, 151–1158 (1991).
138. S. I. Nidich, P. Morehead, R. J. Nidich, D. Sands and H. Sharma, "The Effect of the Maharishi Student Rasayana Food Supplement on Non-verbal Intelligence" *Pers. Ind. Diff.* **15**, 599–601 (1993).
139. L. Schweinhart, H. Barnes and D. Weikart, *Significant Benefits* (High Scope Press, Ypsilanti, MI, 1993).
140. A. Zervigon-Hakes and M. Graham, *Florida's Children: Their Future is In Our Hands* (Florida State University Center for Prevention and Early Intervention Policy, Tallahassee, Florida, 1994).
141. R. Haskins, "Beyond Metaphor: The Efficacy of Early Childhood Education" *Amer. Psychologist* **44**, 274–282 (1989).
142. C. Holden, "Head Start Enters Adulthood" *Science* **247**, 1400–1402 (1990).
143. *Ethyl Corp. v. EPA*, U.S. Court of Appeals for District of Columbia Circuit, Case No. 941-505, U.S. Court of Appeals of the District of Columbia Circuit, 51 F.3d 1053; 1995 U.S. App. LEXIS 8468, January 13, 1995, Argued, April 14, 1995, Decided.
144. *Ethyl Corp. v. Browner*, U.S. Court of Appeals for District of Columbia Circuit, Case No. 94-1516, U.S. Court of Appeals of the District of Columbia Circuit, 1995 U.S. App. LEXIS 29682, Sept. 11, 1995, Argued, October 20, 1995, Decided.
145. J. Zayed, M. Gerin, S. Loranger, P.Sierra, D. Begin and G. Kennedy, "Occupational and environmental exposure of garage workers and taxi drivers to airborne manganese arising from the use of methylcyclopentadienyl manganese tricarbonyl in unleaded gasoline" *J. Amer. Indust. Hygiene Assoc.* **55**, 53–58 (1994).
146. J. Komura and M. Sakarnoto, "Chronic Oral Administration of Methylcyclopentadienyl Manganese Tricarbonyl Altered Brain Biogenic Amines in the Mouse: Comparison with Inorganic Manganese" *Toxicology Letters* **73**, 65–73 (1994).
147. S. Loranger and J. Zayed, "Environmental and Occupational Exposure to Manganese: a Multimedia Assessment" *Int. Arch. Occup. Environ. Health* **65**, 101–110 (1995).
148. S. Loranger, M. C. Bibeau and J. Zayed, "Le manganèse dans l'eau potable et sa contribution à l'exposition humaine" *Rev. Epidém et Santé Public* **42**, 315–321 (1994).
149. *USA Counties on CD-ROM*, based on *State and Metropolitan Area Data Book*, 1982, 1986, and 1991 (Washington, DC, U.S. Census Bureau, 1994).
150. Right-to-Know Network (rtknet.org), operated by OMB Watch and Unison Institute, 1742 Connecticut Ave., NW, Washington, DC 20009; phone 202 234 8494.
151. Data provided courtesy Andrew Gaidurgis, Pelavin Research Institute, 1000 Thomas Jefferson St., NW, Washington, DC 20007.
152. *County Alcohol Problem Indicators*, 1986–1990, U.S. Alcohol Epiderniologic Data Reference Manual, Vol. 3, Fourth edition (1994) U.S. Department of Health and Human Services, NIH Publ. No. 94-3747.
153. R. D. Masters, D. J. Grelotti, B. T. Hone, D. Gonzalez and D. Jones, Jr., "Brain Biochemistry and Social Status: the Neurotoxicity Hypothesis" **In**: *Political Inequality, Intelligence, and Public Policy* (E. White, ed.) (Westview, Boulder, CO., 1997) pp. 141–183.
154. R. D. Masters, B. Way, B. T. Hone, D. J. Grelotti, A. Doshi, D. Gonzaley and D. Jones, Jr. "Environmental Pollution and Crime" *Vermont Law Review* (in press).
155. We thank David Grelotti, Stanley Weinberger, Johanna Blaxall, Sara Tullis, Maura Kelly, Davis Kitchel, Brad Parks, Dr. Robert Perlman, Michael T. McGuire, and Everett L. Hodges for research assistance and helpful comments. Research supported in part by the Gruter Institute for Law and Behavioral Research and Rockefeller center for the Social Sciences, Dartmouth College.

3. CARCINOGENS AND MUTAGENS

DOUGLAS McGREGOR * and CHRISTIANE PARTENSKY *

INTRODUCTION

It is a well recognised property of chemicals that they are toxic and that the division of toxin from non-toxin depends upon the dose. The manner in which this toxicity is expressed is also dose-dependent, so that, for example, the division of carcinogens from non-carcinogens is dependent upon their carcinogenic potency and whether the expression of some lethal, non-neoplastic property of the substance occurs at a dose level that is higher or lower than that required to induce the expression of neoplastic processes. This idea is well-accepted in experimental reproductive toxicology, where an agent may induce malformations, but is not described as a teratogen unless this activity is expressed at a dose at which embryotoxicity or foetotoxicity are not the dominant effects. It is also readily accepted in mutagenesis that dead cells do not mutate. The definition of a carcinogen also depends upon the statistical power of the study to detect differences between an exposed and a nonexposed group, after confounding and other sources of bias have been eliminated or minimized. Such control is much more readily achieved in experimental than in epidemiological studies, although the latter possess the property of greater relevance.

CARCINOGENS

Cancer as a cause of death is not negligible at any age, but it is primarily a terminal illness in aged populations. Carcinogenesis is the multistep, multifactorial process involved in the development of malignant tumours. A carcinogen is a risk factor causally related to an increase in cancer incidence or prevalence. Cancer epidemiologists study the factors by which people develop or die of cancer. To an experimental oncologist, however, a carcinogen is not necessarily an agent causing a pathological phenomenon resulting in death. A carcinogen in this context is an agent which increases the incidence of any neoplasm irrespective of whether it is lethal or potentially lethal. In extreme circumstances, a carcinogen may be defined also as an agent which increases the incidence of certain types of preneoplastic change.

Carcinogens may be physical, chemical or biological agents. Important physical phenomena which are carcinogenic are ultraviolet and ionising radiation. Circumstances involving biological agents which are judged to be carcinogenic are chronic infection with the hepatitis viruses B and C (hepatocellular carcinoma), human papilloma virus types 16 and 18 (cervical cancers), human immunodeficiency virus-1 and, possibly, -2 (Kaposi's sarcoma, non-Hodgkin's lymphoma), human T-lymphotropic virus-I (adult T-cell leukaemia/lymphoma in Japan, Caribbean and West Africa), Epstein–Barr virus (non-Hodgkin's lymphoma) and infection with the blood parasites, *Schistosoma haematobium* (urinary bladder squamous-cell carcinomas in Africa) and *S. japonicum* (colorectal cancers in Japan and China), and the

* International Agency for Research on Cancer, 150 cours Albert Thomas, 69372 Lyon 08 (France).

liver fluke, *Opisthorchis viverrini* (cholangicarcinoma in Thailand). These agents can have a very significant impact upon human health in some populations. In addition, nonmelanocytic skin cancer caused by sunlight is certainly the most common human cancer, although mortality from these skin cancers is far lower than the incidence. Nevertheless, there are many more chemicals which have been recognised as human carcinogens and there are also some human circumstances (related to occupation or life-style) that appear to involve exposures to carcinogens (without, necessarily, identifying the actual carcinogens). The remainder of this section of the chapter will focus on these chemicals.

Those chemicals classified by the International Agency for Research on Cancer as of February 1997 as being human carcinogens (Group 1), probably human carcinogens (Group 2A) or possibly human carcinogens (Group 2B) are listed in the Appendix Table. Summary conclusions regarding their genotoxicity are also listed. Exposure to some of these chemicals would appear not to involve any immediate risk of genotoxicity, a dichotomy that has led to their suggested description as 'directly or indirectly acting carcinogens'[1]. This may be preferred to the more commonly used terms, 'genotoxic and nongenotoxic carcinogens,' since carcinogenesis *always* involves mutation. However, classification as a so-called directly acting carcinogen is not a simple application of results from genotoxicity assays. Where these assay results are positive it may be simpler and a clearer expression of our limited understanding if such agents were to be described as carcinogens that are also genotoxic: for example, the dominant mode of carcinogenic action for an agent that also has genotoxic properties could be receptor-mediated cross-talk reactions rather than its translocation to DNA, intercalation rather than adduction to DNA, production of reactive oxygen species, inhibition of topoisomerase I or II, etc. rather than its potential for the direct induction of genetic damage.

These tables show that a very large *numerical* proportion of all identified carcinogenic agents are industrial chemicals and a very large proportion of exposure circumstances that increase the risk of cancer are occupation-related. This high proportion does not reflect the impact of these risk factors upon population cancer incidence or mortality. Indeed, a conclusion of the Harvard Report on Cancer Prevention is that nearly two-thirds of cancer deaths in the USA can be linked to tobacco use, diet, obesity and lack of exercise[2].

1. *Tobacco*

Unburnt tobacco contains substances that are carcinogenic and genotoxic in rodents and smokeless tobacco products (chewing tobacco, oral snuff) have been shown to increase the risk of oral cancer amongst human users[3]. However, the greater impact of tobacco upon human health occurs when it is burned.

The first clear evidence that tobacco smoking is a major cause of death was published almost simultaneously from studies in the UK[4] and USA[5], that showed a very strong association between the habit and cancer of the lung. When tobacco smoking was evaluated as a carcinogenic hazard by the IARC[6], it was noted that the approximately 90% of lung cancer deaths attributable to smoking applies to men in most western populations, but where women are increasingly using cigarettes the attributable proportion in women is approaching these levels. Cigarettes were the predominant cause of lung cancer, but increased risks were also associated with pipe and/or cigar smoking and there is a strong positive interaction with high-dose exposures to radon daughters or asbestos. Histologically, the risk was greater for squamous cell and small cell carcinomas than for adenocarcinomas of the lung.

Newer evidence has reinforced the conclusions of the IARC evaluation, in which smoking-related diseases were assigned to one of three categories, depending upon how the evidence for confounding factors was assessed. For some diseases, such as lung cancer and peripheral vascular disease, the excess incidence or mortality was almost entirely caused by smoking, whereas confounding factors were considered to be important for a number of other diseases for which dietary or other factors are also known to be causes. High alcohol consumption is a cause of stroke and oesophageal cancer and heavy smokers tend to drink more alcohol than non-smokers; it also predisposes to falls and hip fracture, as does lack of exercise, and heavy smokers tend to exercise less. These factors and categories have been taken into account in Tables I and II, which summarize two sets of the newer data: a prospective study of male British doctors that now has 40 years of follow up and the American Cancer Society study of over one million men and women aged 35 years or more. The results show a remarkable similarity and fully justify the conclusion[9] that "No single measure is known that would have as great an impact on the number of deaths attributable to cancer as a reduction in the use of tobacco or a change to the use of tobacco in a less dangerous way." Although the health effects of tobacco smoke upon nonsmokers is much less than upon smokers, there is now strong and consistent evidence that passive smoking increases the risk of lung cancer and a meta analysis of 34 published prospective and case-control studies indicates a relative risk of 1.24 (95% CI 1.11–1.38)[10]. Data on other smoking-related cancers are too few to permit conclusions.

Tobacco smoke and its condensates are carcinogenic in rodents and are quite clearly genotoxic complex mixtures. The presence of polycyclic aromatic hydrocarbons in these materials (benzo [*a*] pyrene in particular) and the tobacco-specific nitrosamine, 4-(methylnitrosamino)-1-(3-pyridyl)-1-butanone (NNK) has focused attention upon genotoxic mechanisms of carcinogenesis. Such mechanisms are often equated with early stage events, but there are significant health benefits from cessation of smoking. A number of studies have shown how lung cancer mortality risk ratios fall with time and may regain the risk experienced by nonsmokers (reviewed in reference[11]. These results suggest that, in addition to any early genetic events, either there are later mutations that are necessary for cancer development, or there are nonmutagenic processes induced by the complex mixture that are required throughout carcinogenesis, or that both classes of processes must be expressed.

2. *Alcoholic Beverages*

The proportion of cancer deaths attributed to alcohol by Doll and Peto[9] was 3%, with ranges of estimates varying from 2% to 4%. When they were evaluated by the IARC[12], the consumption of alcoholic beverages was considered to be causally related to the occurrence of malignant tumours of the oral cavity, pharynx, oesophagus and liver, the relationship being particularly strong for oesophageal cancers. The evaluation statement is significant because ethanol itself has not been shown to be carcinogenic in experimental animals and there have been no adequate oral studies of any alcoholic beverages. Nevertheless, since all types of alcoholic beverages seem to be implicated, a causal factor that is common to all of them is a reasonable hypothesis. That ethanol might be the factor is supported by elevated risks for oral and pharyngeal cancer among users of high-alcohol mouthwashes[13].

In the last few years, evidence has emerged that the regular consumption of small-to-moderate amounts of alcohol can reduce the risk of ischaemic heart disease and that the U-shaped curve described for this relationship extends to all causes of

Table I Fatal diseases positively associated with smoking—study of male British doctors[7]*

Disease	Standardized mortality per 100,000 men/year		Relative risk (b/a)	Absolute excess risk per 100,000 men/year (b − a)	Attributable proportion† (%)
	Life-long non-smoker (a)	Current cigarette smoker (b)			
(i) *Increased risk largely or entirely caused by smoking*					
Cancer of:					
Lung	14	209	15.0	195	81
Upper respiratory sites	1	24	24.0	23	87
Bladder	13	30	2.3	17	28
Pancreas	16	35	2.2	19	26
Ischaemic heart disease	572	892	1.6	320	15
Respiratory heart disease	0	10	—	10	100
Aortic aneurysm	15	62	4.1	47	48
Chronic obstructive lung disease	10	127	12.7	117	78
(ii) *Increased risk partly caused by smoking*					
Cancer of:					
Oesophagus	4	30	7.5	26	66
Stomach	26	43	1.7	17	17
Kidney	9	13	2.1	4	25
Leukaemia	4	7	1.8	3	19

Stroke	152	203	1.3	51	8
Pneumonia	71	138	1.9	67	21
(iii) *Increased risk due to confounding*					
Cirrhosis of liver	6	32	5.3	26	—
Cancer of liver	7	11	1.6	4	—
Suicide	23	37	1.6	14	—
Poisoning	7	19	2.7	12	—
All diseases excluding those in category (iii)	907	1823	2.0	916	23
All diseases excluding those in category (ii) *and* (iii)	612	1324	2.2	712	26

* Table compiled by[8].

† The proportion of all deaths from the specified disease attributable to smoking, assuming 30% of the population are current smokers and that all the excess risk in smokers is due to smoking. In Group (ii) the actual proportions will be somewhat less than those specified.

Table II Fatal diseases positively associated with smoking—American Cancer Society (CPSII). Men and women aged 35 years or more (Surgeon General Report, 1989)*

Disease		Standardized mortality per 100,000/year†		Relative risk‡	Absolute excess risk per 100,000 per year	Attributable proportion** (%)
		Life-long non-smoker	Current cigarette smoker			
(i) *Increased risk largely or entirely caused by smoking*						
Cancer of:						
Lung	M	24	537	22.4	513	87
	F	18	213	11.9	195	77
Upper respiratory sites	M	1	27	24.5	26	89
	F	2	10	5.6	8	58
Bladder and other urinary organs	M	18	53	2.9	35	36
	F	8	21	2.6	13	32
Pancreas	M	18	38	2.1	20	25
	F	16	37	2.3	21	29
Ischaemic heart disease	M	500	970	1.9	470	22
	F	386	688	1.8	302	19
Aortic aneurysm‡	M	24	98	4.1	74	48
	F	11	52	4.6	41	52
Chronic obstructive pulmonary disease	M	39	378	9.7	339	72
	F	21	216	10.5	195	74
(ii) *Increased risk partly caused by smoking*						
Cancer of:						
Oesophagus	M	9	68	7.6	59	66
	F	4	41	10.3	37	74
Kidney	M	8	23	3.0	15	37
	F	6	8	1.4	2	11

Cerebrovascular lesions	M	147	328	2.2	181	27
	F	236	434	1.8	198	20
(iii) *Increased risk due to confounding*						
Cancer of cervix	F	8	18	2.1	10	—
All diseases excluding those in category (iii)	M	788	2520	3.2	1732	40
	F	708	1720	2.4	1012	30
All diseases excluding those in categories (ii) *or* (iii)	M	588	2010	3.4	1422	42
	F	438	1179	2.7	741	34

* Table compiled by[8].

† Calculated using the published relative risk, the mortality in the population aged $\geqslant 35$ years and assuming that 30% of the population are current smokers.

‡ Taken from the American Cancer Study (Surgeon General Report, 1989).

** The proportion of all deaths from the specified disease attributable to smoking, assuming 30% of the population are current smokers and that all the excess risk in smokers is due to smoking. In Group (ii), the actual proportions will be somewhat less than those specified.

death and persists even for alcohol augmented causes, which include the cancers mentioned, but does not separate them from cirrhosis, alcoholism and external causes[14]. The latter factors would seem unlikely causes of death at low and moderate levels of alcohol intake, but would strongly contribute to the increasing curve at high intake levels.

3. *Food and Diet*

Chemicals occurring in living or once-alive organisms are extremely diverse, numerous and complex with regard to their effects upon other organisms exposed to them. Some may be carcinogenic or anticarcinogenic and frequently the actual component(s) responsible for an effect are difficult to identify. It is not clear, for example, whether the limited human evidence that drinking hot mate is a risk factor for cancers of the upper gastrointestinal tract is due to chemical components of the dried leaves of *Ilex paraguariensis*, a member of the Aquifoliaceae (holly) family, native to Paraguay and Argentina and from which mate is made, physical damage induced by the very high temperature water that is taken into the mouth as it is being drunk, or a combination of these influences. In addition to their concerns regarding tobacco, Doll and Peto[9] considered that it is possible that some nutritional factor(s) may eventually be found to be of comparable importance. The influence of diet on cancer might be a very general one, such as overnutrition, or a more specific one such as providing the means by which powerful carcinogens are ingested, or by affecting the formation of carcinogens in the body, affecting the transport, activation or deactivation of carcinogens, or affecting the promotion of cells that are already transformed to some degree.

(a) *Major Components* Of dietary calories in the so-called average north American diet[15], carbohydrates supply about 46%, lipids supply about 42% and proteins supply about 12%. The simple carbohydrates, sucrose, glucose and fructose supply more than half of the carbohydrate calories in the north American diet, while starch supplies the remainder. This type of diet contrasts with, for example, the so-called Italian-style Mediterranean diet that describes the food patterns typical of Crete, much of the rest of Greece and southern Italy in the early 1960s. At that time, the broad characteristics of the diet were[16]:

(i) An abundance of plant foods (fruit, vegetables, breads, other forms of cereals, potatoes, beans, nuts and seeds);

(ii) Minimally processed, seasonally fresh and locally grown foods;

(iii) Fresh fruit as the typical daily dessert, with sweets containing concentrated sugars or honey consumed a few times per week;

(iv) Olive oil as the principal source of fat;

(v) Dairy products (mainly cheese and yogurt) consumed daily in low to moderate amounts;

(vi) Fish and poultry consumed in low to moderate amounts;

(vii) Zero to four eggs consumed weekly;

(viii) Red meat consumed in low to moderate amounts;

(ix) Wine consumed in low to moderate amounts, mainly with meals.

This diet was low in saturated fat (7–8% of the calories), with total fat ranging from <25% to >35% of the calories from one area to another, and was high in complex carbohydrates and dietary fibre. In addition, the lifestyle of the people included regular physical activity and was associated with low rates of obesity. According to surveys conducted in the late 1970s, daily energy requirements of adult men in central or southern Italy were provided as follows: carbohydrates, 41%; fats, 31–33%; alcohol, 14%; protein, 12–13%[17].

The more obvious benefit of this type of dietary regime is a reduction in risk of coronary disease, but a number of studies have also shown that there is a reduced incidence of breast cancer in the region, in comparison with Northern Europe or North America, and that colorectal cancers are much less common in Greece, Spain and Portugal than in Scotland, Denmark and other Scandinavian countries (reviewed in reference [18]). The mode of expression of these benefits is unclear. It could be as a primary result of protective effects or lower exposure to toxic effects of the diet consumed in other areas. It could be consequential on the diet producing a lower body weight, body mass index (BMI) or weight gain, or perhaps these are only markers for a higher level of physical activity. Studies on migrant populations suggest that environmental factors have a dominating influence over genetic factors in producing geographical patterns of disease distribution (see reference [19]). Thus, Italian migrants to Australia show increases with duration of residence in mortality from cancers of the breast, lung, colon and rectum[20,21].

Excessive body weight as a risk factor for cancer has been investigated in a number of studies, especially of cancers of the colon, breast and uterus. There was evidence of an association of excess body weight with a greater risk for colon cancer in men, but not women, in some of these studies, which included the largest prospective study of almost one million people[22,23]. Four studies did not find any such relationship, except among Japanese men living in Hawaii in one study (reviewed in reference [24]). Excess body weight was associated with higher risk for rectal cancer among men and women combined in one study[23].

For breast cancer in women (reviewed in reference [25]), associations with measures of adiposity vary by age and menopausal status at the time of diagnosis. Heavier women appear to be at decreased risk for developing premenopausal breast cancer; relative risks of about 0.6 were reported for women with BMIs of references. Conversely, heavier women are at increased risk of developing and dying from postmenopausal breast cancer. Although contradictory findings have been observed in cohort studies, modest increases in relative risk in the order of 1.2–1.5 were reported in older postmenopausal women with BMI's of references. Furthermore, adult weight gain and increased central adiposity have been consistently and independently associated with an increased risk of postmenopausal breast cancer. No significant associations have been found between weight loss and postmenopausal breast cancer incidence.

In the case of endometrial cancer, positive and independent associations between body weight or BMI, weight gain and various measures of central adiposity and the incidence of the disease have been consistently found. Increases in relative risks of developing endometrial cancer of 2–3.5 have been reported for women with BMIs of references (reviewed in reference [25]). It also appears that endometrial cancer mortality is increased in heavier and taller women.

On the other hand, a low BMI was associated with increased risks of gastric and lung cancers in the study of Nomura *et al.*[26]. Thus, the above studies are somewhat inconsistent, although suggestive of body weight being a risk factor.

It can be tentatively concluded that the influence of the type of diet, or total dietary intake (perhaps interacting with physical activity level and perhaps expressing in terms of BMI values or body weight gain) may affect the incidence of certain human tumours. This conclusion finds strong support from experiments with rats and mice. Many experiments have shown that dietary/caloric restriction, which results in reduced body weight gain, confers some protection against neoplasia in rodents (reviewed in reference [27]). These effects of dietary restriction have been known since the early years of this century, and perhaps earlier.

Specific major constituents of diet have also been investigated as possible risk factors. These include the high intake of saturated fats, for which geographical correlation studies strongly suggest that diets rich in fats and saturated fats in particular are associated with the high incidence of colorectal cancers observed in western Europe, Australia and North America. This conclusion is supported by most case-control studies. However, the results are not consistent, there being no association in a number of other case-control studies and negative associations in at least two studies (reviewed in reference [28]).

(b) *Minor Components* Of the many thousands of compounds known to occur in unprocessed plants, most have not been subjected to toxicological research. One of the rare, extensively studied compounds is caffeic acid, a metabolic precursor of lignin, a structural polymer in all land plants and ubiquitous in the food supply. It is very largely conjugated, commonly occurring as chlorogenic acids. These can be hydrolysed in the mammalian stomach to caffeic and quinic acids. Caffeic acid causes forestomach tumours in male mice and male and female rats; kidney tumours in female mice and in male rats. It is also clastogenic in mammalian cells *in vitro* but not *in vivo*[29]. Other minor components identified as having carcinogenic activity in rodent experiments are safrole and dihydrosafrole. Oral administration of safrole induced liver tumours in mice and rats, while dihydrosafrole induced liver tumours in mice and oesophageal tumours in rats[30].

One of the functions ascribed to secondary plant compounds is the provision of a defence mechanism against herbivores (insects and other invertebrates) and against attack by pathogens[31]. It is believed that these compounds [phytoalexins] play a decisive role in this context[32] and there is an extensive literature demonstrating their deterrent and/or toxic nature[33,34]. Indeed, some specific examples suggest that the protective role of secondary plant compounds against insects could be their primary function[35–37]. It has been remarked that the natural pesticide concentrations in food may be as much as 10,000 times higher than that of synthetic pesticide residues[38], yet, the concentrations of minor components are normally very low and any one with carcinogenic potential probably would have to be very potent if it was to have an impact upon human cancer incidence. However, if exposure is to several different minor constituents that are carcinogenic, then they may have an additive effect.

(c) *Substances Produced by Cooking and Food Processing* When food juices are subjected to high temperatures, the amino acids, sugars and other constituents also degrade and react to form enormous numbers of new compounds, some of which are desired for their flavour characteristics, e.g. pyrazines. Further heating may produce charring and the emergence of polycyclic aromatic hydrocarbons (PAHs), many of which are carcinogenic and mutagenic. However, almost twenty years ago, Sugimura *et al.*[39] reported that the charred parts of fish and meat have mutagenic activity that cannot be fully accounted for by their PAH content. Later, many new chemicals were discovered in pyrolysates of amino acids and proteins and some of them were in the charred parts of fish and meat (briefly reviewed in reference [40]). Most of these products were derivatives of heterocyclic aromatic amines. Analysis of pyrolysed single amino acids has revealed about 25 polycyclic heterocyclic amines, e.g. Trp-P-1, Trp-P-2, Glu-P-1, Lys-P-1 and Phe-P-1. Pyrolysis of mixtures of amino acids produces still more compounds, such as PhIP, IQ, MeIQ and MeIQx. These compounds are carcinogenic in rodents and are amongst the most potent mutagens known.

Nitrosamines can accumulate in foods as a result of processing. High levels of *N*-nitrosodimethylamine have been reported in some samples of Chinese-style salted fish and low levels are found in vegetables in China, Japan and Korea when traditional methods of pickling involving fermentation of local vegetables—with or without salt—are employed. The human data suggest that vegetables pickled in this way possibly causes oesophageal cancer and that Chinese-style salted fish does indeed cause nasopharyngeal cancer. The latter conclusion finds support from rodent studies, in which the salted fish induced a small number of oral and nasal carcinomas. There appear to have been no adequate rodent studies with pickled vegetables[29].

(d) *Contaminants of Food* Most studies of carcinogenic contaminants have centred upon the mycotoxins. Food storage conditions are without doubt important in determining the levels of many components. Contamination with *Aspergillus flavus* and *A. parasiticus*, particularly in hot, humid climates, can be a severe problem resulting in significant exposures to aflatoxins, which are believed to be human carcinogens[29]. Whether these substances act alone to increase cancer risks, or whether there is an interaction with hepatitis B (and possibly C) is still debated[41].

Other possible mycotoxin contaminants are the toxins derived from *Fusarium moniliforme*: fumonisins B_1 and B_2 and fusarin C. The human evidence for their involvement in cancer is restricted to correlation studies, most of which suggest some relationship between oesophageal cancer rates and the occurrence of *F. moniliforme* or its toxins in maize. This relationship is supported by one carcinogenicity study indicating that fusarin C induces papillomas and carcinomas of the oesophagus and forestomach in female mice and rats. However, a male rat study with cultures of *F. moniliforme* known to contain mainly fusarin C did not produce such tumours, while cultures rich in fumonisins B_1 and B_2 induced hepatocellular carcinomas and cholangiocarcinomas and forestomach papillomas and carcinomas[29].

Contamination of grains by nonedible plants that may contain carcinogenic pyrrolizidine alkaloids can occur in certain communities and chemical contamination may also occur, sometimes with disastrous consequencies, e.g. Japan, in 1968[42], and Taiwan, in 1979[43], but any effect of these upon human cancer rates has not been demonstrated.

4. *Occupational Carcinogens*

Because exposures at work tend to be occupation-specific, then occupation-related cancers might also be expected to be characteristic for a particular occupation. Nevertheless, the first report of an occupational cancer did not appear until the 18th century and other reports did not appear until the latter half of the 19th century.

Estimates of the contribution from occupational exposures to the cancer burden have ranged from 1% to 38% (Table III), although it has been suggested that the highest estimate may have been driven by sociopolitical rather than scientific

Table III Estimates of proportion of occupational cancers (modified from reference [44])

Year	Author	Proportion (%)
1954	Druckrey	1
1967	Clayson	1
1976	Higginson and Muir	1–3
1977	Wynder and Gori	1–10
1977	Epstein	5–15[a]
1977	Cole	10–15
1978	OSHA	23–38
1979	Higginson	1–5
1979	Higginson and Muir	2–6[b]
1981	Doll and Peto	ca. 4
1991	Siemiatycki	no estimate

[a] For males
[b] For the Birmingham area.

considerations[9]. It is interesting to note that Siemiatycki[45] refrained from making any estimate, although he discussed the issues at some length. While the community impact of occupational cancers may be relatively low, occupational cancer tends to be concentrated among relatively small groups of people among whom the risk of developing the disease may be quite large. Therefore, detection of occupational hazards should remain a high priority in cancer prevention, in spite of the low importance that their proportional contribution would suggest. Once the risks have been identified, then they can usually be reduced or even eliminated.

The cancer sites and the typical occupations in which exposure to the established human carcinogens has been shown to constitute a cancer risk are shown in Table IV. Common targets for occupational exposures are the respiratory tract and skin. Reference to Appendix Table will indicate that there are more occupational circumstances listed there than in Table IV. This is because, in some occupations the causal agents for the increased cancer risks have not been identified.

Once human carcinogens have been identified (or even when they have been predicted, as a result of experiments with rodents), focussed efforts can be made to reduce exposure to them. It is improbable that ethylene oxide will ever be identified as a human carcinogen on the basis of epidemiological evidence alone now that sterilization plant workers are better protected than they were some years ago. It seems that concern over the level of chromosomal damage in exposed workers[46] led to the closure of plants in which exposure was particularly high. Similarly, it can be argued that vinyl chloride is now so well regulated and the industry has been so responsive to concerns over the health of workers that the angiosarcomas of the liver, hepatocellular carcinomas, brain tumours, lung tumours and malignancies of the lymphatic and haematopoietic system observed in earlier worker populations[3] are unlikely to be repeated in more recent worker populations. Containment of hazards in this way is much more difficult during primary manufacture and the recovery of raw materials, e.g. mining operations. Nevertheless, there are clear indications that occupational cancer is a topic of reduced scientific interest in Europe and North America. On the other hand, its social impact can remain substantial in the developing countries.

Industrialization in the 20th century has been associated with some populations (mostly in industrialized countries) experiencing improved services and health, while

Table IV Established occupational causes of cancer

Agent	Site of cancer	Occupation
Aromatic amines (4-aminodiphenyl, benzidine, 2-naphthylamine)	Bladder	Dye manufacturers, rubber workers, coal gas manufacturers
Arsenic	Skin, lung	Copper and cobalt smelters, arsenical pesticide manufacturers, some gold miners
Asbestos	Lung, pleura, peritoneum (also probably stomach, large bowel, oesophagus)	Asbestos miners, asbestos textile manufacturers, asbestos insulation workers, certain shipyard workers
Benzene	Marrow, especially erythroleukaemia	Workers with glues and varnishes
Beryllium	Lung	Beryllium extraction, production and fabrication
Bis(chloromethyl) ether and chloromethyl methyl ether	Lung	Makers of ion-exchange resins
Cadmium and cadmium compounds	Lung	Cadimum workers
Chromium VI compounds	Lung	Manufacturers of chromates from chrome ore, pigment manufacturers
Coal-tar pitches	Skin, lung, bladder	Roofers, tar distillers, patent-fuel workers
Crystalline silica (quartz and cristobalite)	Lung	Refractory brick, pottery, quarry and granite workers
Ethylene oxide	Lymphatic and haematopoietic tissues	Sterilization personnel
Ionizing radiations	Lung	Uranium and some other miners
Ionizing radiations	Bone	Luminizers
Ionizing radiations	Marrow, all sites	Radiologists, radiographers
Mineral oils, untreated and mildly treated	Skin	Mule spinners, jute processors, metal machinists
Mustard gas	Larynx, lung	Poison gas manufacturers
Nickel compounds	Nasal sinuses, lung	Nickel refiners
Shale oils	Skin	Shale-oil workers
Soot, tar, oil	Skin, lung	Coal gas manufacturers, roofers, asphalters, aluminium refiners, many groups selectively exposed to certain tars and oils
Strong inorganic acid mists containing sulfuric acid	Nasal sinuses, larynx, lung	Isopropyl alcohol manufacture
UV light	Skin	Farmers, seamen
Vinyl chloride	Liver (angiosarcoma)	PVC manufacturers
	Nasal sinuses	Hardwood furniture manufacturers
	Nasal sinuses	Leather workers
Wood dust (mainly hard woods)	Nasal cavities, paranasal sinuses	Woodworkers

others (mostly in developing countries) are experiencing either slow improvement or even deteriorating conditions[47]. These authors described the latter countries as being characterized by (1) the large size of the informal (work) sector; (2) large numbers of workers, including contract workers, who have little or no support from unions or other worker organizations; (3) the transfer to them of hazardous industries from industrialized countries; (4) the substitution by export (cash) crops for local food crops; (5) the insecure status of the workers, owing to large-scale unemployment, migratory labour and child labour; (6) a lack of legislative protection or the poor enforcement of protective measures; and (7) increasing multinational penetration and decreasing national control over resources. These conditions have been exacerbated by the Third World burden of debt and demonstrate a lack of concern for health issues in which occupation-related cancers and other diseases can be expected to flourish as they did in the currently developed countries during the last two centuries.

MUTAGENS

Mutagenic damage may be caused by the absorption of high energy radiation, reaction with chemicals or an interaction with an invading virus or other parasite. Chemical mutagens come either from the environment of the cell or are produced by the cells themselves as part of their normal metabolism. We now know a great deal about the way in which mutagens can affect the genetic material within cells, but in the early days of human cytogenetics few showed interest in the possibility that pathological conditions might be associated with chromosomal abnormalities. A notable exception was Theodor Boveri (published in 1914), who speculated that chromosomal rearrangements that he observed in cancer cells might be important in the aetiology of malignancy. However, no experimentally demonstrated high concordance of carcinogenic and mutagenic activity emerged until the mutagenicity assays were modified to take account of the metabolic activation of chemicals being tested. This led to such an increase in the sensitivity of *in vitro* assays that, in the early 1970s, it was possible for Ames *et al.*[48] to title a paper, 'Carcinogens are mutagens: a simple test system combining liver homogenates for activation and bacteria for detection.' Before this situation had been reached, concern over mutagen exposure had developed with the discovery of so-called supermutagens: chemicals such as ICR-170, AF-2, hycanthone and β-propriolactone that can induce high levels of mutation at high levels of target cell survival. Several leading geneticists were concerned that supermutagens might be widely distributed[49] because either they had passed through traditional toxicity screens without showing adverse effects or they had never been tested at all. In spite of these concerns, however, the major impetus received by research in mutagenesis came from the belief that carcinogenic activity was predictable by examining the interaction of suspect agents with DNA.

There is no doubt regarding the importance of transmissible mutation in evolution and transmissible damage has been induced experimentally in mice by ionizing radiation and a small number of chemicals. In reality, however, there is little evidence that any agent has increased the frequency of any human mutation which can be transmitted from one generation to the next. That this might have happened was suggested by results from a study of germ-line mutation at minisatellite loci among children born in heavily polluted areas of the Mogilev district of Belarus after the Chernobyl accident[50]. Mutation rates in the 79 Mogilev families examined were higher in areas of high surface ^{137}Cs contamination than in areas where contamination was lower*. However, a similar minisatellite study of children from 50 families exposed to atomic bomb radiation in Hiroshima and Nagasaki (gonadal dose >0.01 Sv) and 50 control families did not show any effect[51]. Minisatellites, or tandem repeat elements, are particularly useful for monitoring human germ-line mutations because the very high rate of spontaneous mutation that alters allele length (repeat copy number) provides a system capable of detecting induced mutations in relatively small populations. The hypervariable loci examined in these two studies have spontaneous mutation rates at least 1000 times higher than in most protein-coding loci. So, even with highly sensitive techniques it has not been possible so far to clearly show any germ-line transmissible effect of potentially very damaging exposures.

Consequently, while adverse effects upon germ-line cells are important to our descendants, we have so far been unable to see what these effects might be and the epidemiologists among future generations will stand little chance of identifying the causes of any increases induced today in diseases with a genetically predisposing component. These are not acceptable reasons for ignoring the possibility of doing harm to our descendants and neglecting to take preventative action. However, it is arguable that the long-term effects of mutagens might be avoided if exposure is controlled on the basis of more immediately hazardous and more easily recognized properties of these agents.

1. *Hierarchy of Assays in Genetic Toxicology*

(a) *Germ-line and Somatic Cell In Vivo Cytogenetic Assays* Our ability to detect DNA damage in germ-line cells is low, particularly in female animals, and is mainly aimed at the identification of clastogenic effects. These are more readily detected in somatic cells, a situation which is reflected in the powerful emphasis placed upon bone marrow studies as compared with studies on spermatocytes, spermatogonia or oocytes. This situation is not as serious as it at first appears.

Holden[52] reviewed the available literature and found 76 compounds which had been tested for chromosomal effects in both somatic and germ-line cells. Of these, concordant results were obtained with 58 chemicals. The remaining 18 chemicals for which there were discordant results were all positive (i.e. induced damage) in somatic cells only. At that time, therefore, the available evidence suggested that a negative somatic cell response is highly predictive of a negative germ-line cell response. Subsequently, it was suggested by the US EPA GeneTox Workshops that six chemicals could be uniquely germ-line cell mutagens[53], but a re-evaluation of the GeneTox Program literature on these compounds indicates that they were misclassified[54] (Table V). Therefore, there was, as of 1989, no reason to change the presumption that all germ-line cell clastogens are also somatic cell clastogens. This kind of thinking led the European Centre for Ecotoxicology and Toxicology of Chemicals (ECETOC) to propose a testing strategy in which agents without somatic cell genotoxicity in vivo could be assumed to have no potential for germ-line cell genotoxicity[55].

It appears that a very high proportion of known human carcinogens induce either micronuclei or chromosomal aberrations in bone marrow[56]. Thus, the real value of

* Although acute exposure was mainly to ^{131}I, this isotope has a half-life of only 8 days, so, when ^{131}I has decayed, more stable isotopes become the main source of radiation exposure; ^{137}Cs is the most important of these isotopes

an *in vivo* cytogenetic test is that a positive response, e.g., in the bone marrow micronucleus or metaphase analysis tests, does help to identify particularly dangerous compounds.

(b) *Cytogenetic Assays In Vivo and In Vitro* Thompson[57] reviewed the literature to find 216 chemicals which had been tested both *in vitro* and in rodent bone marrow tests for clastogenicity. Definitive results were obtained with 181 of them, amongst which there was agreement for 126 chemicals. Of the 55 for which the results did not agree, 53 were positive *in vitro* and negative *in vivo*. Only *d*-ascorbic acid and ethynyloestradiol were negative *in vitro* while inducing significant results in bone marrow. This leads to the conclusion that a chemical that fails to induce a significant response in an *in vitro* clastogenicity assay is unlikely to be clastogenic *in vivo*, in bone marrow assays.

(c) *Germ-line and Somatic Cell In Vivo Mutation Assays* Data relating to gene mutations in germ-line cells are much more limited, but those available support the conclusions reached for clastogens. Data reviewed by Russell *et al.*[58] showed for the mouse specific locus test only eight chemicals as clearly positive and, in addition, they considered one chemical mixture as positive after review of data initially considered negative by formal statistical methods. This number was increased by another five chemicals tested by the US National Toxicology Program (NTP)[59]. These chemicals (Table VI) are all highly reactive and induce a wide variety of toxic effects in man and other animals. Furthermore, all of these individual chemicals that induce point mutations or small deletions in germ-line cells are also clastogenic in germ-line cells, so even germ-line cell damage can be recognized by the easier techniques applied for clastogenic effects.

Table V So-called 'unique' germ-cell mutagens[54]

Chemical	Re-evaluated evidence for clastogenic effects	
	Germ-line cells	Somatic cells
Diethylhexyl phthalate (DEHP)	Questionable	Positive
Ethanol	Positive	Positive
Cyclohexylamine	Unconfirmed positive	Positive
Saccharin, sodium salt	Questionable	Positive
Cadmium chloride	Positive	Positive
Triflupromazine	Questionable	Positive

Table VI Mouse specific locus test-positive chemicals[58,59]

Acrylamide	Melphalan
Chlorambucil	Methyl methane sulphonate
Coal liquid A	Methyl nitrosourea
Cyclophosphamide	Mitomycin C
Ethyl methane sulphonate	Procarbazine
Ethyl nitrosourea	Propyl methane sulphonate
Ethylene oxide	Triethylene melamine

2. *Genetic Toxicology in Carcinogen Identification*

Those agents that do induce heritable mutations in experimental animals will have already been considered to be hazardous chemicals regulatable because of their other toxicological properties which are more easily monitored (usually because they are carcinogenic, but perhaps also because of reproductive effects, neurotoxicity, cardiotoxicity, etc.). It also appears that there are no mutagens or clastogens that are uniquely active in germ-line cells through which their adverse effects could be passed from generation to generation. Therefore, from the view point of toxicological evaluation, if highly reactive, toxic substances and carcinogens are being adequately controlled, then so are mutagens/clastogens with heritable effects. The main toxicological interest in mutagenicity therefore comes down to what mutagenic activity can tell us about carcinogenicity. Indeed, the success of mutagenicity assays developed in the last 30 years has always been measured against the yardstick of carcinogenicity.

A study[60] of four *in vitro* assays (Table VII) for their potential to predict carcinogenic activity indicated a concordance with little variation around 60% for the rodent carcinogenicity results and the results of each of the *in vitro* assays (*Salmonella* mutation assay, chromosomal aberration and sister-chromatid exchange (SCE) assays with chinese hamster ovary (CHO) cells and the *tk* locus mutation assay in mouse lymphoma L5178Y cells). A literature survey of results with the unscheduled DNA synthesis (UDS) assay in primary cultures of hepatocytes (Table VIII) and the *in vivo* rodent bone marrow cell micronucleus test (Table IX) revealed similar concordances. However, the specificity and/or positive predictivity parameters were much better for the *Salmonella* mutation, UDS and micronucleus tests than for the others. Specificity is the proportion of chemicals correctly identified as non-carcinogens within a population of non-carcinogens; while positive predictivity is the proportion of chemicals identified as mutagens which were actually carcinogenic. Specificity is one aspect of how a test is validated, while positive predictivity is one measure of the accuracy of the test in practice. While specificity is the more objective parameter (in that it is not dependent upon the proportion of carcinogens in the test population), positive predictivity is what is actually measured when the test leaves the academic environment and enters the world of unknown carcinogenic activity for which it was designed. For these reasons, these three genotoxicity assays were separated out from other genotoxicity assays in Appendix Table (at the end of the list of references).

Table VII Summary characteristics of four 'short-term tests' for the prediction of carcinogenicity, based upon results with 73 chemicals[60]

	SAL*	CA*	SCE*	ML*tk**
Sensitivity %	45	55	73	70
Specificity %	86	69	43	45
Positive predictivity %	83	73	67	66
Negative predictivity %	51	50	52	50
Concordance %	62	60	62	60

*Abbreviations: SAL, *Salmonella typhimurium* his locus mutation; CA, chromosomal aberrations in CHO cells; SCE, sister-chromatid exchanges in CHO cells; ML*tk*, mouse lymphoma L5178Y cell *tk* locus mutation.

Table VIII Summary characteristics of the primary hepatocyte unscheduled DNA synthesis assay (grain counting) for the prediction of carcinogenicity, based upon published data[61] (Total: 368; Total with carcinogenicity data: 249)

	A		B		C	
	Positive and negative results only		As A and inconclusives* as negatives		As A and inconclusives* as positives	
Sensitivity	96/184	52%	96/194	49%	113/227	50%
Specificity	22/22	100%	48/55	87%	22/22	100%
Positive predictivity	96/96	100%	96/103	93%	113/113	100%
Negative predictivity	22/110	20%	48/146	33%	22/136	16%
Concordance	118/206	57%	144/249	58%	135/249	54%

* Inconclusive results are those not clearly positive or negative in either the carcinogenicity or the UDS assay.

Table IX Summary characteristics of the rodent bone-marrow cell micronucleus test for the prediction of carcinogenicity, based upon published data[61] (Total: 355; Total with carciongenicity data: 205)

	A		B		C	
	Positive and negative results only		As A and inconclusives* as negatives		As A and inconclusives* as positives	
Sensitivity	87/136	64%	87/144	60%	111/173	64%
Specificity	14/30	47%	31/47	66%	14/32	44%
Positive predictivity	87/103	84%	87/117	74%	111/129	86%
Negative predictivity	14/63	22%	31/88	35%	14/76	18%
Concordance	101/166	61%	118/205	58%	125/205	61%

* Inconclusive results are those not clearly positive or negative in either the carciongenicity or the micronucleus assay.

3. *Genetic Events as Biomarkers*

Because human data are most applicable, if they can be obtained, and because of the enormous difficulties in correlating eventual cancer outcome with exposures in what is surmised to be the relevant period during carcinogenesis, there have been attempts to use intermediate endpoints—biomarkers—that may be biomarkers of exposure, e.g. adducts with haemoglobin or DNA, or biomarkers of effect, e.g. polyps in the large bowl; leukoplakia; naevi. 'Biomarker' is an easily used but possibly dangerous term: an easy one because, like a black hole, it attracts and aggregates everything within its vicinity; a dangerous one because causal connections are assumed between one observation and another that may be justified or even correct in some cases, but not in others, yet no distinction between them is made. In environmental and occupational epidemiology it is frequently assumed that the exposure is to mutagens and therefore involves the measurement of protein and nucleic acid adducts in blood and other tissues; altered bases excreted in urine; chromosomal aberrations, micronuclei and SCEs in blood and other tissues. While this assumption may appear reasonable in a particular situation, it is remarkable that so few of these 'biomarkers' have been validated. Consequently, plausible hypotheses are being used directly in human studies without going through any validation step that could be performed with experimental animals.

One rare example of an attempt to validate a marker was a study[62] to ascertain whether or not SCE rates have any association with the risk of cancer at the individual level in rats exposed to *N*-ethyl-*N*-nitrosourea (ENU). SCE frequencies in cultured lymphocytes were measured before as well as 24 hours and 7 days after a single intraperitoneal ENU dose of 0, 25, 50 or 75 mg/kg body weight. The relationship between SCEs and eventual cancer outcome was analysed, i.e., presence or absence of tumour, latency period. ENU induced a maximum of 1.6-fold increase in SCEs/cell observed at both post-treatment sampling times. The mean SCE rates in rats with ENU-specific cancers (various gliomas, thyroid and testicular tumours), or in rats with early or multiple tumours did not differ from those in animals that survived no less than 65 weeks or longer without developing tumours. In the tumour-bearing animals, no relationship was found between the mean SCE rate and survival time. The conclusion from this study was that there is no relationship between SCE frequency and individual susceptibility to ENU. On a group basis, however, rats with high SCE rates did have an increased risk of cancer.

In addition to this single experimental validation study, there is in progress a single follow-up study for subsequent cancer risks of cytogenetic events in peripheral blood lymphocytes in human populations[63]. The cohort consists of 3182 people living in the Nordic countries who were examined between 1970 and 1988 for chromosomal aberrations, sister-chromatid exchanges and micronuclei. The data were collected in 10 different laboratories and so, to standardize the interlaboratory variations, the results for each laboratory were stratified into low (1–33 percentile), medium (34–66 percentile) and high (67–100 percentile) frequency events. At the time of the publication, the analysis was based upon 85 cancers diagnosed between 1970 and 1991. So far, there was a significant association between a cytogenetic event and eventual cancer development only for a high frequency of chromosomal aberrations (standardized incidence ratio 2.1, 95%CI 1.5–2.8, based upon 39 cancers in this group, there being 66 cases for the combined low, medium and high frequency groups). There were no significant trends or increased point estimates in the standardized incidence ratios of sister-chromatid exchanges (with a total of 49 cancer cases) or micronuclei (with a total of only 11 cancer cases). Clearly, the numbers are too small to allow any definite conclusions to be reached at this stage, particularly in the case of micronuclei, but any association of sister-chromatid exchange with eventual cancer outcome is, at best, weak.

CARCINOGENS AND MUTAGENS: CONCLUSIONS

The most important single human carcinogen, in terms of its effects upon human mortality, is tobacco smoke. In comparison, all other fatal carcinogenic risks are small. Sunlight exposure certainly produces more tumours, but a small proportion of these is lethal and sunlight is more difficult to avoid than tobacco smoke should be, and low exposures to sunlight are beneficial. Occupational carcinogens have a much smaller community impact, although they can be significant in specific populations. Exposure controls have improved significantly in the developed countries during the latter half of this century, but the conditions in some developing countries are a growing concern.

Mutagens are important because their activity indicates that they are biologically reactive and may cause damage. Substances that damage germ-line cells experimentally are also highly toxic in other, more readily conducted assays and, therefore, they can be regulated if necessary because of the more immediate harm they can produce. Very frequently this harm is carcinogenesis. Positive results in some genotoxicity assays (particularly the primary hepatocyte UDS, *in vivo* rodent bone marrow cell clastogenicity and the bacterial mutation assays) are informative, while positive results with certain other assays are not. No negative results are informative.

Mutagenic endpoints are used in some circumstances as 'biomarkers' that are interpreted as being important in carcinogenesis, either as indicators of exposure only or also of effect. However, both experimental and human evidence suggest that sister-chromatid exchanges in peripheral blood lymphocytes are not informative with regard to cancer outcome, while the only human follow-up study ever performed suggests that high levels of chromosomal aberrations in lymphocytes may be indicative of an increased risk of developing some form of cancer. No other genetic endpoint has been subjected to similar validation studies.

References

1. D. A. Casciano, G. Talaska and D. Clive, "The potent hepatocarcinogen methapyrilene induces mutations in L5178Y mouse lymphoma cells in the apparent absence of DNA adduct formation" *Mutation Res.* **263**, 127–132 (1991).
2. G. Colditz, W. DeJong, D. Hunter, D. Trichopoulos and W. Willett, Harvard report on cancer prevention. Vol. 1: Causes of human cancer. *Cancer Causes Cont.* **7**, 3–58 (1996).
3. IARC, *IARC Monographs on the Evaluation of Carcinogenic Risks to Humans,* Supplement **7**, *Overall Evaluations of Carcinogenicity: An Updating of IARC Monographs Volumes 1 to 42* (Lyon, 1987).
4. R. Doll and A. B. Hill, "Smoking and carcinoma of the lung" Preliminary report. *Br. Med. J.* **ii**, 739–748 (1950).
5. E. L. Wynder and E. A. Graham, "Tobacco smoking as a possible etiologic factor in bronchiogenic carcinoma. A study of six hundred and eighty-four proved cases" *J. Am. Med. Assoc.* **143**, 329–336 (1950).
6. IARC, *IARC Monographs on the Evaluation of Carcinogenic Risk of Chemicals to Humans,* **Vol. 38**, *Tobacco Smoking* (Lyon, 1986).
7. R. Doll, R. Peto, K. Wheatley, R. Gray and I. Sutherland, "Mortality in relation to smoking: 40 years' observation on male British doctors" *Br. Med. J.* **309**, 901–911 (1994).
8. N. J. Wald and A. K. Hackshaw, "Cigarette smoking: an epidemiological overview" **In**: (R. Doll and J. Crofton, Eds.) *Tobacco and Health. Brit. Med. Bulletin* **52**, 3–11 (1996).
9. R. Doll and R. Peto, "The causes of cancer: quantitative estimates of avoidable risks of cancer in the United States today" *J. Natl. Cancer Inst.* **66**, 1191–1308 (1981).
10. M. R. Law and A. K. Hackshaw, "Environmental tobacco smoke" **In:** (R. Doll and J. Crofton, Eds.) *Tobacco and Health. Brit. Med. Bulletin* **52**, 22–34 (1996).
11. L. Tomatis, A. Aitio, N. E. Day, E. Heseltine, J. Kaldor, A. B. Miller, D. M. Parkin and E. Riboli (Eds.) *Cancer: Causes, Occurrence and Control* (IARC Scientific Publications No. **100**, Lyon, 1990). pp. 303–308.
12. IARC, *IARC Monographs on the Evaluation of Carcinogenic Risks to Humans*, **Vol. 44**, *Alcohol Drinking* (Lyon, 1988).
13. W. J. Blot, "Alcohol and cancer" *Cancer Res.* (Suppl.) **52**, 2119s–2123s (1992).
14. R. Doll, R. Peto, E. Hall, K. Wheatley and R. Gray, "Mortality in relation to consumption of alcohol: 13 years' observations on male British doctors" *Br. Med. J.* **309**, 911–918 (1994).
15. R. L. Whistler and J. R. Daniel, "Carbohydrates" **In**: (O. R. Fennema, Ed.), *Food Chemistry*, 2nd. Edition. (Marcel Dekker, Inc., New York, 1985) pp. 69–137.
16. W. C. Willett, F. Sacks, A. Trichopoulou, G. Drescher, A. Ferro-Luzzi, E. Helsing and D. Trichopoulos, "Mediterranean diet pyramid: a cultural model for healthy eating" *Am. J. Clin. Nutr.* **61(suppl.)**, 1402S–1406S (1995).
17. A. Ferro-Luzzi and S. Sette, "The mediterranean diet: an attempt to define its present and past composition" *Europ. J. Clin. Nutr.* **43(suppl. 2)**, 13–29 (1989).
18. W. P. T. James, G. G. Duthie and K. W. J. Wahle, "The Mediterranean diet: protective or simply not toxic?" *Europ. J. Clin. Nutr.* **43(suppl. 2)**, 31–41 (1989).

19. M. Geddes, D. M. Parkin, M. Khlat, D. Balzi and E. Buiatti (Eds.) *Cancer in Italian Migrant Populations* (IARC Sci. Publ. No. 123. Lyon, 1996).
20. A. J. McMichael, M. G. McCall, J. M. Hartshorne and T. L. Woodings, "Patterns of gastro-intestinal cancer in European migrants to Australia: the role of dietary change" *Int. J. Cancer* **25**, 431–437 (1980).
21. B. K. Armstrong, T. L. Woodings, N. S. Stenhouse and M. G. McCall, *Mortality from Cancer in Migrants to Australia 1962 to 1971* (NH & MRC Research Unit in Epidemiology and Preventive Medicine, Raine Medical Statistics Unit, Perth, WA, University of Western Australia, 1983).
22. E. A. Lew and L. Garfinkel, "Variations in mortality by weight among 750,000 men and women" *J. Chron. Dis.* **32**, 563–576 (1979).
23. R. L. Phillips and D. A. Snowdon, "Dietary relationships with fatal colorectal cancer among Seventh-Day Adventists" *J. Natl. Cancer Inst.* **74**, 307–317 (1985).
24. M. Shike, "Body weight and colon cancer" *Am. J. Clin. Nutr.* **63(suppl.)**, 442S–444S (1996).
25. R. Ballard-Barbash and C. A. Swanson, "Body weight: estimation of risk for breast and endometrial cancers" *Am. J. Clin. Nutr.* **63(suppl.)**, 437S–441S (1996).
26. A. Nomura, L. K. Heilbrun and G. N. Stemmermann, "Body mass index as a predictor of cancer in men" *J. Natl. Cancer Inst.* **74**, 319–323 (1985).
27. D. McGregor, "Tumour incidence reduction in carcinogenicity bioassays" **In**: (B. W. Stewart, D. McGregor and P. Kleihues, Eds.) *Principles of Chemoprevention* (IARC Sci. Publ. No. 139, Lyon, 1996) pp. 277–290.
28. R. Kaaks, "Dietary fat and colorectal cancer: case-control studies" **In**: (P. Sachet, M. Boiron and G. Halpern, Eds.) *Nutrition Cancer* (CERN, Paris, 1995) pp. 25–39.
29. IARC, *IARC Monographs on the Evaluation of Carcinogenic Risks to Humans, Some Naturally Occurring Substances: Food Items and Constituents*, **Vol. 56**, *Heterocyclic Aromatic Amines and Mycotoxins* (IARC, Lyon, 1993).
30. IARC, *IARC Monographs on the Evaluation of Carcinogenic Risk of Chemicals to Man*, **Vol. 10**, *Some Naturally Occurring Substances* (IARC, Lyon, 1976).
31. R. C. Beier, "Natural Pesticides and Bioactive Components in Foods" *Reviews of Environmental Contamination and Toxicology* **113**, 47–137 (1990).
32. M. Wink, "Plant breeding: importance of secondary metabolites for protection against pathogens and herbivores" *Theor. Appl. Genet.* **75**, 225–233 (1988).
33. P. K. Cottee, E. A. Bernays and A. J. Mordue, "Comparisons of deterrence and toxicity of selected secondary plant to an oligophagous and a polyphagous acridid" *Entomol. Exp. Appl.* **46**, 241–247 (1988).
34. G. W. Dawson, D. L. Hallahan, A. Mudd, M. A. Patel, J. A. Pickett, L. J. Wolhams and R. Wallsgrove, "Secondary plant metabolites as target for genetic modification of crop plants for pest resistance" *Pestic. Sci.* **27**, 48–59 (1989).
35. D. McKey, "The distribution of secondary plant compounds within plants" **In**: (G. A. Rosenthal, D. H. Janzen, Eds.) *Herbivores: Their interacting with secondary plant metabolites* (Academic Press., New York, 1979).
36. J. A. Pickett, "Production of behaviour-controlling chemicals by crop plants" *Phil. Trans. R. Soc. Lond. B.* **310**, 235–2329 (1985).
37. J. R. Philogene and J. T. Arnason, "L'influences des composes secondaires des plants sur la biologie des insectes" *Rev. d'entomol. Quebec* **31**, 33–42 (1986).
38. B. N. Ames and L. S. Gold, "Pesticides, risk, and applesauce" *Science* **244**, 755–757 (1989).
39. T. Sugimura, M. Nagao, T. Kawachi, M. Honda, T. Yahagi, Y. Seino, S. Sato, N. Matsukura, T. Matsushima, A. Shirai, M. Sawamura and H. Matsumoto, "Mutagen-carcinogens in food with special reference to highly mutagenic pyrolytic products in broiled foods" **In**: (H. H. Hiatt, J. D. Watson and J. A. Winsten, Eds.) *Origins of Human Cancer* (Book C. Cold Spring Harbor Laboratory, New York, 1977) pp. 1561–1575.
40. T. Sugimura and M. Nagao, "Carcinogenic, mutagenic, and comutagenic aromatic amines in human foods" *Natl. Cancer Inst. Monogr.* **58**, 27–33 (1981).
41. IARC, *IARC Monographs on the Evaluation of Carcinogenic Risks to Humans.* **Vol. 59**, *Hepatitis Viruses* (IARC, Lyon, 1994).
42. Y. Masuda, "The Yusho rice oil poisoning incident" **In**: (A. Schecter, Ed.) *Dioxins and Health* (Plenum Press, New York, 1994), pp. 633–659.
43. C.-C. Hsu, M.-L. M. Yu, Y.-C. J. Chen, Y.-L. L. Guo and W. J. Rogen, "The Yu-cheng rice oil poisoning incident" **In**: (A. Schecter, Ed.) *Dioxins and Health* (Plenum Press, New York, 1994) pp. 661–684.
44. P. Bogovski, "Historical perspective of occupational cancer" **In**: (H. Vainio, M. Sorsa and K. K. Hemminki, Eds.) *Occupational Cancer and Carcinogenesis* (Hemisphere Publishing Corp. Washington, 1981) pp. 1–19.

45. J. Siemiatycki, "Introduction to occupational cancer" **In**: J. Siemiatycki, Ed.) *Risk Factors for Cancer in the Workplace* (CRC Press, Boca Raton, 1991) pp. 1–16.
46. IARC, *IARC Monographs on the Evaluation of Carcinogenic Risks to Humans*. **Vol. 60**, *Some Industrial Chemicals* (Lyon, 1994) pp. 73–159.
47. N. Pearce, E. Matos, M. Koivusalo and S. Wing, "Industrialization and health" **In**: (N. Pearce, E. Matos, H. Vainio, P. Boffetta and M. Kogevinas, Eds.) *Occupational Cancer in Developing Countries* (IARC Scientific Publications No. 129. Lyon, 1994) pp. 7–22.
48. B. N. Ames, W. E. Durston, E. Yamasaki and F. D. Lee, "Carcinogens are mutagens: a simple test system combining liver homogenates for activation and bacteria for detection" *Proc. Natl. Acad. Sci. USA* **70**, 2281–2285 (1973).
49. J. F. Crow, "Rate of genetic change under selection" *Proc. Natl. Acad. Sci. USA* **59**, 655–668 (1968).
50. Y. E. Dubrova, V. N. Nesterov, N. G. Krouchinsky, V. A. Ostapenko, R. Neumann, D. Neil and A. J. Jeffreys, "Human minisatellite mutation rate after the Chernobyl accident" *Nature* **380**, 683–686 (1996).
51. M. Kodaira, C. Satoh, K. Hiyama and K. Toyama, "Lack of effects of atomic bomb radiation on genetic instability of tandem-repetitive elements in human germ cells" *Am. J. Hum. Genet.* **57**, 1275–1283 (1995).
52. H. E. Holden, "Comparison of somatic and germ cell models for cytogenetic screening" *J. Appl. Toxicol.* **2**, 196–200 (1982).
53. A. Auletta and J. Ashby, "Workshop on the relationship between short-term test information and carcinogenicity: Williamsburg, Virginia, January 20–23, 1987" *Environ. Mol. Mutagen.* **11**, 135–145 (1988).
54. I.-D. Adler and J. Ashby, "The present lack of evidence for unique rodent germ-cell mutagens" *Mutation Res.* **212**, 55–66 (1989).
55. P. Arni, J. Ashby, S. Castellino, G. Engelhardt, B. A. Herbold, R. A. J. Priston and W. J. Bontinck, "Assessment of the potential germ cell mutagenicity of industrial and plant protection chemicals as part of an integrated study of genotoxicity *in vitro* and *in vivo*" *Mutation Res.* **203**, 177–184 (1988).
56. M. D. Shelby and E. Zeiger, "Detection of human carcinogens in the Salmonella and rodent bone-marrow cytogenetic tests" *Mutation Res.* **234**, 257–261 (1990).
57. E. D. Thompson, "Comparison of *in vivo* and *in vitro* cytogenetic assay results" *Environ. Mutagen.* **8**, 753–768 (1986).
58. L. B. Russell, P. B. Selby, E. von Halle, W. Sheridan and L. Valcovic, "The mouse specific-locus test with agents other than radiation" *Mutation Res.* **86**, 329–354 (1981).
59. M. D. Shelby, J. B. Bishop, J. M. Mason and K. R. Tindall, "Fertility, reproduction, and genetic disease: studies on the mutagenic effects of environmental agents on mammalian germ cells" *Environ. Health Perspect.* **100**, 283–291 (1993).
60. R. W. Tennant, B. H. Margolin, M. D. Shelby, E. Zeiger, J. K. Haseman, J. Spalding, W. Caspary, M. Resnick, S. Stasiewicz, B. Anderson and B. Minor, "Prediction of chemical carcinogenicity in rodents from in vitro genetic toxicity assays" *Science* **236**, 933–941 (1987).
61. D. McGregor, "A review of some properties of ethylene glycol ethers relevant to their carcinogenic evaluation" *Occupational Hyg.* **2**, 213–235 (1996).
62. A. Aitio, J. R. P. Cabral, A.-M. Camus, D. Galendo, H. Bartsch, M.-L. Aitio, H. Norppa, S. Salomaa, M. Sorsa, K. Husgafvel-Pursiainen and M. Nurminen, "Evaluation of sister chromatid exchange as an indicator of sensitivity to *N*-ethyl-*N*-nitrosourea-induced carcinogenesis in rats" *Teratog. Carcinog. Mutagen.* **8**, 273–286 (1988).
63. L. Hagmar, A. Brøgger, I. L. Hansteen, S. Heim, B. Högstedt, L. Knudsen, B. Lambert, K. Linnainmaa, F. Mitelman, I. Nordenson, C. Reuterwall, S. Salomaa, S. Skerfving and M. Sorsa, "Cancer risk in humans predicted by increased levels of chromosomal aberrations in lymphocytes: Nordic study group on the health risk of chromosome damage" *Cancer Res.* **54**, 2919–2922 (1994).

Appendix Table The IARC evaluations of carcinogenicity to humans of agents, mixtures and exposure circumstances up to December 1996[1] Groups 1, 2A, and 2B only; Genotoxicity not listed for mixtures or exposure circumstances

	SAL	UDS	MN	OTHER
Group 1—Carcinogenic to humans				
Agents and groups of agents				
Aflatoxins, naturally occurring [1402-68-2] (*56*; 1993)	+	+	+	+
4-Aminobiphenyl [92-67-1] (*1*, *Suppl. 7*; 1987)	+	+	+	+
Arsenic [7440-38-2] and arsenic compounds[2] (*23*, *Suppl. 7*; 1987)	−			+
Asbestos [1332-21-4] (*14*, *Suppl. 7*; 1987)	−	−		
Azathioprine [446-86-6] (*26*, *Suppl. 7*; 1987)	+		+	
Benzene [71-43-2] (*29*, *Suppl. 7*; 1987)	−	−	+	?
Benzidine [92-87-5] (*29*, *Suppl. 7*; 1987)	+	+	+	+
Beryllium [7440-41-7] and beryllium compounds[3] (*58*; 1993)	−	−	−	+
N,N-Bis(2-chloroethyl)-2-naphthylamine (Chlornaphazine) [494-03-1] (*4*, *Suppl. 7*; 1987)	+	+	+	+
Bis(chloromethyl)ether [542-88-1] and chloromethyl methyl ether [107-30-2] (technical-grade) (*4*, *Suppl. 7*; 1987)	+			
1,4-Butanediol dimethanesulfonate (Busulphan; Myleran) [55-98-1] (*4*, *Suppl. 7*; 1987)	+		+	+
Cadmium [7440-43-9] and cadmium compounds[3] (*58*; 1993)	−		−	+
Chlorambucil [305-03-3] (*26*, *Suppl. 7*; 1987)	+		+	
1-(2-Chloroethyl)-3-(4-methylcyclohexyl)-1-nitrosourea (Methyl-CCNU; Semustine) [13909-09-6] (*Suppl. 7*; 1987)	+			+
Chromium [VI] compounds[3] (*49*; 1990)	+		+	+
Ciclosporin [79217-60-0] (*50*; 1990)				
Cyclophosphamide [50-18-0] [6055-19-2] (*26*, *Suppl. 7*; 1987)	+	+	+	+
Diethylstilboestrol [56-53-1] (*21*, *Suppl. 7*; 1987)	−	−	(+)	+
Erionite [66733-21-9] (*42*, *Suppl. 7*; 1987)				
Ethylene oxide[4] [75-21-8] (*60*; 1994)	+		+	+
Helicobacter pylori (infection with) (*61*; 1994)				
Hepatitis B virus (chronic infection with) (*59*; 1994)				
Hepatitis C virus Chronic infection with) (*59*; 1994)				
Human immunodeficiency virus type 1 (infection with) (*67*; 1996)				
Human papillomavirus type 16 (*64*; 1995)				
Human papillomavirus type 18 (*64*; 1995)				
Human T-cell lymphotropic virus type I (*67*; 1996)				
Melphalan [148-82-3] (*9*, *Suppl. 7*; 1987)			+	+
8-Methoxypsoralen (Methoxsalen) [298-81-7] plus ultraviolet A radiation (*24*, *Suppl. 7*; 1987)	+	+		
MOPP and other combined chemotherapy including alkylating agents (*Suppl. 7*; 1987)				
Mustard gas (Sulfur mustard) [505-60-2] (*9*, *Suppl. 7*; 1987)				+
2-Naphthylamine [91-59-8] (*4*, *Suppl. 7*; 1987)	+	+	+	+
Nickel compounds[3] (*49*; 1990)	−	−	−	+

SAL, mutation in *Salmonella typhimurium* liver microsome assay; UDS, unscheduled DNA synthesis in primary cultures of rodent hepatocytes; MN, micronucleus induction in rodent bone marrow cells; OTHER, other mammalian cells in *in vitro* or *in vivo* assays; +, positive; (+), weak positive; −, negative; ?, inconclusive.

[1] Chemical Abstract Number is given in square brackets; the IARC Monograph volumes and year in which the latest evaluation was published are given in parentheses.

[2] This evaluation applies to the group of chemicals as a whole and not necessarily to all individual chemicals within the group.

[3] Evaluated as a group.

[4] Overall evaluation upgraded from 2A to 1 with supporting evidence from other data relevant to the evaluation of carcinogenicity and its mechanisms.

Appendix Table (*Continued*)

	SAL	UDS	MN	OTHER
Oestrogen replacement therapy (*Suppl. 7*; 1987)				
Oestrogens, nonsteroidal[2] (*Suppl. 7*; 1987)	–			
Oestrogens, steroidal[2] (*Suppl. 7*; 1987)	–	–		–
Opisthorchis viverrini (infection with) (*61*; 1994)				
Oral contraceptives, combined[5] (*Suppl. 7*; 1987)				
Oral contraceptives, sequential (*Suppl. 7*; 1987)				
Radon [10043-92-2] and its decay products (*43*; 1988)				
Schistosoma haematobium (infection with) (*61*; 1994)				
Silica [14808-60-7], crystalline (inhaled in the form of quartz or cristobalite from occupational sources) (*68*; 1997)			–	+
Solar radiation (*55*; 1992)				
Talc containing asbestiform fibres (*42*, *Suppl. 7*; 1987)				
Tamoxifen[6] [10540-29-1] (*66*; 1996)		–		+
2,3,7,8-Tetrachlorodibenzo-*para*-dioxin (TCDD) [1746-01-6] (*69*; 1997)	–	–	–	?
Thiotepa [52-24-4] (*50*; 1990)			+	+
Treosulfan [299-75-2] (*26*, *Suppl. 7*; 1987)			+	
Vinyl chloride [75-01-4] (*19*, *Suppl. 7*; 1987)	+	+	+	
Mixtures				
Alcoholic beverages (*44*; 1988)				
Analgesic mixtures containing phenacetin (*Suppl. 7*; 1987)				
Betel quid with tobacco (*37*, *Suppl. 7*; 1987)				
Coal-tar pitches [65996-93-2] (*35*, *Suppl. 7*; 1987)				
Coal-tars [8007-45-2] (*35*, *Suppl. 7*; 1987)				
Mineral oils, untreated and mildly treated (*33*, *Suppl. 7*; 1987)				
Salted fish (Chinese-style) (*56*; 1993)				
Shale-oils [68308-34-9] (*35*, *Suppl. 7*; 1987)				
Soots (*35*, *Suppl. 7*; 1987)				
Tobacco products, smokeless (*37*, *Suppl. 7*; 1987)				
Tobacco smoke (*38*, *Suppl. 7*; 1987)				
Wood dust (*62*; 1995)				
Exposure circumstances				
Aluminium production (*34*, *Suppl. 7*; 1987)				
Auramine, manufacture of (*Suppl. 7*; 1987)				
Boot and shoe manufacture and repair (*25*, *Suppl. 7*; 1987)				
Coal gasification (*34*, *Suppl. 7*; 1987)				
Coke production (*34*, *Suppl. 7*; 1987)				
Furniture and cabinet making (*25*, *Suppl. 7*; 1987)				
Haematite mining (underground) with exposure to radon (*1*, *Suppl. 7*; 1987)				
Iron and steel founding (*34*, *Suppl. 7*; 1987)				
Isopropanol manufacture (strong-acid process) (*Suppl. 7*; 1987)				
Magenta, manufacture of (*57*; 1993)				
Painter (occupational exposure as a) (*47*; 1989)				
Rubber industry (*28*, *Suppl. 7*; 1987)				
Strong-inorganic-acid mists containing sulfuric acid (occupational exposure to) (*54*; 1992)				

[5] There is also conclusive evidence that these agents have a protective effect against cancers of the ovary and endometrium.

[6] There is also conclusive evidence that this agent (tamoxifen) reduces the risk of contralateral breast cancer.

Appendix Table (*Continued*)

	SAL	UDS	MN	OTHER
Group 2A—Probably carcinogenic to humans				
Agents and groups of agents				
Acrylamide[7] [79-06-1] (*60*; 1994)	−	−	+	+
Acrylonitrile [107-13-1] (*19*, *Suppl. 7*; 1987)	+	−	−	+
Adriamycin[7] [23214-92-8] (*10*, *Suppl. 7*; 1987)	+	−	+	+
Androgenic (anabolic) steroids (*Suppl. 7*; 1987)				
Azacitidine[7] [320-67-2] (*50*; 1990)	+			+
Benz[*a*]anthracene[7] [56-55-3] (*32*, *Suppl. 7*; 1987)	+	+		+
Benzidine-based dyes[7] (*Suppl. 7*; 1987)	+	+	+	+
Benzo[*a*]pyrene[7] [50-32-8] (*32*, *Suppl. 7*; 1987)	+	+	+	+
Bischloroethyl nitrosourea (BCNU) [154-93-8] (*26*, *Suppl. 7*; 1987)	+		+	+
1,3-Butadiene [106-99-0] (*54*; 1992)	+		+	−
Captafol[7] [2425-06-1] (*53*; 1991)				+
Chloramphenicol[7] [56-75-7] (*50*; 1990)	−			+
1-(2-Chloroethyl)-3-cyclohexyl-1-nitrosourea[7] (CCNU) [13010-47-4] (*26*, *Suppl. 7*; 1987)	+			+
para-Chloro-*ortho*-toluidine [95-69-2] and its strong acid salts[3] (*48*; 1990)	−			+
Chlorozotocin[7] [54749-90-5] (*50*; 1990)				
Cisplatin[7] [15663-27-1] (*26*, *Suppl. 7*; 1987)	+		+	+
Clonorchis sinensis (infection with)[7] (*61*; 1994)				
Dibenz[*a*,*h*]anthracene[7] [53-70-3] (*32*, *Suppl. 7*; 1987)	+	−		+
Diethyl sulfate [64-67-5] (*54*; 1992)	+	+		+
Dimethylcarbamoyl chloride[7] [79-44-7] (*12*, *Suppl. 7*; 1987)	+	−	+	+
Dimethyl sulfate[7] [77-78-1] (*4*, *Suppl. 7*; 1987)	+	+		+
Epichlorohydrin[7] [106-89-8] (*11*, *Suppl. 7*; 1987)	+	−	−	+
Ethylene dibromide[7] [106-93-4] (*15*, *Suppl. 7*; 1987)		+	−	+
N-Ethyl-*N*-nitrosourea[7] [759-73-9] (*17*, *Suppl. 7*; 1987)	+		+	+
Formaldehyde [50-00-0] (*62*; 1995)	+	+	−	+
Human papillomavirus type 31 (*64*; 1995)				
Human papillomavirus type 33 (*64*; 1995)				
IQ[7] (2-Amino-3-methylimidazo[4,5-*f*]quinoline) [76180-96-6] (*56*; 1993)	+	+	−	+
5-Methoxypsoralen[7] [484-20-8] (*40*, *Suppl. 7*; 1987)	(+)			
4,4′-Methylene bis(2-chloroaniline) (MOCA)[7] [101-14-4] (*57*; 1993)	+	+	+	+
N-Methyl-*N*′-nitro-*N*-nitrosoguanidine[7] (MNNG) [70-25-7] (*4*, *Suppl. 7*; 1987)	+	+	+	+
N-Methyl-*N*-nitrosourea[7] [684-93-5] (*17*, *Suppl. 7*; 1987)	+	+		+
Nitrogen mustard [51-75-2] (*9*, *Suppl. 7*; 1987)	+	+	+	+
N-Nitrosodiethylamine[7] [55-18-5] (*17*, *Suppl.7*; 1987)	+	+		+
N-Nitrosodimethylamine[7] [62-75-9] (*17*, *Suppl. 7*; 1987)	+	+	+	+
Phenacetin [62-44-2] (*24*, *Suppl. 7*; 1987)	?		+	−
Procarbazine hydrochloride[7] [366-70-1] (*26*, *Suppl. 7*; 1987)	+	−	+	+
Styrene-7,8-oxide[7] [96-09-3] (*60*; 1994)	+		−	+
Tetrachloroethylene [127-18-4] (*63*; 1995)	−	−		+
Trichloroethylene [79-01-6] (*63*; 1995)	+	+	+	+
1,2,3-Trichloropropane [96-18-4] (*63*; 1995)	+	−		+
Tris(2,3-dibromopropyl)phosphate[7] [126-72-7] (*20*, *Suppl. 7*; 1987)	+	(+)	+	+
Ultraviolet radiation A[7] (*55*; 1992)	+			+
Ultraviolet radiation B[7] (*55*; 1992)	+			+
Ultraviolet radiation C[7] (*55*; 1992)				+

[7] Overall evaluation upgraded from 2B to 2A with supporting evidence from other data relevant to the evaluation of carcinogenicity and its mechanisms.

Appendix Table (*Continued*)

	SAL	UDS	MN	OTHER
Vinyl bromide[7] [593-60-2] (*39*, *Suppl. 7*; 1987)	+	+		
Vinyl fluoride [75-02-5] (*63*; 1995)			+	+
Mixtures				
Creosotes [8001-58-9] (*35*, *Suppl. 7*; 1987)				
Diesel engine exhaust (*46*; 1989)				
Hot mate (*51*; 1991)				
Non-arsenical insecticides (occupational exposures in spraying and application of) (*53*; 1991)				
Polychlorinated biphenyls [1336-36-3] (*18*, *Suppl. 7*; 1987)				
Exposure circumstances				
Art glass, glass containers and pressed ware (manufacture of) (*58*; 1993)				
Hairdresser or barber (occupational exposure as a) (*57*; 1993)				
Petroleum refining (occupational exposures in) (*45*; 1989)				
Sunlamps and sunbeds (use of) (*55*; 1992)				
Group 2B—Possibly carcinogenic to humans				
Agents and groups of agents				
A-α-C (2-Amino-9*H*-pyrido[2,3-*b*]indole) [26148-68-5] (*40*, *Suppl. 7*; 1987)		+		
Acetaldehyde [75-07-0] (*36*, *Suppl. 7*; 1987)	−			+
Acetamide [60-35-5] (*7*, *Suppl. 7*; 1987)	−		(+)	
AF-2 [2-(2-Furyl)-3-(5-nitro-2-furyl)acrylamide [3688-53-7] (*31*, *Suppl. 7*; 1987)	+	+		+
Aflatoxin M_1 [6795-23-9] (*56*; 1993)	+	+		
para-Aminoazobenzene [60-09-3] (*8*, *Suppl. 7*; 1987)	+	+		+
ortho-Aminoazotoluene [97-56-3] (*8*, *Suppl. 7*; 1987)	+	+		
2-Amino-5-(5-nitro-2-furyl)-1,3,4-thiadiazole [712-68-5] (*7*, *Suppl. 7*; 1987)				
Amitrole [61-82-5] (*41*, *Suppl. 7*; 1987)	−		−	−
ortho-Anisidine [90-04-0] (*27*, *Suppl. 7*; 1987)	+	−		+
Antimony trioxide [1309-64-4] (*47*; 1989)				
Aramite® [140-57-8] (*5*, *Suppl. 7*; 1987)				
Atrazine[8] [1912-24-9] (*53*; 1991)	−			−
Auramine [492-80-8] (technical-grade) (*1*, *Suppl. 7*; 1987)	+		−	+
Azaserine [115-02-6] (*10*, *Suppl. 7*; 1987)		+		
Benzo[*b*]fluoranthene [205-99-2] (*32*, *Suppl. 7*; 1987)	?			
Benzo[*j*]fluoranthene [205-82-3] (*32*, *Suppl. 7*; 1987)				
Benzo[*k*]fluoranthene [207-08-9] (*32*, *Suppl. 7*; 1987)	+			
Benzofuran [271-89-6] (*63*; 1995)	−			+
Benzyl violet 4B [1694-09-3] (*16*, *Suppl. 7*; 1987)	+			
Bleomycins[8] [11056-06-7] (*26*, *Suppl. 7*; 1987)	+		+	+
Bracken fern (*40*, *Suppl. 7*; 1987)				
Bromodichloromethane [75-27-4] (*52*, 1991)	?		−	+
Butylated hydroxyanisole (BHA) [25013-16-5] (*40*, *Suppl. 7*; 1987)	−	−	−	
β-Butyrolactone [3068-88-0] (*11*, *Suppl. 7*; 1987)	+			
Caffeic acid [331-39-5] (1993) (*56*; 1993)	−		−	+

[8] Overall evaluation upgraded from 3 to 2B with supporting evidence from other data relevant to the evaluation of carcinogenicity and its mechanisms.

Appendix Table (*Continued*)

	SAL	UDS	MN	OTHER
Carbon black [1333-86-4] (*65*; 1996)	+			+
Carbon tetrachloride [56-23-5] (*20*, *Suppl. 7*; 1987)	−	−	−	+
Ceramic fibres (*43*; 1988)				
Chlordane [57-74-9] (*53*; 1991)	−	−		
Chlordecone (Kepone) [143-50-0] (*20*, *Suppl. 7*; 1987)	−			−
Chlorendic acid [115-28-6] (*48*; 1990)	−			+
α-Chlorinated toluenes (benzyl chloride, benzal chloride, benzotrichloride) (*Suppl. 7*; 1987)	+		−	+
para-Chloroaniline [106-47-8] (*57*; 1993)	+			+
Chloroform [67-66-3] (*20*, *Suppl. 7*; 1987)	−	−		
1-Chloro-2-methylpropene [513-37-1] (*63*; 1995)	+			+
Chlorophenols (*41*, *Suppl. 7*; 1987)	−			
Chlorophenoxy herbicides (*41*, *Suppl. 7*; 1987)	−			
4-Chloro-*ortho*-phenylenediamine [95-83-0] (*27*, *Suppl. 7*; 1987)	+		+	+
CI Acid Red 114 [6459-94-5] (*57*; 1993)	+	−		−
CI Basic Red 9 [569-61-9] (*57*; 1993)	+	+		+
CI Direct Blue 15 [2429-74-5] (*57*; 1993)	+	−		−
Citrus Red No. 2 [6358-53-8] (*8*, *Suppl. 7*; 1987)	+			
Cobalt [7440-48-4] and cobalt compounds[3] (*52*; 1991)				
para-Cresidine [120-71-8] (*27*; *Suppl. 7*; 1987)	+	−		
Cycasin [14901-08-7] (*10*, *Suppl. 7*; 1987)	+	+	−	
Dacarbazine [4342-03-4] (*26*, *Suppl. 7*; 1987)	+			+
Dantron (Chrysazin; 1,8-Dihydroxyanthraquinone) [117-10-2] (*50*; 1990)	−	?		
Daunomycin [20830-81-3] (*10*, *Suppl. 7*; 1987)				+
DDT [*p,p′*-DDT, 50-29-3] (*53*; 1991)	−	−	−	−
N,N′-Diacetylbenzidine [613-35-4] (*16*, *Suppl. 7*; 1987)				
2,4-Diaminoanisole [615-05-4] (*27*, *Suppl. 7*; 1987)	+			+
4,4′-Diaminodiphenyl ether [101-80-4] (*29*, *Suppl. 7*; 1987)				
2,4-Diaminotoluene [95-80-7] (*16*, *Suppl. 7*; 1987)	+	+		+
Dibenz[*a,h*]acridine [226-36-8] (*32*, *Suppl. 7*; 1987)				−
Dibenz[*a,j*]acridine [224-42-0] (*32*, *Suppl. 7*; 1987)	+	−		+
7*H*-Dibenzo[*c,g*]carbazole [194-59-2] (*32*, *Suppl. 7*; 1987)				
Dibenzo[*a,e*]pyrene [192-65-4] (*32*, *Suppl. 7*; 1987)				
Dibenzo[*a,h*]pyrene [189-64-0] (*32*, *Suppl. 7*; 1987)		−		
Dibenzo[*a,i*]pyrene [189-55-9] (*32*, *Suppl. 7*; 1987)	+			
Dibenzo[*a,l*]pyrene [191-30-0] (*32*, *Suppl. 7*; 1987)				
1,2-Dibromo-3-chloropropane [96-12-8] (*20*, *Suppl. 7*; 1987)	+		+	+
para-Dichlorobenzene [106-46-7] (*29*, *Suppl. 7*; 1987)	−		+	−
3,3′-Dichlorobenzidine [91-94-1] (*29*, *Suppl. 7*; 1987)	+		+	
3,3′-Dichloro-4,4′-diaminodiphenyl ether [28434-86-8] (*16*, *Suppl. 7*; 1987)				
1,2-Dichloroethane [107-06-2] (*20*, *Suppl. 7*; 1987)	+	+		
Dichloromethane (methylene chloride) [75-09-2] (*41*, *Suppl. 7*; 1987)	+		−	+
1,3-Dichloropropene [542-75-6] (technical-grade) (*41*, *Suppl. 7*; 1987)	+			+
Dichlorvos [62-73-7] (1991) (*53*; 1991)	+			+
Diepoxybutane [1464-53-5] (*11*, *Suppl. 7*; 1987)	+	−		+
Di(2-ethylhexyl)phthalate [117-81-7] (*29*, *Suppl. 7*; 1987)	−	−		−
1,2-Diethylhydrazine [1615-80-1] (*4*, *Suppl. 7*; 1987)				
Diglycidyl resorcinol ether [101-90-6] (*36*, *Suppl. 7*; 1987)				+
Dihydrosafrole [94-58-6] (*10*, *Suppl. 7*; 1987)				
Diisopropyl sulfate [2973-10-6] (*54*; 1992)				

Appendix Table (*Continued*)

	SAL	UDS	MN	OTHER
3,3′-Dimethoxybenzidine (*ortho*-Dianisidine) [119-90-4] (*4*, *Suppl. 7*; 1987)	+	+		+
para-Dimethylaminoazobenzene [60-11-7] (*8*, *Suppl. 7*; 1987)	+	+	+	+
trans-2-[(Dimethylamino)methylimino]-5-[2-(5-nitro-2-furyl)-vinyl]-1,3,4-oxadiazole [25962-77-0] (*7*, *Suppl. 7*; 1987)	+			
2,6-Dimethylaniline (2,6-Xylidine) [87-62-7] (*57*; 1993)	–		–	+
3,3′-Dimethylbenzidine (*ortho*-Tolidine) [119-93-7] (*1*, *Suppl. 7*; 1987)				
Dimethylformamide [68-12-2] (*47*; 1989)	–	–	–	–
1,1-Dimethylhydrazine [57-14-7] (*4*, *Suppl. 7*; 1987)	?			+
1,2-Dimethylhydrazine [540-73-8] (*4*, *Suppl. 7*; 1987)	+	–	+	+
3,7-Dinitrofluoranthene [105735-71-5] (*65*; 1996)	+		+	+
3,9-Dinitrofluoranthene [22506-53-2] (*65*; 1996)	+		+	+
1,6-Dinitropyrene [42397-64-8] (*46*; 1989)	+	+		+
1,8-Dinitropyrene [42397-65-9] (*46*; 1989)	+	+		+
2,4-Dinitrotoluene [121-14-2] (*65*; 1996)	+	–		+
2,6-Dinitrotoluene [606-20-2] (*65*; 1996)	+	–		+
1,4-Dioxane [123-91-1] (*11*, *Suppl. 7*; 1987)	–			–
Disperse Blue 1 [2475-45-8] (*48*; 1990)	+			+
Ethyl acrylate [140-88-5] (*39*, *Suppl. 7*; 1987)	?		?	+
Ethylene thiourea [96-45-7] (*7*, *Suppl. 7*; 1987)	+	–	–	–
Ethyl methanesulfonate [62-50-0] (*7*, *Suppl. 7*; 1987)	+	+	+	+
2-(2-Formylhydrazino)-4-(5-nitro-2-furyl)thiazole [3570-75-0] (*7*, *Suppl. 7*; 1987)				
Furan [110-00-9] (*63*; 1995)	–	–		+
Glasswool (*43*; 1988)				
Glu-P-1 (2-Amino-6-methyldipyrido[1,2-*a*:3′,2′-*d*]imidazole) [67730-11-4] (*40*, *Suppl. 7*; 1987)		+		+
Glu-P-2 (2-Aminodipyrido[1,2-*a*:3′,2′-*d*]imidazole) [67730-10-3] (*40*, *Suppl. 7*; 1987)		–		+
Glycidaldehyde [765-34-4] (*11*, *Suppl. 7*; 1987)	+			+
Griseofulvin [126-07-8] (*10*, *Suppl. 7*; 1987)	–	–		
HC Blue No. 1 [2784-94-3] (*57*; 1993)	–	+	?	+
Heptachlor [76-44-8] (*53*; 1991)	–	–		+
Hexachlorobenzene [118-74-1] (*20*, *Suppl. 7*; 1987)	–			–
Hexachlorocyclohexanes (*20*, *Suppl. 7*; 1987)	–	–		–
Hexamethylphosphoramide [680-31-9] (*15*, *Suppl. 7*; 1987)	+	+	+	+
Human immunodeficiency virus type 2 (infection with) (*67*; 1996)				
Human papillomaviruses: some types other than 16, 18, 31 and 33 (*64*; 1995)				
Hydrazine [302-01-2] (*4*, *Suppl. 7*; 1987)	+	–	–	+
Indeno[1,2,3-*cd*]pyrene [193-39-5] (*32*, *Suppl. 7*; 1987)	+			
Iron-dextran complex [9004-66-4] (*2*, *Suppl. 7*; 1987)				–
Isoprene [78-79-5] (*60*; 1994)	–		+	+
Lasiocarpine [303-34-4] (*10*, *Suppl. 7*; 1987)	+	+		+
Lead [7439-92-1] and lead compounds, inorganic[3] (*23*, *Suppl. 7*; 1987)	–		+	–
Magenta [632-99-5] (containing CI Basic Red 9) (*57*; 1993)				
MeA-α-C (2-Amino-3-methyl-9*H*-pyrido[2,3-*b*]indole) [68006-83-7] (*40*, *Suppl. 7*; 1987)		+		
Medroxyprogesterone acetate [71-58-9] (*21*, *Suppl. 7*; 1987)				
MeIQ (2-Amino-3,4-dimethylimidazo[4,5-*f*]quinoline) [77094-11-2] (*56*; 1993)	+	+		+

Appendix Table (*Continued*)

	SAL	UDS	MN	OTHER
MeIQx (2-Amino-3,8-dimethylimidazo[4,5-*f*]quinoxaline) [77500-04-0] (*56*; 1993)	+	+	−	+
Merphalan [531-76-0] (*9*, *Suppl. 7*; 1987)				
2-Methylaziridine (Propyleneimine) [75-55-8] (*9*, *Suppl. 7*; 1987)				
Methylazoxymethanol acetate [592-62-1] (*10*, *Suppl. 7*; 1987)	−	+		+
5-Methylchrysene [3697-24-3] (*32*, *Suppl. 7*; 1987)	+	+		
4,4′-Methylene bis(2-methylaniline) [838-88-0] (*4*, *Suppl. 7*; 1987)				
4,4′-Methylenedianiline [101-77-9] (*39*, *Suppl. 7*; 1987)	+	−		+
Methylmercury compounds[3] (*58*; 1993)	−			+
Methyl methanesulfonate [66-27-3] (*7*, *Suppl. 7*; 1987)	+	+	+	+
2-Methyl-1-nitroanthraquinone [129-15-7] (uncertain purity) (*27*, *Suppl. 7*; 1987)	+			
N-Methyl-*N*-nitrosourethane [615-53-2] (*4*, *Suppl. 7*; 1987)				
Methylthiouracil [56-04-2] (*7*, *Suppl. 7*; 1987)	−			
Metronidazole [443-48-1] (*13*, *Suppl. 7*; 1987)	+	−	−	
Mirex [2385-85-5] (*20*, *Suppl. 7*; 1987)	−	−		
Mitomycin C [50-07-7] (*10*, *Suppl. 7*; 1987)	+	+	+	+
Monocrotaline [315-22-0] (*10*, *Suppl. 7*; 1987)				
5-(Morpholinomethyl)-3-[(5-nitrofurfurylidene)amino]-2-oxazolidinone [3795-88-8] (*7*, *Suppl. 7*; 1987)				
Nafenopin [3771-19-5] (*24*, *Suppl. 7*; 1987)		−		
Nickel, metallic [7440-02-0] and alloys (*49*; 1990)				
Niridazole [61-57-4] (*13*, *Suppl. 7*; 1987)				
Nitrilotriacetic acid [139-13-9] and its salts[3] (*48*; 1990)	−	−	−	?
5-Nitroacenaphthene [602-87-9] (*16*, *Suppl. 7*; 1987)	+	+		
2-Nitroanisole [91-23-6] (*65*; 1996)	+			+
Nitrobenzene [98-95-3] (*65*; 1996)	−	−		−
6-Nitrochrysene [7496-02-8] (*46*; 1989)	+			
Nitrofen [1836-75-5] (technical-grade) (*30*, *Suppl. 7*; 1987)	+			
2-Nitrofluorene [607-57-8] (*46*; 1989)	+	+		+
1-[(5-Nitrofurfurylidene)amino]-2-imidazolidinone [555-84-0] (*7*, *Suppl. 7*; 1987)				
N-[4-(5-Nitro-2-furyl)-2-thiazolyl]acetamide [531-82-8] (*7*, *Suppl. 7*; 1987)	+			
Nitrogen mustard *N*-oxide [126-85-2] (*9*, *Suppl. 7*; 1987)				
2-Nitropropane [79-46-9] (*29*, *Suppl. 7*; 1987)	+	+	−	−
1-Nitropyrene [5522-43-0] (*46*; 1989)	+	+		+
4-Nitropyrene [57835-92-4] (*46*; 1989)				
N-Nitrosodi-*n*-butylamine [924-16-3] (*17*, *Suppl. 7*; 1987)				−
N-Nitrosodiethanolamine [1116-54-7] (*17*, *Suppl. 7*; 1987)	+			+
N-Nitrosodi-*n*-propylamine [621-64-7] (*17*, *Suppl. 7*; 1987)	+	+		+
3-(*N*-Nitrosomethylamino)propionitrile [60153-49-3] (*37*, *Suppl. 7*; 1987)		+		
4-(*N*-Nitrosomethylamino)-1-(3-pyridyl)-1-butanone (NNK) [64091-91-4] (*37*, *Suppl. 7*; 1987)		+		
N-Nitrosomethylethylamine [10595-95-6] (*17*, *Suppl. 7*; 1987)				
N-Nitrosomethylvinylamine [4549-40-0] (*17*, *Suppl. 7*; 1987)				
N-Nitrosomorpholine [59-89-2] (*17*, *Suppl. 7*; 1987)	+	+	+	
N′-Nitrosonornicotine [16543-55-8] (*37*, *Suppl. 7*; 1987)		+		
N-Nitrosopiperidine [100-75-4] (*17*, *Suppl. 7*; 1987)	+			
N-Nitrosopyrrolidine [930-55-2] (*17*, *Suppl. 7*; 1987)	+	+		+
N-Nitrososarcosine [13256-22-9] (*17*, *Suppl. 7*; 1987)				
Ochratoxin A [303-47-9] (*56*; 1993)	−	+		+
Oil Orange SS [2646-17-5] (*8*, *Suppl. 7*; 1987)	+			

Appendix Table (*Continued*)

	SAL	UDS	MN	OTHER
Oxazepam [604-75-1] (*66*; 1996)	–	–		+
Palygorskite (attapulgite) [12174-11-7] (long fibres, > 5 μm) (*68*; 1997)		–		
Panfuran S (containing dihydroxymethylfuratrizine [794-93-4]) (*24*, *Suppl. 7*; 1987)				
Pentachlorophenol [87-86-5] (*53*; 1991)	–			(+)
Phenazopyridine hydrochloride [136-40-3] (*24*, *Suppl. 7*; 1987)	?	–		+
Phenobarbital [50-06-6] (*13*, *Suppl. 7*; 1987)	(+)	–	+	(+)
Phenoxybenzamine hydrochloride [63-92-3] (*24*, *Suppl. 7*; 1987)	+			
Phenyl glycidyl ether [122-60-1] (*47*; 1989)				–
Phenytoin [57-41-0] (*66*; 1996)	–		?	+
PhIP (2-Amino-1-methyl-6-phenylimidazo[4,5-*b*]pyridine) [105650-23-5] (*56*; 1993)	+		?	+
Ponceau MX [3761-53-3] (*8*, *Suppl. 7*; 1987)	(+)			
Ponceau 3R [3564-09-8] (*8*, *Suppl. 7*; 1987)	–			
Potassium bromate [7758-01-2] (*40*, *Suppl. 7*; 1987)	+		+	+
Progestins (*Suppl. 7*; 1987)	–	–		–
1,3-Propane sultone [1120-71-4] (*4*, *Suppl. 7*; 1987)	+			+
β-Propiolactone [57-57-8] (*4*, *Suppl. 7*; 1987)	+			+
Propylene oxide [75-56-9] (*60*; 1994)	+		+	+
Propylthiouracil [51-52-5] (*7*, *Suppl. 7*; 1987)				
Rockwool (*43*; 1988)				
Saccharin [81-07-2] (*22*, *Suppl. 7*; 1987)	–		–	+
Safrole [94-59-7] (*10*, *Suppl. 7*; 1987)	+	–	–	+
Schistosoma japonicum (infection with) (*61*; 1994)				
Slagwool (*43*; 1988)				
Sodium *ortho*-phenylphenate [132-27-4] (*30*, *Suppl. 7*; 1987)				–
Sterigmatocystin [10048-13-2] (*10*, *Suppl. 7*; 1987)		+		
Streptozotocin [18883-66-4] (*17*, *Suppl. 7*; 1987)	+	+		+
Styrene[8] [100-42-5] (*60*; 1994)	?		–	+
Sulfallate [95-06-7] (*30*, *Suppl. 7*; 1987)	+			
Tetranitromethane [509-14-8] (*65*; 1996)	+			+
Thioacetamide [62-55-5] (*7*, *Suppl. 7*; 1987)	–		+	+
4,4′-Thiodianiline [139-65-1] (*27*, *Suppl. 7*; 1987)	+			
Thiourea [62-56-6] (*7*, *Suppl. 7*; 1987)	–			?
Toluene diisocyanates [26471-62-5] (*39*, *Suppl. 7*; 1987)	+		–	+
ortho-Toluidine [95-53-4] (*27*, *Suppl. 7*; 1987)	?	?	–	+
Trichlormethine (Trimustine hydrochloride) [817-09-4] (*50*; 1990)				
Trp-P-1 (3-Amino-1,4-dimethyl-5*H*-pyrido[4,3-*b*]indole) [62450-06-0] (*31*, *Suppl. 7*; 1987)		+		+
Trp-P-2 (3-Amino-1-methyl-5*H*-pyrido[4,3-*b*]indole) [62450-07-1] (*31*, *Suppl. 7*; 1987)		–		+
Trypan blue [72-57-1] (*8*, *Suppl. 7*; 1987)	+	?		
Uracil mustard [66-75-1] (*9*, *Suppl. 7*; 1987)	+			+
Urethane [51-79-6] (*7*, *Suppl. 7*; 1987)	+		+	–
Vinyl acetate [108-05-4] (*63*; 1995)	–			+
4-Vinylcyclohexene [100-40-3] (*60*; 1994)	–			
4-Vinylcyclohexene diepoxide [107-87-6] (*60*; 1994)	+			+
Mixtures				
Bitumens [8052-42-4], extracts of steam-refined and air-refined (*35*, *Suppl. 7*; 1987)				
Carrageenan [9000-07-1], degraded (*31*, *Suppl. 7*; 1987)				

Appendix Table (*Continued*)

	SAL	UDS	MN	OTHER
Chlorinated paraffins of average carbon chain length C_{12} and average degree of chlorination approximately 60% (*48*; 1990)				
Coffee (urinary bladder)[9] (*51*; 1991)				
Diesel fuel, marine[8] (*45*; 1989)				
Engine exhaust, gasoline (*46*; 1989)				
Fuel oils, residual (heavy) (*45*; 1989)				
Gasoline[8] (*45*; 1989)				
Pickled vegetables (traditional in Asia) (*56*; 1993)				
Polybrominated biphenyls [Firemaster BP-6, 59536-65-1] (*41*, *Suppl. 7*; 1987)				
Toxaphene (Polychlorinated camphenes) [8001-35-2] (*20*, *Suppl. 7*; 1987)				
Toxins derived from *Fusarium moniliforme* (*56*; 1993)				
Welding fumes (*49*; 1990)				
Exposure circumstances				
Carpentry and joinery (*25*, *Suppl. 7*; 1987)				
Dry cleaning (occupational exposures in) (*63*; 1995)				
Printing processes (occupational exposures in) (*65*; 1996)				
Textile manufacturing industry (work in) (*48*; 1990)				

[9] There is some evidence of an inverse relationship between coffee drinking and cancer of the large bowel; coffee drinking could not be classified as to its carcinogenicity to other organs.

4. GEOCHEMISTRY, METAL TOXINS AND DEVELOPMENT PLANNING

FREDERIC R. SIEGEL*

INTRODUCTION

Geochemistry deals with the physical, chemical and biological principles that influence and/or control the fractionation, migration, deposition and distribution of chemical elements (and their isotopes) and compounds in the spectrum of materials that make up the Earth's crust. Surface and near-surface crustal matter are the focus of applied studies in environmental geochemistry as well as in geochemical exploration for mineral and hydrocarbon resources. Samples used in such studies comprise a varied group of solids, liquids, gases, and biological forms[1]. Each sample is unique as to what its chemical signatures represent in terms of volume and area as well as natural and/or contaminant components. A rock outcrop sample gives the chemistry at the point the rock was collected whereas a stream sediment or water sample carries chemical signals of earth materials from the entire upstream watershed including mineral deposits and/or contaminants. The stream sediment also signals low concentrations of chemical elements which could suggest areas with weathered products deficient in one or more than one essential nutrient. Similarly, a soil sample has a chemistry representive of the pit from which it is collected (perhaps 700–1000 cm^3). It reflects the natural chemistry of the underlying rock from which the soil formed plus contaminants that may have been added to it at the sample site (e.g., from atmospheric deposition). Chemical signals from vegetation samples derive from a volume of soil that is as extensive as their root systems or of an aquifer system where roots tap groundwater to sustain growth (to millions of cm^3)[2].

Environmental geochemistry and mineral exploration research are linked by principles and processes although they focus on contrasting problems and solutions. Samples used to target regional or local areas in mineral exploration programs are the same ones used in environmental geochemistry assessments of the health of an ecosystem and its inhabitants. The chemical elements and compounds that are analysed in exploration geochemistry projects coincide with many which are potentially toxic to life forms. These include the heavy metals Cd, Co, Cr, Cu, Fe, Mo, Ni, Pb, V and Zn, plus As and Hg, and selected process determinant elements (Al, Ca, Mn and S). Some of the compounds studied are NO_3^-, CN^-, SO_2, CO_2, metal oxides and hydroxides, humus, and humic and fulvic acids. They have drawn the attention of research that relates the health status of life forms to metal toxins in the food web. Most originate as effluent discharge and atmospheric emission from industrial and manufacturing sources and smelter-refining operations and as drainage from mines and mine wastes. Coal- and oil-fired energy generating facilities add to the atmospheric loading of heavy metal and metalloid contaminants. Earth surface and near-surface environments are subject to additional contamination from

* Department of Geology, George Washington University, Washington, DC 20052 (USA)

imperfect isolation, containment and disposal of potentially toxic wastes from sources cited above. These may contaminate large areas. Other sources that often affect fairly localized areas originate from agricultural activity and domestic and business wastes. Discharge into the environment and poor waste disposal controls of effluents carrying metal toxins is a global problem inherited from the past (in industrialized societies) and contemporary to some degree (in developing countries). Metal toxins pose health risks to plant, animal and human populations via ingestion through the atmosphere, the food web (food crops, food fish, and food animals) and water.

The environmental geochemist who works on epidemiological problems does not pretend to establish a medical relationship between potentially toxic elements (PTEs) or compounds in the food web and prevalence or incidence of disease(s). He/she reports the results of observations and measurements (geographic distributions and concentrations) on the metals and/or compounds in a study area and (physical–chemical) conditions there. In addition, the environmental geochemist may propose possible sources for anomalous metal concentrations to medical–epidemiological–toxicological researchers. It is they who establish the statistically confident toxin/disease relations and then follow through with the hows and whys of mechanisms of disease development in an environment. Ultimately this might lead to a therapy for an affected population and a plan to alleviate the intrusion of the toxin on other inhabitants of the ecosystem.

1. *Chemical Elements and Health in Humans*

There are several metals (and non-metals) that are essential or probably essential to life forms. Table I categorizes these for humans into macronutrients for which 100 or more mg/day are necessary for the effective and efficient functioning of the body and into micronutrients which require only a few mg/day. Groups are also given for possibly essential micronutrients and for some potentially toxic metals (PTMs). Ingestion of higher or lower than optimum ranges of concentrations of essential nutrients over time can result in excesses or deficiencies that can be toxic to humans. This can manifest itself via chronic maladies, marked symptoms, and sickness that diminishes the quality of life and can ultimately result in incapacitation or death. For some PTMs, ingestion of excess concentrations are simply bioevacuated

Table I The role of trace elements in humans

Essential macronutrients (100 mg/day or more): Ca, Cl, Mg, P, K, Na, **S**
Essential micronutrients (no more than a few mg/day) **Cr**, **Co**, **Cu**, F, I, **Fe**, **Mn**, Se, **Zn**
Possibly essential micronutrients: **Ni**, Si, **Sn**, **V**
Trace pollutants: Al, **As**, **Cd**, **Pb**, **Hg** (Some may possibly be essential)
Others: **Sb**, Ba, Be, B, Br, Li, Rb, **Ag**, Sr, **Ti**

R. G. Crounse, W. J. Pories, J. T. Bray and R. L. Mauger, "Geochemistry and Man: health and disease; 1. essential elements; 2. elements possibly essential, those toxic and others" **In**: (I. Thornton, Ed.) *Applied Environmental Geochemistry* (Academic Press, London 1983) pp. 267–308; pp. 309–333.

and toxicity is not a problem. There is no natural compensation for deficiency concentrations in the body unless foodstuffs with the necessary nutrient(s) or pill supplements are added to the dietary intake.

SOME CHEMICAL ELEMENT PATHWAYS AND MEDIATED DISEASES IN HUMANS

There are many examples of the impact of lack of nutrient ingestion (non-metals and metals) on human health as well as of the ingestion of too much of an essential nutrient or other chemical elements over time. The non-metallic element F is a classic example of both deficient and excess ingestion response modes. With the ingestion of deficiency concentrations of F over time, bone development will suffer and manifest itself by increased caries and susceptibility to osteoporosis. Ingestion of excess F over time can result in mottled teeth and excessive bone growth and ligament calcification. A natural deficiency of I (non-metallic) in the food chain was determined many years ago to be the principal cause of goitre in humans worldwide. In 1980 the link was made between a lack of Se (non-metallic) in the earth materials (rocks, soils, waters) in an area of China and a deficiency of Se in the food web. This deficiency in Se in the human diet resulted in a childrens' cardiomyopathy condition (Keshan disease) that was responsible for a great number of deaths over the years[3]. In the case of I, the diet was supplemented by adding this nutrient to foodstuff (e.g. bread) or by iodizing salt, and goitre from a deficiency of I was eliminated. A weekly supplement of Se of 0.5–1.0 mg (dose is a function of age) by pill greatly reduced the incidence of Keshan disease in China[4].

A lack of absorption of Fe and probably Zn (endocrine abnormalities) through the food web was thought to be the cause of deficiency-related anaemia, dwarfism, hypogonadism and marked hepatosplenomegaly in an area of Iran[5]. In some segments of Iranian society, geophagia (soil mastication) is practiced. Where the diseases were prevalent, the soils were rich in clay minerals. It was proposed that because of cation exchange reactions, Fe and possibly Zn in food and water were adsorbed on the ingested clay minerals and were not available for absorption by the body although these essential micronutrient elements were present in good concentrations in the soils and water. The disease problem was resolved after a period of Fe therapy. A supplement of Zn alleviated the affected population's endocrine problems.

Ingestion of naturally occurring excesses over time of As and essential micronutrients such as Cr and Cu, for example, have also caused human health problems or symptoms that mimic a disease. Excess As in well waters has led to sickness and deaths on the Indian subcontinent, Asia, and South and North America. Ingestion of excess Cr can lead to false symptoms for diabetes. Ingestion of excess Cu through natural sources in the food web can cause gastrointestinal problems. Where areas with high natural contents of potentially toxic elements are identified, there are options to the land-use. The areas can be avoided for cultivation of food crops that will take the PTEs up in proportion to their concentrations in soils and water (for crop irrigation, aquaculture or mariculture) or even hyperaccumulate them. Instead, cultivation of non-food crops (e.g. cotton) or crops that discriminate against the uptake of PTMs in the growth environments can be encouraged (or genetically engineered). There is the possibility of soil amendment to immobilize target PTMs (e.g. liming to increase the pH). However, determinations must be made that the remediation process will not mobilize other PTMs in the growth environment that could present a health risk if they enter the food web. If viable solutions are not found, the land can be taken out of agricultural production and used for other purposes such a recreation. Water as a carrier of PTMs to an agricultural area is subject to treatment before use, but this option is a function of the cost for the volume of water required.

There may be only empirical data to suggest a relationship between natural geochemical parameters and a disease in an area. Nonetheless, these relations can be relevant to medical scientists. For example, clusters of multiple sclerosis incidence have been identified in Henribourg, Saskatchewan, Canada. A geochemical survey of the soils and waters showed that the area has calcareous, chloride and Zn-rich soils and waters, that Cu, Mo and V are deficient in soils but not in water, that Se and S are deficient in soil and water, that the upper soil layer is enriched in Pb, and that nitrate and nitrite are in the water[6]. Nitrate concentrations were consistently and markedly higher in areas with a greater incidence of multiple sclerosis. There are many questions that can then be asked: 'is the abundance of a toxic element or the deficiency of an element harmful to normal biological functions?'; or 'can the excess or deficiency provide biological conditions for growth of infective or parasitic organisms?'; 'how important is synergism with respect to an excess state or antagonism to a deficiency state?'; or 'do any of the above conditions increase the probability of triggering the disease?'[6]. If parallel investigations are carried out in other areas with clusters of multiple sclerosis, similarities or differences in geochemical parameter-disease relationships can be useful in setting a direction for continuing research.

Human activities that result in the addition of high concentrations of PTMs to an environment and food web can be dealt with once the source(s) of pollutants is identified. The ease of remediation depends on several factors including areal magnitude of the problem, mode of dispersion of the PTMs, accessibility to the source(s) and responses of the PTMs to immobilization procedures. The identification mode is one in which the role of the environmental geochemist is paramount.

This can be explored best by initially reviewing a common human activity that inadvertently resulted in the development of a health hazard because active processes and the response of earth materials beneath the earth's surface were not evaluated. During the 1960s and thereafter, thousands of tube (water) wells were dug in West Bengal to give year-round support for an irrigation-intensive new form of (Green Revolution) rice. The wells reached from 20 to 150 m and penetrated three aquifers. Water is extracted mainly from the middle aquifer at about 80 m or more. During the 1980s arsenic poisoning was diagnosed in the human population. The afflicted population had high concentrations of arsenic in urine, hair, skin, nail and liver tissue samples. This is characteristic of people drinking arsenic-polluted water for a long time. Tube well waters were analysed and 62% of 20,000 wells sampled had arsenic concentrations above the World Health Organization permissible limit of 0.01 mg/l with some as high as 3.7 mg/l. The U.S. EPA limit is 0.05 mg/l. At present, about 200,000 people have arsenical skin lesions and many also have hyperkeratoses, hardened patches of skin that may develop into cancer. Liver effects and respiratory problems have also been associated with the arsenic poisoning in India[7,8]. In neighboring western Bangladesh 700 arsenic poisoning cases have recently been reported. About 1/3 of the 300 tube wells tested gave waters with contaminant levels higher than the safe limit set by the World Health Organization[9]. Several million Indians and Bangladeshis are at risk where well waters have not yet been analysed. Previous to the large scale extraction of water, village wells used for domestic and limited irrigation needs did not cause any large scale and regular fluctuations in the aquifer water table and no cases of arsenic poisoning afflicted West Bengali populations.

The source of the high (natural) arsenic concentrations was found in a few clay-silt rich sediment layers in the aquifer that contained the sulfide mineral pyrite (FeS_2) in the aquifer. Arsenic is a trace impurity in the pyrite. Decomposition of the pyrite and release of its arsenic contents is related to the large scale extraction of groundwater. Strong seasonal fluctuations in the water table result in rapid and regular influx of aerated (oxygenated) water into sediment pore spaces[7]. The sulfide mineral degrades by oxidation, forms ferric sulfate (found in the sediment), releases H^+ into the groundwater (decreasing pH) and mobilizes arsenic in the water as arsenite ($HAsO_3^{2-}$). The principal pathway of this most toxic form of arsenic to humans is through drinking water. The possible uptake, bioaccumulation and pathway of arsenic to humans through the rice irrigated with the polluted water is not yet known. With ongoing research, any other mineral source of arsenic will be identified.

There are some possible solutions to this problem. One is to import arsenic-free piped water from the Ganges River but the cost would be $200 million and would take 20 years when a rather immediate solution is necessary. Deep well water might be an answer for the human population[8]. The rice could be replaced with less water intensive food crops which would discriminate against the uptake of arsenic already in the soil as the result of past irrigation. A short-term response is necessary. The use of a ferric oxide/alkaline low cost filter treatment of the water to render it safe for human consumption may be feasible using a similar approach as was proposed for the chemical fixation of arsenic in contaminated soils[10]. Water treatment kits made in India cost about $14 but this is a major part of the income for inhabitants in the area affected. The World Bank, other international lending organizations and granting agencies and national AID programs would do well to donate filter kits for the low income people of the region. If kits are needed for five million families, the cost of $70 million would be a small investment within the context of what is being spent worldwide to raise health standards and the quality of life for people. This provides time until long-term solutions are put into place. China has a major problem with arsenic poisoning as well and will benefit by the data and solution(s) that evolve from the Indian experience.

In another case, Cd draining into river water from Zn mine tailings over time was related to an outbreak of osteomalacia (the "itai-itai" disease) in Japan during the post-WW II period. Geochemically, because of ionic size–charge similarities with Zn, Cd can enter into the crystal structure of the Zn ore mineral sphalerite (ZnS; Zn [Cd]S) so that a mine source was proposed. Analyses of stream sediments targeted the source mine. A knowledge of physical-chemical conditions (acid mine drainage and oxidizing conditions) explained how the Cd was mobilized from mine tailings into the fluvial system[11]. The Cd entered the food web through drinking and cooking water drawn directly from the river and in rice irrigated with river water. A lag time of several years before the onset of notable symptoms delayed the identification of the problem. Once identified, however, the pollutant source was eliminated and after a natural cleansing period the water could be used again for domestic and agricultural purposes. A Cd load remained in the rice field soils so that alternate discriminator crops were sought. Given the state of plant genetic engineering, it is possible that given a similar situation today, a rice could be developed that would not take up the soil Cd. This case served as a signal event where ore being mined contained Cd and was basic to the elimination of Cd sources in other areas where the itai-itai disease was diagnosed. It should have stimulated the investigation of other than mining areas where Cd could get into the food web.

This has not been the case, however. Rice grown in soils formed on a uraniferous black shale (contains 6.3 ppm Cd) in the Deog-Pyoung area of Korea contains

0.6 ppm Cd. The dietary intake of 574 gm/day of rice translates to an ingestion of 344 μg/day of Cd which is comparable with the ingestion rate in the itai-itai disease area of Japan just described. The Korean dietary load of Cd is increased by leafy and other vegetables grown in these soils (e.g. lettuce, cabbage, red pepper) as well as tobacco which contains an average of 14 ppm Cd in dry matter (with a maximum of 46 ppm). Each pack of cigarettes adds 42 μg to the ingested Cd load[12,13]. Although the "itai-itai" disease has not yet afflicted the Deog-Pyoung population, the potential is not diminished. The obvious use of light liming can amend the soil and decrease the bioavailability of Cd but may not be desireable for black shales. Liming could facilitate the mobility of Mo and its uptake in forage for ruminants that could result in a hypocuprosis condition.

Neurotoxin problems associated with Pb ingestion is well documented as a major danger to young children. A principal source is dust in homes that have Pb-based paints or soils from paint scraping mixed with soils that children play on or transfer from hand to mouth[14,15]. Vegetables grown in the contaminated soils provide another pathway to the human population. In Romania, Pb and Cd intoxication is found in populations living in the Baia Mare and Copsa Mica smelter areas. This affected mainly children. Almost 60% of the 14,554 people investigated had excessive (150–250 μg/l for 51.9%) and harmful (>250 μg/l) levels of Pb in the urine[16]. Other diseases that have been attributed to the smelter sources are saturnine encephalopathy (slowly developing brain disease), radial nerve paralysis and saturnine colic. Similarly, the Bukowno smelter area, Poland has topsoils contaminated with Pb, Zn and Cd[17]. Again, children were impacted most with 75% of blood samples in the 5–8 age group at Pb poisoning concentrations (>10 μg/dl). Urine contents of Cd averaged 0.85 μg/dl. This exceeds the health hazard limit of 0.5 μg/dl[18]. The pathway of pollutants to humans at Bukowno has not yet been determined. This is a global problem that has existed and still emits from smelter and mining operations where economic development goals supercede environmental sensitivity and the health of at risk populations[19–22].

Contaminant Pb in the atmosphere and in soils near heavily travelled roads has been eliminated as a source in many countries by the use of unleaded gasoline. However, contaminated soils juxtaposed to these roads is still in place. Much smelter-originated atmospheric Pb at mining operations in many parts of the world has been eliminated from the atmosphere by terminating operation of a smelter and by the use of chemical scrubber–electrical precipitator systems in chimneys in modern or modified operations. Older smelter operations can still load the atmosphere and poison downwind populations and soils not only with Pb but also with As, Cd, Cr, Cu, Hg, Ni, V and Zn[23,24]. The soils retain this loading and can be a foodcrop web path for ingestion with time of metal toxins and are mainly associated with old mining operations, heavy metal industries, oil-refineries and energy generation facilities. Clean-up of topsoil by excavation is possible but costly. Bioremediation is being studied and practiced in some areas but is a slow although effective process.

Neurotoxin poisoning from Hg discharged into the environment from industrial activities or as agricultural biocide runoff is well documented. In the classic Minamata Bay, Honshi Island, Japan case, inorganic Hg, a less dangerous form of the element, was discharged into the bay in 1953 from a plastics plant. Under the aerobic conditions of the bay waters the Hg^{2+} (98%) was converted by biomethylation to methyl mercury (CH_3Hg) which was accumulated by organisms at the beginning of the food web. This was biomagnified by food fish along the web to as high as 50 ppm. The maximum permissible concentration of Hg in food is 0.5 ppm. The population proximate to the bay depended on fish as a principal food source. Over time there was a progressive poisoning of the population and by 1960, 43 people died and another 116 were permanently incapacitated. Whether there was genetic damage has not yet been determined. Since 1960 many others have died or have been disabled and still others have developed various symptoms of Hg poisoning and the progressive deterioration of the nervous system. The final legal settlements were completed during 1996.

Atmospheric deposition of Hg from operations where it is volatilized can be an important source of Hg contamination. This may be in placer gold mining areas where Hg serves as an amalgam for Au and is burned off for Au recovery without regard to environmental pollution as in the Brazilian Amazon[19,25,26]. This may be downwind from smelters where Hg is associated with sulfide ore deposits. In the Siberian city of Norilsk, the principle Ni–Cu sulfide ore smelter, built in 1934, has toxified the atmosphere with various PTMs. Atmospheric deposition of Hg downwind in the drainage basin of the Pyasina River may be responsible for high Hg contents in Kara Sea sediments (to 2045 ppb)[27]. If this Hg accesses the marine food web, it could ultimately bioaccumulate in food fish and higher order life forms.

Although not metal toxins, the compounds NO_3^- from agricultural runoff and CN^- from mineral processing of Au ore have invaded aquifer systems and caused poisoning of human and animal populations. When aquifers feed streams and rivers, the impact on wildlife drinking the CN^- contaminated waters can be marked.

The functions of essential trace elements for humans and animals and the symptoms that could develop from deficiency in essential nutrient intake over time are the subject of continuous investigation[28].

RELATIONSHIP TO HEALTH OF OTHER LIFE FORMS

Other life forms have their own essential nutrient intake requirements. If there are excesses or deficiencies in nutrient intake or the ingestion of potentially toxic metals over time, vegetation, animals, and fish, for example, can respond to degrading ecosystem conditions in varying ways. Some life forms may tolerate excess concentrations of metals in their growth environment or discriminate against uptake; others will not. Some will bioaccumulate and biomagnify metal toxins. There may be subclinical (asymptomatic) health and development problems, sickness (e.g. manifested by weight loss or lack of weight gain), diminished yield of product (e.g. beef, milk, fieldcrops, fruit), development of morphological abnormalities (e.g. tumors in fish), failure to reproduce and others. For example, corn (maize) and the coffee plant are highly sensitive to a deficiency of Zn, barley to a deficiency of Fe, rice to deficiencies of Fe and Cu, soya bean to deficiencies of Fe and Mn, sugar beet to deficiencies of Fe, Mn and Cu and wheat to deficiencies of Mn and Cu[28–30]. These are all important crops for human populations. Potentially toxic metals bioaccumulated in the food web can translocate to humans over time and sickness and disabling effects or death can follow.

INDUSTRIAL ACTIVITIES AS A SOURCE OF METAL TOXINS

The bolded element symbols in Table I indicate those that are targeted in mineral exploration. The addition of Be, Mo and Tl to the bolded grouping gives the elements on the 1993 EPA priority pollutant list and the PTMs followed by the Arctic Monitoring and Assessment Program initiated in 1992. They can contaminate

Table II Potentially toxic trace metals in waste effluents from various industries

Mining and ore preparation	As, Be, Cd, Pb, Hg, Mn, U, Zn
Metallurgy	As, Be, Bi, Cd, Cr, Cu, Pb, Hg, Ni, V, Zn
Chemical industry	As, Ba, Cd, Cr, Cu, Pb, Hg, Sn, U, V, Zn
Glass manufacture	As, Ba, Pb, Ni
Pulp and paper factories	Cr, Cu, Pb, Hg, Ni
Textiles	As, Ba, Cd, Cu, Pb, Hg, Ni
Fertilizers	As, Cd, Cr, Cu, Pb, Hg, Mn, Ni, Zn
Oil refining	As, Cd, Cr, Cu, Pb, Ni, V, Zn

J. E. Fergusson, *Inorganic Chemistry and the Earth*. (Pergamon Press, Oxford, 1982).

Table III Industries for which EPA is developing specific regulations for pretreatment of effluents

Timber processing	Laundries
Leather taning	Soaps and detergents
Thermal electrical generation	Machinery
Petroleum refining	Copper working
Iron and steel working	Aluminum working
Paving and roofing	Plastics
Batteries	Paint and ink manufacture
Printing	Coated coil fabrication
Mining	Enameled products
Foundaries	Adhesives
Plastics/synthetics manufacture	Explosives fabrication
Pulp and paper	Gum and wood
Pharmaceutical manufacture	Textiles
Rubber manufacture	Electroplating

P. A. Vesilind, J. J. Peirce and R. F. Weiner, *Environmental Engineering* (Third Edition) (Butterworth-Heinemann, Boston, 1994) Table 11-2. p. 244.

the environment from mining and related activities cited above and from industrial effluent, atmospheric emission and waste disposal sources. Some specific examples of environmentally damaging metals that can intrude the ecosystem from various industries are given in Table II. The EPA has given a listing of 34 industries for which there must be a pre-treatment of waste-bearing effluents before discharge (Table III).

CHEMISTRY OF THE ENVIRONMENT AND DEVELOPMENT PLANNING

There are a myriad of development sectors that can be added to an already thriving area. In a new area, the installation of a major project (e.g. hydroelectric dam) will serve as a node for additional ones (e.g. power-based industries, agriculture). Each project is unique in its physical, chemical and biological requirements. Water volume and quality, soil chemistry and waste disposal (solids, liquids and gases) as well as safe siting from natural hazards such as earthquakes, volcanoes, floods, landslides and others are among the most important requirements to be met[30].

Without planning on the basis of the geochemical environment a new venture, especially agricultural, can be short-lived. For example, a food project for the cultivation of a dairy herd with natural forage would be at risk if forage is grown in soils formed from the weathering (disintegration and decomposition) of black shales. Black shales are sedimentary rocks that formed under reducing conditions. They commonly have **natural** high concentrations of Cd and Mo and other heavy metals (e.g. As, Sb, Cu, Zn, Ni) and a mildly acid pH. Copper is an essential nutrient for cattle and is found in the soils and in the forage. The Mo can be mobilized and accumulated by forage grasses in proportion to its concentration in soils. The presence of high Mo concentrations in the animal diet effectively blocks the absorption of Cu with the result that hypocuprosis develops from Cu deficiency. In sheep, a Cu deficiency can cause swayback or enzootic ataxia. Symptoms of hypocuprosis disease may initially be subclinical so that the deficiency toxicity is not readily obvious. The symptoms that become evident are diharrea, coat dullness and discoloration, loss of appetite and weight, drop in milk production, anemia and emaciation[31]. This can be an economic disaster. Recognition of the problem can lead to a chemical solution for the soils that lowers pH thus diminishing the mobility of Mo, but attention must be given to what potentially toxic elements might be present and be made bioavailable by decreasing to lower levels of pH (e.g. Cd). An agricultural solution might be used by planting forage that discriminates against Mo accumulation. A dietary solution with injected Cu, or abandonment of a foraging site for "at risk" animals are other options for dealing with the problem. As important is the fact that definition of the problem can serve as a signal event that alerts similar development sector projects in geologically similar environments to the possibility of metal toxicity before subclinical symptoms become obvious symptoms. Such a case was described and did serve as a signal event for dairy herds in the UK[31]. The same environmental problem may be developing in the Deog-Pyoung area of Korea cited previously with regard to Cd loading in humans in an area where soils formed from uraniferous black shales. These soils also have high Mo contents (136 ppm) which give a low Cu : Mo ratio (2.65) in rice stalks that are the principal forage for cattle[13]. A low Cu : Mo ratio suggests the possibility of hypocuprosis such as in the UK although no clinical symptoms are yet obvious in the Korean ruminants. Certainly, testing for Cu in animals foraging in the area is called for to find out if there is a deficiency of this essential element so that if therapy is required it can be administered to the animals to maintain their health status and productivity. Table IV gives examples of excesses and deficiencies that might be expected in different rock types and which may be reflected in soils formed from them or in waters moving through them.

Each project is specific in terms of human needs for people who work at or service projects and for their families. This translates to a clean environment secure from natural and anthropogenic hazards, water suitable in quantity and quality for drinking, personal hygiene and sanitation, secure waste disposal sites that will not leak metal toxins into water supplies or soils, uncontaminated foodstuffs and clean air for the physical and living environment[30]. Failure to meet one or a combination of the human requirements can lead to the ingestion of metal toxins. With time this can cause a drop in productivity on the job from chronic illness or from time lost to absenteeism from sickness. Ultimately, this can lead to permanently lost productivity from life-lived with disability and years of life lost because of premature mortality[32].

A global burden of disease study categorized the fraction of disability and mortality attributable to 10 major risk factors. The burden is given in terms of DALYs (disability-adjusted life years) for the 17 major diseases or injuries that affect human populations. The top three of these for 1990, lower respiratory infections, diarrheal diseases, and conditions arising during the pre-natal period have

Table IV Chemical element excesses and deficiencies that may be expected in different rock types and soils formed from them and waters in contact with them

Rock type/environment	Potential environment deficiencies	Potential environment excesses
Limestones	Mo, Sr, Mg	P, K
Shales	P	Se, Mo
Black shales		Heavy metals, As, Sb, Hg, U
Phosphorites		V, U, Ni, Cr, Zn, Mn, P
Peat	Cu (not available)	Zn, Cu, Pb, Cd
Serpentine	N, P, Ca, Mo, Mn	Ni, Cr, Mg, Fe
Complex ores		Cu, Pb, Zn, Au, As
Coastal sands	Fe, Mn, Cu, Se, Co, B, Cr	
Evaporite basins		Sr, Cl, Na, B, K, Li, Fe, Mo

H. L. Cannon, "Trace element deficiencies in some geochemical provinces of the US" **In**: (D. D. Hemphill, Ed.) *Trace Substances in Environmental Health—III* (Univ. of Missouri, Columbia, Missouri, 1970) pp. 38–40.

associations to environmental geochemistry to a greater or lesser degree. Another three (ischemic heart disease, cerebrovascular disease, and iron-deficiency anemia) are likely related to some degree to geochemical characteristics of the ecosystem. Because of increasing environmental awareness and sensitivity to the causes of diseases and injuries and because of investment to alleviate or eliminate obvious causes, the top three major diseases are expected to drop in ranking. By the year 2020 lower respiratory infections percent of total DALYs is predicted to drop from 8.2 to 3.1, diarrheal diseases from 7.2 to 2.7, and pre-natal period conditions from 6.7 to 2.5. On the other hand, ischemic heart disease and cerebrovascular disease will rise in their percent of total DALYs from 3.4 to 5.9 and 2.8 to 4.4, respectively. Of the 10 major risk factors in 1990, three that have some relation to environmental geochemistry are malnutrition, poor water supply, hygiene and sanitation, and air pollution; respectively, these are responsible for 15.9, 6.8 and 0.5 percent of the total DALYs. Globally almost 80 percent of the burden of disease is concentrated in developing countries. Diminishing the risk factors and hence lowering the DALYs lost to diseases that are in some way related to geochemistry and environmental toxicity will improve the quality of peoples' lives and productivity for their well being. Interestingly, the summary study[32] made no mention of metal toxins as a cause of disease. However, metal toxification is a global problem and can affect important populations. For example, as previously noted **naturally** high As levels in well waters used for drinking, cooking and irrigation of rice have been related to increased risk for skin and internal cancers that have impacted about 200,000 people in India and Bangladesh and signaled risk areas elsewhere in the world such as China[7–9].

In the sections that follow, the so-called "natural" metal concentrations in earth materials will be discussed as will be the factors that influence metal mobility, dispersion and deposition. Observations and measurements that are available to identify natural and human-influenced PTMs-rich geochemical environments will be discussed especially where metals have been or can be mobilized into an ecosystem via atmosphere, water, and soil and present a problem. The possible effects of these on the sustainable use and yield of an ecosystem to benefit human and other life forms are considered. There is much global pollution now in areas which have large and dense human populations or where population growth (natural and from changing demographics) will take place. There is no question that remediation is required for these sites world-wide, large and small, especially for soils, aquifers, surface waters and sediments (including wetlands) and for the general ecosystem. Approaches to remediation are described and followed by a consideration of optimal development and the evaluation of metal toxins in environmental development planning.

NATURAL CONCENTRATIONS OF METAL TOXINS IN EARTH MATERIALS

1. *Earth Materials*

There are three major classes of rocks exposed at the earth's surface or hidden beneath a soil-vegetation cover. These are called igneous, sedimentary and metamorphic rocks. Each has many sub-classes and forms under different environmental conditions within the Earth and at or close to the Earth's surface. The rocks are categorized on the basis of several observations and measurements. One of these is mineralogic and comprises the minerals present and their weight percent contributions to a rock. Mineral content can vary markedly between rock classes and within them. There are special situations during rock-forming processes where the metals have been concentrated and form ore minerals and metal deposits. Hydrothermal (hot, metal-bearing) fluids with high metal concentrations have been injected into rocks posterior to their crystallization or lithification also creating ore minerals and metal deposits. Great changes in metal contents are found where ore deposits occur and in rocks proximate to them.

Minerals define the chemistry of a rock, the chemistry of soils that form from rock disintegration and decomposition (the weathering process) and the chemistry of the waters that flow over them or through them (in aquifers). If a parent rock from which a soil forms contains natural high concentrations of metals or metal ores, these metals will often show increased contents in soil overlying it and in water active in the weathering process. Whether PTMs are mobilized from soils or into waters in potentially toxic concentrations is controlled by several factors: pH, reduction–oxidation potential, adsorption, organic matter, and others[33]. Both soils and waters are foundations of the food web. Foodcrops or forage metal profiles may reflect the chemistry of the soils in which they grow and of rainfall. Vegetation may bioaccumulate a metal (or more than one metal) if there is a bioaffinity for an element. Also, some vegetation will grow only where there is a given concentration of a bioavailable metal. This observation and the chemical analysis of leaves, twigs, pine needles and bark, for example, are the basis for geobotanical and biogeochemical prospecting[34]. Similarly, land food animal chemistry may reflect that of the forage and of drinking water (soil and rock influenced) whereas food fish may have a metal composition influenced by that of stream, lake or ocean waters and sediments. Bioaccumulation can be an important process in these two settings as well. Particulates ingested from an uncontaminated atmosphere add to the natural metal loading in an organism. This can be considered primary metal loading in the food web. The average contents of some PTMs and their ranges in the various media are given in Table V.

2. *General Concentration Data on PTMs in Ecosystem Matter*

Concentrations in rocks of metals that are potentially toxic to inhabitants of the ecosystem are generally low. Table V gives average contents of potentially toxic metals of selected rock types in contact with the living environment. It complements

Table V The average natural concentrations in selected earth materials of some metals that are potentially toxic to humans and other life forms. Compilation from several sources.

Metal	Earth materials						
	Granite	Basalt	Shale	Ocean clay	Limestone	Deep sea carbonate	Streams (ppb)
Al%	7.2	7.8	8.0	8.4	0.42	2.0	50
As	2	2	13	13	1	1	2
Cd	0.13	0.22	0.3	0.42	0.03	0.0X	
Co	4	48	19	74	0.1	7	0.2
Cr	10	170	90	90	11	11	1
Cu	20	87	45	250	4	30	7
Fe%	1.42	8.65	4.72	6.50	0.38	0.9	40
Hg	0.03	0.01	0.4	0.03	0.04	0.0X	0.07
Mn%	0.045	0.15	0.09	0.67	0.11	0.1	8
Mo	1	1.5	2.6	27	0.4	3	0.5
Ni	10	130	68	225	20	30	2
Pb	17	6	20	80	9	9	1
V	50	250	130	120	20	20	1
Zn	50	105	95	165	20	35	30

Metal	Soil (with range)	Vegetation (ash)	
		Natural	On mineralized terrain
Al		8610	
As	6 (0.1–40)		
Cd	0.06 (0.1–0.7)		
Co	8 (1–40)	To 30	>50
		9	
Cr	100 (5–3000)	9	
Cu	20 (2–100)	15–30	50–60 to 100–200
		183	sometimes >1000
Fe	3.8% (0.7–55%)	6740	
Hg	0.03 (0.01–0.3)		
Mn	850 (100–4000)	4815	
Mo	2 (0.2–5)	13	
Ni	40 (10–1000)	5–30	>100 sometimes 2000
		65	
Pb	10 (2–100)	0.2–30	
		70	
V	100 (20–500)	22	
Zn	50 (10–300)	20–100	300 grasses
		100–1000	500–1000 wood
		1400	

Table IV cited earlier which identified rock types/environments that have high or low PTEs concentrations that might translocate to the ecosystem where they occur as excesses and deficiencies. It is obvious from Table V that some elements have greatly different values within one rock class (e.g. Cr in igneous—granite and basalt) and between rock classes (e.g. As, Cr, Co, and Ni in granite and shale). Published average analyses of potentially toxic metals are only guides to what one might expect to find in areas of a single or of multiple rocks types and derived earth materials and in life forms that inhabit an ecosystem. For environmental planning pre-development, specific site analyses should be made and evaluated.

NATURAL BASELINE METAL CONCENTRATIONS IN EARTH MATERIALS

Global averages can not be used for decision making with regards to environmental geochemistry and potentially toxic chemical components. Each study area has its own natural geochemical baseline values in the various ecosystem media. These baselines should be established before assessments are made of health risks to humans and other organisms in an area being considered for development. In addition to the natural systems loading one must add a supplementary loading that takes place through physical, chemical and biological factors that enhance metal concentrations in soils, sediments, waters, vegetation and the atmosphere. Together these provide a rather complete assessment of probable pathways for natural metal exposure, the metals involved and their concentrations, and the health risk they pose individually or in assemblages to human habitation and use of an environment.

1. *Establishing the Natural Geochemical Baselines for PTMs*

The concept of natural background or baseline concentrations of chemical elements is based on the fact that average concentrations can be generated for areas of any size for any ecosystem medium, inorganic or organic, solid, liquid or gas. However, statistical distributions and variations in these concentrations should be considered when calculating a realistic assessment of natural contents of ecosystem components. In geochemistry, most distributions are not normal but rather are skewed and the skew is generally positive. Many geochemical distributions approach lognormal distributions. Thus geochemists do not define a natural background concentration for a chemical element as the arithmetic mean value $(X) \pm$ one standard deviation (σ) but rather use the geometric mean value $(X) \pm$ the values representing the (skewed) 84th $(+\sigma)$ and 16th $(-\sigma)$ percentiles. These values reflect the distributions of the natural populations and can be determined readily from a cumulative frequency plot prepared using lognormal probability paper[35]. They give a realistic natural background "range" for PTMs in earth materials being evaluated in environmental planning both for development planning and for setting remediation target values.

Changes in rock type, climate, vegetation and topography can affect sample chemistry in small and large areas. This causes changes in background values that are the basis for establishing chemical environmental intrusion. Trend surface analysis is a statistical method of polynomial surface fitting by which the general trend in an area (the natural background) is separated from random local variations in a series of data in the form of a calculated trend surface and residuals. The precise form of the calculated surface is that in which the sums of the squares of the residuals (or the differences between the observed and computed values) have minimum values. The trend surface contains the systematic or regional component of the variations in the geochemical data and the residuals include the local geochemical values not related to the principle element distribution. It is best applied when there is a rather uniform sampling grid so that extrapolation to unsampled areas is kept to a minimum. Trend surface analysis maps rural areas with anomalous positive (potential for excess in the ecosystem) and negative (potential for deficiency in the ecosystem) residuals[35] and can be important to development planning. The residuals can alert environmental assessment teams to the possibility for excess and/or deficiency concentrations of metals in an area being studied for capital investment that may be sensitive to the environmental chemistry.

The reliability of natural baseline "range" data for an ecosystem being studied is dependent on several factors. These include the samples or subsamples analysed, the density of sampling with respect to the area of interest, the detection limits of the analytical method(s), and the accuracy and precision of the data.

2. *Geochemical Atlases*

Sources for existing data on element concentrations and distributions, and their statistical evaluation are available in many national and regional atlases. These atlases have been prepared for use in agriculture and pollution studies to assess potential problems and predict high risk areas for environmentally-related diseases. Many include epidemiological data. Sample types that have been used in analyses reported in these publications include rock, soil, stream sediment, overbank (and floodplain) sediment, lake sediment, estuarine sediment, marine sediment, wetlands sediment, surface water, vegetation and other life forms.

The first detailed geochemical atlas and the one that set the standard and model for those that followed was the 1978 publication of the Wolfson Geochemical Atlas of England and Wales[36], planned and prepared by John Webb of Imperial College, one of the important personages in the field of applied geochemistry, and his team of geoscientists. In 1995, a global geochemical database was published in which systematic geochemical maps of the world for environmental and resource management were compiled and published[37]. In addition, data for most PTMs abound and thematic geochemical maps continue to be presented in publications on exploration geochemistry and general geochemical surveys on various sample media from federal and state geological surveys, universities, and private companies. For example, geochemical maps of Finland and Sweden were published together in 1993[38], a geochemical atlas for Austria was completed in 1988[39] and one for Poland was completed in 1995[40], and during 1996 geochemical maps showing the distribution and concentration of heavy metals in Jamaican surface soils were published[41]. In recent years there has been increasing emphasis on integrated geochemical studies in high resolution geochemical mapping[42]. These encompass environmental and mineral exploration factors that focus on a broad spectrum of elements and compounds that can help to identify and resolve chemical environmental problems or highlight potential problems as well as serve to give direction to exploration programs for ore minerals and other mineral commodities.

Element concentrations may be lower or higher than the natural baseline range suggesting the possibility of deficiencies or excesses of essential nutrients and the existence of areas at risk because of PTMs. The final evaluation falls then on planning (1) to deal with the deficiency (amending a soil or common foods, for example), and (2) determining the accessability of one or more than one PTMs to the food web to assess risk. The latter would of necessity include how element mobility would be affected by a change in land use policies and practices as a result of planned development. For example, there could be changes in soil or water pH, in earth materials reduction-oxidation conditions, in hydrology, (e.g. greatly increased groundwater extraction, draining wetlands, inundation), in cation exchange and complexation possibilities and other factors. In a futuristic mode, the influence of global climate change has to be assessed as well.

ENVIRONMENTAL INFLUENCE ON ELEMENT CONCENTRATION AND DISTRIBUTION

Element concentrations and distributions in surface and near-surface earth systems and the atmosphere are the result of mobility and precipitation of free ions, ionic compounds and anionic complexes. These may be in matrices of gases or vapours that evaporate or sublime from liquids or solids, and as dust or fumes or as solid metal or metal oxide particulates that condense from vapours, and in aqueous combination or in solids. The mobilized form determines the route(s) of exposure to life form populations. For humans these include ingestion, inhalation and through the skin[43]. Other life forms have similar routes of exposure modified by the forms themselves such as via roots for vegetation with accumulation (or discrimination) according to species, cultivars and duration of exposure. Table VI shows food crops that accumulate some PTMs in edible portions and in their leaves to greater or lesser degrees.

Whether the ions or ionic and anionic complexes are mobilized or precipitated and then perhaps remobilized is dictated by changes in physical, chemical and biological conditions in source areas and depositional environments. In soils, surface aquatic environments, and their associated sediments, food crops, food animals (including fish and foul), and aquifer waters, the controlling parameters that influence mobility and deposition of PTMs are many. These include some already discussed: pH, redox potential, concentrations, chemical gradients, cation exchange capacity, clay minerals, Fe/Mn oxyhydroxides, ad sorption-desorption, and bioaccumulation. The interaction of these parameters and varying responses of PTMs to them (e.g. synergistic or antagonistic) can make interpretations of environmental influences complex.

1. *Important Influences*

The array of factors that by themselves or combined influence the mobility of PTMs and their potential for release from soils or sediments are formidable. Perhaps foremost of these are the ways in which PTEs are found physically in these media (Fig. 1) and how some specific PTMs are distributed in various phases of a soil or sediment as defined by their reaction to chemical attack (Fig. 2). The element Cr for example is tightly bound in the residual phase and not easily mobilized whereas Cd has more distribution in easily soluble and exchangeable phases than the other elements considered. Other factors that affect mobility, some already cited, include (1) chemical parameters such as pH, Eh (redox potential), organic matter content, salinity, concentrations of competing ions, ionic or anionic complexes, and moisture;

Table VI General metal accumulation of the PTMs Cd and Pb in edible portions of food crops and Cu, Ni and Zn in their leaves

PTM	High accumulations	Low accumulations
Cd	Lettuce, spinach, celery, cabbage	Potato, maize, french beans, peas
Pb	Kale, ryegrass, celery	Some barley cvs, potato, maize
Cu	Sugar beet, certain barley cvs	Leek, cabbage, onion
Ni	Sugar beet, ryegrass, mangold, turnip	Maize, leek, barley cv, onion
Zn	Sugar beet, mangold, spinach, beetroot	Potato, leek, tomato, onion

B. J. Alloway, "Soil processes and the behaviour of metals." **In**: (B. J. Alloway, (Ed.) *Heavy Metals in Soils*, 2nd Ed., Blackie Academic & Professional, London, 1995) pp. 11–37.

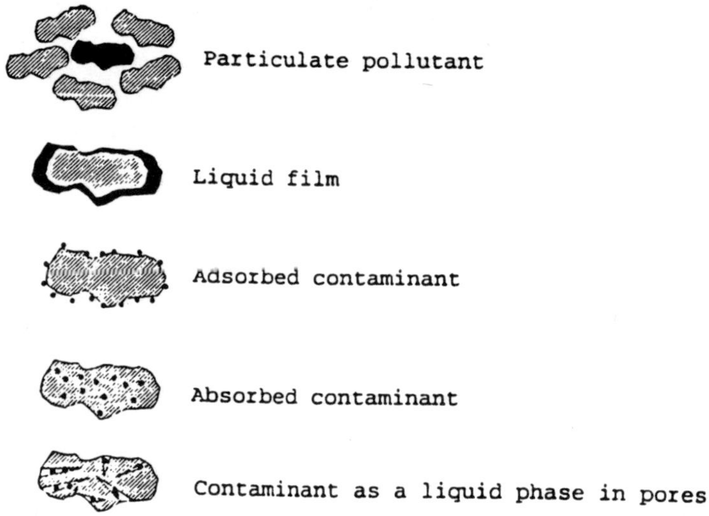

Figure 1 Physical states of contaminants in soils and sediments[57]. These affect potentially toxic metals chemical reactivity, mobility and bioavailability.

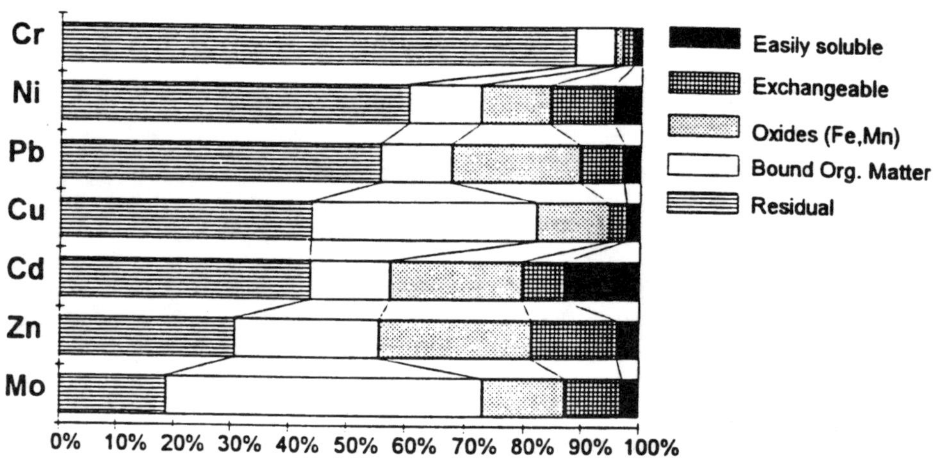

Figure 2 Distribution of some potentially toxic metals in solid phases of soils in Poland. Distributions in sediments are similar. A. Kabata-Pendias. "Trace metals in soil of Poland—occurrence and behaviour", *Trace Substances in Environmental Health* **25**, 53–70 (1992); A. Kabata-Pendias and H. Pendias, *Trace Elements in Soils and Plants*, 2nd Ed. (CRC Press, Boca Raton, Florida, 1991).

(2) physical parameters such as texture (size distribution) plus structure and penetratability, presence of one or an assemblage of clay minerals, their surface areas and cation exchange capacities, presence of oxyhydroxides of Fe, Mn and Al and carbonate minerals, climate (temperature); and (3) biological parameters such as microbial activity, vegetative root extension and penetration, and burrowing organisms. Micro-organism mediated PTMs oxidation–reduction reactions and bio-methylation–demethylation reactions are listed in Table VII. Groupings of many of

Table VII Micro-organism mediated PTMs oxidation–reduction and biomethylation–demethylation reactions

Reaction	PTM
Reduction	As^{5+}, Cr^{6+}, Fe^{3+}, Hg^{+}, Mn^{4+}, Se^{4+}
Oxidation	As^{3+}, Cr^{3+}, Fe^{0}, Fe^{2+}, Mn^{2+}, Sb^{3+}
(Bio)-methylation	As^{5+}, Cd^{2+}, Hg^{2+}, Se^{4+}, Sn^{2+}
Demethylation	R Hg^{2+}

P. Doelman, "Microbiology of soil and sediments" **In**: (W. Salomons and W. M. Stigliani, Eds.) *Biogeodynamics of Pollutants in Soils and Sediments* (Springer, Berlin 1995) pp. 31–52.

the same factors affect chemical element mobility and precipitation in wetland, lake, fluvial (streams and rivers), estuarine and oceanic (marine) sediments, and surface and aquifer waters.

Under acidic conditions, metals such as Cd, Co, Cr, Cu, Mn, Ni, Pb and Zn become more soluble and mobile. This enhances their uptake by grasses and other vegetation. Increased acidity (decreased pH) also increases the amount of dissolved Al and C_{org} in a soil, swamp or wetland environment. Under oxidizing conditions, decomposing organic matter releases associated PTMs and causes greater acidity[44,45]. This increases mobility for the metals cited above. Conversely, increases in pH to basic conditions establishes a geochemical barrier and many PTMs with low K_{sp} are immobilized and their uptake by vegetation is restricted. For example with a pH rise to 7.4 to 9 in aquatic systems favors the adsorption of Zn, Cr and Cd onto suspended matter[46]. Indeed, liming of agricultural soils increases pH and dissolved organic carbon and can immobilize metals (see acidic mobilized group above) that are potentially harmful to crop growth or because of their translocation to fodder or food crops. At the same time, however, other metals such as Mo can be mobilized and made bioavailable to fodder. As previously noted, ruminants including food and dairy cattle can develop hypocuprosis because the Mo contents of fodder interferes with the absorption of Cu by the animals.

The PTMs may have different mobility responses when a redox factor is combined with pH. This has been represented in a pH–Eh graph for some of the metals[47]. In Fig. 3, the head and width of an arrow for an element shows the pH–Eh conditions under which mobility increases; the tail of an arrow shows the same for decreased mobility. It is clear that As and Mn increase in mobility with decreasing pH and greater reduction potential (to As^{3+} and Mn^{2+}); Fe as Fe^{2+} and Mo as Mo^{2+} react similarly. The diagram also shows that with weakly acidic conditions to about pH 6 and increasing oxidation potential from $> +0.2$ to $> +0.6$, the elements Cd, Ni and Hg have increased mobility. With decreasing pH to 4 or less and increasing oxidation potentials from $+0.2$ to $> +0.6$ Zn, Cu, Pb are mobilized as are Fe and Mn at pH of <2. With marine conditions of pH about 8.1 and oxidation, Fe and Mn are precipitated as oxyhydroxides.

Under reducing conditions in the marine environment and in the presence of sulfide, chalcophile elements such as Zn, Cu, Fe, Hg, Ni, Pb and others will precipitate as sulfide mineral phases. Changes in oxidation or reduction potentials in a marine environment, for example, can lead to the separation of elements. The separation in the oceans may be geographic and related to bathymetry and restricted circulation that leads to changing reduction–oxidation potential. It may be vertical in cores due to changing reduction potential from decaying organics and bacterial activity. In both environments some PTMs are immobilized and precipitated under

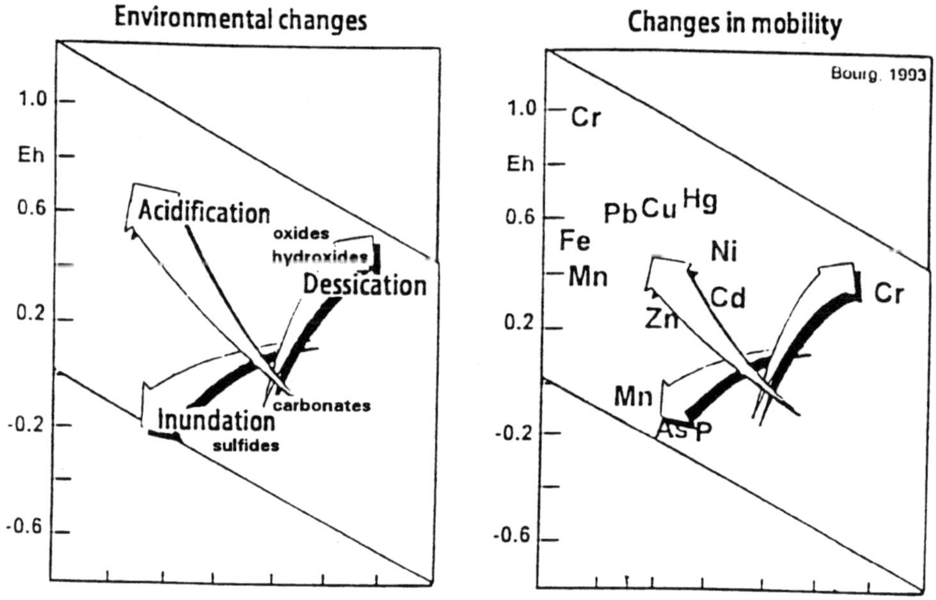

Figure 3 pH–Eh (oxidation–reduction potential) diagrams showing environmental changes and main mineral types controlling the solubilities of potentially toxic metals and the resulting trends of the increasing mobilities[67].

conditions where others are still mobile and being moved by currents or driven by geochemical gradients to precipitate elsewhere. Natural pH values for most soils are in the range of about 3 to 8.5 and redox potentials from -0.35 to $+0.8$. Shallow subsurface waters have more restricted ranges with pH from 5.5 to 8.5 and redox potentials from -0.08 to $+0.6$[48]. Elements respond to changes in these redox conditions in ways similar to the responses in the oceanic environment.

Adsorption is a function of the density of unsatisfied surface charges on colloids and clay size particulates. The charges may be negative in the case of clay minerals, manganese oxyhydroxides and organic colloids (especially humic compounds) thus providing adsorption sites for cations including PTMs[49,50]. The unsatisfied charges may be positive as for iron oxyhydroxides and provide capture sites for anionic complexes such as arsenate, molybdate, chromate, vanadate, and phosphate. PTMs that are associated with adsorption or precipitation phases are shown in Table VIII. Adsorption capacity varies between sorbents and with environmental conditions. For example, Fe and Mn oxyhydroxides have the strongest adsorption sites followed by particulate organics and clay minerals[51–53]. Among the clay minerals, whether in soils or sediments, kaolinite has the least adsorption capacity, illite and chlorite have intermediate capacities and montmorillonite (smectite) has the highest adsorption capacity. This is a function of the mineral structure and the surface areas per unit mass. Montmorillonite has by far the greatest surface area (Table IX). In addition, adsorption of most PTMs increases with increasing pH as is predictable from the consideration of mobilization of PTMs from soils or sediments.

There is a finite capacity of sorbents to store a contaminant under favored conditions. When this capacity is exceeded, the contaminant may diffuse into the surrounding medium[44,45]. If the conditions change there may be a major release of

Table VIII PTMs normally found with secondary minerals in soils

Mineral	Sorbed or precipitated PTMs
Fe oxides	V, Mn, Ni, Cu, Zn, Mo, As
Mn oxides	Fe, Co, Ni, Zn, Pb
Carbonates of Ca	V, Mn, Fe, Co, Cd
Clay minerals	
Smectites aka	
Montmorillonites	Ti, V, Cr, Mn, Fe, Co, Ni, Cu, Zn, Pb
Illites	V, Ni, Co, Cr, Zn, Cu, Pb
Vermiculites	Ti, Mn, Fe

G. Sposito, "The chemical forms of trace metals in soils" **In**: (I. Thornton, Ed.) *Applied Environmental Geochemistry* (Academic Press, London, 1983) pp. 123–170.

Table IX Typical surface properties of sorbents that tend to decrease metal solubility[52]

Soils and soil minerals	Surface area (m^2/g)
Kaolinite	7–30
Illite	65–100
Montmorillonite	700–800
Mn oxides	30–300
Fe oxides	40–80
Clays and loams	20–270
Silty loams	20–200
Sandy loams and loamy sands	10–70
Rendzinas	15–170
Calcareous acquifer sand	0.5–5

contaminants into the ecosystem damaging its productivity. This is the concept of the chemical time bomb (CTB) which refers to a chain of events resulting in the delayed and sudden occurrence of harmful environmental effects due to the mobilization of chemical stored in soils (decades or centuries) in response to slow alterations in environmental conditions[54]. The CTBs differ in the rate of the occurrence and intensity in response to climate (temperature and amount of daylight). For example, peat (reducing environment with sulfide present) is a known accumulator of PTMs. In Nordic countries peat bogs have been drained and reclaimed for harvestable forests. Exposure of new soils to oxidation conditions led to acidification from the decomposition of pyrite and increased bioavailability of PTMs. This was harmful to root growth of trees and resulted in declines of forests and of fish stocks in adjacent wetlands. In addition, the impact of metal contamination on some parasites and predators lessened the natural control of forest pests[55].

Biomethlyation or the microbial mediated formation of organometallic compounds from inorganic precursors is an important process that affects the toxicity potential of several PTMs. It creates toxic species of Hg that enter the food web as CH_3Hg^+ or $(CH_3)Hg_2$ after its release from the microbial system. Other PTMs which undergo biomethylation in minor amounts include As, Cr, Sb, Se, Sn and Tl. Biomethylation is stimulated in warm waters which are slightly acidic, rich in organic matter and have high bacterial activity[46]. A global focus has been on the gold rush areas of the Amazon which encompass parts of Brazil, Colombia, Guyana, Peru and Venezuela. Without the enforcement of environmental legislation that is in place to protect the ecosystem, the amalgam excess from the placer mining operations severely contaminates water and sediment downstream from placer mining sites. The inorganic Hg is biomethylated and high Hg concentrations are found in fish consumed by indigenous populations that occupy the river bank areas. Urine and blood samples from miners and in the indigenous populations are high in Hg and will, in the near future lead to adverse health effects of Hg poisoning[19,56].

These and other physical, chemical and biological influences on element concentration and distribution in the ecosystem are inexorably linked. The understanding of how they affect individual PTMs or the response of PTMs in synergistic or antagonistic systems is essential in planning to maintain the health and productivity of the living environment.

REMEDIATION OR MANAGEMENT OF METAL-CONTAMINATED ENVIRONMENTS

Remediation of human-originated metal contamination can make property and land useable again and protect ecosystem inhabitants from future metal toxicity. Remediation of metal pollution at its source or at contaminated sites is possible in most situations. However, the potential for remediation is limited by two factors: (1) economic resources available for the "curative" process(es); and (2) **impossible to achieve concentration targets** set by regulatory agencies on the basis of average global or regional values for the earth material(s) being studied. These regulatory concentration standards can be quite different from area specific values which are determined by geology, geological processes and environmental conditions at a study site.

For example, an area underlain by igneous rocks that are ultrabasic in their composition may have natural background concentrations of Cr, Ni and Co that far exceed a global standard value. The metal concentrations in soils formed from these rocks will carry correspondingly high chemical signals. Any polluted soils (e.g. from smelter or industrial operations) in such geological regions can be identified and selected for remediation. Targeted remediation values can not be less than natural concentrations in unpolluted soils in the area although these natural concentrations may be in excess of regulatory standards derived from global or regional average values. A realistic remediation standard value for metal concentrations in an area being evaluated can, in many cases, be calculated from existing geochemical data on rocks, stream (lake, oceanic) sediments, soils, waters, and the atmosphere. Lacking this, an environmental assessment team can require controlled scientific sampling and chemical analysis of the medium under investigation in the study area.

Remediation of earth materials containing metal toxin pollutants that could enter a food web and put human populations at risk can be approached using physical, chemical or biological methods or a combination of these. The method(s) used depends to a large part on the size of the problem (area and volume), the medium to be remediated (water [surface, subsurface/aquifer], soil, air), and the time constraints for alleviating or eliminating the pollution problem. Ultimately, however, remediation possibilites are based on the medium that is contaminated, its properties and capacity to hold contaminant metals, and the metals that are the clean-up targets. This then extends to include the concentrations of target metal(s), the presence of other contaminants and their concentrations, the physical state of the pollutant metal(s) (e.g. in solution), the metal(s) association and relative mobility in the medium (e.g. chemical speciation), the properties (e.g. physical and chemical stability), and contaminant response to biological processes.

Physical, chemical and biological methods have been used to remediate soils. If the area and depth of metal-contaminated soil is not large ($<\frac{1}{2}$ hectare and <1 m respectively), excavation can cleanse an environment, and with soil replacement where necessary, allow its use for projects sensitive to soil chemistry. This presumes that the excavated soil will be treated and/or disposed of in a secure manner. There is no social, economic or political benefit in resolving one problem and creating another. For example, the highly toxic metals Hg and As can be volatilized from contaminated soils at relatively high temperatures[57]. If the volatiles are not captured before emission to the atmosphere, atmospheric dispersion and deposition can pollute soils and fluvial systems in downwind areas as has been the case with smelter operations. Ultimately, runoff can carry the Hg and As into marine environments where they can enter into the food web to fish, sea and land mammals, and humans. Most other heavy metals are immobilized in slag if excavation and incineration is the remediation method of choice so that secure disposal of the slag is a necessity as well.

In situ chemical extraction from soils has been used where the texture of the soil (generally sand) lends itself to good hydraulic (high permeability) properties in the direction that parallels the metal pollutant distribution. The target metal(s) should be soluble in a pressure-injected infiltrating aqueous phase from an input well or solubility should increase in response to the addition of chemical conditioners such as acids, peptizers, or agents that alter redox conditions[57]. Extraction of Cd from soils in Holland using a 3.5 pH HCl aqueous phase conditioner brought the Cd concentration from 10 ppm to 1 ppm. The cost was less than would have been the case if excavation had been used. The pumped extracted phase has then to be treated before its return to the environment. Research is going on in microbiological leaching as a method to solubilize metal contaminants in soils and make them available to an acidic aqueous extractant phase[57].

Where PTMs in soils are bioavailable, phytoremediation or the cleansing of soils by removal of PTMs using vegetation has much promise[58]. The potential for this method of remediation is greatest where (hyper-)accumulator plants native to a climate can be found that are rapid-growing and have large masses. With current plant genetics technologies, native species might be engineered to have or improve on the desired properties for an efficient phytoremediator. It may then be possible to harvest the vegetation for its biomass energy potential and metal contents for recycling (or secure disposal).

A promising new electro-remediation technique is being field-tested to evaluate its effectiveness in extracting PTMs from clay-rich (low permeabiity) soils[59]. This first involves the creation of highly permeable zones (e.g. by hydraulic fracturing) in the contaminated soil zones which are used as sorption/degradation sites by injecting agents such as sorbents, chemicals, and microbes. Secondly, electro-osmosis is used as a liquid pump for flushing pollutants from the soil into the treatment sites. Thirdly, liquid flow can be reversed by switching electrical polarity to increase the efficiency of pollutant removal as well as allow multiple passes of the pollutants through the treatment layers for thorough sorption/degradation. As with the other remediation technologies, the extracted pollutant-bearing solution has to be properly treated and/or disposed of in a secure manner.

Contaminated sediments in ponds (e.g. industrial), lakes, streams, rivers, estuaries, inland seas, oceans, and associated wetlands present special problems in that there is much water to deal with and the contamination is often widespread and diffuse. Geochemical engineering solutions can be sought by identifying factors that can accelerate mobility from PTMs contaminated sediments. These include lowering the pH, changing reduction–oxidation conditions, inorganic and organic complexation and microbial mediation. The PTMs are then assessed for their response to physical/chemical barriers to mobility (e.g. adsorption, controlled sedimentation and precipitation) to determine the most secure way to deal with the contaminants and their carriers[46,60]. Although removal of contaminant-laden sediments by dredging has been used in the past, research shows that it should no longer be a viable option if the dredging changes environmental conditions that lead to PTMs mobilization (Table X) and bioavailability. This is especially important if dredging exposes reduced sediment to oxidizing conditions. In experimental studies, the oxidative release of As from industrial solid wastes at a pH of 5 was complete in 5 weeks and the release of Zn began after 5 weeks and was continually enhanced. In other experiments, 20–90% of pyrite bound metals were released in a day or less by exposure to oxidation conditions in sea water where they would be bioavailable[46]. Thus, where applicable, environmental conditions under which PTMs are immobile and can be "stored" under permanent anoxic conditions should not be disturbed. In environments where natural bioturbation of (oxidized) surface aquatic sediments (by fish and, invertebrates such as crinoids, echinoderms and sponges and burrowing forms such as pogonophorans) can expose reduced sediment layers to aeration (oxidation) to depths of 20 cm and lead to mobilization of some PTMs[61], a different solution to insure immobilization would be applicable. On the sea floor in harbors and bays or specific aquatic disposal sites that have received PTMs loading this could be done by capping areas that would be susceptible to changing conditions with clean sediment[46]. The situation is so different for each PTMs-contaminated aquatic sediment that each requires its own monitoring and evaluation of how to best prevent the interaction of the contaminants with the living environment.

Remediation of PTMs in surface and subsurface waters presents two different problems. Surface waters can be self-cleansing if the source of metal contaminants is anthropogenic and can be eliminated. This would include industrial effluent which can be treated to remove metal contaminants before discharge into surface waters. Acid mine drainage from abandoned mines, laden with metal contaminants and seeping into streams and lakes is a problem still seeking a definitive relatively short-term solution. In Finland, the acid mine drainage problem is compounded by

Table X Mobilization of PTMs during dredging. Concentrations in mg/l except Hg in μg/l

PTM	Channel sediment pore water (a)	River water conc. (b)	a/b	Effluent at a man-created marsh		
				Expected conc.	Measured conc.	%change
Zn	0.12	0.052	2	0.065	5.30	+8069
Cu	0.012	0.004	3	0.0055	0.051	+827
Pb	0.077	0.002	38	0.016	0.142	+788
Cd	0.009	0.001	9	0.0025	0.019	+660
Ni	0.054	0.001	54	0.011	0.035	+218
Hg	3.2	0.26	12	0.82	2.0	+144
Mn	6.94	0.03	230	1.34	1.19	−11
Fe	57.3	0.26	220	11.12	6.01	−46

D. A. Darby, D. D. Adams and W. T. Nivens, "Early sediment changes and element mobilization in a man-made estuary marsh." **In**: (P. G. Sly, Ed.) *Sediment and Water Interactions* (Springer, Berlin, 1986) pp. 343–351.

atmospheric deposition of Cd, Cr, Cu, Pb, Ni and V that originated in Eastern and Central Europe and added to the load of As, Cd, Cu, Fe, S and Zn in dust and seepage from mining[24]. Sediments deposited from waters such as these may have served as sinks for water-borne contaminants and present still another contaminant source to deal with.

Two methods of managing engineered wetlands have been used to alleviate metal pollution problems in surface waters. In the first, contaminated effluent and entrained sediment is directed to neutralizing ponds or engineered wetlands where, in various stages, the flow is slowed, sediment is deposited and the contaminant is chemically neutralized and continuously monitored before water flow continues to the natural drainage[62]. The second uses the same engineered wetlands stage method but metal toxin(s) are cleansed from the contaminated waters by rhyzofiltration using accumulator plants such as the water hyacinth. The water hyacinth grows in wetlands, has a large biomass, is rapid-growing and is an efficient accumulator of a broad assemblage of PTMs[63]. As in phytoremediation, the biomass can be harvested for energy generation and its metal contents can be recycled or disposed of in a secure manner.

Subsurface or aquifer waters can be contaminated by leakage from poorly designed landfills or chemical dumps. Rain water filtering through the waste can over a period of time leach PTMs from wastes (loose and from breached containers) and move them through porous, permeable earth materials into an unconfined aquifer. Populations downflow from the input area are then subject to metal toxicity if the water is ingested, used for cooking or for irrigating food crops that can bioaccumulate metal toxins. When the source of metal toxins is identified, it can be physically removed to a secure site and remediation can begin. Since water in aquifers moves very slowly, generally less than 0.3–1.0 m/day, metal toxins will remain in the aquifer downflow system for a long time unless the contaminated water is removed and treated. This has been done at Love Canal, New York, where released buried chemical wastes caused major human health problems. Because of the tragedy of Love Canal, the Superfund was created in the United States in 1980 to deal with sites that are highly contaminated and have clearly been responsible for major health problems or could lead to them. In such a case, after removal, treatment and secure disposal of the pollutant generating wastes, exploratory wells defined the nature and extent of soil pollution and the hydraulic and directional characteristics of subsurface water flow. An impermeable barrier is put in place to keep contaminants on site. Wells are used to inject fluid under pressure to drive the contaminated water to collecting wells from where it is pumped to a treatment plant. The process is slow and may take 10 years or more depending on the areal extent and volume to be extracted and the hydraulics of the aquifer. This method is costly but is effective and can return the earth system to natural, liveable conditions. Housing is again being sold and population is returning to the Love Canal site[30].

In cases where a remediation solution is not feasible because of the lack of economic resources or the extent and nature of a pollution intrusion (whether naturally occurring or anthropogenic), management can help alleviate/eliminate or control the development of health hazards. This is the solution being sought in the case of the arsenic poisoning from polluted groundwater in West Bengal, India, described previously[7,8]. In the case of soils, management would include amending soils chemically to immobilize heavy metals which might otherwise access the food web. For example, in controlled experiments, the addition of soluble inorganic phosphate to mine-waste contaminated soils aided immobilization of the Pb and Zn and might be useful to immobilize other PTMS[64]. A site for chemical management must be evaluated thoroughly lest an additive to immobilize some PTMs will mobilize others. As already noted, altering the redox potential in a soil or sediment towards reduction can stimulate the mobility of Fe, As, Mo and Mn whereas other PTMs remain immobile. Under oxidizing conditions the reverse can take place with mobilization of PTMs other than the four cited above. Whether mobilization occurs or not is also affected by changes of pH. Management can maintain a good living environment and the productivity of terrain if pre-management evaluations are thorough and managed systems are monitored so that a PTM release can be dealt with before a major environmental intrusion occurs.

Whether by remediation or by management, environments can be brought back to "health" as living systems for habitats, water, food and other natural resources. If they are maintained and other PTMs-contaminated environments managed and monitored effectively, they will be able to serve the needs of humankind and other living forms now and for future generations.

OPTIMAL DEVELOPMENT WITH SUSTAINABLE USE AND YIELD OF ECOSYSTEMS

Sustainable use and yield of project environments may be achievable in some development sectors such as fisheries and forestry. How long this can be maintained is a function of several global factors. First and foremost is the reality of population growth. The annual rate of global population increase is 1.6%. Most of this in developing areas of the world (Asia, Africa, Latin America and the Indian sub-continent). The present world population of about 5.9 billion could double in 43 years. Because growth rate appears to be decreasing, estimates are for a stabilized world population of between 8.5 and 9.5 billion. Renewable natural resources will surely be steady or even increased whereas nonrenewable resources will decrease; arable land is being lost (to urbanization, to water projects, to industrialization) even as there is a world-wide drive to conserve and improve environmental systems. The reality of the situation with respect to a growing global population is that it will continue to be an uphill struggle to alleviate poverty and improve the quality of life especially in developing regions. If, at the least, metal toxins from human activities can be controlled so that they do not enter into the food web in contaminant concentrations, one phase of improving the environment for human and other life forms is achievable and sustainable.

Sustainable development implies the promotion of growth and expansion. Clearly, if sustainability implies non-negative changes in availability of natural resources (metal ores, soil, soil quality, water and water quality, land biomass, water biomass and waste assimilation capacity of the receiving environments), it is an impossible goal[65].

Sustainable development defined as "a process of change in which the exploitation of natural resources, the direction of investments, the orientation of technological development, and institutional change are all focused on enhancing current and future potential to meet human needs and aspirations" is reasonable[66]. It is perhaps better to use the concept of optimal development in order to try to achieve the generation to generation pass through of environmental quality and natural resources and their inherent social and economic benefits[67].

Optimal development is possible although interruptible to greater or lesser degrees depending on a project (e.g. agricultural, industrial) and the application of available defenses against natural and anthropogenic hazards in development planning.

In the past, decisions to use them or not have often made on the basis of economics. This is no longer acceptable. Metal toxin capture at a source should be regulated by laws that are strictly enforced as should also be disposal or recycling options. There are remediation techniques that can be used in areas that were contaminated by metal toxins from natural sources and/or from human activities before disposal controls were proposed and instituted. In any development planning where potentially toxic metals will be from point sources or where they are already *in situ* from natural or anthropogenic sources, defenses must be put in place or remediation must take place. Lacking this, metal toxins will enter the ecosystem food web and pose long-term health risks to consumer populations[68–70].

References

1. F. R. Siegel, "Applied geochemistry" **In**: *Encyclopedia of Applied Geology* (Van Nostrand Reinhold, New York, 1984) **XIII**, pp. 207–215.
2. F. R. Siegel and A. P. Segura, "Biogeochemistry for future mineral resource exploration programs in the Central American–Caribbean region" **In**: (R. L. Miller, G. Escalante, J. A. Reinemund and M. L. Bergin, Eds.) *Energy and Mineral Potential of the Central American–Caribbean Regions* (Springer, Berlin, Earth Science Series (1995) **16**, pp. 329–334.
3. X. Chen, G. Yang, J. Chen, X. Chen, Z. Wem and K. Ge, "Studies on the relations of selenium and Keshan disease" *Biol. Trace Elem. Res.* **2**, 91–107 (1980).
4. Z. Yin, S. Yue and R. He, "Characteristics of the environmental geochemistry and its relationship with 'Keshan disease' in the plateau sandy soil of Keshiketengqi, Inner Mongolia" *Scientia Geographica Sinica* **3**, 175–182 (1983) (in Chinese).
5. J. C. Smith and J. A. Halsted, "Clay ingestion (geophagia) as a source of zinc for rats" *Jour. Nutr.* **100**, 973–980 (1970).
6. D. G. Irvine, H. B. Schiefer and W. J. Hader, "Geotoxicology of multiple sclerosis: the Henribourg, Saskatchewan cluster focus: II. The soil" *Sci. Total Environ.* **77**, 176–188 (1988).
7. D. Dipankar and 7 co-authors, "Arsenic in groundwater in six districts of West Bengal, India" *Envir. Geochem. and Health* **18**, 5–15 (1996).
8. P. Bagla and J. Kaiser, "India's spreading health crisis draws global arsenic experts" *Science* **274**, 174–175 (1996).
9. P. Salopek, "Arsenic poisoning cases plague hundreds of villages in Bangladesh" *Washington Post* A17 (4 January, 1997).
10. D. E. Voight, S. L. Brantley and R. J.-C. Hennet, "Chemical fixation of arsenic in contaminated soils" *Applied Geochem.* **11**, 633–642 (1996).
11. J. Kobayashi and N. Hagino, "Strange osteomalacia by pollution from cadmium mining" *Progress Rept.* WP 00359, (Okayama Univ.) 10–24 (1965).
12. K. W. Kim and I. Thornton, "Influence of Ordovician uraniferous black shales on the trace element composition of soils and food crops, Korea" *Applied Geochem.* **Suppl. 2**, 249–255 (1993).
13. K. W. Kim and I. Thornton, "Influence of uraniferous black shales on cadmium, molybdenum and selenium in soils and crop plants in the Deog-Pyoung area of Korea" *Envir. Geochem. and Health*, **15**, 119–133 (1993).
14. I. Thornton, "Environmental geochemistry and health in the 1990s: a global perspective" *Applied Geochem.* **Suppl. 2**, 203–210 (1993).
15. J. Watt, I. Thornton and J. Cotter-Howells, "Physical evidence suggesting the transfer of soil Pb into young children via hand-to-mouth activity" *Applied Geochem.* **Suppl. 2**, 269–272 (1993).
16. R. Lacatusu, C. Rauta, S. Carstea and I. Ghelase, "Soil–plant–man relationships in heavy metal polluted areas in Romania" **In**: (R. Fuge, M. Billet and O. Selinus, Eds.) *Environmental Geochemistry, 3rd Int'l Symp.*, Krakow, Poland, *Applied Geochem.* **11**, 105–107 (1996).
17. J. F. Verner and M. H. Ramsey, "Heavy metal contamination of soils around a Pb–Zn smelter in Bukowno, Poland" **In**: (R. Fuge, M. Billet and O. Selinus, Eds.) *Environmental Geochemistry, 3rd Int'l Symp.*, Krakow, Poland, *Applied Geochem.* **11,** 11–16 (1996).
18. J. E. Fergusson, *The Heavy Elements: Chemistry, Environmental Impact and Health Effects* (Pergamon Press, Oxford, England, 1990).
19. I. Thornton, "Impacts of mining on the environment: some local, regional and global issues" **In**: (R. Fuge, M. Billet and O. Selinus, Eds.) *Environmental Geochemistry, 3rd Int'l Symp.*, Krakow, Poland, *Applied Geochem.* **11**, 355–361 (1996).

20. B. J. Alloway, "The origins of heavy metals in soils" **In**: (B. J. Alloway Ed.) *Heavy Metals in Soils* (Blackie Academic & Professional, 1995) pp. 38–57.
21. M. C. Jung and I. Thornton, "Heavy metal contamination of soils and plants in the vicinity of a lead–zinc mine, Korea" **In**: (R. Fuge, M. Billet and O. Selinus, Eds.) *Environmental Geochemistry, 3rd Int'l* Symp., Krakow, Poland, *Applied Geochem.* **11**, pp. 53–59 (1996).
22. S. Ustya and V. Petrikova, "Heavy metal pollution of soils and crops in Northern Bohemia" **In**: (R. Fuge, M. Billet and O. Selinus, Eds.) *Environmental Geochemistry, 3rd Int'l Symp.*, Krakow, Poland, *Applied Geochem.* **11**, 77–80 (1996).
23. L. Matschullat and E. Bozau, "Atmospheric element input in the Eastern Ore Mountains" **In**: (R. Fuge, M. Billet and O. Selinus, Eds.) *Environmental Geochemistry, 3rd Int'l* Symp., Krakow, Poland, *Applied Geochem.* **11**, 149–154 (1996).
24. E. Kubin and H. Lippo, "Atmospheric heavy metal deposition in Finland from 1985–1990" **In**: (R. Fuge, M. Billet and O. Selinus Eds.) *Environmental Geochemistry, 3rd Int'l Symp.*, Krakow, Poland, *Applied Geochem.* **11**, 155–161 (1996).
25. D. Cleary, I. Thornton, N. Brown, G. Kazantis, T. Delves and S. Worthington, "Mercury in Brazil" *Nature* **369**, 613–614 (1994).
26. I. Thornton, D. Cleary and G. Kazantis, "Mercury contamination from gold mining in the Brazilian Amazon: health implication" **In**: (P. Varnavas, Ed.) *Environmental Contamination, Proceedings of the 6th Int'l Conf.*, Delphi, Greece (1994) 7–12.
27. F. R. Siegel, J. H. Kravitz and H. H. Hopkins, "Arsenic and mercury in Kara Sea-Barents Sea sediment cores (abs.) *Annual meeting., Geol. Soc. America, Denver, in Abstracts Book* **27**, A43 (1996).
28. W. Mertz, "The essential trace elements" *Science* **213**, 1332–1338 (1981).
29. B. J. Alloway, "Soil processes and the behaviour of metals": **In**: (B. J. Alloway, Ed.), *Heavy Metals in Soils*, 2nd Ed. (Blackie Academic and Professional, London, 1995) pp. 11–37.
30. F. R. Siegel, *Natural and Anthropogenic Hazards in Development Planning* (R. G. Landes Inc., Austin, and Academic Press, New York and London, 1996).
31. J. S. Webb, "Regional geochemical reconnaissance in medical geography" **In**: H. L. Cannon and H. C. Hopps, Eds.), *Environmental Geochemistry and Health, Geol. Soc. America Memoir* (1971) **123**, pp. 31–42.
32. C. J. L. Murray and A. D. Lopez, "Evidence-based health policy—lessons from the global burden of disease study" *Science* **274**, 740–743 (1996).
33. A. I. Perel'man, "Geochemical barriers: theory and practical applications" *Applied Geochem.* **1**, 669–680 (1986).
34. R. R. Brooks, *Biological Methods of Prospecting for Minerals* (Wiley-Interscience, New York (1983).
35. F. R. Siegel, *Applied Geochemistry* (Wiley-Interscience, New York, 1974).
36. J. S. Webb, I. Thornton, M. Thornton, R. J. Howarth and P. L. Lowenstein, *The Wolfson Geochemical Atlas of England and Wales* (Oxford Univ. Press, London, 1978).
37. A. G. Darnley, A. Bjorklund, B. Bolviken, N. Gustavsson, P. V. Koval, J. A. Plant, A. Steenfelt, M. Tauchid and X. Xie, with contributions by R. G. Garrett and G. E. M. Hall, "A Global Geochemical Database: For Environmental and Resource Management" UNESCO *Earth Sciences* **19** (1995).
38. N. Gustavsson, E. Lampio, B. Nilsson, G. Norblad, F. Ros and R. Salminen, "Geochemical Maps of Finland and Sweden" *Geol. Surv. of Finland, Espoo, Finland and Swedish Geol. Co.*, Uppsala, Sweden (1993).
39. F. Thalmann, O. Schermann, E. Schroll and G. Hausberger, "Geochemischer Atlas der Republik Osterreich; Boehmische Masse und Zentralzone der Ostalpen; Bachsedimente" *Geol. Bundesanst.*, Vienna, 35 maps at 1:1000 (1988).
40. J. Lis and A. Pasieczna, *Geochemical Atlas of Poland* (Polish Geological Institute, Warsaw, (72 plates, 15 tables, 1995).
41. A. H. M. Johnson, G. C. Lalor, J. Preston, H. Robotham, C. Thompson and M. Vutchkov, "Heavy metals in Jamaican surface soils" *Envir. Geochem. and Health* **18**, 113–121 (1996).
42. P. R. Simpson, N. Breward, D. M. A. Flight, T. R. Lister, J. M. Cook, B. Smith and G. E. M. Hall, "High resolution regional hydrogeochemical baseline mapping of stream water of Wales, the Welsh borders and West Midland region" **In**: (R. Fuge, M. Billet and O. Selinus, Eds.) *Environmental Geochemistry, 3rd Int'l Symp.*, Krakow, Poland, *Applied Geochem.* **11**, 621–632 (1996).
43. S. E. Manahan, *Toxicological Chemistry* (2nd Ed.) (Lewis Publishers, Boca Raton, FL, 1992).
44. M. J. Wilson and N. Bell, "Acid deposition and heavy metal mobilization" **In**: (R. Fuge, M. Billet and O. Selinus, Eds.) *Environmental Geochemistry, 3rd Int'l Symp.*, Krakow, Poland, *Applied Geochem.* **11**, 133–137 (1996).
45. A. C. M. Bourg and J. P. G. Loch, "Mobilization of heavy metals as affected by pH and redox conditions" **In**: (W. Salomons and W. M. Stigliani, Eds.), *Biodynamics of Pollutants in Soils and Sediments* (Springer, Berlin 1995) pp. 87–102.

46. U. Förstner, "Contaminated aquatic sediments and waste sites: geochemical engineering solutions" **In**: (H. E. Allen, Ed.) *Metal Contaminated Aquatic Sediments* (Ann Arbor Press, Chelsea, Michigan, 1995) pp. 237–256.
47. U. Förstner, "Changes in metal mobilities in aquatic and terrestrial cycles" **In**: (J. W. Patterson and R. Passino, Eds.), *Metals Speciation, Separation and Recovery* (Lewis Publishers, Chelsea, Michigan, 1987, pp. 3–26.
48. L. G. M. Bass Becking, I. R. Kaplan and D. Moore, "Limits of the natural environment in terms of pH and oxidation–reduction potentials *Jour. Geol.* **68**, 243–284 (1960).
49. G. W. Brummer, *The Importance of Chemical Speciation in Environmental Processes* (Springer, Berlin, 1986) pp. 169–192.
50. Y. Chen and F. J. Stevenson, "The role of organic matter in modern agriculture" **In**: Y. Chen and Y. Avinmelech, Eds.) (Martinus Nijhoff, Dordrecht, 1986) pp. 73–112.
51. V. Novotny, "Diffuse sources of pollution by toxic metals and impact on receiving waters" **In**: (W. Salomons, U. Förstner and P. Mader, Eds.) *Heavy Metals: Problems and Solutions* (Springer, Berlin, 1995) pp. 33–52.
52. A. C. M. Bourg, "Speciation of heavy metals in soils and groundwater and implications for their natural and provoked mobility" **In**: (W. Salomons, U. Förstner and P. Mader, Eds.) *Heavy Metals: Problems and Solutions* (Springer, Berlin, 1995) pp. 19–31.
53. A. Kabata-Pendias, "Behavioural properties of trace metals in soils" *Applied Geochem.*, **Suppl. 2**, 3–9 (1993).
54. W. M. Stigliani, P. Doelman, W. Salomons, R. Schulin, G. R. B. Smidt and S. E. A. T. M. Van der Zee, "Chemical time bombs: predicting the unpredictable" *Environment* **33**, 4–9; 26–30 (1991).
55. C. J. Konsten, G. R. B. ter Mueulen-Smidt, W. M. Stigliani, W. Salomons and H. Eusackers, "Summary of the workshop on delayed effects of chemical in soils and sediments (Chemical Time Bombs) with emphasis on the Scandinavian region" *Applied Geochem.* **Suppl. 2**, 295–299 (1993).
56. D. Lacerda, O. Malm, J. R. D. Guimaraes, W. Salomons and R.-D. Wilken, "Mercury and the new gold rush in the south" **In**: (W. Salomons and W. M. Stigliani, Eds.), *Biogeodynamics of Pollutants in Soils and Sediments* (Springer, Berlin, 1995) pp. 213–245.
57. W. H. Rulkens, J. T. C. Grotenhuis and R. Tichy, "Methods for cleaning contaminated soils and sediments" **In**: (W. Salomons, U. Förstner and P. Mader, Eds.), *Heavy Metals: Problems and Solutions* (Springer, Berlin, 1955) pp. 169–191.
58. R. M. Atlas, "Bioremediation" *Chem. and Eng. News* 32–42 (April 3, 1995).
59. Monsanto, "Development of an integrated *in-situ* remediation technology" *Technology Development Data Sheet* (1995).
60. R. D. Schuiling, "Geochemical engineering—some thoughts on a new research field" *Applied Geochem.* **5**, 251–262 (1990).
61. G. M. Matisoff, "Effects of bioturbation on solute and particle transport in sediments" **In**: (H. E. Allen, Ed.) (Ann Arbor Press, Chelsea, Michigan 1995) pp. 201–272.
62. K. Fytas and J. Hadjigeorgiou, "An assessment of acid rock drainage continuous monitoring technology" *Envir. Geol.* **25**, 36–42 (1995).
63. F. R. Siegel, "Environmental geochemistry in development planning: an example from the Nile delta, Egypt" *Jour. Geochem. Explor.* **55**, 265–273 (1995).
64. J. Cotter-Howells and S. Capron, "Remediation of contaminated land by formation of heavy metal phosphates" **In**: (R. Fuge, M. Billet and O. Selinus, Eds.) *Environmental Geochemistry, 3rd Int'l Symp*, Krakow, Poland, *Applied Geochem.* **11**, 335–342 (1996).
65. P. Dasgupta, "Optimal versus sustainable development" **In**: (I. Serageldin and A. Steer, Eds.) *Valuing the Environment* (Washington, D.C. World Bank, 1994) 35–46.
66. G. L. Brady and P. C. F. Geets, "Sustainable development: the challenge of implementation" *Int. Jour. Sustainable Development* **1**, 189–197 (1994).
67. T. C. Koopmans, *On the Concept of Optimal Economic Growth* (Pontificiae Academiae Scientiarium Script Varia, 1965).
68. E. Merian (Ed.), *Metals and Their Compounds in the Environment: Occurrence, Analysis, and Biological Relevance* (VCH, Weinhem, 1991).
69. B. E. Davies, "Trace elements in the human environment: problems and risks," *Envir. Geochem. and Health* **16**, 97–106 (1994).
70. D. J. Thomas, "Arsenic toxicity in humans: research problems and prospects" *Envir. Geochem. and Health* **16**, 107–111 (1994).

5. THE ENVIRONMENTAL TOXICOLOGY OF VOLCANIC GASES IN THE EUROPEAN ENVIRONMENT

J. P. GRATTAN *

INTRODUCTION

Many volcanic gases and resultant aerosols are toxic and may have severe impacts upon environments which are exposed to these emissions. Studies of these phenomena have emphasised the toxic impact upon environments in close vicinity to the volcano[1,2] or even in close proximity to an active vent[3]. However, the impact of volcanic gases may be felt over a much larger area, one need only consider the impact of the Laki fissure eruption, 1783–1784 AD, upon the vegetation, livestock and people of Iceland[4–6], to realise the magnitude of the environmental disaster brought about by the gases emitted at that time. Similar phenomena are now known to occur on a continental scale. In 1783 AD much of Europe was covered in a dry fog composed of volcanic gases emitted from eruptions in Iceland and Italy[7–9]. The environmental impact of this event across Europe was dramatic. The hazard posed by this and similar phenomena will be assessed and their potential role in the contemporary European environment will be discussed.

THE TOXICOLOGY OF VOLCANIC GASES

Volcanoes emit a wide range of potentially toxic gases via eruptive and non-eruptive processes. The most common compounds which result are presented in Table I, and include ammonia, carbon dioxide, carbon monoxide, fluorine, hydrochloric acid, hydrofluoric acid, hydrogen sulphide, sulphur dioxide and sulphuric acid[2,10,11]. The impact of volcanic gases, as with gases emitted by anthropogenic pollution, depends primarily upon the macro and micro climatic processes which may concentrate or dilute them, and upon the subsequent oxidation processes which may occur in the atmosphere as gases become aerosols. The sensitivity of the receptor to pollution must also be considered[12]. Acid gases and halogens of volcanogenic origin or otherwise may be transported considerable distances[13–16] and be deposited as dry particles[17], as acid rain and snow[18–20] or as an acid fog, mist or dew[21,22].

1. *Non-Eruptive Emissions and Plant Tolerance*

Volatile output from volcanoes is not limited to eruptive activity. For example Mt. Etna, Sicily, emits 30 tons of fluorine daily through non-eruptive degassing, as well as annual outputs of sulphur dioxide which are the equivalent of the total industrial emissions of modern France[23]. Despite this output, only small concentrations

* The University of Wales, Aberystwyth, The Institute of Geography and Earth Sciences, Aberystwyth SY23 3DB (UK). E mail: jpg@aber.ac.uk.

Table I Toxicology of volcanic gases[2,10,11,65]

Ammonia		
Animal thresholds	100 ppm in air	Irritation of mucous membranes, coughing, dyspnea and vomiting, conjunctivitis, retention of urine, lung oedema.
Plant thresholds	10 ppm in air	"Cooked" appearance becoming brown on drying
Carbon dioxide		
Animal thresholds	5000 ppm in air	Asphyxiation, drowsiness, weakness
Plant thresholds	No marked effect	
Carbon monoxide		
Animal thresholds	100 ppm in air	Shortness of breath, impairment of hearing and vision; dizziness, at high concentrations unconsciousness and death
Plant thresholds	500 ppm in air	Epinasty, chlorosis, shedding of leaves
Fluorine		
Animal thresholds	0.1 ppm in air	Powerful caustic irritant, conjunctivitis, skin irritation, bone degeneration, mottling of teeth, lung inflammation
Plant thresholds	0.1 ppm in air	Marginal leaf lesions, shedding of leaves
Hydrochloric acid		
Animal thresholds	5 ppm in air	Irritation of mucous membranes, spasm of the larynx, higher levels can result in pulmonary oedema
Plant thresholds	10 ppm in air	Marginal leaf lesions, shedding of leaves
Hydrofluoric acid		
Animal thresholds	3 ppm in air	Ulcers in respiratory tract. Skin burns, degeneration of subcutaneous tissues, gangrene, eye irritation, blindness, lung oedema
Plant thresholds	Variable	Marginal leaf lesions, shedding of leaves
Hydrogen sulphide		
Animal thresholds	20 ppm in air	Skin irritation at 20 ppm, respiratory paralysis at 500 ppm, photophobia, inflammation of cornea and respiratory tract
Plant thresholds	1 ppm in air	Chronic degeneration depending on concentration
Sulphur dioxide		
Animal thresholds	10 ppm in air	Eye irritation, burning of trachea, lung oedema, lung paralysis
Plant thresholds	1 ppm in air	Chronic degeneration depending on concentration. Chronic, yellowing and bleaching of leaves. Acute, collapse of marginal and inter-coistal areas, drying to ivory or brownish red
Sulphuric acid		
Animal thresholds	0.1 ppm in air	Burning, rapid disintegration of tissue, lung damage
Plant thresholds	Variable	Scorching or charring, depending on moisture content and size of aerosol droplets

of hydrogen fluoride have been detected in surveys of the leaves of plants downwind of the emissions[1,2]. Synergistic relationships between volatiles may lessen the intensity of their impact. The emission of sulphur dioxide in train with the fluorine may have caused the stomata of the plants to close and thus hinder the absorption of fluorine[24]. Alternatively the vegetation growing in the vicinity may develop tolerance by natural selection and long exposure to these emissions. Austrian pine [*Pinus nigra*] growing on the slopes of Mt. Etna were first planted at the turn of the century and initially had a mortality rate of 80%. The trees growing today are the descendants of the surviving 20% and have a high tolerance to fluorosis and acidification. Austrian pines planted from fresh stock had the same 80% mortality rate as the original planting[25].

Fluorine is particularly toxic and is frequently a component of gases emitted in Icelandic volcanic eruptions. Following the eruption of Eldfjell, 23–25 of June 1973, the concentration of fluorine sixty kilometres away in Myraldur was 3000 ppm in air: where the threshold for plant damage is only 0.1 ppm[1,4]. The pine needles of mature conifers turned brown, young conifers were killed and mosses were severely affected or eradicated. That this scale of damage should occur in Iceland is perhaps not surprising. However, documentary evidence will be presented below which describes similar environmental pollution phenomena which took place in Europe in 1783 AD, over 600 miles away from the likely volcanic source in Iceland, the Laki fissure eruption.

VOLCANIC GASES AND THE LAKI FISSURE ERUPTION

1. *Eruption Dynamics*

This classic basalt fissure eruption[26], took place between 8 June 1783–February 1784, and produced 250 Mt of sulphuric acid, 60% of which was released during the first 6 weeks of the eruption[27,28], a quantity rarely matched by other Holocene eruptions in the northern hemisphere[29–33]. The relatively low explosive force of this eruption failed to transmit the majority of the volcanic gases produced directly into the stratosphere. As a result perhaps 60% of the total emitted pollutants were confined in the troposphere[27,28,34]. Regional atmospheric circulation patterns then transported volatile gases retained within the troposphere to Europe where their environmental impact, described below, was dramatic.

2. *Volatile Dispersal and Concentration*

Evidence assembled by Kington[35] suggests that the prevailing weather conditions in Europe during the early summer of 1783, a deep low pressure system over Iceland and a virtually stationary high pressure cell over Europe, facilitated the dispersal of these toxic volcanic gases and aerosols from Laki over a large area of western and central Europe. Thus it seems that the huge quantities of gas ejected by the Laki fissure were drawn to a high altitude in the troposphere and then transported by high velocity winds directed towards the high pressure cell stationed over the continent, where they descended towards the land surface[36]. Heavily charged with the acid aerosols and other volatiles produced by the Laki eruption and eruptions in Italy[9] the presence of the anti-cyclone undoubtedly ensured that pollution, in greater or lesser amounts, was delivered to virtually the entire continent in June and July 1783. The proximal and distal impacts of the gases emitted will be considered below.

3. *Eruptive Emissions: Proximal Damage*

An outline account of the proximal damage is presented here to allow the reader to compare the environmental impacts which occurred in Iceland with those which occurred on the European mainland. Readers wishing more detail should turn to Thorarinsson[4], and other references made below.

"The grass was singed and seemed to wither and stopped growing" The Sheriff of Rangárvallsýsla[6].

"The air has smelt strongly of sulphur... The mouths of sheep and cattle have also been yellow with sulphur... and all the animals have been restless" The Sheriff of snæfellsness[6].

"Leaves on trees withered so that in June it looked as though it was far into October. The grass in the homefields became pale and in some places stopped growing." The Sheriff of Suður Múlasýsla[6].

"The pestilential effect of the fire caused death and damage to horses sheep and cattle after the following fashion. The horses lost all flesh; on some the hide rotted all along the back; manes and tails decayed and came off at a sharp pull. Knotty growths appeared about the joints particularly around the fetlock. The head swelled inordinately, whereupon followed a paralysis of the jaw, so that the beasts could not graze or feed, for what they were able to chew dropped from their mouths again. The entrails corrupted, the bones withered, quite drained of marrow... Sheep suffered yet more grievous harm; there was hardly a member but knotted, particularly the jaws, so that the knots pushed out through the skin close to the bone. Brisket, hips and legs—around those parts large bony excrescences developed which caused the leg bones to curve or else deformed them in various ways. Bones and knuckles as soft as if chewed lungs, liver and heart, in some swollen, in some shrivelled, the entrails rotten and soft, full of sand and worms. The shred of flesh that remained was after the same fashion. What passed for meat was both rank and bitter, and thereto full of strong poison, wherefore the eating of it proved the death of many a man, notwithstanding that people tried to cure it, clean it and salt it, according as their skill and means permitted" Jon Steingrimsson, Sida district[4].

Translated in Thorarinsson[4] this account continues at length and makes grim reading.

In Iceland during the summer of 1783 grassland was poisoned and, as a consequence, heavy mortality was sustained among herds and flocks of grazing animals. Moreover, it has been estimated that 24% of the human population eventually died[4–6]. The brief palimpsest of Icelandic material presented above suggests that the gases emitted by the Laki fissure were extremely toxic and included fluorine and sulphur. The impact of the volcanic gases upon the relatively proximal environment of Iceland was severe, as will be seen below the environmental impact of the toxic gases in Europe was hardly less so.

4. *Eruptive Emissions: Distal Damage*

Beyond Iceland, in the summer of 1783, a persistent dry fog was reported from many areas of Europe (Fig. 1). This fog, which is described in detail below, possessed a strong fetid smell and affected human health, damaged plants and killed insects. The material presented below graphically illustrates the contribution of toxic volcanic gases to the European environment.

(a) Descriptions of the Dry Fog

(i) France

"This year there appeared, in the month of June, a thick fog, which hung between the sky and the earth, which was often called a dry fog because it did not moisten the earth, and was often mistaken for a thick smoke" M. Oueilhe, Curé of Larmon, near Toulouse[37].

"...a fog which arose at the beginning of June and which turned the sun to the colour of blood, This fog lasted until the last days of July and then returned at the end of August and continued well into September. It was thick, dry and had a sulphurous odour. We could see no break in this cloud and one man could not see another at 300 paces" M. Picard Curé of Oinville[37].

"The fog was accompanied by an very intense odour of sulphur and creosote"
"The odour of sulphur was very strong"
"Similar to the smell of coal, others say the fog smelt like a forge or burning horn"

Descriptions of the fog in France assembled by Harreaux[38].

(ii) Germany

Embden, July 12th: "The thick and dry mist which has continued so long seems spread over the whole of Europe... during the day it veils the sun and in the evening there is a tainted odour"[39].

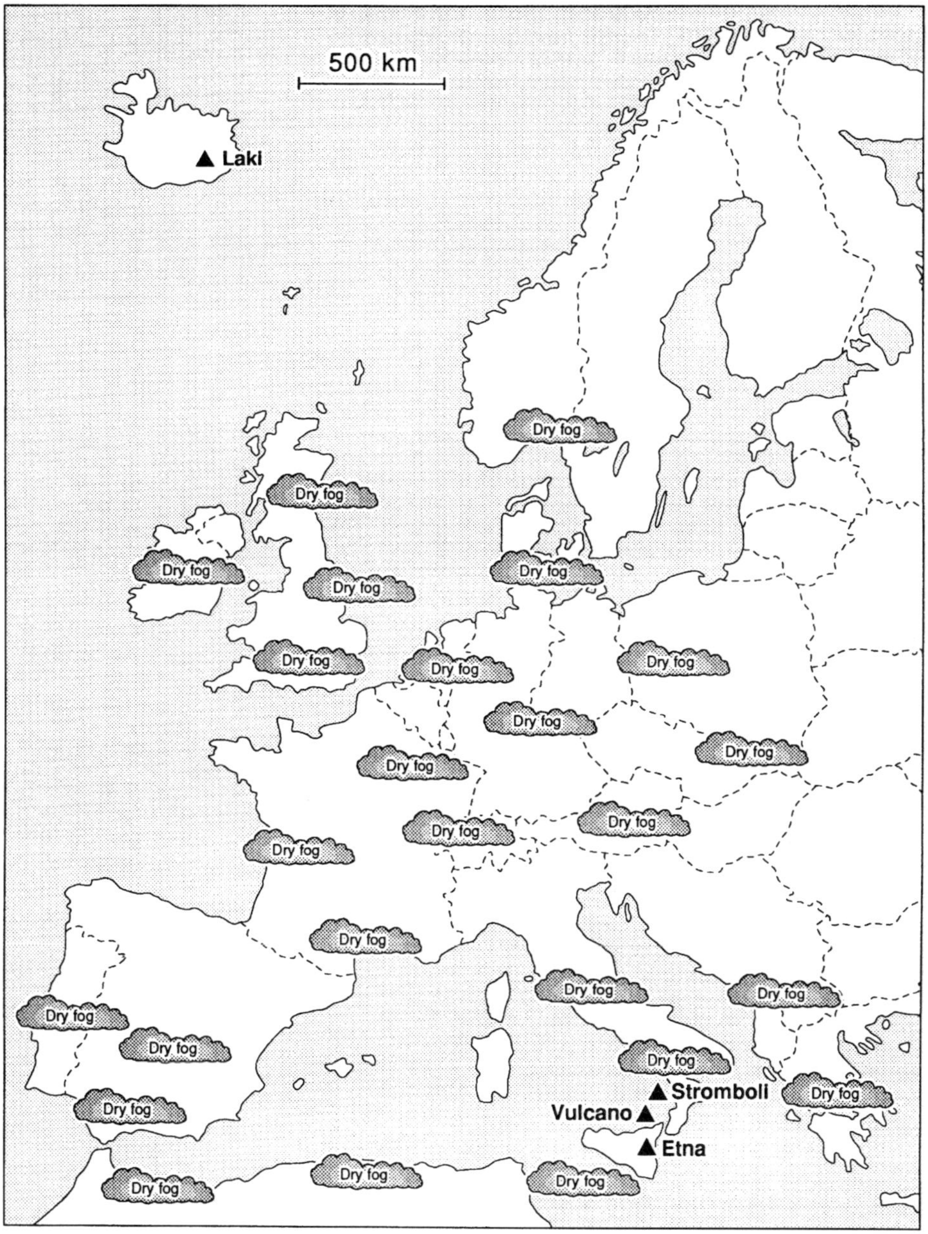

Figure 1 The extent of the dry fog in Europe during the summer of 1783.

(iii) Great Britain

"The state of the atmosphere for this week past has been more remarkably close and thick than was ever observed at this season. Such a haziness has prevailed, that the hills, at two or three miles distance, have not been discernible, and the appearance of the sun has been like that of a faint ball of fire, without a ray darting from it"[40]

"...the peculiar haze or smoky fog, that prevailed for many weeks in this island and in every part of Europe, and even beyond its limits, was a most extraordinary appearance, unlike anything known within the memory of man. By my journal I find I had noticed this strange occurrence from June 23 to July 20 inclusive"[41].

(iv) Italy

"...the moon appeared ruddy and... the sun could be looked at without being blinded. The fog was hot, dry and dense, and this phenomenon was observed not only by us, but also elsewhere in Italy, Germany and France"[9].

(v) Netherlands

"On many days after the 24th June, in both the town and countryside there was a strong, persistent fog... the fog was very dense and accompanied by a very strong smell of sulphur"[42]

"The month of June was not noteworthy until the 19th when a remarkable fog contaminated all of Europe... was different from other fogs in its constancy, its density and its extreme dryness. The hygrometer indicated an excessive dryness after the 23rd when the fog continued to increase in strength. These conditions persisted for the whole month. The sun, seen through the fog appeared to have a red face and was without strength,... nor was it possible to make out objects viewed at a distance without making a considerable effort. ...from the morning of the 24th it was accompanied by a very perceptible odour of sulphur, which even penetrated into houses"[43].

(vi) Mediterranean and North Africa

"By the late mails from Africa it appears that the fogs in summer were thicker and more suffocating all along their coasts than with us in England, and that in the archipelago and along the Mediterranean sea they were so thick as to render communication dangerous"[44].

From Scandinavia in the north to the southern coast of the Mediterranean the descriptions of the fog exhibit a remarkable consistency. The fog was thick and dry, it was persistent and it possessed a fetid sulphurous smell. It was observed over both land and sea and reduced visibility dramatically. This consistency of these accounts suggests that the fog was spread over much of Europe, further descriptions are presented in Brayshay and Grattan[45].

(b) Impacts on Animals and Human Health

The geographical extent of descriptions of damage to human and animal health are less widespread than the descriptions of the dry fog, but they are consistent, severe, and are very similar to the impacts described today, of high intensity pollution events of anthropogenic origin[46,47].

(i) Netherlands

"After the 24th, (June) many people in the open air experienced an uncomfortable pressure, headaches and experienced a difficulty breathing exactly like that encountered when the air is full of burning sulphur, asthmatics suffered to an even greater degree: horses, cattle and sheep were not affected but the fog brought about a great extermination of insects, particularly amongst leaf aphids"[42].

"Those people with weak chests experienced a similar sensation to that experienced when exposed to burning sulphur"[43].

(ii) France

"While the sun was obscured there was a sickness which caused innumerable deaths". The Curé of Broué[37].

"To the beginning of March 1784 and through the previous summer the parish of Champersu has been afflicted by an pestilence which afflicted the throat. One believes that the dry fogs of May, June, July and August (1783), that turned the sun as red as blood delivered this curse" M. Dreux, Curé of Umpeau[37].

"The fogs have been followed by great storms and sicknesses which have driven a third of the men in many parishes to their tombs". The Curé of Landelles[37].

"... the fog ... is the cause of the tingling felt in the hands, eyes, lips and throat"[38].

(iii) Italy

Gennari, clearly associated the fog with an increase in deaths:

"This month ... relentless diseases raged and many people died within a few days"[9].

Asthma, skin and eye irritation, irritation of mucous membranes and even death are indicated above. The impact of the fog on human health will be discussed in detail below. However, the symptoms described suggest that the sulphur dioxide concentration in the fog was in excess of 250 μg/m^3 which is the threshold for the inducement or intensification of asthma[48,49]. The increase in mortality described in France and northern Italy is less readily ascribed to the influence of the toxic gases within the fog than the asthma outbreak in the Netherlands. However, it is well known that severe air pollution events often result in short lived dramatic increases in mortality such as have occurred in Athens[50], Detroit[51], and London[52].

(c) Impacts on Vegetation

The descriptions of damage to vegetation presented here are also consistent with a very intense episode of environmental pollution which extended from Scandinavia to the Mediterranean sea (Fig. 2). The phenomena described below are very similar to present day results of industrial air pollution. However, in 1783 the source was volcanic, not anthropogenic, and on a scale beyond the current ability of Europe's industries to generate in the Eighteenth Century.

(i) France

"This fog was nearly always dry and damaged the corn which yielded hardly any crop" M. Picard Curé of Oinville[37].

"The fog had dried the wheat so badly that there was scarcely any yield" The Curé of Oinville St. Liphard[37].

(ii) Germany

"all the trees on the borders of the (river) Ems have been stripped of their leaves in one night"[53].

(iii) Great Britain

"On Monday night last [June 23rd], a very sudden and extraordinary alteration in the appearance of the grass and corn growing in this neighbourhood ... in so much that the grazing land, which only the day before was full of juice and had upon it the most delightful verdure, did, immediately after this uncommon event, look as if it had dried up by the sun, and was to walk on like hay. The beans were turned to a whitish colour, the leaf and blade appearing as if dead"[54].

"On Wednesday June 25th it was first observed here, and in this neighbourhood, that all the different species of grain, viz, wheat, barley, and oats, were very yellow, and in general to have had all their leaves but their upper ones in particular, withered, within two or three inches at their ends; the forward barley and the oats most so ... their awns appeared ... withered also. Many of the oats' ... chaff husks

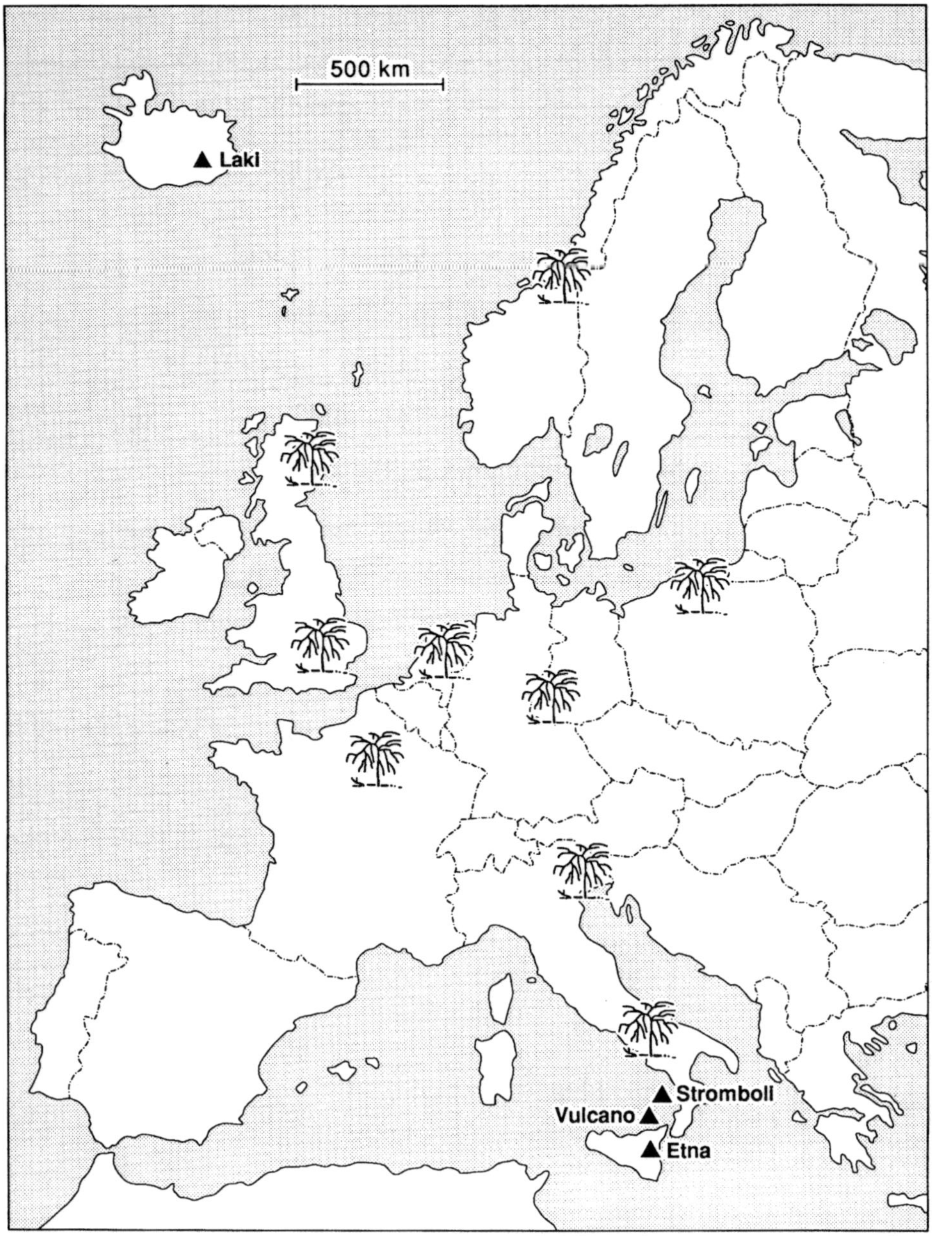

Figure 2 Extent of environmental damage reported during the dry fog of 1783.

were withered in like manner; ... About this time, and for 3 days both before and after, there was an uncommon gloom in the air, with a dead calm. The dews were very profuse. The sun was scarce visible even at mid-day, and then entirely shorn of its beams so as to be viewed by the naked eye without pain"[55].

"The aristæ of the barley, which was coming into ear, became brown and weathered at their extremities, as did the leaves of the oats; the rye had the appearance of being mildewed; so that the farmers were alarmed for those crops. ... The Larch, Weymouth Pine, and hardy Scotch fir, had the tips of their leaves withered. The leaves of some ashes very much sheltered in my garden suffered greatly. ... Cherry-trees, a standard peach tree, filbert and hasel-nut-trees, shed their leaves plentifully, and littered the walks as in autumn. ... All these vegetables appeared exactly as if a fire had been lighted near them, that had shrivelled and discoloured their leaves"[56].

(iv) Netherlands

"On the morning of the 28th, the leaves of many trees were faded, grass and vegetables appeared likewise. Leaves and fruits fell as if in autumn: afterwards the whole countryside looked desolate"[43].

"On the morning of the 25th the land offered an aspect of severe desolation, the green colour of the plants had disappeared and everywhere the leaves were dry, just as in October or November; ... This affected a wide variety of plants: some were covered in spots, others changed gradually while some leaves dried up completely. Some leaves did not entirely deteriorate and these continued to grow, but their leaf tips were decayed. Another noticeable change was that in a moment the colour could change from green to brown, black, grey or white. Others kept their natural colour but overnight on the 25th their tips were wizened. Afterwards a great quantity of leaves fell"[42].

(v) Norway

"Here we could only smell the smoke and yet we got sick. It was the fog which approached the country, which must have diluted the poison, it fell on the leaves of various vegetation which then withered. It is, therefore, no wonder that the valleys and fields in Iceland were quite destroyed and became useless as pastures for the cattle" Brun 1786[7].

Brugmans[42] studied the effects of the "sulphuric smog" on over 200 species of plants in the Netherlands. Highly susceptible species included trees such as *Betula alba*, *Corylus avellana*, *Fagus castanaea*, *Pinus sylvestris*, *Populus alba*, *Salix*, *Tilia europea*, *Cedrus*, *Juglans regia* and garden flowers such as roses, *Calendula*, *Centaurea* and *Paeonia*. The shedding of leaves, chlorosis, leaf lesions, bleaching, burning and chronic leaf deterioration are all described above in graphic detail and suggest that ammonia, carbon monoxide, fluorine, hydrochloric acid, hydrofluoric acid, hydrogen sulphide, sulphur dioxide and sulphuric acid may have been present in the dry fog.

TOXIC COMPONENTS OF THE 1783 DRY FOG

We may infer much about the nature and composition of the fog via table one and modern studies of the impact of air pollution on visibility, human health and vegetation.

A persistent reduction in visibility has been observed to occur when increased levels of sulphates are present in the lower atmosphere[57]. Stothers[58] has calculated that visibility at the surface was reduced to as low as 2 km and that the optical depth of the atmosphere may have been as high as 4 in some areas of Europe (for a detailed discussion see Stothers[58]). Many of the symptoms described above, the shedding of leaves, chlorosis and dramatic changes of colour suggests that a suite of toxic volatile gases were present in the dry fog. A study of table one indicates that these were primarily sulphur and fluorine. These deductions are in broad agreement with the emissions of the Laki fissure reported by Devine *et al.*[29].

All the reported symptoms suggest that a broad suite of acid volatiles were present in sufficient concentration to cause serious plant damage[22,59]. Chlorosis and the shedding of leaves is a classic response to concentrations of fluorine and hydrofluoric acid, and charring is typical of damage caused by a sulphuric acid aerosol[11]. The damage to the Scotch fir reported by Sir John Cullum[56] is typical of the damage caused by the absorption of sulphur dioxide[60]. Leaf lesions may be observed at a pH < 3.5 and serious leaf damage will occur if pH ⩽ 2.8[61].

If we refer to modern analogues[48–52] the shortness of breath experienced by many people and the worsened condition of asthma sufferers it may be deduced that the 24 hour mean concentrations of sulphur dioxide exceeded 250 μg/m^3.

DISCUSSION

It is clear that during June and July 1783, Europe suffered an environmental pollution event of a scale and magnitude hitherto unimagined, and certainly beyond the efforts of the fledgling industries of Britain and Europe to then generate. The only conceivable source for this pollution was volcanic activity which occurred at this time, the Laki Fissure eruption in Iceland and perhaps smaller eruptions of Vesuvius, Stromboli and Vulcano in Italy.

The documentary material presented above demonstrates that erupted volcanic gases may be concentrated within the troposphere by regional air circulation and concentrated to a sufficient degree to cause severe environmental pollution at great distances from the eruptive source. The volcanic gases emitted in 1783 were concentrated in an atmosphere which contained very low quantities of gas which were anthropogenic in origin, yet the environmental impact of these gases were severe. The exceedence of safety guidelines for atmospheric pollution in European cities is now common, particularly under stable atmospheric conditions[50,62–64], precisely the conditions which favour the concentration of volcanic gases in toxic concentrations. Were the events of 1783 to be repeated today, the volcanic material transported to Europe would necessarily enhance the already polluted air found in European conurbations. It should not be assumed that volcanogenic dry fogs are only generated by rare high magnitude low-frequency events as the Laki fissure eruption. Camuffo and Enzi[9] have documented frequent and severe environmental impacts following the eruption of relatively minor Italian volcanoes and nineteen episodes of crop damage and dry fogs have been documented between 1374 and 1819.

This paper has outlined the historical impact of volcanic gases in the European environment and has suggested that volcanic eruptions may be a major episodic point source for air pollution. The concentration of toxic volcanic gases, at great distances from the eruption, may exceed the threshold for plant and animal damage. The environmental impact of volcanic gases may therefore be severe and notable impacts on human health and vegetation may be anticipated. Should a sulphur producing volcanic eruption occur in Iceland or Italy, at a time when human activity and atmospheric circulation have already created a stressed or marginal urban air environment then a major health hazard may occur.

References

1. J. P. Garrec, A. Lounowski and R. Plebin, “The influence of volcanic fluoride emissions on the surrounding vegetation” *Fluoride* **10**, 153–156 (1977).
2. R. H. Mandl, L. H. Weinstein and M. Keveny, “Effects of hydrogen fluoride and sulphur dioxide alone and in combination on several species of plants” *Environmental Pollution* **9**, 133–143 (1975).
3. T. Takamatsu, J. Boratynski, and K. Satake, “Effects of volcanic acid deposition on Soil Chemistry: I. Status of exchangeable cations and sulfur” *Soil Science* **154**(6), 435–448 (1992).
4. S. Thórarinsson, “On the damage caused by volcanic eruptions with special reference to tephra and gases” **In**: (D. K. Grayson and P. D. Sheets, eds.) *Volcanic Activity and Human Ecology* (London, Academic Press, 1979) pp. 125–159.

5. G. A. Gunnlaugsson, G. M. Gudbergsson, S. Thórarinsson, S. Raffnson and T. Einarsson, (eds.), *Skáftareldar 1783–1784* (Rekyavik. Mal Og Menning, 1984) (In Icelandic with English summaries).
6. A. E. J. Ogilvie, "The climate of Iceland 1701–1784" *Jökull* **36**, 57–73 (1986).
7. S. Thórarinsson, "Greetings from Iceland. Ash falls and volcanic aerosols in Scandinavia" *Geografiska Annaler* **63A**, 109–118 (1981).
8. J. P. Grattan and F. B. Pyatt, "Acid damage in Europe caused by the Laki Fissure eruption—an historical review" *The Science of the Total Environment* **151**, 241–247 (1994).
9. D. Camuffo, and S. Enzi, "Impacts of the clouds of volcanic aerosls in Italy during the last seven centuries" *Natural Hazards* **11**, 135–161 (1995).
10. R. E. Wilcox, "Some effects of the recent volcanic ash falls with special reference to Alaska" *US Geological Survey Bulletin* **1028-N**, 409–476 (1959).
11. R. J. Blong, *Volcanic Hazards: a source book on the effects of eruptions* (Sydney, Academic Press, 1984).
12. K. Bull, "Critical load maps for the U.K." *NERC news* 31–32 (July, 1991).
13. B. Fisher, "Acid rain—the long range transport of air pollutants" *Weather* **36**, 367–369 (1981).
14. J. P. Grattan, and D. J. Charman, "Non-climatic factors and the environmental impact of volcanic volatiles: implications of the Laki Fissure eruption of AD 1783" *The Holocene* **4**(1), 101–106 (1994).
15. N. Z. Heidam, P. Wahlin and K. Kemp, "Arctic Aerosol studies in Greenland" *Abstracts of papers of the American Chemical Society* **201**, 34–35 (1991).
16. B. C. Halter, J. M. Harris and K. A. Rahn, "A study of winter variability in carbon dioxide and arctic haze aerosols at Barrow, Alaska" *Atmospheric Environment* **19**, 2011–2024 (1985).
17. P. D. Fowler, "Transfer to terrestrial surfaces" *Phil. Trans. R. Soc. Lond.* **B305**, 281–297 (1984).
18. T. D. Davies, P. W. Abrahams, M. Tranter, I. Blackwood, P. Brimblecombe and C. E. Vincent, "Black acidic snow in the remote Scottish Highlands" *Nature* **312**, 58–61 (1984).
19. R. M. Koerner and D. Fisher, "Acid snow in the Canadian high arctic" *Nature* **295**, 137 (1982).
20. N. Oskarsson, "Aluminium Fluoride in a volcanic cloud: Evidence from the 1991 Hekla eruption" *Proceedings of the Nordic Geological Conference* **1991**, 129 (1992).
21. J. P. Grattan, "The distal impact of volcanic gases and aerosols in Europe: A reveiw of the phenomenon and assessment of vulnerability in the late Twentieth Century" **In**: (J. G. Maund, S. Penn and M. G. Culshaw, eds.) *Geohazards and Engineering Geology. Proceedings of the 31st Annual conference of the Engineering group of the Geological Society*, 1995) pp. 123–136.
22. J. Wisniewski, "The potential acidity associated with dews, frosts and fogs" *Water, Air, and Soil Pollution* **17**(4), 361–377 (1982).
23. P. Allard, J. Carbonelle, D. Dajlevic, J. Le Bronec, P. Morel, M. C. Robe, J. M. Maurenas, R. Faivre-Pierret, D. Martin, J. C. Sabroux and P. Zettwoog, "Eruptive and diffuse emissions of CO2 from Mt Etna" *Nature* **351**, 387–391 (1991).
24. R. A. Reinert, "Plant responses to air pollutant mixtures" *Ann. Rev. Phytopath.* **22**, 421–442 (1984).
25. F. Le Guern, R. Faivre-Pierret and J. P. Garrec, "Atmospheric contribution of volcanic sulphur vapour and its influence on the surrounding vegetation" *J. Volcanol. Geotherm. Res.* **35**, 173–178 (1988).
26. S. Thórarinsson, "The Lakagigar eruption of 1783" *Bulletin Volcanologique* **33**(3), 910–929 (1969).
27. Th. Thordarson and S. Self, "The Laki [Skaftár Fires] and Grímsvötn eruptions in 1783–85" *Bulletin Volcanologique* **55**, 233–263 (1993).
28. Th. Thordarson, S. Self, N. Oskarsson and T. Hulsebosch, "Sulfur, chlorine, and fluorine degassing and atmospheric Loading by the 1783–1784 AD Laki (Skaftar fires) eruption in Iceland" *Bulletin Of Volcanology* **58**, 205–225 (1996).
29. J. D. Devine, H. Sigurdsson, A. N. Davis, and S. Self, "Estimates of Sulfur and Chlorine yield to the Atmosphere from volcanic eruptions and potential climatic effects" *Journal of Geophysical Research* **89**, 6309–6325 (1984).
30. Á. Hjartanson, "Environmental changes in Iceland following the great Þjórsá Lava eruption 780014C years BP' *Münchener Geographische Abhandlungen* **B12**, 147–156 (1994).
31. G. C. Newhall and S. Self, "The volcanic explosivity index (VEI): An estimate of explosive magnitude for volcanism" *Journal of Geophysical Research* **87**(C2), 1231–1238 (1982).
32. J. M. Palais and H. Sigurdsson, "Petrologic evidence of volatile emissions from major historic and pre-historic volcanic eruptions" **In**: (A. Berger, R. E. Dickinson and J. W. Kidson, eds.) *Understanding Climate Change. Geophysical Monograph 52* (IUGG 7, 1989) pp. 31–53.
33. H. Sigurdsson, J. D. Devine and A. N. Davis, "The petrologic estimate of volcanic degassing" *Jökull* **35**, 1–8 (1985).
34. R. J. Fiacco, Th. Thordarson, M. S. Germani, S. Self, J. Palais, S. Whitlow and P. Groutes, "Atmospheric aerosol loading and transport due to the 1783–84 Laki fissure eruption in Iceland, interpreted from Ash palticles and acidity in the GISP 2 ice-core" *Quaternary Research* **42**(3), 231–244 (1994).

35. J. A. Kington, *The Weather for the 1780s over Europe* (Cambridge, C.U.P. 1988).
36. J. P. Grattan and M. B. Brayshay, "An Amazing and Portentous summer: Environmental and social responses in Britain to the 1783 eruption of an Iceland Volcano" *The Geographical Journal* **161**(2), 125–134 (1995).
37. R. Rabartin and P. Rocher, *Les Volcans et la Révolution Française* (Paris, L'Association Volcanologique Européene 1993).
38. M. Harreax, "Du Brouillard Sec" *Mémoirs, Société Archéologique d'Eure et Loire* **4**, 30–31 (1858).
39. *The Morning Herald and Daily Advertiser* (August 5th, 1783).
40. *The Bristol Journal* (July 19th, 1783).
41. G. White, *The Natural History of Selbourne* (1789) (Reprinted London, Penguin, 1977).
42. S. J. Brugmans, *Natuurkundige verhandeling over een zwavelagtigen nevel den 24 Juni 1783 innde provincie van stad en lande en naburige landen waargenomen.* (A physical treatise on a sulphuric smog as observed on the 24th of July 1783 in the province of Groningen and neighbouring countries): Leyden, 58pp. (1787).
43. Mr. V. Swinden, "Observations sur quelques particularités météorologiques de l'année 1783" *Memoires de l 'Académie Royale des Sciences, Turin*, **Anneés 1784–1785**, 113–140 (1786).
44. *The Gentleman's Magazine* (September, 1783).
45. M. Brayshay and J. P. Grattan. "Environmental and social responses in Europe to the 1783 eruption of the Laki Fissure volcano in Iceland: A consideration of contemporary documentary evidence" *Volcanoes in the Quaternary* (Geological Society of London, In Press).
46. J. G. Ayres, "Health effects of air pollution" *Chemistry and Industry* **21**, 827–830 (1996).
47. M. T. Krishna and A. J. Chauhan, "Air pollution and health" *Journal of the Royal College of Physicians of London* **30**, 448–452 (1996).
48. J. Koenig, W. E. Pierson and R. Frank, "Acute effects of inhaled sulphur dioxide plus sodium chloride aerosol on pulmonary function in asthmatic adolescents" *Environmental Research*, **22**, 145–153 (1979).
49. World Health Organization, *Environmental Health Criteria Volume 8, Sulfur oxides and suspended particulate matter* (Geneva, WHO, 1979).
50. Y. Klidonas, "The quality of the atmosphere in Athens" *The Science of the Total Environment* **129**, 83–94 (1993).
51. J. Schwartz and R. Morris, "Air pollution and hospital admissions for cardiovascular disease in Detroit, Michigan" *American Journal of Epidemiology* **142**, 23–35 (1995).
52. E. T. Wilkins, "Air pollution aspects of the London fog of December 1952" *Quarterly Journal of the Royal Meteorological Society* **80**, 267–271 (1954).
53. *The Ipswich Journal* (August 9th, 1783).
54. *Cambridge Chronicle and Journal* (July 5th, 1783).
55. *The Ipswich Journal* (July 12th, 1783).
56. Rev. Sir J. Cullum, "Of a remarkable frost on the 23rd of June, 1783" *Philosophical Transactions of the Royal Society*. **Abridged volume 15**, 604 (1785).
57. B. D. Clarkson and B. R. Clarkson, "Vegetation decline following recent eruptions on White Island, Bay of Plenty, New Zealand" *New Zealand Journal of Botany* **32**, 21–36 (1994.).
58. R. B. Stothers, "The Great Dry Fog of 1783" *Climatic Change* **32**, 79–89 (1996).
59. J. N. Cape, "Direct damage to vegetation caused by acid rain and polluted cloud" *Environmental Pollution* **82**, 167–180 (1993).
60. C. Caput, Y. Belot, D. Auclair and N. Decourt, "Absorption of Sulphur Dioxide by Pine needles leading to acute injury" *Environmental Pollution* **16**, 3–15 (1978).
61. Watt Committee On Energy, *Report number 14. Acid Rain* (London, The Watt Committee on Energy Limited, 1984).
62. J. S. Bower, K. J. Stevenson, G. F. Broughton, J. Vallance-Plews, J. E. Lampert, B. P. Sweeney, S. W. Eaton, P. G. Willis, B. Stacey, G. S. Driver, S. E. Laight, R. Berwick and M. S. Jackson, *Air Pollution in the UK 1992/93* (Stevenage, Warren Spring Laboratory, 1994).
63. UNEP/WHO. *Urban Air Pollution in the Mega cities of the World* (Oxford, Blackwell, 1993).
64. M. Shahgedanova and T. P. Burt, "New Data on Air pollution in the former Soviet Union" *Global Enviqronmental change* **4**, 201–227 (1994).
65. W. F. Giggenbach, "Chemical Composition of volcanic gases" **In**: (R. Scarpa and R. I. Tilling, eds.) *Monitoring and Mitigation of Volcanic Hazards* (Berlin, Springer Verlag, 1996) pp. 221–256.

6. ECONOMIC AND ENVIRONMENTAL COSTS OF PESTICIDE USE*

DAVID PIMENTEL, ANTHONY GREINER and TAD BASHORE**

INTRODUCTION

Approximately three million tons of pesticides are applied throughout the world each year[1] at a current purchase price of $29 billion[2]. These three million tons are comprised of roughly 1600 different chemicals[3]. In 1995, sales of pesticides in the United States reached 625,000 tons of active ingredients and a conservatively estimated $6.7 billion[4]. Approximately 600 different types of pesticides are used annually in the U.S.[5].

Despite the widespread use of pesticides in the United States, pests such as insects, plant pathogens, and weeds destroy 37% of all potential food and fiber crops[6]. It is estimated that pest levels would increase 10% if no pesticides were used at all, while specific crop losses would range from zero to nearly 100%[7]. Pesticides, therefore, make a significant contribution to maintaining world food production. In general, each dollar invested in pesticide control returns about $4 in crops saved[8].

The attitude towards pesticides and other chemicals has changed, however, since the publication of Rachel Carson's *Silent Spring* in 1962. Although pesticides are generally profitable, their use does not always decrease crop losses. For example, even with the 10-fold increase in insecticide use in the United States from 1945 to 1989, total crop losses from insect damage have nearly doubled from 7% to 13%[9]. This rise in crop losses to insects is, in part, caused by changes in agricultural practices. For example, the practice of rotating corn with other crops has been replaced with the continuous production of corn on about half of its total hectarage and has resulted in nearly a fourfold increase in corn losses to insects despite a more than 1000-fold increase in insecticide use in corn production[10].

Most benefits of pesticides are based only on direct crop returns. Such assessments do not include the indirect environmental and economic costs associated with pesticides. It has been estimated that only 0.1% of applied pesticides reach the target pests, leaving the bulk of the pesticides (99.9%) to impact the environment[11]. To facilitate the development and implementation of a balanced and sound policy of pesticide use, these costs must be examined. Twenty years ago the U.S. Environmental Protection Agency identified the need for such a risk investigation[12], but only a few papers concerning this difficult topic have been published; recently, however, the EPA, FDA, and USDA have jointly announced that emergency exemptions for "cancer-causing" pesticides will no longer be given[13]. Additionally, in 1995 the EPA and the chemical manufacturer DuPont reached an agreement which will result in the elimination of the potentially carcinogenic herbicide cyanazine by the year 2002[14]. Perhaps these are examples of the first major steps towards severely limiting hazardous pesticide use.

---

* Portions of this chapter are reprinted with the permission of Chapman & Hall, Inc., New York, NY (USA). Their portions appeared in *The Pesticide Question* (Routledge, Chapman & Hall, 1993).

** College of Agriculture and Life Sciences, Cornell University, Ithaca, New York 14853-0901 (USA).

This chapter is an updated version of Pimentel *et al.*'s comprehensive investigation[15] of the effects of the nation's dependence on pesticides.

PESTICIDES AND PUBLIC HEALTH

Human poisonings and their related illnesses are clearly the highest price paid for pesticide use. About 110,000 pesticide poisonings are reported each year in the U.S.[16] resulting in approximately 20 accidental fatalities per year, the majority of which are children[17]. Due to gaps in the demographic data, however, this figure may represent only 80% of the total number of poisonings[18]. Worldwide, an estimated 3 million severe human pesticide poisonings occur each year of which approximately 220,000 are fatal[19].

Although it is impossible to place a precise monetary value on human life, the "cost" of human pesticide poisonings has been estimated. Insurance industry studies have computed monetary ranges between $1.6 and $8.5 million for the value of a "statistical life"[20]. Alternatively, the conservative estimate of $2.2 million per human life—the average value that the surviving spouse of a slain New York City policeman receives—may be used[21]. Based on this figure and the available data, human pesticide poisonings and related illnesses in the U.S. are estimated to total about $933 million each year (Table I).

The situation is even worse in other regions of the world. Approximately 80% of the pesticides produced annually in the world are used in developed countries[22], but less than half of all pesticide-induced deaths occur in these countries[23]. A higher proportion of pesticide poisonings to deaths occur in developing countries where there tend to be inadequate occupational safety standards, protective clothing, and washing facilities; insufficient enforcement; poor labeling of pesticides; illiteracy; and

Table I Estimated economic costs of human pesticide poisonings and other pesticide-related illnesses in the United States each year

Effects of pesticides on human health	Total costs ($)
Cost of hospitalized poisonings	
2380[a] × 2.84 days @ $1000/day	6,759,000
Cost of outpatient-treated poisonings	
27,000[c] × $630[b]	17,010,000
Lost work due to poisonings	
4680[a] workers × 4.7 days × $80/day	1,760,000
Pesticide cancers	
<12,000[d] cases × $70,700[c]/case	848,400,000
Cost of fatalities	
27 accidental fatalities[c] × $2.2 million	59,400,000
Total	933,329,000

[a] T. J. Keefe, E. P. Savage, S. Munn and H. W. Wheeler, *Evaluation of Epidemiological Factors from Two National Studies of Hospitalized Pesticide Poisonings, U.S.A.* (Exposure Assessment Branch, Hazard Evaluation Division, Office of Pesticides and Toxic Substances, U.S. Environmental Protection Agency, Washington, DC, 1990).
[b] Includes hospitalization, foregone earnings, and transportation.
[c] J. Blondell, EPA, Washington, DC (personal communication, 1991).
[d] See text for details.

insufficient knowledge of pesticide hazards[24]. In China, for example, reports from the Chinese National Statistics Bureau revealed the occurrence of 48,377 pesticide poisonings resulting in 3204 fatalities in 1995 (R. Roush, University of Adelaide, Australia, personal communication, 1996).

Additionally, average pesticide residue levels in food are often higher in developing countries than in developed nations. For example, a study in Egypt reported that a majority of assayed milk samples, when tested for fifteen different pesticides, contained residue levels between 60–80%[25]. By way of contrast, 50% of the milk samples analyzed in a U.S. milk study had pesticide residues, all in trace quantities well below EPA and FDA regulatory limits[26].

About 35% of the foods purchased by American consumers, however, do have detectable levels of pesticide residues[27]. Between 1–3% of these foods have pesticide residue levels that are above the legal tolerance level[28]. Residue levels may be even higher than this because the analytical methods now employed in the U.S. detect only about one-third of the more than 600 pesticides in use[29]. Current U.S. regulations allow the residues of 325 of these pesticides to remain in the food supply[30]. The contamination rate is undoubtedly higher for fruits and vegetables because these foods receive the highest dosage of pesticides. In fact, one USDA study has shown that some pesticide residue remains in fruits and vegetables even after they have been washed, peeled, or cored[31]. Consequently, 97% of the public is concerned about pesticide residues in its food[32].

Throughout the world, the highest levels of pesticide exposure are found in farm workers, pesticide applicators, and people who live adjacent to heavily treated agricultural land. Because farmers and farm workers directly handle 70–80% of the pesticides they use, they are at the greatest risk of exposure[33]. The epidemiological evidence suggests a significantly higher rate of cancer incidence among farmers and farm workers in the U.S. and Europe than among non-farm workers in some areas[34]. In these high-risk populations, there is strong evidence for associations between lymphomas and soft-tissue sarcomas and certain herbicides[35], as well as between lung cancer and exposure to organochlorine insecticides[36].

Consequently, both the acute and chronic health effects of pesticides warrant attention and concern. While the acute toxicity of most pesticides is well documented[37], information on chronic human illnesses such as cancer is not as sound[38]. For example, based on animal studies, the International Agency for Research on Cancer found "sufficient" evidence of carcinogenicity in eighteen pesticides and "limited" evidence in an additional sixteen pesticides[39]. According to Wargo[40], EPA toxicologists estimate that one-third of the 325 pesticides which are allowed to remain on food as residues produce "some tumorigenic evidence" when tested on laboratory animals at the maximum-tolerated dosage. Similarly, studies have reported an increased prevalence of certain cancers in farmers[41]. However, a recent study in Saskatchewan found no significant difference in non-Hodgkin's lymphoma mortality between farmers and non-farmers[42], and D. Schottenfeld of the University of Michigan (personal communication, 1991) estimates that fewer than 1% of the human cancer cases in the U.S. are attributable to pesticide exposure. Since there are approximately 1.2 million new cancer cases annually[43], Schottenfeld's assessment suggests that less than 12,000 cases of cancer per year are due to pesticides.

However, there is evidence to suggest that many other acute and chronic maladies are associated with pesticide use[44]. For example, the recently banned pesticide dibromochloropropane (DBCP), which is used for plant pathogen control, was found to cause testicular dysfunction in animal studies[45] and was linked to infertility in human workers who had been exposed to the chemical[46]. Also, a large body of evidence obtained from animal studies suggests that pesticides can produce immune dysfunction[47]. In a study of women who had chronically ingested groundwater contaminated with low levels (mean of 16.6 ppb) of aldicarb, Fiore *et al.*[48] reported evidence of significantly reduced immune response, although these women did not exhibit any overt health problems.

There is also growing evidence of sterility in humans and various other animals, particularly in males, due to various chemicals and pesticides in the environment. Sperm counts in Europe have declined by about 50% and continue to decrease an additional 2% per year. Young male river otters in the lower Columbia River and male alligators in Florida's Lake Apopka have smaller reproductive organs than males in unpolluted regions of their respective habitats[49].

Although it is often difficult to determine the impact of individual pesticides and other chemicals, the chronic health problems associated with organophosphorus pesticides—which have largely replaced the banned organochlorines—are of particular concern[50]. The malady Organophosphate Induced Delayed Polyneuropathy (OPIDP) is well-documented and is marked by irreversible neurological defects[51]. The deterioration of memory, moods, and the capacity for abstract thought have been observed in some cases[52], while other cases indicate that persistent neurotoxic effects may result even after the termination of an acute organophosphorus poisoning incident[53].

Chronic conditions such as OPIDP constitute an important public health issue because of their potential cost to society. For example, the effect of pesticides on children has become a growing concern[54]. Children can be exposed to pesticides on a daily basis in a variety of ways: through the foods they eat, in the houses where they live, or in the communities where they play[55]. With the increased realization of the distinct physiological differences between adults and children, it has become obvious that the present pesticide tolerance and regulatory system, as it relates to children, is severely lacking. All of the regulations to date have been based on adult tolerances. Children have much higher metabolic rates than adults, and their ability to activate, detoxify, and excrete xenobiotic compounds is different than that of adults. Also, because of their smaller physical size, children are exposed to higher levels of pesticides per unit of body weight. Evidence of this is found in a study which reported that 50% of all pesticide poisonings in England and Wales involved children under the age of ten[56]. The use of pesticides in the home has also been linked to childhood cancer[57]. In general, the realization that children's sensitivities to toxins are much different than those of adults has provided the impetus for the movement towards setting specific pesticide regulations with children in mind[58]. Wargo has suggested the need for increased investigations into the effects of pesticides on other age groups such as the elderly, as well as on different ethnic groups[59].

DOMESTIC ANIMAL POISONINGS AND CONTAMINATED ANIMAL PRODUCTS

Thousands of domestic animals are poisoned by pesticides each year. The majority of these poisonings involve dogs and cats (Table II)[60]. For example, nearly 40% of the 25,000 calls made in 1987 to the National Animal Poison Control Center at the University of Illinois concerned pesticide poisonings in dogs and cats[61]. Similarly, Kansas State University reported that 67% of all animal pesticide poisonings involved dogs and cats[62]. These figures are not surprising because dogs and cats

Table II Estimated domestic-animal pesticide poisonings in the United States

Livestock	Number × 1000	$ per head	Number of illnesses[f]	$ Cost per poisoning[g]	$ Cost of poisonings	Number of deaths[e]	$ Cost of deaths × 1000[h]	Total $ × 1000
Cattle	100,987[a]	659[a]	101	131.80	13,312	8	5,272	18,584
Dairy Cattle	9,528[a]	1,160[a]	10	232.00	2,320	1	1,160	3,480
Dogs	52,500[c]	125[i]	53	25.00	1,325	4	500	1,825
Horses	5,480[a]	1,000[d]	6	200.00	1,200	1	1,000	2,200
Cats	57,000[c]	20[i]	57	4.00	228	5	100	328
Swine	57,904[a]	74.90[a]	58	14.98	869	5	375	1,244
Chickens	6,689,110[a]	1.55[a]	6,690	0.31	2,074	535	830	2,904
Turkeys	287,220[a]	8.72[c]	288	1.75	504	23	201	705
Sheep	9,079[a]	70.30[a]	9	14.06	127	1	71	198
Total	7,268,808				21,959		9,509	31,468

[a] *Agricultural Statistics* (U.S. Department of Agriculture, Government Printing Office, Washington, DC, 1994).
[b] *1993 Production Yearbook* (Food and Agriculture Organization of the United Nations, Rome, 1994).
[c] *Statistical Abstract of the United States 1994* (U.S. Bureau of the Census, U.S. Government Printing Office, Washington, DC, 1994).
[d] *Statistical Abstract of the United States 1990* (U.S. Bureau of the Census, U.S. Dept. of Commerce, Washington, DC, 1990).
[e] Based on a 0.008% mortality rate (see text).
[f] Based on a 0.1% illness rate (see text).
[g] Based on each animal illness costing 20% of total production value of that animal.
[h] The death of the animal illness costing 20% of total production value of that animal.
[i] Estimated.

usually wander freely about the home and farm and, therefore, have a greater opportunity to come into contact with pesticides than other domesticated animals.

Such animals, however, are also susceptible to pesticide poisoning. For example, the American Association of Poison Control Centers reported in one study that 11,166 of 28,198 reported cases (39.6%) of exposure to non-drug products were due to pesticide poisonings. Similarly, 163 of 425 (38.4%) reported fatalities were due to pesticides[63].

The best estimates indicate that about 20% of the total monetary value of animal production, or about $4.2 billion, is lost to all animal illnesses, including pesticide poisonings[64]. Colvin reported that 0.5% of animal illnesses and 0.04% of all animal deaths reported to a veterinary diagnostic laboratory were due to pesticide toxicosis[65]. Thus, an estimated $31 million is lost to pesticide poisonings (Table II).

This estimate is conservative because it is based only on poisonings reported to veterinarians. Many animal pesticide poisonings that occur in the home and on farms go undiagnosed and are attributed to factors other than pesticides. In addition, when farm animals are poisoned and little can be done for the animals, the farmer seldom calls a veterinarian but rather waits for the animal to recover or destroys it (G. Maylin, Cornell University, personal communication, 1977). Such cases are usually unreported.

Additional economic losses occur when meat, milk, and eggs are contaminated with pesticides. Animals can become contaminated from pesticides that are either applied to crops or farm buildings for pest control, or from pesticides applied directly to the animal for veterinary purposes. While studies to date have shown that these residues do build-up in the fatty tissues of livestock, the residues are well below federal tolerance limits[66].

In the United States, all animals slaughtered for human consumption, if shipped interstate, and all imported meat and poultry must be inspected by the U.S. Department of Agriculture. This inspection is to ensure that the meat and products are wholesome, properly labeled, and do not present a health hazard. One part of this inspection, which involves monitoring meat for pesticide and other chemical residues, is the responsibility of the National Residue Program (NRP). The samples taken are intended to insure the detection of a chemical if it is present in 1% of the slaughtered animals[67]. Generally, violative residues are found less frequently in chickens than in livestock and most frequently in swine[68].

However, of the more than 600 pesticides now in use, NRP tests are made for only the 41 which have been determined by the FDA, EPA, and FSIS to be a public health concern (D. Beermann, Cornell University, personal communication, 1991)[69]. While the monitoring program records the number and type of violations, there is no significant cost to the animal industry because the meat, including poultry, is generally sold and consumed before the test results are available. About 3% of chickens with illegal pesticide residues are sold in the market[70].

Compliance sampling is designed to prevent the contamination of meat and milk. When a producer is suspected of marketing contaminated livestock, the carcasses are detained until the residue analyses are reported. If illegal residues are present, the carcasses or products are condemned and the producer is prohibited from marketing other animals until it is confirmed that all the livestock are safe[71]. If carcasses are not suspected of being contaminated, then by the time the results of the residue tests are reported the carcasses have been sold to consumers. This reliance on the suspicion of contamination is a major deficiency in the surveillance program.

In addition to animal carcasses, pesticide-contaminated milk cannot be sold and must be disposed of. In certain incidents these losses are substantial. In Hawaii in

1982, 80% of the milk supply on the island of Oahu (at a value of more than $8.5 million) was condemned by public health officials because it had been contaminated with the insecticide heptachlor[72]. This incident had immediate and far-reaching effects on the island's milk industry: reduced milk sales due to the contaminated milk were estimated at $39,000 per dairy farmer. Subsequently, the lack of consumer confidence changed the structure of the island's milk industry. Because consumers considered the island's milk to be unsafe, most of the milk supply began to be imported.

When the costs attributable to domestic animal poisonings and contaminated meat, milk, and eggs are combined, the economic value of all livestock products in the United States lost to pesticide contamination is estimated to be at least $31.5 million annually (Table II).

Similarly, other nations lose significant numbers of livestock and large amounts of animal products each year due to pesticide-induced illness or death. Exact data concerning these livestock losses do not exist and the available information comes only from reports of the incidence of mass destruction of livestock. For example, when the pesticide leptophos was used by Egyptian farmers on rice and other crops, 1300 draft animals were fatally poisoned[73]. The estimated economic losses were significant, but exact figures are not available.

In addition, countries that export meat to the United States can experience tremendous economic losses if the meat is found to be contaminated with pesticides. In a 15-year period, the beef industries in Guatemala, Honduras, and Nicaragua lost more than $1.7 million due to the pesticide contamination of exported meat[74]. In these countries, meat which is too contaminated for export is sold in local markets. Obviously such policies contribute to public health problems.

BENEFICIAL NATURAL PREDATORS AND PARASITES

Many species—especially predators and parasites—control or help to control herbivorous populations in both natural and agro-ecosystems. Indeed, these natural beneficial species theoretically make it possible for ecosystems to remain green and foliated. With the parasites and predators maintaining herbivore populations at low levels, only a relatively small amount of plant biomass is removed each growing season[75].

Natural enemies play a major role in keeping the populations of many insect and mite pests under control[76], but they can be adversely affected by pesticides[77]. For example, bollworm, tobacco budworm, cotton aphid, spider mites, and cotton loopers have reached outbreak levels in cotton crops following the destruction of their natural enemies by pesticides[78], while European red mites, red-banded leafroller, San Jose scale, oystershell scale, rosy apple aphid, wooly apple aphid, white apple leafhoppers, two-spotted spider mites, and apple rust mites have reached outbreak levels in apple crops for the same reason[79]. Significant pest outbreaks also have occurred in other crops[80]. In addition, because parasitic and predaceous insects often have complex searching and attack behaviors, sublethal insecticide dosages may alter this behavior, thus disrupting effective biological control strategies (L. E. Ehler, University of California, personal communication, 1991).

Fungicides can also contribute to pest outbreaks by reducing fungal pathogens that are naturally parasitic on many insects. For example, the use of benomyl for plant pathogen control reduces populations of entomopathogenic fungi and results in the increased survival of velvet bean caterpillars and cabbage loopers in soybeans. The increased number of insects eventually leads to reduced soybean yields[81].

When outbreaks of secondary pests occur because their natural enemies are destroyed by pesticides, additional and sometimes more expensive pesticide treatments have to be made in an effort to sustain crop yields. This raises overall costs and contributes to pesticide-related problems.

An estimated $520 million can be attributed to the cost of additional pesticide applications and increased crop losses, both of which result from the destruction of natural enemies by pesticides (Table III).

Although no reliable estimates are available concerning the impact of the loss of natural enemies in terms of increased pesticide use and/or reduced yields, general observations by entomologists indicate that the impact of the loss of natural enemies is severe in many parts of the world. For example, from 1980 to 1985 insecticide use in rice production in Indonesia dramatically increased[82]. This caused the destruction of beneficial natural enemies of the brown planthopper, and pest populations subsequently exploded. Rice yields dropped so significantly that rice had to be imported into Indonesia for the first time in many years. The estimated loss in rice in just a two-year period was $1.5 billion[83].

Following that incident, entomologist Dr. I. N. Oka and his associates (who previously had developed a successful, low insecticide program for rice pests in Indonesia) were consulted by Indonesian President Suharto's staff to determine what should be done to resolve the problem (I. N. Oka, Bogor Food Research Institute,

Table III Losses due to the destruction of beneficial natural enemies in U.S. crops ($ millions)

Crops	Total expenditures for insect control with pesticides ($)[a]	Amount of added control costs ($)
Cotton	320	160
Tobacco	5	1
Potatoes	31	8
Peanuts	18	2
Tomatoes	11	2
Onions	1	0.2
Apples	43	11
Cherries	2	1
Peaches	12	2
Grapes	3	1
Oranges	8	2
Grapefruit	5	1
Lemons	1	0.2
Nuts	160	16
Other	500	50
Total		$257.4 ($520)[b]

[a] D. Pimentel, L. McLaughlin, A. Zepp, B. Lakitan, T. Kraus, P. Kleinman, F. Vancini, W. J. Roach, E. Graap, W. S. Keeton and G. Selig, "Environmental and economic impacts of reducing U.S. agricultural pesticide use" **In**: (D. Pimentel, ed.) *Handbook on Pest Management in Agriculture* (CRC Press, Boca Raton, FL, 1991) pp. 679–718.

[b] Because the added pesticide treatments do not provide as effective control as the natural enemies, we estimate that at least an additional $260 million in crops are lost to pests. Thus, the total loss due to the destruction of natural enemies is estimated to be at least $520 million/year.

Indonesia, personal communication, 1990). Their advice was to substantially reduce insecticide use and return to a sound "treat-when-necessary" program that protected the natural enemies. Following Oka's advice, President Suharto in 1986 mandated that 57 of the 64 pesticides used on rice would be withdrawn and pest management practices improved. Pesticide subsidies were also eliminated. Subsequently rice yields increased to levels well above those recorded during the period of heavy pesticide use[84]. The Indonesian Minister of Agriculture has recently banned twenty-eight of the 64 pesticides originally banned for use on rice for use on *all* crops (R. Roush, University of Adelaide, Australia, personal communication, 1996).

D. Rosen (Hebrew University of Jerusalem, personal communication, 1991) estimates that natural enemies of pests account for up to 90% of the control of pest species achieved in agro-ecosystems and natural systems; we estimate that about half of the control of pest species is due to natural enemies. Pesticides provide an additional control of 10%[85], while the remaining 40% is due to host-plant resistance and other limiting factors present in the agro-ecosystem[86].

Parasites, predators, and host-plant resistance are estimated to account for about 80% of the non-chemical control of pest insects and plant pathogens in crops[87]. Many cultural controls such as crop rotations, soil and water management, fertilizer management, awareness of planting times, crop-plant density, trap crops, and polyculture provide additional pest control. Together these non-chemical controls can be used effectively to reduce U.S. pesticide use by as much as one-half without any reduction in crop yields[88].

PESTICIDE RESISTANCE IN PESTS

In addition to destroying natural enemy populations, the extensive use of pesticides has often resulted in the development of pesticide resistance in insect pests, plant pathogens, and weeds. In a report by the United Nations Environment Programme, pesticide resistance was ranked as one of the top four environmental problems in the world[89]. More than 500 insect and mite species, a total of 150 plant pathogen species, and more than 270 weed species are now resistant to pesticides[90]. In fact, more than 1600 insect species have developed resistance to pesticides since the 1940s[91].

Increased pesticide resistance in pest populations frequently results in the need for several additional applications of commonly used pesticides to maintain anticipated crop yields. These additional pesticide applications compound the problem by increasing environmental selection for resistance traits. Despite attempts to deal with the problem, pesticide resistance continues to develop[92].

The impact of pesticide resistance is felt in the economics of agricultural production. A striking example of this occurred in northeastern Mexico and the Lower Rio Grande of Texas[93]. Over time, extremely high pesticide resistance had developed in the tobacco budworm population until, finally, in the early 1970s approximately 285,000 ha of cotton had to be abandoned because pesticides were ineffective and the crop could not be protected from the budworm. The economic and social impacts on the Texan and Mexican farming communities dependent upon cotton were devastating.

A study by Carrasco-Tauber indicates the extent of costs attributed to pesticide resistance[94]. The study reported a yearly loss of \$45–120 per ha to pesticide resistance in California cotton. Thus, approximately \$348 million of the California cotton crop was lost to resistance. Since \$3.6 billion of U.S. cotton was harvested in 1984, the loss due to resistance for that year was approximately 10%. Assuming a 10% loss in other major crops that receive heavy pesticide treatments in the United States, crop losses due to pesticide resistance are estimated to be $1.4 billion per year.

A detailed study by Archibald further demonstrated the hidden costs of pesticide resistance in California cotton [95]. She reported that 74% more organophosphorus insecticides were required in 1981 to achieve the same kill of pests like *Heliothis* spp. than were required in 1979. Her analysis demonstrated that the diminishing effect of pesticides in addition to intensified pest control reduced the economic return per dollar of pesticide invested to only $1.14.

Furthermore, efforts to control resistant *Heliothis* spp. exact a cost on other crops when large, uncontrolled populations of *Heliothis* and other pests disperse. In addition, the cotton aphid and the whitefly exploded as secondary cotton pests because of both their own and their natural enemies' exposure to high concentrations of insecticides. They also have developed resistance to insecticides.

The total external cost attributed to the development of pesticide resistance is estimated to range between 10–25% of current pesticide treatment costs, or approximately $400 million each year in the United States alone[96]. In other words, at least 10% of the pesticides used in the United States is applied just to combat the increased resistance that has developed in various pest species.

In addition to plant pests, a large number of insect and mite pests of both livestock and humans has become resistant to pesticides[97]. Although a relatively small quantity of pesticide is applied for control of livestock and human pests, the cost of resistance has become significant. Based on available data, we estimate the yearly cost of resistance in insect and mite pests of livestock and humans to be about $30 million in the United States.

Although the costs of pesticide resistance are high in the United States, the costs in developing countries are significantly greater because pesticides are not only used to control agricultural pests but are also vital for the control of disease vectors[98]. One of the major costs of resistance in tropical countries is associated with malaria control. By 1961, the incidence of malaria in India (after early pesticide use) had declined to only 41,000 cases. Because mosquitoes developed resistance to pesticides and malarial parasites developed resistance to drugs, the incidence of malaria in India now has exploded to about 59,000,000 cases per year[99]. Similar problems are occurring not only in India but also in the rest of Asia, Africa, and South America. The total world incidence of malaria is estimated to be 100–120 million cases resulting in 1–2 million deaths[100].

BEE POISONINGS AND REDUCED POLLINATION

Honey and wild bees are vital for the pollination of fruits, vegetables, and other crops. It has been estimated that the production of approximately one-third of all human food is dependent upon bee pollination[101]. The direct and indirect benefits of bees to agricultural production range from $10–33 billion each year in the United States (E. L. Atkins, University of California, personal communication, 1990)[102]. Estimates of their benefits in Canada and Australia are $1.2 billion[103] and $196 million[104] respectively.

Because most insecticides used in agriculture are toxic to bees[105], pesticides have a major impact on both honey bee and wild bee populations. Buchmann and Nabhan report that increased use of organochlorine pesticides in the U.S. has resulted in a 43% decline in honey bee colonies[106]. D. Mayer (Washington State University, personal communication, 1990) estimates that approximately 20% of all honey bee colonies are adversely affected by pesticides. He includes the roughly 5% of U.S. bee colonies that are killed outright or die during the winter because of pesticide exposure. Mayer calculates that these direct annual losses reach $13.3 million (Table IV). Another 15% of the bee colonies either are seriously weakened by pesticides or suffer losses when apiculturists have to move colonies to avoid pesticide damage.

According to Mayer, the yearly estimated loss from partial bee kills, reduced honey production, and the cost of moving colonies totals about $25.3 million. Also, as a result of heavy pesticide use on certain crops, beekeepers are excluded from 4–6 million ha of otherwise suitable apiary locations. It is estimated that the yearly loss in potential honey production in these regions is about $27 million (D. Mayer, Washington State University, personal communication, 1990).

In addition to these direct losses caused by damage to bees and honey production, many crops are lost because of the lack of pollination. In California, for example, approximately one million colonies of honey bees are rented annually at the rate of $20 per colony in order to augment the natural pollination of almonds, alfalfa, melons, and other fruits and vegetables (R. A. Morse, Cornell University, personal communication, 1990). Since California produces nearly 50% of the bee-pollinated crops in the U.S., the total cost for bee rental for the entire country is estimated to be $40 million. We estimate at least one-tenth ($4 million) of this cost is attributable to the effects of pesticides (Table IV).

Estimates of annual agricultural losses due to the reduction in pollination by pesticides may range as high as $4 billion per year (J. Lockwood, University of Wyoming, personal communication, 1990). For most crops, both crop yield and quality are enhanced by effective pollination. For example, several studies have demonstrated that effective pollination by bees resulted in yield increases from 20–30% in several cotton varieties[107]. Assuming that a conservative 10% increase in cotton yield would result from more efficient pollination, after deducting charges for bee rental the net annual gain for cotton alone could be as high as $400 million. However, using bees to enhance cotton pollination is impossible at present because of the intensive use of insecticides on cotton[108].

Several studies emphasize that pollination is required for increased crop yields, and more importantly, it will increase the quality of fruits and vegetables[109]. In experiments with melons, E. L. Atkins (University of California, personal communication, 1990) reported that with adequate pollination melon yields were increased 10% and quality was raised 25% as measured by the dollar value of the crop.

Based on the analysis of honey bee and related pollination losses caused by pesticides, pollination losses attributable to pesticides are estimated to represent about 10% of pollinated crops and have a yearly cost of about $200 million. After adding the cost of reduced pollination to the other environmental costs of pesticides on

Table IV Estimated honey bee losses and pollination losses from honey bees and wild bees

Colony losses from pesticides	$13.3 million/year
Honey and wax losses	$25.3 million/year
Loss of potential honey production	$27.0 million/year
Bee rental for pollination	$4.0 million/year
Pollination losses	$200.0 million/year
Total	$319.6 million/year

honey and wild bees, the total annual loss is calculated to be about $320 million (Table IV). Clearly, the available evidence confirms that the yearly cost of direct honey bee losses—together with reduced yields resulting from poor pollination—is significant.

CROP AND CROP PRODUCT LOSSES

Pesticides are applied to protect crops from pests in order to increase yields, but sometimes the crops are damaged by the pesticide treatments themselves. This occurs when the recommended dosages suppress crop growth, development, and yield; pesticides drift from the targeted crop to damage adjacent nearby crops; residual herbicides either prevent chemical-sensitive crops from being planted in rotation or inhibit the growth of the crops that are planted; and/or excessive pesticide residues accumulate on crops, necessitating the destruction of the harvest. Crop losses translate into financial losses for growers, distributors, wholesalers, transporters, retailers, food processors, and others. Potential profits as well as investments are lost. The costs of crop losses increase when the related costs of investigations, regulation, insurance, and litigation are added to the equation. Ultimately the consumer pays for these losses in higher market prices.

Data on crop losses due to pesticide use are difficult to obtain. Many losses are never reported to state and federal agencies because the parties often settle privately (B. D. Berver, Office of Agronomy Services, South Dakota, personal communication, 1990; R. Batteese, Board of Pesticide Control, Maine Department of Agriculture, personal communication, 1990; J. Peterson, Pesticide/Noxious Weed Division, North Dakota Department of Agriculture, personal communication, 1990; E. Streams, EPA, Region VII, personal communication, 1990). For example, in the State of North Dakota, only an estimated one-third of the pesticide-induced crop losses are reported to the State Department of Agriculture (Peterson, personal communication, 1990). Furthermore, according to the Federal Crop Insurance Corporation, losses due to pesticide use are not insurable because of the difficulty of determining pesticide damage (E. Edgeton, Federal Crop Insurance Corporation, Washington, DC, personal communication, 1990).

Damage to crops may occur even when recommended dosages of herbicides and insecticides are applied to crops under normal environmental conditions (J. Neal, Chemical Pesticides Program, Cornell University, personal communication, 1990)[110]. Heavy dosages of insecticides used on crops have been reported to suppress growth and yield in both cotton and strawberry crops[111]. The increased susceptibility of some crops to insects and diseases following normal use of 2,4-D and other herbicides has been demonstrated in several studies[112]. Furthermore, when weather and/or soil conditions are inappropriate for pesticide application, herbicide treatments may cause yield reductions ranging from 2–50%[113].

Crops are lost when pesticides drift from the target crops to non-target crops, some of which may be as far as several miles downwind[114]. Drift occurs with almost all methods of pesticide application, including both ground and aerial equipment. The potential problem is greatest when pesticides are applied by aircraft[115]. With aircraft application, 50–75% of the applied pesticides miss the target area[116]. In contrast, 10–35% of the pesticide applied with ground application equipment misses the target area[117]. The most serious drift problems are caused by "speed sprayers" and "mist-blower sprayers" because with these application technologies approximately 35% of the pesticide drifts away from the target area, and larger amounts of pesticide are applied by these sprayers than by aircraft (E. L. Atkins, University of California, personal communication, 1990).

Crop injury and subsequent loss due to drift is particularly common in areas planted with diverse crops. For example, in southwest Texas in 1983 and 1984, nearly \$20 million of cotton was destroyed from drifting 2,4-D herbicide when adjacent wheat fields were aerially sprayed with the herbicide[118].

Clearly, drift damage, human exposure, and widespread environmental contamination are inherent in the process of pesticide application and add to the cost of using pesticides. As a result, commercial applicators are frequently sued for damage inflicted during or after treatment. Therefore, most U.S. applicators now carry liability insurance at an estimated cost of about \$245 million per year (D. Witzman, U.S. Aviation Underwriters, Tennessee, personal communication, 1990; H. Collins, National Agricultural Aviation Association, Washington, DC, personal communication, 1990)[119].

When residues of some herbicides persist in the soil, crops planted in rotation may be injured[120]. In 1988–89, an estimated \$25–30 million of Iowa's soybean crop was lost due to the persistence of the herbicide Sceptor in the soil (R. G. Hartzler, Cooperative Extension Service, Iowa State University, personal communication, 1990).

Herbicide persistence can sometimes prevent growers from rotating their crops—a situation that may force them to continue planting the same crop (T. Tomas, Nebraska Sustainable Agriculture Society, Hartington, Nebraska, personal communication, 1990)[121]. For example, the use of Sceptor in Iowa, as mentioned above, has prevented farmers from implementing their plan to plant soybeans after corn (Hartzler, personal communication, 1990). Unfortunately, the continuous planting of some crops in the same field often intensifies insect, weed, and pathogen problems[122]. Such pest problems not only reduce crop yields but often require added pesticide applications.

Although crop losses caused by pesticides seem to be a small percentage of total U.S. crop production, their total value is significant. For example, an average of 0.14% of San Joaquin County, California's total crop production was lost to pesticides from 1986 to 1987 (Agricultural Commissioner, San Joaquin County, California, personal communication, 1990)[123]. Similarly, in Yolo County, California, approximately 0.18% of the county's total crop production was lost in 1989 (Agricultural Commissioner, Yolo County, California, personal communication, 1990)[124]. Estimates indicate that less than 0.05% of Iowa's annual soybean crop is lost to pesticides (Hartzler, personal communication, 1990).

An average 0.1% loss in the annual U.S. production of corn, soybeans, cotton, and wheat—four crops which together account for about 90% of the herbicide and insecticide used in U.S. agriculture—was valued at \$40.9 million in 1993[125]. Assuming that only one-third of the incidents involving crop losses due to pesticides are reported to authorities, the total value of all crops lost because of pesticides could be as high as three times this amount, or \$123 million annually.

This \$123 million, however, does not take into account either other crop losses or major but recurrent events such as the large-scale losses that occurred in Iowa in 1988–1989 (\$25–30 million); Texas in 1983–84 (\$20 million), and California's aldicarb/watermelon crisis in 1985 (\$8 million, see below). These recurrent losses alone represent an average of \$30 million each year, raising the estimated average crop loss value from the use of pesticides to approximately \$153 million.

Additional losses are incurred when food crops are disposed of because they exceed the EPA regulatory tolerances for pesticide residue levels. Assuming that all

Table V Estimated loss of crops and trees due to the use of pesticides

Impacts	Total costs (in millions of $)
Crop losses	153
Crop applicator insurance	245
Crops destroyed because of excess pesticide contamination	550
Investigations and testing	
Governmental	10
Private	1
Total	959

of the crops and crop products that exceeded the EPA's regulatory tolerances (reported to be at least 1%) were disposed of as required by law, about $550 million in crops would be destroyed annually because of excessive pesticide contamination[126]. Because most of the crops with pesticide residues above the tolerance levels are neither detected nor destroyed, they are consumed by the public, thereby avoiding financial loss but creating public health risks.

A well publicized incident in California in 1985 illustrates this problem. Excess pesticides in food generally go undetected unless a large number of people become ill after the food is consumed. Thus, when more than 1000 people became ill after eating contaminated watermelons in California, approximately $1.5 million worth of watermelons were destroyed (R. Magee, State of California Department of Food and Agriculture, Sacramento, personal communication, 1990). After the public became ill, it was learned that several California farmers had treated watermelons with the insecticide aldicarb (Temik) which is not registered for use on watermelons[127]. Following this crisis, the California State Assembly appropriated $6.2 million to be awarded to claimants affected by state seizure and freeze orders[128]. According to the California Department of Food and Agriculture, an estimated $800,000 in investigative costs and litigation fees resulted from this one incident (Magee, personal communication, 1990). The California Department of Health Services was assumed to have incurred similar expenses, placing the total cost of the incident at nearly $8 million.

To avoid other dangerous and costly incidents like the California watermelon crisis, many private distributors and grocers are testing their produce for the presence of pesticides to reassure themselves and consumers of the safety of the food they handle (C. Merrilees, Consumer Pesticide Project, National Toxics Campaign, San Francisco, personal communication, 1990). In the U.S., this testing is currently estimated to cost $1 million per year (Merrilees, personal communication, 1990), but if all of the retail grocers nationwide were to undertake such testing, the calculated cost would be approximately $66 million per year based on data from California.

The special investigations of crop losses due to pesticide use that are conducted by state and federal agencies are also costly. From 1987–89, the State of Montana Department of Agriculture conducted an average of 80 pesticide-related investigations per year at an average cost of $3500 per investigation (S. F. Baril, State of Montana, Department of Agriculture, personal communication, 1990). Also, the State of Hawaii conducts approximately five investigations per year at a cost of nearly $10,000 each (R. Boesch, Pesticide Programs, Department of Agriculture, State of

Hawaii, personal communication, 1990). By averaging the number of investigations from seven states (Arkansas, Hawaii, Idaho, Iowa, Louisiana, Mississippi, and Texas) and by using the conservative Montana state figure of $3500 per investigation, we estimate that the average state conducts 70 investigations per year at a cost of $246,000 annually. Using these data, investigations in the U.S. are estimated to total $10 million annually. This figure does not include investigation costs at the federal level.

When crop seizures, insurance, and investigation costs are added to the costs of direct crop losses due to the use of pesticides in commercial crop production, the total loss is estimated to be about $959 million annually in the United States (Table V).

GROUND AND SURFACE WATER CONTAMINATION

Certain pesticides applied to crops eventually end up in ground and surface water and have even been detected in rain[129] and fog[130]. It has been estimated that 1% of all pesticides from non-irrigated farmland runs off the land; 4% of pesticides run off of irrigated farmland; and 33% of pesticides run off farmland when they are applied by airplane[131].

Pesticide contamination of surface waters—lakes, rivers, streams, and reservoirs—is an important concern due to the extensive use of such water for drinking and recreation. Particularly disturbing is the fact that conventional drinking water treatment is not designed for, and subsequently does not remove, pesticides. One study demonstrated that even after conventional treatment, 90% of drinking water samples contained at least one pesticide, while 58% of the samples contained at least four different pesticides[132]. After studying the drinking water in 29 U.S. cities over a four month period, Cohen *et al.* observed that the herbicide atrazine was present in the tap water of 28 of the cities (97%) and cyanazine was present in 25 of the cities (86%)[133]. In addition, federal health levels for atrazine and cyanazine were exceeded in 17% and 35% of all samples taken, respectively.

At present, water treatment plants do not regularly monitor the water they treat for pesticides. An estimate for the daily cost of monitoring the triazine herbicides is $1500 per city per year, or $0.10 per person[134]. There are over 11,000 surface water treatment systems of which approximately 4000 are large systems and 7000 are small systems[135]. Assuming a cost of $1500 per year for the large systems and $3000 per year for the small systems, the total cost of surface water monitoring would be approximately $27 million per year.

The three most common pesticides found in groundwater are aldicarb (an insecticide), and alachlor and atrazine (both herbicides)[136]. Estimates indicate that nearly one-half of the groundwater and well water in the United States is, or has the potential to be, contaminated[137]. EPA reported that 10.4% of community wells and 4.2% of rural domestic wells have detectable levels of at least one of the 127 pesticides tested in a national survey[138]. It would cost an estimated $1.3 billion annually for the United States to monitor its well and groundwater for pesticide residues[139].

The fact that approximately one-half of the population obtains its water from wells[140] and the fact that pesticide residues remain for long periods of time in contaminated groundwater are both major concerns[141]. Not only are there just a few microorganisms that have the potential to degrade pesticides[142], but the groundwater recharge rate averages less than 1% per year[143].

Monitoring pesticides in groundwater is only a portion of the total cost of U.S. groundwater contamination. Also, there is the high cost of cleanup. For instance, at the Rocky Mountain Arsenal near Denver, Colorado, the removal of pesticides from groundwater and soil was estimated to cost approximately $2 billion[144]. If all pesticide-contaminated groundwater were cleared of pesticides before human consumption, the cost would be about $500 million[145]. Note, the cleanup process requires a water survey to target the contaminated water for cleanup. Thus, the total monitoring and cleaning costs of pesticide-polluted groundwater are estimated to be approximately $1.8 billion annually.

FISHERY LOSSES

Pesticides are washed into aquatic ecosystems by water runoff and soil erosion. About 18 t/ha/yr of soil are washed and/or blown from pesticide treated cropland into adjacent locations including streams and lakes[146]. Pesticides also drift into streams and lakes and contaminate these aquatic systems[147]. Some soluble pesticides are easily leached into streams and lakes[148] and are readily taken up by aquatic organisms[149]. A nationwide survey of fish in the United States showed pesticide residues present in almost all of the 119 fish species that were examined[150].

Once in aquatic systems, pesticides cause fishery losses in several ways. These include high pesticide concentrations in water; low level doses that may kill highly susceptible fish fry; the elimination of essential fish foods like insects and other invertebrates; or the reduction of dissolved oxygen levels in the water due to the decomposition of aquatic plants killed by pesticides. In addition, because government safety restrictions ban the catching or sale of fish contaminated with pesticide residues, such unmarketable fish are considered an economic loss[151]. Furthermore, health advisories restrict the amount of sportfish that can be eaten which, in turn, impacts the revenues gained from tourism[152].

Each year large numbers of fish are killed by pesticides. Based on EPA data, we calculate that from 1977–87 the number of fish kills due to all factors was 141 million fish per year; 6–14 million fish per year are killed by pesticides[153]. These estimates of fish kills are considered to be low for the following reasons: first, in 20% of fish kills, no estimate is made of the number of fish killed; second, fish kills frequently cannot be investigated quickly enough to accurately determine the primary cause of death. In addition, fast moving waters in rivers dilute pollutants so that causes of kills frequently cannot be identified. Moving waters also wash away some of the poisoned fish, while other poisoned fish sink to the bottom and cannot be counted. Perhaps most important is the fact that few, if any, of the widespread and more frequent low-level pesticide poisonings are dramatic enough to be observed; therefore, they tend to go unrecognized and unreported[154].

Studies have shown that high dosages of aquatic herbicides for weed control are toxic to fish[155], however, several studies have shown that at actual application dosage rates, these herbicides have no effect on the fish which live in the water. By using radio telemetry techniques coupled with laboratory work, researchers have demonstrated that the application and presence of herbicides in the water does not drive the fish away or inhibit their feeding habits[156]. These studies did not, however, investigate any potential short- or long-term biochemical effects the herbicides may have on the fish in the aquatic ecosystem.

The average value of a fish was estimated in 1982 to be about $1.70, using the guidelines of the American Fisheries Society[157], but it was reported that the Coors Beer Company might be "fined up to $10 per dead fish, plus other penalties" for an accidental beer spill in a creek for which they were held liable[158]. Using the estimated value of $4.00, the total value of the conservatively estimated 6–14 million fish killed per year ranges from $24–56 million. This calculation only takes into account direct impacts; indirect impacts, such as tourism, are more difficult to define but can be substantial[159]. For instance, the revenue generated in Massachusetts by marine recreational fishing has been estimated to be $637 million annually[160]. So, the actual loss is probably several times the $24–56 million estimate when all of the indirect impacts are taken into account.

WILD BIRDS AND MAMMALS

Wild birds and mammals are also damaged by pesticides, and consequently make excellent "indicator species". Deleterious effects of pesticides on wildlife include death from direct exposure or secondary poisonings from consuming contaminated prey; reduced survival, growth, and reproductive rates from exposure to sublethal dosages; and habitat reduction through elimination of food sources and refuges[161]. In the United States, approximately 3 kg of pesticide per hectare are applied to about 160 million ha of land per year[162]. With such a large portion of the land area treated with heavy dosages of pesticide, it is to be expected that the impact of pesticides on wildlife is significant.

The full extent of bird and mammal destruction is difficult to determine because these animals are often secretive, camouflaged, highly mobile, and live in dense grass, shrubs, and trees. Typical field studies of the effects of pesticides often obtain extremely low estimates of bird and mammal mortality[163]. This is because bird carcasses disappear quickly, well before the dead birds can be found and counted. Studies show that only 50% of birds are recovered even when the birds' location is known[164]. Furthermore, when known numbers of bird carcasses were placed in identified locations in the field, 62–92% disappeared overnight due to vertebrate and invertebrate scavengers[165]. Also, field studies seldom account for birds that die a distance from the treated areas. Finally, birds often hide and die in inconspicuous locations. A recent study suggested an approach to solving this problem: designating the bird population in question with bands before the application of pesticides and establishing the survival rate of the banded birds after the pesticide has been applied[166].

Nevertheless, many bird casualties caused by pesticides have been reported. For instance, White *et al.* reported that 1200 Canada geese were killed in one wheat field that was sprayed with a 2:1 mixture of parathion and methyl parathion at a rate of 0.8 kg per ha[167]. Carbofuran applied to alfalfa killed more than 5000 ducks and geese in 5 incidents, while the same chemical applied to vegetable crops killed 1400 ducks in a single incident[168]. Carbofuran is estimated to kill 1–2 million birds each year[169]. Another pesticide, diazinon, killed 700 of the wintering population of 2500 Atlantic Brant Geese after it was applied to three golf courses[170].

Several studies report that the use of herbicides in crop production results in the total elimination of the weeds that harbor some insects (R. Beiswenger, University of Wyoming, personal communication, 1990)[171]. This has led to significant reductions in the grey partridge in the United Kingdom and the common pheasant in the United States. In the case of the partridge, population levels have decreased more than 77% because partridge chicks (also pheasant chicks) depend on insects to supply them with the protein needed for their development and survival (R. Beiswenger, University of Wyoming, personal communication, 1990)[172].

Frequently the form of a pesticide influences its toxicity to wildlife[173]. For example, treated seed and insecticide granules—including carbofuran, fensulfothion, fonofos, and phorate—are particularly toxic to birds when consumed. Many birds will ingest these granules either on purpose or by accident, thereby consuming the pesticide directly. Some recent research has focused on the spraying of the pesticide treated area with a solution of a naturally occurring plant substance which is unpalatable to many types of birds[174]. Another approach includes treating the granules with taste repellents before field application[175]. Despite these measures, estimates are that 0.23–1.5 birds per ha are killed in Canada, while in the United States 0.25–8.9 birds per ha per year are estimated to have been killed by pesticides[176].

Pesticide toxicity can also be species-specific. For instance, in the United Kingdom, about 1500 greylag and pink-footed geese died over a four-year period due to exposure to carbophenothion. This occurred despite pre-approval toxicity tests using Canada geese. Further tests revealed that the greylag and pink-footed geese were much more sensitive to carbophenothion than were the Canada geese[177].

Pesticides also adversely affect the reproductive potential of many birds and mammals. Exposure of birds, especially predatory birds, to chlorinated insecticides has caused reproductive failure and has been attributed to eggshell thinning[178]. Most of the affected populations have recovered since the banning of DDT in the United States[179]. However, DDT and its metabolite DDE remain a concern because DDT continues to be used in some South American countries which are the wintering areas for numerous bird species[180].

Several pesticides—especially DBCP, dimethoate, and deltamethrinare—are reported to reduce sperm production in certain mammals[181]. Clearly, when this occurs the capacity of certain wild mammals to survive is reduced.

Habitat alteration and destruction can be expected to reduce mammal populations. For example, when glyphosate was applied to forest clearcuts to eliminate low growing vegetation, the southern red-backed vole population was greatly reduced because its food source and cover were practically eliminated[182]. In another vivid example, the decline of the alligator population in Florida's Lake Apopka was linked to a large spill of DDT[183]. Similarly, the effects of herbicides have also been observed in other mammals[184], including sixteen predator species in the Great Lakes region[185]. These findings have raised concerns about how much humans are being affected by these chemicals, especially since the average male sperm count has decreased by more than 40% since 1938[186]. Overall, however, the impacts of pesticides on mammals have been inadequately investigated.

Although the gross values for wildlife are not available, expenditures involving wildlife are one measure of its monetary value. Non-consumptive users of wildlife spent an estimated $14.3 billion in 1985[187]. Bird watchers in the U.S. spend an estimated $600 million annually on their sport and an additional $500 million on birdseed—a total of $1.1 billion[188]. The money spent by bird hunters to harvest 5 million game birds was $1.1 billion, or approximately $216 per bird[189]. In addition, estimates of the value of all types of birds ranged from $0.40 to more than $800 per bird. The $0.40 per bird was based on the costs of birdwatching and the $800 per bird was based on the replacement costs of the affected species[190].

If it is assumed that the damages pesticides inflect on birds occur primarily on the 160 million hectares of cropland that receive most of the pesticide, and the bird population is estimated to be 4.2 birds per hectare of cropland[191], then 672 million birds are directly exposed to pesticides. If it is conservatively estimated that only 10% of the bird population is killed, then the total number of birds killed is 67 million. Note that this estimate is at the low end of the range of 0.25 to 8.9 birds per ha killed annually by pesticides mentioned earlier in this section. Also, this is considered a conservative estimate because secondary losses to pesticide reductions in invertebrate prey poisonings were not included in the assessment. Assuming that the average value of a bird is $30, then an estimated $2 billion worth of birds are destroyed annually.

Also, a total of $102 million is spent yearly by the U.S. Fish and Wildlife Service on their Endangered Species Program which aims to re-establish species such as the bald eagle, peregrine falcon, osprey, and brown pelican that in some cases were reduced by pesticides[192].

When all of the above costs are combined, we estimate that the bird losses associated with pesticide use in the U.S. total approximately $2.1 billion per year.

MICROORGANISMS AND INVERTEBRATES

Pesticides easily find their way into soils where they may be toxic to arthropods, earthworms, fungi, bacteria, and protozoa. Small organisms are vital to ecosystems because they dominate both the structure and function of natural systems.

For example, an estimated 4.5 tons per ha of fungi and bacteria exist in the upper 15 cm of soil[193]. Together with arthropods, they make up 95% of all species and 98% of all biomass (excluding vascular plants). Microorganisms are essential to the proper functioning of the ecosystem because they break down organic matter and enable vital chemical elements to be recycled[194]. Equally important is their ability to "fix" nitrogen to make it available for plants[195]. The role of microorganisms cannot be over-emphasized because in natural, agricultural, and forestry ecosystems they are essential agents in the biogeochemical recycling of vital elements[196].

Earthworms and insects aid in bringing new soil to the surface at a rate of up to 500 t/ha/yr (D. Pimentel, C. Wilson, C. McCullum, R. Huang, P. Dwen, J. Flack, Q. Tran, T. Saltman, and B. Cliff, *Economic and Environmental Benefits of Biodiversity*, unpublished manuscript, 1996). This action improves soil formation and structure for plant growth and makes various nutrients more available for absorption by plants. The holes in the soil made by earthworms and insects (up to 10,000 holes per square meter) also facilitate the percolation of water into the soil[197], thereby slowing rapid water runoff from the land and preventing soil erosion.

Unfortunately, our understanding of the effects of pesticides on these soil organisms is minimal. Studies have shown a wide range of effects from pesticides, ranging from drastically lowered survival rates to, in some instances, increased growth[198]. This is not surprising since this is a large and varied group of organisms in which each species reacts differently[199]. However, insecticides, fungicides, and herbicides generally reduce both species diversity and the total biomass of the biota in the soil[200].

Earthworms have been researched to a much greater extent, and most pesticides have been shown to be toxic to them[201]. Stringer and Lyons reported that apple tree leaves had accumulated on the surface of soil where earthworms had been killed by pesticides[202]. Apple scab, a disease carried over from season to season on fallen leaves, is commonly treated with fungicides. Some fungicides, insecticides, and herbicides can be toxic to the earthworms that would otherwise remove and recycle the surface leaves[203].

On golf courses and other lawns, the destruction of earthworms by pesticides results in the accumulation of dead grass or thatch in the turf[204]. Special equipment must be used, at considerable expense, to remove this thatch.

Although these invertebrates and microorganisms are essential to the vital structure and function of all ecosystems, it is impossible to place a dollar value on the damage caused by pesticides to this large group of organisms. To date no relevant quantitative data on the value of microorganism destruction by pesticides has been collected.

GOVERNMENT FUNDS FOR PESTICIDE POLLUTION REGULATION

The cost of carrying out state and federal regulatory actions, as well as the pesticide monitoring programs that are needed to control pesticide pollution, are major environmental costs associated with pesticide use. These funds are spent to reduce the hazards of pesticides and to protect the integrity of the environment and public health.

Approximately 1.3 million individuals have been certified to apply restricted pesticides in the U.S.[205]. At least $1 million is spent each year by the state and federal governments to train and register these pesticide applicators (D. Rutz, Cornell University, personal communication, 1991). Also, more than $40 million is spent each year by the EPA for the registration and re-registration of pesticides[206]. Based on these known expenditures, we estimate that the federal and state governments spend approximately $180 million per year for the regulation and monitoring of pesticide pollution (Table VI)[207]. Some estimates for the combined public and private-sector costs of regulating pesticides, however, range as high as $1 billion per year[208].

Although enormous amounts of government money are being spent to reduce pesticide pollution, many of the costs of pesticides are not taken into account. Also, many serious environmental and social problems have yet to be corrected by improved government policies.

CONCLUSION

An investment of about $6.7 billion in pesticide control saves approximately $26 billion in U.S. crops, based on direct costs and benefits[209]. The indirect costs of

Table VI Total estimated environmental and social costs of pesticides in the United States

Costs	Million $/year
Public health impacts	933
Domestic animal deaths and contamination	31
Loss of natural enemies	520
Cost of pesticide resistance	1400
Honey bee and pollination losses	320
Crop losses	959
Surface water monitoring	27
Groundwater contamination	1800
Fishery losses	56
Bird losses	2100
Government regulations/monitoring expenditures	180
Total	8326

pesticide use to the environment and public health, however, need to be balanced against these benefits. Based on the available data, the environmental and social costs of pesticide use total approximately $8.3 billion each year (Table VI). Users of pesticides in agriculture pay for only about $3.2 billion of this cost which includes problems arising from pesticide resistance and destruction of natural enemies. Society eventually pays this $3.2 billion plus the remaining $5.1 billion in environmental and public health costs (Table VI).

Our assessment of the environmental and health problems associated with pesticides faced problems of scarce data which made this assessment of the complex pesticide situation incomplete. For example, what is an acceptable monetary value for a human life lost or a cancer illness due to pesticides? Equally difficult is placing a monetary value on killed wild birds and other wildlife, invertebrates, microbes, or contaminated food and groundwater.

In addition to the costs that cannot be accurately measured, there are additional costs that have not been included in the $8.3 billion per year figure. A complete accounting of the indirect costs should include accidental poisonings like the "aldicarb/watermelon" crisis; domestic animal poisonings; unrecorded losses of fish, wildlife, crops, trees, and other plants; losses resulting from the destruction of soil invertebrates, microflora, and microfauna; the true costs of human pesticide poisonings; water and soil pollution; and human health effects like cancer and sterility. If the full environmental and social costs could be measured as a whole, the total cost would be significantly greater than our estimate of $8.3 billion per year. Such a complete long-term cost/benefit analysis of pesticide use would surely reduce the perceived profitability of pesticides.

Human pesticide poisonings, reduced natural enemy populations, increased pesticide resistance, and honey bee poisonings account for a substantial portion of the calculated environmental and social costs of pesticide use in the United States. Fortunately, some losses of natural enemies and pesticide resistance problems are being alleviated through carefully planned use of integrated pest management (IPM) practices. But a great deal remains to be done to reduce these important environmental costs[210].

This investigation not only underscores the serious nature of the environmental and social costs of pesticides, but emphasizes the great need for a more detailed investigation of their environmental and economic impacts. Pesticides are, and will continue to be, a valuable pest control tool. Meanwhile, more accurate and realistic cost/benefit analyses will enable agriculturists and other scientists to minimize the risks of pesticides and to develop and increase the use of non-chemical pest controls in order to maximize the benefits of pest control for all of society.

References

1. J. Wargo, *Our Children's Toxic Legacy: How Science and Law Fail to Protect Us from Pesticides* (Yale University Press, New Haven, CT, 1996).
2. C. M. Benbrook, *Pest Management at the Crossroads* (Consumers Union, Yonkers, NY, 1996).
3. T. Colborn, D. Dumanoski and J. P. Myers, *Our Stolen Future: How We Are Threatening Our Fertility, Intelligence, and Survival—A Scientific Detective Story* (Dutton, New York, NY, 1996).
4. A. L. Aspelin, *Pesticide Industry Sales and Usage: 1994 and 1995 Market Estimates, Preliminary* (U.S. Environmental Protection Agency, Washington, DC, 1996); *Statistical Abstract of the United States 1995* (U.S. Government Printing Office, Washington, DC, 1995).
5. D. Pimentel, L. McLaughlin, A. Zepp, B. Lakitan, T. Kraus, P. Kleinman, F. Vancini, W. J. Roach, E. Graap, W. S. Keeton and G. Selig, "Environmental and economic impacts of reducing U.S.

agricultural pesticide use" **In**: (D. Pimentel, ed.) *Handbook on Pest Management in Agriculture* (CRC Press, Boca Raton, FL, 1991) pp. 679–718.
6. Pimentel *et al*. (1991). See ref. 5.
7. D. Pimentel, J. Krummel, D. Gallahan, J. Hough, A. Merrill, I. Schreiner, P. Vittum, F. Koziol, E. Back, D. Yen and S. Fiance, "Benefits and costs of pesticide use in U.S. food production" *BioScience* **28**, 778–784 (1978).
8. C. Carrasco-Tauber, "Pesticide Productivity Revisited" *M.S. Thesis* (University of Massachusetts, Amherst, MA, 1989); Pimentel *et al*., 1991.
9. Pimentel *et al*. (1991). See ref. 5.
10. Pimentel *et al*. (1991). See ref. 5.
11. D. Pimentel, "Amounts of pesticides reaching target pests: environmental impacts and ethics" *J. Agri. Environ. Ethics* **8**, 17–29 (1995).
12. *Minutes of Administrator's Pesticide Policy Advisory Committee* (U.S. Environmental Protection Agency, Washington, DC, March, 1977).
13. FDA memo from Division of Federal-State Relations (7 May, 1993).
14. "DuPont, US EPA reach accord on phasing out cyanazine" *J. Am. Water Works Assoc*. **87**, 16 (1995).
15. Pimentel *et al*. (1991). See ref. 5.
16. Benbrook (1996). See ref. 2.
17. J. Blondell, *Pesticide Poisoning Summary* (Health Effects Division, Office of Pesticide Programs, U.S. Environmental Protection Agency, Washington, DC, 1992).
18. Blondell (1992). See ref. 17.
19. "World Health Organization Press Release" *Tropical Diseases News* **31**, 3 (March 28, 1990).
20. A. Fisher, L. G. Chestnut and D. M. Violette, "The value of reducing risks of death: a note on new evidence" *J. Policy Anal. and Mgt*. **8**, 88–100 (1989); E. P. Nash, *What's a life worth?* (New York Times, New York, 1994).
21. Nash (1994). See ref. 20.
22. World Resources Institute/UN Environment Programme/UN Development Programme, *World Resources: 1994–1995* (Oxford University Press, Oxford, UK, 1994).
23. Committee, House of Commons Agriculture, *The Effects of Pesticides on Human Health: Report and Proceedings of the Committee*, 2nd Special Report **Vol. 1** (Her Majesty's Stationery Office, London, 1987).
24. D. Bull, *A Growing Problem: Pesticides and the Third World Poor* (Oxfam, Oxford, 1982); F. Forget, "Pesticides and the third world" *J. Toxicol. Environ. Health* **32**, 11–31 (1991).
25. S. M. Dogheim, E. N. Nasr, M. M. Almaz and M. M. El-Tohamy, "Pesticide residues in milk and fish samples collected from two Egyptian Governorato" *J. Assoc. Off. Anal. Chem*. **73**, 19–21 (1990).
26. W. J. Trotter and R. Dickerson, "Pesticide residues in composited milk collected through the U.S. pasteurized milk network" *J. AOAC Intl*. **76**, 1220–1225 (1993).
27. FDA, "Food and Drug Administration Pesticide Program Residues in Foods, 1989" *J. Assoc. Off. Anal. Chem*. **73**, 127A–146A (1990).
28. H. K. Hundley, T. Cairns, M. A. Luke and H. T. Masumoto, "Pesticide residue findings by the Luke method in domestic and imported foods and animal feeds for the fiscal years 1982–1986" *J. Assoc. Off. Anal. Chem*. **71**, 875–877 (1988); FDA, 1990.
29. *Pesticide Residues in Food: Technologies for Detection* (Office of Technology Assessment, U.S. Congress, Washington, DC, 1988); J. P. Minyard and W. E. Roberts, "A state data resource on toxic chemicals in foods" **In**: (B. G. Tweedy, H. J. Dishburger, L. G. Ballantine and J. McCarthy, eds.) *Pesticide Residues and Food Safety: A Harvest of Viewpoints* (American Chemical Society, Washington, DC, 1991) pp. 151–161.
30. Wargo (1996). See ref. 1.
31. R. Wiles and C. Campbell, *Washed, Peeled—Contaminated: Pesticides Residues in Ready-To-Eat Fruits and Vegetables* (Environmental Working Group, Washington, DC, 1994).
32. "Food and Drug Administration Pesticide Program Residues in Foods, 1988" *J. Assoc. Off. Anal. Chem*. **72**, 133A–152A (1989).
33. M. Moses, "Pesticide-related health problems and farmworkers" *J. Am. Assoc. Occup. Health Nurs*. **37**, 115–130 (1989); R. McConnell, A. F. P. Anton and R. Magnotti, "Crop duster aviation mechanics: high risk for pesticide poisoning" *Am. J. Public Health* **80**, 1236–1239 (1990); A. Blair and S. H. Zahm, "Cancer among farmers" *Occup. Med*. **6**, 335–354 (1991); M. T. Al-Jaghbir, A. S. Salhab and F. A. Hamarsheh, "Dermal and inhalation exposure to dimethoate" *Arch. Environ. Contam. Toxicol*. **22**, 358–361 (1992); C. Wesseling, L. Castillo and C. Elinder, "Pesticide poisonings in Costa Rica" *Scand. J. Work Environ. Health* **19**, 227–235 (1993); S. Ciesielski, D. P. Loomis, S. R. Mims and A. Auer, "Pesticide exposures, cholinesterase depression, and symptoms among North Carolina migrant farmworkers" *Am. J. Public Health* **84**, 446–451 (1994); H. H. McDuffie, "Women at work: agriculture and pesticides" *J. Occup. Med*. **36**, 1240–1246 (1994).

34. D. S. Sharp, B. Eskenazi, R. Harrison, P. Callas and A. H. Smith, "Delayed health hazards of pesticide exposure" *Ann. Rev. Public Health* **7**, 441–471 (1986); A. Blair, H. Malker, K. P. Cantor, L. Burmeister and K. Wirklund, "Cancer among farmers: a review" *Scand. J. Work. Environ. Health* **11**, 397–407 (1985); L. M. Brown, A. Blair, R. Gibson, G. D. Everett, K. P. Cantor, L. M. Schuman, L. F. Burmeister, S. F. Van Lier and F. Dick, "Pesticide exposures and other agricultural risk factors for leukemia among men in Iowa and Minnesota" *Cancer Res.* **50**, 6585–6591 (1990); K. P. Cantor, A. Blair, G. Everett, R. Gibson, L. F. Burmeister L. M. Brown, L. Schuman and F. R. Dick, "Pesticides and other agricultural risk factors for non-Hodgkin's lymphoma among men in Iowa and Minnesota" *Cancer Research* **52**, 2447–2455 (1992).
35. S. K. Hoar, A. Blair, F. F. Holmes, C. D. Boysen, R. J. Robel, R. Hoover and J. F. Fraumeni, "Agricultural herbicide use and risk of lymphoma and soft-tissue sarcoma" *J. Amer. Med. Assoc.* **256**, 1141–1147 (1986); A. Blair and H. Zahm, "Methodologic issues in exposure assessment for case-control studies of cancer and herbicides" *Am. J. Indus. Med.* **18**, 285–293 (1990); S. H. Zahm, D. D. Weisenburger, P. A. Babbitt, R. C. Saal, J. B. Vaught, K. P. Cantor and A. Blair, "A case-control study of non-Hodgkin's lymphoma and the herbicide 2,4-dichlorophenoxyacetic acid (2,4-D) in Eastern Nebraska" *Epidemiology* **1**, 349–356 (1990).
36. A. Blair, O. Axelson, C. Franklin, O. E. Paynter, N. Pearce, De. Stevenson, J. E. Trosko, H. Vaubui, G. Williams, J. Woods and S. H. Zahm, "Carcinogenic effects of pesticides" **In**: (C. F. Wilkinson and S. R. Baker, eds.) *The Effect of Pesticides on Human Health* (Princeton Scientific Pub. Co., Princeton, NJ, 1990) pp. 201–260; A. C. Pesatori, J. M. Sontag, J. H. Lubin, D. Consonni and A. Blair, "Cohort mortality and nested case-control study of lung cancer among structural pest control workers in Florida (United States)" *Cancer Causes & Control* **5**, 310–318 (1994).
37. D. J. Ecobichon, J. E. Davies, J. Doull, M. Ehrich, R. Joy, D. McMillan, R. MacPhail, L. W. Reiter, W. Slikker and H. Tilson, "Neurotoxic effects of pesticides" **In**: (C. F. Wilkinson and S. R. Baker, eds.) *The Effect of Pesticides on Human Health* (Princeton Scientific Pub. Co., Princeton, NJ, 1990) pp. 131–199.
38. C. F. Wilkinson, "Introduction and overview" **In**: (C. F. Wilkinson and S. R. Baker, eds.) *The Effect of Pesticides on Human Health* (Princeton Scientific Pub. Co., Princeton, NJ, 1990) pp. 5–33.
39. Statement by D. William Lijinsky before the Committee on Labor and Human Resources, U.S. Senate, June 6, 1989, Washington, DC; WHO/UNEP, *Public Health Impact of Pesticides Used in Agriculture* (World Health Organization/United Nations Environment Programme, Geneva, 1990).
40. Wargo (1996). See ref. 1.
41. J. S. Woods, L. Polissar, R. K. Severson, L. S. Heuser and B. G. Kulander, "Soft tissue sarcoma and non-Hodgkin's lymphoma in relation to phenoxy herbicide and chlorinated phenol exposure in western Washington, USA" *J. Nat. Cancer Inst.* **78**, 899–910 (1987); WHO/UNEP, 1989; Cantor *et al.*, 1992.
42. D. T. Wigle, R. M. Samenciw, K. Wilkins, D. Riedel, L. Ritter, H. I. Morrison and Y. Mao, "Mortality study of Canadian male farm operators: non-Hodgkin's lymphoma mortality and agricultural practices in Saskatchewan" *J. Nat. Canc. Inst.* **82**, 575–582 (1990).
43. *Statistical Abstract of the United States 1995* (U.S. Bureau of the Census, U.S. Government Printing Office, Washington, DC, 1995).
44. Colborn *et al.* (1996). See ref. 3.
45. R. H. Foote, E. C. Schermerhorn and M. E. Simkin, "Measurement of semen quality, fertility, and reproductive hormones to assess dibromochloropropane (DBCP) effects in live rabbits" *Fund. and Appl. Tox.* **6**, 628–637 (1986); D. S. Sharp, B. Eskenazi, R. Harrison, P. Callas and A. H. Smith, "Delayed health hazards of pesticide exposure" *Ann. Rev. Public Health* **7**, 441–471 (1986); I. Shaked, U. A. Sod-Moriah, J. Kaplanske, G. Potashnik and O. Buckman, "Reproductive performance of dibromochloropropane-treated female rats" *Int. J. Fert.* **33**, 129–133 (1988); D. Shemi, Z. Marx, J. Kaplanski, G. Potashnik and U. A. Sod-Moriah, "Testicular damage development in rats injected with dibromochloropropane DBCP" *Andrologia* **20**, 331–337 (1988); D. Shemi, U. A. Sod-Moriah, M. Abraham, M. Friedlaender, G. Potashnik and J. Kaplanski, "Ultrastructure of testicular cells in rats treated with dibromochloropropane DBCP" *Andrologia* **21**, 229–236 (1989).
46. D. Whorton, R. M. Krauss, S. Marshall and T. H. Milby, "Infertility in male pesticide workers" *The Lancet* 1259–1261 (Dec. 17, 1977); G. Potashnik and I. Yanai-Inbar, "Dibromochloropropane (DBCP): an 8-year reevaluation of testicular function and reproductive performance" *Fertility and Sterility* **47**, 317–323 (1987).
47. B. H. Devens, M. H. Grayson, T. Imamura and K. E. Rodgers, "O,O,S-trimethyl phosphorothioate effects on immunocompetence" *Pestic. Biochem. Physiol.* **24**, 251–259 (1985); L. J. Olson, B. J. Erickson, R. D. Hinsdill, J. A. Wyman, W. P. Porter, R. C. Bidgood and E. V. Norheim, "Aldicarb immunomodulation in mice: an inverse dose-response to parts per billion in drinking water" *Arch. Environ. Contam. Tox.* **16**, 433–439 (1987); M. I. Luster, J. A. Bland and J. H. Dean, "Molecular and cellular basis of chemically induced immunotoxicity" *Ann. Rev. Pharmacol. Tox.* **27**,

23–49 (1987); P. T. Thomas and R. V. House, "Pesticide-induced modulation of immune system" **In**: (N. N. Ragsdale and R. E. Menzer, eds.) *Carcinogenicity and Pesticides: Principles, Issues, and Relationships*, Am. Chem. Soc. Symposium, Ser. 414, Washington, DC (1989) pp. 94–106.
48. M. C. Fiore, H. A. Anderson, R. Hong, R. Golubjatnikov, J. E. Seiser, D. Nordstrom, L. Hanrahan and D. Belluck, "Chronic exposure to aldicarb-contaminated groundwater and human immune-function" *Environ. Res.* **41**, 633–645 (1986).
49. Colborn *et al.* (1996). See ref. 3.
50. Ecobichon *et al.* (1990). See ref. 37.
51. M. Lotti, "The pathogenesis of organophosphate polyneuropathy" *Crit. Rev. Toxicol.* **21**, 465–487 (1992).
52. E. P. Savage, T. J. Keefe, L. W. Mounce, R. K. Heaton, A. Lewis and P. J. Burcar, "Chronic neurological seqelae of acute organophosphate pesticide poisoning" *Arch. Environ. Health* **43**, 38–45 (1988).
53. Ecobichon *et al.* (1990).
54. Wargo (1996). See ref. 1.
55. Wiles and Campbell (1994); J. R. Davis, R. C. Brownson and R. Garcia, "Family pesticide use in the home, garden, orchard and yard" *Arch. Environ. Contam. Toxicol.* **22**, 260–266 (1992); M. Moses, E. S. Johnson, W. K. Anger, V. W. Burse, S. W. Horstman, R. J. Jackson, R. G. Lewis, K. T. Maddy, R. McConnell, W. J. Meggs and S. H. Zahm, "Environmental equity and pesticide exposure" *Toxicol. Ind. Health* **9**, 913–959 (1993).
56. P. Casey and J. A. Vale, "Deaths from pesticide poisoning in England and Wales: 1945–1989" *Human and Exp. Toxicol.* **13**, 95–101 (1994).
57. J. R. Davis, R. C. Brownson, R. Garcia, B. J. Bentz and A. Turner, "Family pesticide use and childhood brain cancer" *Arch. Environ. Contam. Toxicol.* **24**, 87–92 (1993); S. Cordier, M. J. Iglesias, C. Le Goaster, M. M. Guyot, L. Mandereau and D. Hemon, "Incidence and risk factors for childhood brain tumors in the Ile de France" *Intl. J. of Cancer* **59**, 776–782 (1994); J. K. Leiss and D. A. Savitz, "Home pesticide use and childhood cancer: a case-control study" *Am. J. Public Health* **85**, 249–252 (1995).
58. Wiles and Campbell (1994). See ref. 31.
59. Wargo (1996). See ref. 1.
60. J. C. Haliburton and W. B. Buck, "Animal poison control center: summary of telephone inquiries during first three years of service" *J. Am. Vet. Med. Assoc.* **182**, 514–515 (1983); M. J. Murphy, "Toxin exposures in dogs and cats: pesticides and biotoxins" *J. Am. Vet. Med. Assoc.* **205**, 414–421 (1994).
61. V. R. Beasley and H. Trammel, "Insecticide" **In**: (R. W. Kirk, ed.) *Current Veterinary Therapy: Small Animal Practice* (W. B. Saunders, Philadelphia, 1989) pp. 97–107.
62. J. Barton and F. Oehme, "Incidence and characteristics of animal poisonings seen at Kansas State University from 1975 to 1980" *Vet. and Human Tox.* **23**, 101–102 (1981).
63. C. S. Hornfeldt and M. J. Murphy, "1990 report of the American Association of Poison Control Centers: poisonings in animals" *Journal of the American Veterinary Medical Association* **200**(8), 1077–1080 (1992).
64. S. M. Gaafar, W. E. Howard and R. Marsh, *World Animal Science B: Parasites, Pests, and Predators* (Elsevier, Amsterdam, 1985).
65. B. M. Colvin, "Pesticide uses and animal toxicoses" *Vet. and Human Tox.* **29** (Suppl. 2), 15 (1987).
66. I. D. Robertson, A. Naprasnik and D. Morrow, "The sources of pesticide contamination in Queensland livestock" *Aus. Vet. J.* **67**, 152–153 (1990); A. Herrera, A. A. Arino, M. P. Conchello, R. Lazaro, S. Bayarri and C. Perez, "Organochlorine pesticide residues in spanish meat products and meat of different species" *J. Food Prot.* **57**, 441–444 (1994); A. Spiric and S. Raicevic, "Residues of lindane in the adipose tissue of different animal species over a three year period (1991–1993)" *Acta Veterinaria (Beograd)* **44**, 303–308 (1994); M. D. Garrido, M. Jodral and R. Pozo, "Organochlorine pesticides in Spanish sterilized milk and associated health risks" *J. Food Prot.* **57**, 249–252 (1994).
67. *Meat and Poultry Inspection. The Scientific Basis of the Nation's Program* (National Academy of Sciences, Washington, DC, 1985); L. M. Crawford and D. M. Schor, "Pesticides from a regulatory perspective" **In**: (B. G. Tweedy, H. J. Dishburger, L. G. Ballantine and J. McCarthy, eds.) *Pesticide Residues and Food Safety: A Harvest of Viewpoints* (American Chemical Society, Washington, DC, 1991) pp. 308–312.
68. M. K. Cordle, "USDA regulation of residues in meat and poultry products" *J. Anim. Sci.* **66**, 413–433 (1988).
69. NAS (1985). See ref. 67.
70. *Regulating Pesticides in Food* (National Academy of Sciences, Washington, DC, 1987).
71. NAS (1985). See ref. 67.
72. E. van Ravenswaay and E. Smith, "Food contamination: consumer reactions and producer losses" *National Food Review* (Spring), 14–16 (1986).
73. A. H. Sebae, "Incidents of local pesticide hazards and their toxicological interpretation" *Proceedings of UC/AID University of Alexandria Seminar Workshop in Pesticide Management*, Alexandria, Egypt (1977) in Bull (1982) pp. 137–152.

74. ICAITI, *An Environmental and Economic Study of the Consequences of Pesticide Use in Central American Cotton Production* (Central American Research Institute for Industry, United Nations Environment Programme, Guatemala City, Guatemala, 1977).
75. N. G. Hairston, F. E. Smith and L. B. Slobodkin, "Community structure, population control and competition" *Amer. Nat.* **94**, 421–425 (1960).
76. P. DeBach, *Biological Control of Insect Pests and Weeds* (Rheinhold, New York, 1964); C. B. Huffaker, *Biological Control* (Plenum, New York, 1977); D. Pimentel, "Herbivore population feeding pressure on plant host: feedback evolution and host conservation" *Oikos* **53**, 289–302 (1988).
77. R. van den Bosch and P. S. Messenger, *Biological Control* (Intext Educational, New York, 1973); P. L. Adkisson, "Alternatives to the unilateral use of insecticides for insect pest control in certain field crops" **In**: (L. F. Seatz, ed.) *Symposium on Ecology and Agricultural Production* (Univ. of Tennessee, Knoxville, TN, 1977) pp. 129–144; D. N. Ferro, "Insect pest outbreaks in agroecosystems" **In**: (P. Barbosa and J. C. Schultz, eds.) *Insect Outbreaks* (Academic Press, San Diego, 1987) pp. 195–215; D. M. Jackson and J. J. Lam, Jr. "Jalysus-wickhami hemiptera berytidae toxicity of pesticides applied to the soil or in the transplant water of flue-cured tobacco" *J. Econ. Entomol.* **82**, 913–918 (1989); B. A. Croft, *Arthropod Biological Control Agents and Pesticides* (Wiley, New York, 1990); L. E. Wills, B. A. Mullens and J. D. Mandeville, "Effects of pesticides on filth fly predators coleoptera histeridae staphylinidae acarina macrochelidae uropodidae in caged layer poultry manure" *J. Econ. Entomol.* **83**, 451–457 (1990).
78. Adkisson (1977); *Pest Management Strategies* (Office of Technology, U.S. Congress, Washington, DC, 1977); D. L. Murray, *Cultivating Crisis: The Human Cost of Pesticides in Latin America* (Univ. of Texas Press, Austin, TX, 1994).
79. B. E. Tabashnik and B. A. Croft, "Evolution of pesticide resistance in apple pests and their natural enemies" *Entomophaga* **30**, 37–49 (1985); R. H. Messing, B. A. Croft and K. Currans, "Assessing pesticide risk to arthropod natural enemies using expert system technology" *App. Nat. Resour. Management* **3**, 1–11 (1989); Croft, 1990; J. Kovach and A. M. Agnello, "Apple pests-pest management system for insects" **In**: (D. Pimentel, ed.) *Handbook of Pest Management in Agriculture* (CRC Press, Boca Raton, FL, 1991) pp. 107–116.
80. C. B. Huffaker and C. E. Kennett, "Experimental studies on predation: predation and cyclamen mite populations on strawberries in California" *Hilgardia* **26**, 191–222 (1956); Huffaker (1977); OTA (1979); Croft (1990); Pimentel *et al.* (1991); Murray (1994).
81. C. M. Ignoffo, D. L. Hostetter, C. Garcia and R. W. Pinnelle, "Sensitivity of the entomopathogenic fungus *Nomuraea rileyi* to chemical pesticides used on soybeans" *Environ. Ent.* **4**, 765–768 (1975); D. W. Johnson, L. P. Kish and G. E. Allen, "Field evaluation of selected pesticides on the natural development of the entomopathogen, *Nomuraea rileyi*, on the velvetbean caterpillar in soybean" *Environ. Ent.* **5**, 964–966 (1976).
82. I. N. Oka, "Success and challenges of the Indonesian national integrated pest management program in the rice-based cropping system" *Crop Protection* **10**, 163–165 (1991).
83. *Integrated Pest Management in Rice in Indonesia* (Food and Agriculture Organization, United Nations, Jakarta, May 1988).
84. FAO (1988).
85. Pimentel *et al.* (1991). See ref. 5.
86. Pimentel *et al.* (1991). See ref. 5.
87. Pimentel *et al.* (1991). See ref. 5.
88. Pimentel *et al.* (1991). See ref. 5.
89. *The State of the Environment: Selected Topics* (United Nations Environment Programme, Governing Council, Seventh Session, Nairobi, Kenya, 1979).
90. Benbrook (1996).
91. *The State of Food and Agriculture 1989* (Food and Agriculture Organization of the United Nations, Rome, 1990).
92. T. J. Dennehy, J. P. Nyrop, R. T. Roush, J. P. Sanderson and J. G. Scott, "Managing pest resistance to pesticides: a challenge to New York's agriculture" *New York's Food and Life Sciences Quarterly* **17**, 4 and 13–17 (1987); Murray, 1994.
93. P. L. Adkisson, "The integrated control of the insect pests of cotton" *Tall Timbers Conf. Ecol. Control Habitat Mgmt.* **4**, 175–188 (1972); *Pest Control: An Assessment of Present and Alternative Technologies* (National Academy of Sciences, Washington, DC, 1975).
94. Carrasco-Tauber (1989).
95. S. O. Archibald, A Dynamic Analysis of Production Externalities: Pesticide Resistance in California, *Ph. D. Thesis*, (University of California, Davis, CA, 1984).
96. A. M. LaFarge, "The persistence of resistance" *Agrichemical Age* **29**, 10–12 (1985); C. R. Harper and D. Zilberman, "Pesticide regulation: problems in trading off economic benefits against health risks" **In**: (D. Zilberman and J. B. Siebert, eds.) *Economic Perspectives on Pesticide Use in California* (University of California, Berkeley, CA, 1990) pp. 181–208.

97. G. P. Georghiou, "The magnitude of the resistance problem" *Pesticide Resistance, Strategies and Tactics for Management* (National Academy of Sciences, Washington, DC, 1986) pp. 18–41.
98. D. Faber, *Environment Under Fire* (Monthly Review Press, New York, 1993); Murray (1994).
99. R. Reuben, "Obstacles to malaria control in India-the human factor" **In**: (W. W. Service, ed.) *Demography and Vector-Borne Disease* (CRC Press, Boca Raton, FL, 1989) pp. 143–154.
100. *Our Planet, Our Health: Report of the WHO Commission on Health and Environment* (World Health Organization, Geneva, 1992); *Malaria Prevention and Control* (National Academy of Sciences, Washington, DC, 1991); *Malaria: Obstacles and Opportunities* (National Academy of Sciences, Washington, DC, 1991).
101. K. W. Richards, "Non-apis bees as crop pollinators" *Revue Suisse de Zoologie* **100**, 807–822 (1993); C. S. Williamson, "Conserving Europe's bees: why all the buzz?" *TREE* **10**, 309–310 (1995).
102. W. E. Robinson, R. Nowogrodzki and R. A. Morse, "The value of honey bees as pollinators of U.S. crops" *Am. Bee J.* **129**, 477–487 (1989).
103. M. L. Winston and S. C. Scott, "The value of bee pollination to Canadian apiculture" *Can. Beekeep.* **11**, 134 (1984).
104. R. A. Gill, "The value of honeybee pollination to society" *Acta Hort.* **288**, 62–68 (1991).
105. K. F. Haynes, "Sublethal effects of neurotoxic insecticides on insect behavior" *Ann. Rev. Entom.* **33**, 149–168 (1988); K. MacKenzie and M. L. Winston, "Effects of sublethal exposure to diazinon and temporal division of labor in the honeybee" *J. Econ. Entom.* **82**, 75–82 (1989); S. L. Buchmann and G. P. Nabhan, *The Forgotten Pollinators* (Island Press, Washington, DC, 1996).
106. Buchmann and Nabhan (1996). See ref. 105.
107. S. E. McGregor, C. Rhyne, S. Worley and F. E. Todd, "The role of honeybees in cotton pollination" *Agron. Jour.* **47**, 23–25 (1955); V. Mahadevan and K. C. Chandy, "Preliminary studies on the increase in cotton yield due to honeybee pollination" *Madras Agr. Jour.* **46**, 23–26 (1959); A. S. Tanda, "Bee pollination increases yield of 2 interplanted varieties of asiatic cotton (*gossypium arboreum l.*) (*apis cerana indica, apis mellifera*, India)" *Am. Bee J.* **124**, 539–540 (1984).
108. S. E. McGregor, *Insect pollination of cultivated crop plants*, Agricultural Handbook No. 496 (U.S. Dept. of Agr., Agr. Res. Ser., Washington, DC, 1976).
109. R. C. Sihag, "Effects of pesticides and bee pollination on seed yield of some crops in India" *J. Apicultural Res.* **27**, 49–54 (1988); E. Mussen, "California crop pollination" *Cleanings in Bee Culture* **118**, 646–647 (1990); R. W. Currie, M. L. Winston and K. N. Slessor, "Effect of synthetic queen mandibular pheromone sprays on honey bee hymenoptera apidae pollination of berry crops" *J. Econ. Entomol.* **85**, 1300–1306 (1992); D. M. Cribb, D. W. Hand and R. N. Edmondson, "A comparative study of the effects of using the honeybee as a pollinating agent of glasshouse tomato" *J. Hortic. Sci.* **68**, 79–88 (1993).
110. W. L. Chan, "Comparative study of weed control methods in rice" *Jour. Taiwan Agr. Res.* **14**, 1–13 (1965).
111. ICAITI, 1977; V. R. Reddy, D. N. Baker, F. D. Whisler and R. E. Fye, "Application of GOSSYM to yield decline in cotton I. Systems analysis of effects of herbicides on growth, development, and yield" *Agron. Jour.* **79**, 42–47 (1987); J. T. Trumbel, W. Carson, H. Nakakihara and V. Voth, "Impact of pesticides for tomato fruitworm (*Lepidoptera: Noctuidae*) suppression on photosynthesis, yield, and nontarget arthropods in strawberries" *J. Econ. Entomol.* **81**, 608–614 (1988).
112. I. N. Oka and D. Pimentel, "Herbicide (2,4-D) increases insect and pathogen pests on corn" *Science* **193**, 239–240 (1976); J. Altman, "Impact of herbicides on plant diseases" **In**: (C. A. Parker, A. D. Rovia, K. J. Moore and P. T. W. Wong, eds.) *Ecology and Management of Soilborne Plant Pathogens* (Am. Phytopathological Soc., St. Paul, MN, 1985) pp. 227–231; A. D. Rovira and H. J. McDonald, "Effects of the herbicide chlorsulfuron on rhizoctonia bare patch and take-all of barley and wheat" *Plant Disease* **70**, 879–882 (1986).
113. R. von Rumker and F. Horay, *Farmers' Pesticide Use Decisions and Attitudes on Alternate Crop Protection Methods* (U.S. Environmental Protection Agency, Washington, DC, 1974); B. R. Elliot, J. M. Lumb, T. G. Reeves and T. E. Telford, "Yield losses in weed-free wheat and barley due to post-emergence herbicides" *Weed Res.* **15**, 107–111 (1975); M. B. Akins, L. S. Jeffery, J. R. Overton and T. H. Morgan, "Soybean response to preemergence herbicides" *Proc. S. Weed Sci. Soc.* **29**, 50 (1976).
114. J. Henderson, *Legal Aspects of Crop Spraying*, Univ. Ill. Agr. Exp. Sta. Circ. 99 (Univ. of Illinois, Urbana, IL, 1968); C. J. Barnes, T. L. Lavy and J. D. Mattice, "Exposure of non-applicator personnel and adjacent areas to aerially applied propanil" *Bul. Environ. Contam. and Tox.* **39**, 126–133 (1987).
115. G. W. Ware, B. J. Estesen, W. P. Cahill, P. D. Gerhardt and K. R. Frost, "Pesticide drift. I. High-clearance vs aerial application of sprays" *J. Econ. Entomol.* **62**, 840–843 (1969).
116. G. W. Ware, W. P. Cahill, P. D. Gerhardt and J. W. Witt, "Pesticide drift IV: on-target deposits from aerial application of insecticides" *J. Econ. Entomol.* **63**, 1982–1983 (1970); ICAITI (1977);

G. W. Ware, *Reducing Pesticide Application Drift-Losses* (University of Arizona, College of Agriculture, Cooperative Extension, Tucson, AZ, 1983); G. W. Ware, W. P. Cahill, B. J. Estesen, W. C. Kronland and N. A. Buck, "Pesticide drift deposit efficiency from ground sprays on cotton" *J. Econ. Entomol.* **68**, 549–550 (1975); N. B. Akesson and W. E. Yates, "Physical parameters affecting aircraft spray application" **In**: (W. Y. Garner and J. Harvey, eds.) *Chemical and Biological Controls in Forestry* (Am. Chem. Soc. Ser. 238, Washington, DC, 1984) pp. 95–111; F. Mazariegos, *The Use of Pesticides in the Cultivation of Cotton in Central America* (United Nations Environment Programme, Industry and Environment, Guatemala, July/August/September, 1985).

117. Ware *et al.* (1975); F. R. Hall, "Pesticide application technology and integrated pest management (IPM)" **In**: (D. Pimentel, ed.) *Handbook of Pest Management in Agriculture*, **Vol. II**. (CRC Press, Boca Raton, FL, 1991) pp. 135–170.
118. D. Hanner, "Herbicide drift prompts state inquiry" *Dallas (Texas) Morning News* (July 25, 1984).
119. *Census of Civil Aircraft* (U.S. Federal Aviation Administration, Washington, DC, 1988).
120. H. V. Nanjappa and N. M. Hosmani, "Residual effect of herbicides applied in transplanted fingermillet (*Eleusive coracana Gaertu.*) on the succeeding crops" *Indian Jour. of Agron.* **28**, 42–45 (1983); C. B. Rogers, "Fluometuron carryover and damage to subsequent crops" *Dissertation Abstract International* **45**(8), 2375B (1985); J. W. Keeling, R. W. Lloyd and J. R. Abernathy, "Rotational crop response to repeated applications of korflurazon" *Weed Tech.* **3**, 122–125 (1989).
121. J. Altman, "Impact of herbicides on plant diseases" **In**: (C. A. Parker, A. D. Rovia, K. J. Moore and P. T. W. Wong, eds.) *Ecology and Management of Soilborne Plant Pathogens* (Am. Phytopathological Soc., St. Paul, MN, 1985) pp. 227–231.
122. *Restoring the Quality of Our Environment* (President's Science Advisory Committee, The White House, Washington, DC, 1965); National Academy of Sciences, 1975; Pimentel *et al.*, 1991.
123. *San Joaquin County Agricultural Report 1989* (San Joaquin County Department of Agriculture, San Joaquin County, CA, 1990).
124. *Yolo County 1989 Agricultural Report* (Office of the Agricultural Commissioner, Yolo County, CA, 1990).
125. U.S. Department of Agriculture, *Agricultural Statistics* (U.S. Department of Agriculture, Government Printing Office, Washington, DC, 1994).
126. Calculated based on data from: "Food and Drug Administration pesticide program residues in foods, 1989" *Jour. Assoc. Off. Anal. Chem.* **73**, 127A–146A (1990); and U.S. Department of Agriculture, *Agricultural Statistics* (U.S. Department of Agriculture, Government Printing Office, Washington, DC, 1989).
127. R. B. Taylor, "State sues three farmers over pesticide use on watermelons" *Los Angeles Times* I (CC) 3–4 (1986); K. Kizer, *California's Fourth of July Food Poisoning Epidemic from Aldicarb-Contaminated Watermelons* (State of California Department of Health Services, Sacramento, 1986).
128. *Legislative Counsel's Digest*, State of California, Legislative Assembly Bill No. 2755 (1986).
129. P. R. Richards, J. W. Kramer, D. B. Baker and K. A. Drieger, "Pesticides in rainwater in the northeastern United States" *Nature* **327**, 129–131 (1987); Natural Resources Defense Council, *After Silent Spring* (NRDC Publications, New York, NY, 1993).
130. D. E. Glotfelty, "Pesticides in fog" *Nature* **325**, 602–603 (1987); Natural Resources Defense Council (1993).
131. N. D. Ananyeva, N. N. Naumova, J. Rogers and W. C. Steen, "Microbial transformation of selected organic chemicals in natural aquatic systems" **In**: (J. L. Schnoor, ed.) *Fate of Pesticides and Chemicals in the Environment* (John Wiley & Sons, New York, 1992) pp. 275–294.
132. R. D. Kelley, "Pesticides in Iowa's drinking water" *Proceedings of a Conference: Pesticides and Groundwater: A Health Concern for the Midwest*, Freshwater Foundation, Navarre, MN (Oct. 16–17, 1986) pp. 121–122.
133. B. Cohen, B. Wiles and E. Bondoc, *Weed Killers by the Glass* (Environmental Working Group, Washington, DC, 1995).
134. Cohen *et al.* (1995). See ref. 133.
135. J. Auerbach, "Costs and benefits of current SDWA regulations" *J. Am. Water Works Assoc.* **86**, 69–78 (1994); U. Natarajan and R. Rajagopal, "Economics of screening for pesticides in ground water" *Water Resour. Bull.* **30**, 579–588 (1994).
136. C. D. Osteen and P. I. Szmedra, *Agricultural Pesticide Use Trends and Policy Issues*, Agr. Econ. Report No. 622 (U.S. Dept. of Agr., Econ. Res. Ser., Washington, DC, 1989).
137. T. Holmes, E. Nielsen and L. Lee, *Managing Groundwater Contamination in Rural Areas: Rural Development Perspectives*. (U.S. Dept. of Agr., Econ. Res. Ser., Washington, DC, 1988); Natural Resources Defense Council, 1993.
138. *National Pesticide Survey—Summary* (U.S. Environmental Protection Agency, Washington, DC, 1990).

139. E. G. Nielsen and L. K. Lee, *The Magnitude and Costs of Groundwater Contamination from Agricultural Chemicals. A National Perspective*, Econ. Res. Ser. Staff Report AGES870318 (U.S. Dept. of Agr., Econ. Res. Ser., Natural Resour. Econ. Div., Washington, DC, 1987).
140. P. L. McCarty, P. V. Roberts, M. Reinhard and G. Hopkins, "Movement and Transformation of Halogenated Aliphatic Compounds in Natural Systems" **In**: (J. L. Schnoor, ed.) *Fate of Pesticides and Chemicals in the Environment* (John Wiley & Sons, New York, 1992) pp. 191–210; H. Beitz, H. Schmidt and F. Herzel, "Occurrence, toxicological and ecotoxicological significance of pesticides in groundwater and surface water" **In**: (H. Borner, ed.) *Pesticides in Ground and Surface Water* (Springer-Verlag, Berlin, 1994) pp. 1–56.
141. D. I. Gustafson, *Pesticides in Drinking Water* (Van Nostrand Reinhold, New York, 1993).
142. R. J. Larson and R. M. Ventullo, "Biodegradation potential of groundwater bacteria" **In**: (D. M. Nielsen, ed.) *Proceedings of the Third National Symposium on Aquifer Restoration and Groundwater Monitoring*, May 25–27, National Water Well Association, Worthington, OH (1983) pp. 402–409; V. Pye and J. Kelley, "The extent of groundwater contamination in the United States" **In**: (NAS, ed.) *Groundwater Contamination* (National Academy of Sciences, Washington, DC, 1984).
143. *The Global 2000 Report to the President* (Council on Environmental Quality and the U.S. Department of State, Washington, DC, 1980).
144. "Shell Loses Suit on Cleanup Cost" *New York Times*, New York, NY, A24 (1988).
145. Based on the costs of cleaning water: R. M. Clark, "Water supply regionalization: a critical evaluation" *Proc. Am. Soc. Civil Eng.* **105**, 279–294 (1979); F. van der Leeden, F. L. Troise and D. K. Todd, *The Water Encyclopedia*, 2nd ed. (Lewis Pub., Chelsea, MI, 1990).
146. *The Second RCA Appraisal: Soil, Water, and Related Resources on Nonfederal Land in the United States, Analysis of Conditions and Trends* (U.S. Department of Agriculture, Washington, DC, 1989).
147. R. B. Clark, *Marine Pollution* (Clarendon Press, Oxford, 1989).
148. Nielsen and Lee (1987). See ref. 139.
149. T. Tsuda, S. Aoki, M. Kojima and H. Harada, "Pesticides in water and fish from rivers flowing into Lake Biwa" *Toxicol. Environ. Chem.* **34**, 39–55 (1991); A. G. Miskiewicz and P. J. Gibbs, "Organochlorine pesticides and hexachlorobenzene in tissues of fish and invertebrates caught near a sewage outfall" *Environ. Poll.* **84**, 269–277 (1994); J. K. Buttner, J. C. Makarewicz and T. W. Lewis, "Concentration of selected priority organic contaminants in fish maintained on formulated diets in Lake Ontario water" *The Prog. Fish-Culturist* **57**, 141–146 (1995); M. Chevreuil, A. Carru, A. Chesterikoff, P. Boet, E. Tales and J. Allardi, "Contamination of fish from different areas of the river Seine (France) by organic (PCB and pesticides) and metallic (Cd, Cr, Cu, Fe, Mn, Pb, Zn) micropollutants" *Sci. of the Total Environ.* **162**, 31–42 (1995); Y. Wang, C. Jaw and Y. Chen, "Accumulation of 2,4-D and glyphosate in fish and water hyacinth" *Water, Air and Soil Poll.* **74**, 397–403 (1994).
150. D. W. Kuehl and B. Butterworth, "A national study of chemical residues in fish III: study results" *Chemosphere* **29**, 523–535 (1994).
151. *Fish Kills Caused by Pollution, 1977–1987*, Draft Report (U.S. Environmental Protection Agency, Office of Water Regulations and Standards, Washington, DC, 1990); B. A. Knuth, "Implementing chemical contamination policies in sport-fisheries: agency partnerships and constituency influence" *Jour. Mgt. Sci. and Policy Anal.* **6**, 69–81 (1989); *Guide to Eating Ontario Sport Fish* (Ministry of Environment and Ministry of Natural Resources, Ontario, 1990).
152. New York State Department of Environmental Conservation, *New York Fishing Regulations Guide 1994–1995* (WJF Marketing Services, Inc.).
153. EPA, *Fish Kills Caused by Pollution, 1977–1987* (1990).
154. EPA, *Fish Kills Caused by Pollution, 1977–1987* (1990).
155. N. K. Neskovic, V. Karan, I. Elezovic, V. Poleksic and M. Budimir "Toxic effects of 2,4-D herbicide on fish" *J. Environ. Sci. Health* **29**, 265–279 (1994).
156. M. B. Bain and S. E. Boltz, "Effect of aquatic plant control on the microdistribution and population characteristics of largemouth bass" *Trans. of the Amer. Fish. Soc.* **121**, 94–103 (1992); R. Montgomery, "Fishing not hurt by weed killers, research shows" *Bassmaster Magazine* **28**, 17 (1995).
157. "Monetary values of freshwater fish and fish-kill counting guidelines" *Special Pub. No. 13* (Amer. Fisheries Society, Bethesda, MD, 1982).
158. "Too much beer kills thousands" *Oregon State University Barometer* (May 14, 1991).
159. A. R. Graefe and R. B. Ditton, "Bay and offshore fishing in the Galveston Bay area: a comparative study of fishing patterns, fishermen characteristics, and expenditures" *N. Am. J. Fish. Man.* **6**, 192–199 (1986); N. A. Connelly and T. L. Brown, "Net economic value of the freshwater recreational fisheries of New York" *Trans. Amer. Fish. Soc.* **120**, 770–775 (1991).
160. D. A. Storey and P. G. Allen, "Economic impact of marine recreational fishing in Massachusetts" *N. Am. J. Fish. Man.* **13**, 698–708 (1993).

161. F. L. McEwen and G. R. Stephenson, *The Use and Significance of Pesticides in the Environment* (John Wiley & Sons, Inc., New York, 1979); C. E. Grue, W. J. Fleming and E. F. Hill, "Assessing hazards of organophosphate pesticides to wildlife" *Transactions of the North American Wildlife and Natural Resources Conference (48th)*, Washington, DC, Wildlife Management Institute (1983); R. W. Risebrough, "Pesticides and bird populations" R. F. Johnston, ed., *Current Ornithology* (Plenum Press, New York, 1986) pp. 397–427; G. J. Smith *Pesticide Use and Toxicology in Relation to Wildlife: Organophosphorus and Carbamate Compounds*, Resource Publication 170 (U.S. Dept. of Interior, Fish and Wildlife Service, Washington, DC, 1987); K. M. Fluetsch and D. W. Sparling, "Avian nesting success and diversity in conventionally and organically managed apple orchards" *Env. Tox. Chem.* **13**, 1651–1659 (1994).
162. Pimentel *et al.* (1991). See ref. 5.
163. P. Mineau and B. T. Collins, "Avian mortality in agro-ecosystems: 2, Methods of detection" **In**: (M. P. Greaves, B. D. Smith and P. W. Greig-Smith, eds.) *Field Methods for the Study of Environmental Effects of Pesticides, Proceedings of a Symposium Organized by the British Crop Protection Council*, Churchill College, Cambridge, UK (1988) pp. 13–27.
164. P. Mineau, "Avian mortality in agroecosystems: I. The case against granule insecticides in Canada" **In**: (M. P. Greaves, B. D. Smith and P. W. Greig-Smith, eds.) *Field Methods for the Study of Environmental Effects of Pesticides*, British Crop Protection Council (BPCP) Monograph 40 (BPCP, Thornton Heath, London, 1988) pp. 3–12.
165. R. Balcomb, "Songbird carcasses disappear rapidly from agricultural fields" *Auk* **103**, 817–821 (1986).
166. R. W. Knapton and P. Mineau, "Effects of granular formulations of terbufos and fonofos applied to cornfields on mortality and reproductive success of songbirds" *Ecotoxicology* **4**, 138–153 (1995).
167. D. H. White, C. A. Mitchell, L. D. Wynn, E. L. Flickinger and E. J. Kolbe, "Organophosphate insecticide poisoning of Canada geese in the Texas Panhandle" *J. Field Ornithology* **53**, 22–27 (1982).
168. E. L. Flickinger, K. A. King, W. F. Stout and M. M. Mohn, "Wildlife hazards from furadan 3G applications to rice in Texas" *J. Wildlife Mgt.* **44**, 190–197 (1980); E. L. Flickinger, G. Juenger, T. J. Roffe, M. R. Smith and R. J. Irwin, "Poisoning of Canada geese in Texas by parathion sprayed for control of Russian wheat aphid" *J. Wildlife Diseases* **27**, 265–268 (1991).
169. *Carbofuran: A Special Review Technical Support Document* (U.S. Environmental Protection Agency, Office of Pesticides and Toxic Substances, Washington, DC, 1989).
170. W. B. Stone and P. B. Gradoni, "Wildlife mortality related to the use of the pesticide diazinon" *Northeastern Environ. Sci.* **4**, 30–38 (1985).
171. G. R. Potts, *The Partridge: Pesticides, Predation and Conservation* (Collins, London, 1986).
172. Potts (1986). See ref. 171.
173. A. R. Hardy, "Estimating exposure: the identification of species at risk and routes of exposure" **In**: (L. Somerville and C. H. Walker, eds.) *Pesticide Effects on Terrestrial Wildlife* (Taylor & Francis, London, 1990) pp. 81–97.
174. C. Chen, "Flavoring against fowl" *Cornell Countryman* (April/May, 1995) p. 24.
175. F. N. Mastrota and J. A. Mench, "Evaluation of taste repellents with northern bobwhites for deterring ingestion of granular pesticides" *Env. Tox. Chem.* **14**, 631–638 (1995).
176. Mineau (1988). See ref. 164.
177. P. W. Greig-Smith, "Understanding the impact of pesticides on wild birds by monitoring incidents of poisoning" **In**: (R. J. Kendall and T. E. Lacher, eds.) *Wildlife Toxicology and Population Modeling* (Lewis Publishers, Boca Raton, FL, 1990) pp. 301–320.
178. W. H. Stickel, L. F. Stickel, R. A. Dyrland and D. L. Hughes, "DDE in birds: lethal residues and loss rates" *Arch. Environ. Contam. and Tox.* **13**, 1–6 (1984); Risebrough (1986); L. M. Gonzalez and F. Hiraldo. "Organochlorines and heavy metal contamination in the eggs of the Spanish Imperial Eagle (*Aquila [heliaca]adaberti*) and accompanying changes in eggshell morphology and chemistry" *Environ. Poll.* **51**, 241–258 (1988); J. E. Elliot, R. J. Norstrom and J. A. Keith, "Organochlorines and eggshell thinning in Northern Gannets (*Sula bassanus*) from Eastern Canada 1968–1984" *Environ. Poll.* **52**, 81–102 (1988); P. Mineau, D. C. Boersma and B. Collins, "An analysis of avian reproduction studies submitted for pesticide registration" *Ecotox. Env. Safety* **29**, 304–329 (1994).
179. J. C. Bednarz, D. Klem, L. J. Goodrich and S. E. Senner, "Migration counts of raptors at Hawk Mountain, Pennsylvania as indicators of population trends, 1934–1986" *Auk* **107**, 96–109 (1990).
180. Stickel *et al.* (1984). See ref. 178.
181. M. H. Salem, Z. Abo-Elezz, G. A. Abd-Allah, G. A. Hassan and N. Shakes, "Effect of organophosphorus (dimethoate) and pyrethroid (deltamethrin) pesticides on semen characteristics in rabbits" *J. Environ. Sci. Health* **B23**, 279–290 (1988); Foote *et al.* (1986).
182. P. D'Anieri, D. M. Leslie and M. L. McCormack, "Small mammals in glyphosphate-treated clearcuts in northern Maine" *Canadian Field Nat.* **101**, 547–550 (1987).
183. S. Begley and D. Glick, "The estrogen complex" *Newsweek* 76 77 (March 21, 1994).

184. D. Pimentel, *Ecological Effects of Pesticides on Non-Target Species* (U.S. Govt. Printing Office, Washington, DC, 1971).
185. W. Fantle, "The incredible shrinking man" *The Progressive* 12–13 (October, 1994).
186. E. Carlsen, A. Giwercman, N. Keiding and N. E. Skakkebaek, "Evidence for decreasing quality of semen during past 50 years" *Brit. Med. Jour.* **305**, 609–613 (1992).
187. *1985 Survey of Fishing, Hunting, and Wildlife Associated Recreation* (U.S. Fish and Wildlife Service, U.S. Dept. of Interior, Washington, DC, 1988).
188. U.S. Fish and Wildlife Service (1988). See ref. 187.
189. U.S. Fish and Wildlife Service (1988). See ref. 187.
190. F. E. Walgenbach, *Economic Damage Assessment of Flora and Fauna Resulting from Unlawful Environmental Degradation* (California Department of Fish and Game, Sacramento, CA, 1979); R. T. Tinney, *The Oil Drilling Prohibitions at the Channel Islands and Pt. Reyes-Fallallon Island National Marine Sanctuaries: Some Costs and Benefits*, Report to Center for Environmental Education, Washington, DC, 1982; J. Dobbins, *Resources Damage Assessment of the T/V Puerto Rican Oil Spill Incident*, Report to NOAA, Sanctuary Program Division (James Dobbins Associates, Inc., Washington, DC, 1986); P. C. James, "Internalizing externalities: granular carbofuran use on rapeseed in Canada" *Ecol. Econ.* **13**, 181–184 (1995).
191. J. H. Blew, "Breeding bird census. 92 conventional cash crop" *Farm. Jour. Field Ornithology* **61**, Suppl., 80–81 (1990).
192. *Federal and State Endangered Species Expenditures* (U.S. Fish and Wildlife Service; Washington, DC, 1991).
193. R. Stanier, M. Doudoroff and E. Adelberg, *The Microbial World* (Prentice Hall, London, 1970).
194. R. M. Atlas and R. Bartha, *Microbial Ecology: Fundamentals and Applications* 2nd ed. (Benjamin Cummings Co., Menlo Park, CA, 1987).
195. T. Brock and M. Madigan, *Biology of Microorganisms* (Prentice Hall, London, 1988).
196. Brock and Madigan (1988).
197. F. D. Hole, "Effects of animals on soil" *Geoderma* **25**, 75–112 (1981); C. A. Edwards and J. R. Lofty, "Nitrogenous fertilizers and earthworm populations in agricultural soils" *Soil Biol. Biochem.* **14**, 515–521 (1982).
198. L. Somerville and M. P. Greaves, *Pesticide Effects on Soil Microflora* (Taylor and Francis, London, 1987); C. A. Edwards, "Impact of herbicides on soil ecosystems" *Crit. Rev. in Plant Sci.* **8**, 221–257 (1989); A. L. Jones, D. B. Johnson and D. L. Suett, "Effects of soil treatments with aldicarb, carbofuran and chlorfenvinphos on the size and composition of microbial biomass" **In**: (A. Walker, ed.) *Pesticides in Soils and Water: Current Perspectives* (United Kingdom, British Crop Protection Council, 1991) pp. 75–82.
199. I. Ahmad and D. Malloch, "Interaction of soil microflora with the bioherbicide phosphinothricin" *Agri. Ecosys. and Envir.* **54**, 165–174 (1995).
200. Pimentel (1971). See ref. 184.
201. T. C. Kuo and Y. T. Huang, "Lethal effects of the five commonly used pesticides on the earthworm *Bimastus parvus" Eisen. J. Agri. Assoc. China New Series* **162**, 33–42 (1993); H. Kula, "Comparison of laboratory and field testing for the assessment of pesticide side effects on earthworms" *Acta Zoologica Fennica* **196**, 338–341 (1995); A. I. Mohamed, G. A. Nair, H. H. Kassem and M. Nuruzzaman, "Impacts of pesticides on the survival and body mass of the earthworm, *Aporrectodea caliginosa* (Annelida: Oligochaeta)" *Acta Zoologica Fennica* **196**, 344–347 (1995); S. A. Reinecke, A. J. Reinecke, M. L. Froneman, "The effects of dieldrin on the sperm ultrastructure of the earthworm *Eudrilus eugeniae* (Oligochaeta)" *Environmental Toxicology and Chemistry* **14**(6), 961–965 (1995).
202. A. Stringer and C. Lyons, "The effect of benomyl and thiophanate-methyl on earthworm populations in apple orchards" *Pesticide Science* **5**, 189–196 (1974).
203. C. Edwards and J. Lofty, *Biology of Earthworms* (Chapman and Hall, London, 1977).
204. D. A. Potter and S. K. Braman, "Ecology and management of turfgrass insects" *Ann. Rev. Entom.* **36**, 383–406 (1991).
205. Benbrook (1996). See ref. 2.
206. *Pesticides: EPA's Formidable Task to Assess and Regulate Their Risks* (U.S. General Accounting Office, Washington, DC, 1986).
207. USBC (1995). See ref. 43.
208. Benbrook (1996). See ref. 2.
209. Pimentel *et al.* (1991); USBC (1995). See ref. 5, ref. 43.
210. Pimentel *et al.* (1991). See ref. 5.

7. SECONDARY EFFECTS OF PESTICIDE EXPOSURES

MARC LAPPÉ *

INTRODUCTION

Many second-order effects are not anticipated in either the primary toxicity testing regime or the toxicity profiles generated by knowledge of the pesticide's classification. The US Environmental Protection Agency (EPA) typically works with a substantial regulatory burden, making consideration of secondary impacts of pesticides highly impractical. The EPA reviews over 5000 registration submissions each year, including a review of at least 20 applications for a new registration of product[1]. While it is axiomatic that all pesticides currently in use must fulfill certain toxicity testing requirements prior to their registration[2], such testing fails to anticipate all potential adverse effects stemming from exposure, particularly if such exposure is protracted and at low doses. There is presently a window of opportunity for expanding the net of tests required for registration to anticipate secondary effects since international agencies are presently working towards an integrated management and risk assessment program under the International Programme on Chemical Safety[3]. The current impetus to harmonize risk assessment across national borders is presently concentrating on the health risk assessment of pesticides.

CURRENT PRACTICE

Present toxicity testing typically includes acute and chronic toxicity multigenerational studies of reproductive effects, neurotoxicity, genotoxicity, and high-dose carcinogenicity testing. Formal requirements do not presently embrace assays to determine if a suspect pesticide contributes to endocrinologic disruption, immunotoxicity, or more subtle neurotoxic effects as measured by performance or behavior. Practical limitations also preclude testing of combinations of pesticides for potential additive, potentiating or synergistic effects.

SOURCES OF EXPOSURE

Because many of the more subtle secondary effects of pesticide exposure occur long after exposure has ceased and some subclinical findings have been made after only low level exposure it is important to review possible unforeseen sources for exposure. This is particularly important in light of the fact that the effective doses of pesticides in children's diets often exceeds that of adults because of dietary habits or body weight parameters. In some Asia-Pacific countries, pesticide exposure of the general

* Director, CETOS (Center for Ethics and Toxics), 39141 S. Highway One PO Box 673, Gualala CA 95445 (USA).

population is estimated by a team of Sri Lankan investigators to be unusually high compared to other regions[4]. In the United States, the average dietary exposure to certain pesticides, notably dieldrin and possibly lindane, currently (circa 1996–97) exceeds the health-based standards established by the US Environmental Protection Agency[5].

Occupational exposures provide a more consistent though still varied dosage spectrum. A study by the University of California based Institute for Health Policy Studies in San Francisco found that estimated absorbed daily doses for mixers, loaders and applicators of pesticides ranged from less than 0.0001% to 48% of the LD50 values for any given pesticide. In spite of the high end exposures, the major concern was for chronic rather than acute effects[6]. Data from Minnesota on chronic illness supports this proposition, since chronic but not acute illness was statistically elevated among a group of 1000 state licensed pesticide applicators[7].

A SYNTHETIC VIEW

Many secondary toxic effects are predictable consequences of the known interactions of pesticides with key receptors. Neurological effects often depend on the binding of the pesticide to a receptor site, notably the acetylcholinesterase receptor or others in the central nervous system. Other toxic effects are associated with induction or inhibition of key metabolic enzymes, binding to aryl hydrocarbon receptors, cytochrome P450 enzymes, and flavin-containing monooxygenases[8]. Less well known are the close interactions between the nervous and the immune and reproductive systems. The key organs in these systems, notably the spleen and pituitary–adrenal axis are innervated and many of the mediating cells of the immune system carry receptors for neurotransmitters such as dopamine and endorphins. These observations suggest that an effect on the nervous system may produce a cascade of secondary effects that can affect reproductive and immune function. By binding with critical receptors, different pesticides may produce cumulative effects, even at doses lower than those now appreciated to produce toxicity.

IMMUNOLOGICAL EFFECTS

Studies of pesticide manufacture workers have suggested that job-related exposures can be linked to hematological abnormalities linked to immune function. A study in Poland of 19 male organophosphate pesticide workers vs 18 unexposed female controls in the same plant showed an elevation in leukocyte counts, coupled with elevations of two of three antibody fractions (IgG and IgA)[9]. In addition to this evidence of immunostimulation, the same study found evidence of immune depression for helper lymphocyte and neutrophils. Since both of these latter indices are correlated with immune function, these results can be interpreted to reflect an impairment of immunity more generally. While full-blown human evidence of immunosuppression is currently based on case reports, a large body of animal literature strongly suggests that such effects, even following chronic low level-exposure, may be found[10].

It is particularly relevant that exposure to organophosphorous pesticides (dimethoate and methylparathion) in rats leads to immune system damage following three generations of exposure to 1/75 of the LD50 dose[11]. Other organophosphate pesticides, notably phosalone, have also been shown to produce damage to the cellular and humoral response of rodents[12]. At least one research team believes that pesticides may directly damage the humoral arm of immunity, generally sparing cellular immune functions[13].

HEMATOLOGICAL MALIGNANCIES

Studies showing paradoxical stimulation of white cell counts suggest that under some circumstances pesticide exposure can lead to a stimulation and/or provocation of certain immune cell lingeages such as specific antibody forming cells or leukocyte populations generally. In keeping with this possibility, some research has shown a temporal spatial link between rural farming activites and lymphoproliferative malignancies generally[14]. Several tumors of this type are consistently elevated among pesticide-using farmers[15]. The credibility of this association is increased by the demonstrated ability of specific organophosphate pesticides to break chromosomes[16], or, in the case of malathion, to increase mutation rates at particular loci in lymphocyte target cells[17]. Additional support for this linkage is given by the consistency of the findings: several independent studies have found a reproducible association of pesticide exposure to lymphoproliferative tumors[18]. The ability of several pesticides to provoke immune stimulation might explain in part the otherwise problematic association of pesticides without proven carcinogenicity with tumors of the immune system, notably chronic lymphocytic leukemia (CLL), non-Hodgkin's lymphoma (NHL) and myeloma. In the latter instance, pesticide-exposed workers have been found to be at an elevated risk for a condition thought to be a precursor to myeloma, namely monoclonal gammopathy[19]. This condition reflects the proliferation of a single clone of B cell antibody producing lymphocytes. Should one or more such cell lines become immortalized, the malignancy known as multiple myeloma develops.

The possible causal association of prior pesticide exposure to CLL and NHL is strongly suggested by studies which identify prior work-place conditions involving pesticides with a heightened risk of disease. A case in point is the study of 187 cases of CLLs and NHLs compared to 977 population controls. Occupational history was significantly linked to a raised odds ratio for both tumor types where those occupations entailed use of carbamate and organophosphate pesticides[20]. Pesticide exposure was also strongly linked to certain types of brain cancer (notably non-astrocytic neuroepithelial tumors) in a Norwegian study, particularly among children age 0–14 years who were offspring of parents with heavy pesticide usage[21].

These observations suggest that unanticipated effects may occur more commonly than recognized. By binding with cellular receptors, some pesticides may work through secondary pathways, producing subtle chronic toxicity. Delayed neurotoxicity may be an example of such an effect. A number of organophosphate pesticides (but not all, chlorpyrifos is a notable exception) can cause delayed neurotoxicity, a syndrome that includes ataxia, tremors and peripheral neuropathy.

CHRONIC AND DELAYED NEUROTOXICITY

A delayed effect from high-level exposure to organophosphate-containing compound has been recognized since the 1930s when triorthocresylphosphate contaminated ginger drinks produced thousands of cases of "ginger jake" in the southern United States. In this state, affected individuals show characteristic spasticity. Other data have shown that long term exposure to organophosphates can produce losses in vibrotactile sensitivity, impairment of neurobehavioral abilities, and peripheral neuropathies. Using the hen model, one research team has found that combinations of inhibitors, pesticides and repellants can greatly exacerbate their individual effects[22]. Evidence of peripheral nerve damage is clearest in workers who have had chronic exposure to multiple organophosphate pesticides. For instance, peripheral nerve function was found to be significantly impaired as measured by conduction velocities in 131 Dutch flower bulb farmers following a 20 year history of low-level exposure to various pesticides[23].

Vibration threshold sensitivity tests performed on 90 male pesticide applicators and farmers who excreted urinary metabolites of guthion demonstrated a clear decrement, particularly in the dominant hand[24]. In another study, researchers who examined 68 operators during a period when they were not applying pesticides found deficits in their vibrotactile abilities in comparison with a matched group of controls. A related study of 128 men with frank poisoning compared to 90 controls found deficits in cognitive abilities and evidence of peripheral neuropathy in both fingers and toes (as measured by vibrotactile sensitivity) as long as two years after last exposure[25]. A study of 146 sheep dippers who used organophosphates that included diazinon, propetamphos and chlrofenvinphos for an average of 15 years compared with 143 controls showed that the exposed men had deficits on measures of attention and information processing[26]. In each study doses were not calculated but were likely to be sub-acute.

Other studies have shown only in each study doses were not calculated but were likely to be sub-acute marginal effects of sub-clinical intoxication with organophosphate pesticides. For instance, Ames *et al.*, found that only one measure of neurological function (serial digit performance) was affected following a cholinesterase-lowering, but sub-toxic exposure[27]. However, the overwhelming majority of such studies reveals a pattern of deficits that suggest frank, delayed polyneuropathy may be only the tip of the iceberg of a wide spectrum of impairment. At least one research group believes neurological sequelae from exposure may be much more common than previously thought[28]. These and related investigations have established that prior low level exposure to organophosphate pesticides can produce lasting, subclinical deficits in neurological function.

Of great concern is the possibility that chronic exposure may potentiate certain neurological abnormalities long after exposure ceases. Certain of the above cited studies examined workers two or more years after exposure, yet still found effects. The possibility of a persistent neurological effect of chronic low level exposure has been examined in an experimental setting in which rats were exposed to lindane at 10 mg/kg orally. Two to four weeks after the last dose, the exposed rats still showed a clear cut increase in enhanced behavioral responses, notably myolonic jerks and clinic seizures. Direct demonstration of electrical "kindling" of the amygdala (a prodromal sign associated with convulsions) could be found commencing 4–6 weeks after the last dose[29]. These data suggest that long-term abnormalities in central neurological function may follow even short-lived exposure to certain pesticides.

In support of this proposition, workers who had sprayed fenthion, an organophosphate pesticide, demonstrated significant alterations in memory as measured by visual retention and "passalong" tests in the absence of any overt evidence of clinical toxicity[30]. Given the persistent reports of central nervous system abnormalities following exposure to such pesticides, including anxiety, depression and cognitive impairments, the need for psychological evaluation following low-level exposures is great[31].

REPRODUCTIVE TOXICITY

Some pesticides, most notably nematocides like dibromochloropropane (DBCP) and its analogs, have a long history of disrupting reproductive performance. More recently, pesticides like carbofuran and glyphosate with generally "acceptable" toxicity profiles have been found to produce abnormalities in sperm quality[32].

These and related effects may be explained by an indirect effect of pesticides on the hypothalamic–pituitary axis. For instance, anti-androgenic fungicides, like the classic xenoestrogens tetrachlorodibenzo-*p*-dioxin and PCBs (notably PCB #169), all produce dramatic abnormalities in rodent sex differentiation[33]. Other data suggests a direct toxic effect of pesticides on the target organ. This pattern of toxicity, for instance, applies to lindane's reproductive damage to the testis[34], and other fetal germ cell toxicants. While the available human data for direct reproductive toxicity is still scanty, strong inferential evidence for a male mediated effect can be gleaned from studies that show increased birth defects linked to occupation. A case in point is the study of birth defects among children of pesticide applicators in Minnesota. The study in question found a highly significant increase in birth defects among offspring of licensed private applicators, as well as a general increase in defects among the population who live in areas with high herbicide and fungicide usage[35].

DISCUSSION

Rather than considering these reports of adverse secondary effects of pesticides as coincidental and unrelated, it is reasonable to consider their possible linkage to common underlying toxicity mechanisms. The neurological, immunological and reproductive systems have intimate linkages through direct innervation, common receptors and influence by the hypothalamic–pituitary–adrenal axis. Several pesticides, notably 2,4 D and trifluralin and the fungicides Maneb and Mancozeb are endocrine disruptors[37] and hormone levels can directly influence immune function. Elevated levels of testosterone may depress immune function and disturbances in testosterone may follow certain pesticide exposures[38]. It is well known that environmental estrogens initiate their physiological actions in target tissues predominantly through binding with certain cell nucleus receptors. The most recent thinking in this area suggests that environmental estrogens derive their endocrine activity from their differential activation of certain estrogen response elements in the genome[39]. Following this activation, specific areas of the affected genetic material are activated and protein transcription is enhanced. These data comport with previous findings that show estrogenic pesticides bind to estrogen receptors in many estrogen-responsive cells[40].

These observations underscore the likelihood that additional secondary effects of low level pesticide exposure will be uncovered. This is particularly likely, since binding ability to estrogen receptors, immune interactions via cytokine receptors and neurological effects all have the capacity for greatly magnifying the effective dose of a given pesticide. In the interim, an expanded testing regime would clearly be helpful for anticipating secondary effects. One such model is the newly developed screen for estrogenic activity developed by Tufts University researchers. This test employs an *in vitro* model in which proliferation of cells with estrogen receptors is measured against the dose of a given pesticide[41]. Other assays that are in development include an electrotretinogram to anticipate neurotoxicity[42], new genotoxicity assays[43], and a quantitative vibrotactile model for measuring peripheral neuropathy[44].

The legislative mandate to anticipate and preclude major secondary health impacts of pesticides is embodied in the Federal Insecticide Fungicide and Rodenticide Act (FIFRA) which directs the Office of Pesticide Programs within the US EPA to ensure that pesticide use in commerce will not result in unreasonable adverse effects to both humans and the environment. Presently, much effort is being directed at "regulatory relief", i.e. the reduction of the burden of reviewing health effects[36]. To address the mandate of FIFRA, such efforts must be counterbalanced with an expanded and sustainable effort to reduce the health consequences of chronic exposure to pesticides, an effort that is presently not being undertaken.

The need for expanding such testing to incorporate secondary effects is great. Many more subtle effects from environmental exposure to neurotoxic pesticides have become known in the last decade, including but not limited to reduction in intelligence, impairment in reasoning ability, shortening of attention span and alteration of behavior. Many of the same pesticides have been associated with immune depression, hematological malignancies and reproductive damage[45]. An ideal assay which anticipated some or all of these effects might be developed based on the common receptors or endocrine systems affected by the pesticides in question. For this to occur, a more holistic view of toxicity is needed, one that embraces the commonality of key toxicological pathways and recognizes the need to limit or prevent secondary impacts that differentially affect children and reduce their ability to realize their full potential.

References

1. L. E. Culleen, "Pesticide registration in the United States: overview and new directions" *Quality Assurance* **3**, 291–299 (1994).
2. H. A. Kuiper, "The role of toxicology in the evaluation of new agrochemicals" *J. Environmental Science and Health B* **31**, 353–363 (1996).
3. C. Dragula and G. Burin, "International harmonization for the risk assessment of pesticides: results of an IPCS survey" *Regulatory Toxicology and Pharmacology* **20**, 337–353 (1994).
4. R. Fernando, "Pesticide poisoning in the Asia-Pacific Region and the role of a regional information network" *J. Toxicology and Clinical Toxicology* **33**, 677–682 (1995).
5. D. L. MacIntosh, J. D. Spengler, H. Ozkaynak *et al.*, "Dietary exposures to selected metals and pesticides" *Environmental Health Perspectives* **104**, 202–209 (1996).
6. T. J. Woodruff, A. D. Kyle and F. Y. Bois, "Evaluating health risks from occupational exposure to pesticides and the regulatory response" *Environmental Health Perspectives* **102**, 108801096 (1994).
7. V. G. Farry, J. T. Kelloy, J. M. Sprafka *et al.*, "Survey of health and use characterization of pesticide appliers in Minnesota" *Archives of Environmental Health* **49**, 337–343 (1994).
8. E. Hodgson and P. E. Levi, "Pesticides: an important but underused model for the environmental health sciences" *Environmental Health Perspectives* **104**, Suppl. 1, 97–196 (1996).
9. P. Klucinski, A. Hrycek, H. Stasjiura-Zielniska *et al.*, "Humoral and cellular immunity rates in chemical plant workers employed in the production of liquid pesticides" *Int. J. Occupational Medicine and Environmental Health* **9**, 103–110 (1996).
10. T. Vial, B. Nicolas and J. Descotes, "Clinical immunotoxicity of pesticides" *J. Toxicology and Environmental Health* **48**, 215–229 (1996).
11. L. Institoris, O. Siroki and I. Desi, "Immunotoxicity study of repeated small doses of dimethoate and methylparathion administered to rats over three generations" *Human and Experimental Toxicology* **14**, 879–883 (1995).
12. S. H. Kowalczyk-Bronisz, J. Kotz and B. Bubak *et al.*, "Studies on the effect of organophosphorous pesticide phosalone on mouse immune system" *Archives of Veterinary Medicine* (Poland) **34**, 221–230 (1994).
13. B. D. Banerjee, B. C. Koner and A. Ray, "Immunotoxicity of pesticides: perspectives and trends" *Indian J. Experimental Biology* **34**, 723–733 (1996).
14. D. Waterhouse, W. J. Carman, D. Schottenfeld *et al.*, "Cancer incidence in the rural community of Tecumseh, Michigan: a pattern of increased lymphopoietic neoplasms" *Cancer* **77**, 763–770 (1996).

15. A. Blair and S. H. Zahm, "Agricultural exposures and cancer" *Environmental Health Perspectives* **103**, Suppl. 8, 205–208 (1995).
16. O. Hohammad, A. A. Walid and K. Ghada, "Chromosomal aberrations in human lymphocytes from two groups of workers occupational exposed to pesticides in Syria" *Environmental Research* **70**, 24–29 (1995).
17. J. M. Pluth, J. A. Nicklas, J. P. O'Neill and R. J. Albertini, "Increased frequency of specific genomic deltions resulting from *in vitro* malathion exposure" *Cancer Research* **56**, 2393–2399 (1996).
18. See, for example, K. Wiklund and J. Dich, "Cancer risks among male farmers in Sweden" *European J. Cancer Prevention* **4**, 81–90 (1995).
19. P. Pasqualetti, A. Collacciani and R. Casale, "Risk of monoclonal gammopathy of undetermined significance: a case-referent study" *Amer. J. Hematology* **52**, 217–220 (1996).
20. O. Nanni, D. Amadori, C. Lugaresi *et al.*, "Chronic lymphocytic leukemias and non-Hodgkin's lymphoas by histological type in farming-animal breeding workers: a population case-control study based on a priori exposure matrices" *Occupational and Environmental Medicine* **53**, 652–657 (1996).
21. P. Kristnesen, A. Andersen and L. M. Irgens *et al.*, "Cancer in offspring of parents engaged in agricultural activities in Norway" *Int. J. Cancer* **65**, 39–50 (1996).
22. M. B. Abou-Donia, K. R. Wilmarth, A. A. Abdel-Rahman *et al.*, "Increased neurotoxicity following concurrent exposure to pyridostigmine bromide, DEET and chlropyrifos" *Fundamentals of Applied Toxicology* **34**, 201–202 (1996).
23. M. W. Rutjten, H. J. Salle, M. M. Verberk and M. Smink, "Effect of chronic mixed pesticide exposure on peripheral and autonomic nerve function" *Archives of Environmental Health* **49**, 188–195 (1994).
24. L. Stokes, A. Stark, E. Marshall and A. Narang, "Neurotoxicity among pesticide applicators exposed to organophosphates" *Occupational and Environmental Medicine* **52**, 648–653 (1995).
25. K. Steenland, B. Jenkins, R. Ames *et al.*, "Chronic neurological sequelae to organosphate pesticide poisoning, *Amer. J. Public Health* **84**, 731–736 (1995).
26. R. Stephens, A. Spurgeon, I. Calvert *et al.*, "Neurophyschological effects of long-term exposure to organophoshpates in sheep dip" *Lancet* **345,** 1135–1139 (1995).
27. R. G. Ames, K. Steenland, B. Jenkins *et al.*, "Chronic neurologic sequelae to cholinesterase inhibition among agricultural pesticide applicators, k" *Archives of Environmental Health* **50**, 440–444 (1995).
28. R. McConnell, M. Keifer and L. Rosenstock, "Elevated quantitative vibrotactile threshold among workers previously poisoned with methamidiphos and other organophosphate pesticides" *American J. Industrial Medicine* **25**, 325–334 (1994).
29. M. E. Gilbert, "Repeated exposure to lindane leads to behavioral sensitization and facilitates electrical kindling" *Neurotoxicology and Toxicology* **17**, 131–141 (1995).
30. U. K. Misra, M. Prasad and C. M. Pandey, "A study of cognitive functions and event related potentials following organophosphate exposure" *Electromyographic and Clinical Neurophysiology* **34**, 197–203 (1994).
31. J. Mearns, J. Dunn and P. R. Lees-Haley, "Psychological effects of organophosphate pesticides: a review and call for research by psychologists" *J. Clinical Psychology* **50**, 286–294 (1994).
32. M. I. Yousef, M. H. Salem and H. Z. Ibrahim *et al.*, "Toxic effects of carborfuran and glyphoasate on semen characteristics in rabbits" *J. Environmental Science and Health B* **30**, 513–534 (1995).
33. L. E. Gray Jr and W. R. Kelce, "Latent effects of pesticides and toxic substances on sexual differentiation of rodents" *Toxicology and Industrial Health* **12**, 515–531 (1996).
34. P. R. Dalsenter, A. S. Faqui, J. Webb *et al.*, "Reproductive toxicity and tissue concentrations of lindane in adult male rats" *Human and Experimental Toxicology* **15**, 406–410 (1996).
35. V. F. Garry, D. Schreinemachers, M. E. Harkins and J. Griffith, "Pesticide appliers, biocides and birth defects in rural Minnesota" *Environmental Health Perspectives* **104**, 394–399 (1996).
36. J. T. McClintock, J. L. Kough and R. D. Sjoblad, "Regulatory oversight of biochemical pesticides by the U.S. EPOA: health effects considerations" *Regulatory Toxicology and Pharmacology* **19**, 115–124 (1994).
37. T. Colborn, F. S. von Saal and A. M. Soto, "Developmental effects of endocrine-disrupting chemicals in wildlife and humans" *Environmental Health Perspectives* **101**, 378–384 (1993).
38. H. S. James, *Cited* in 26 above.
39. G. M. Stancel, H. L. Boettger-Tong and C. Chiappetta *et al.*, "Toxicity of endogenous and environmental estrogens: what is the role of elemental interactions" *Environmental Health Perspectives* **103**, Suppl. 7, 29–33 (1995).
40. F. S. Vom Saal, S. C. Nagel, P. Palaza *et al.*, "Estrogenic pesticides: binding relative to estradiol in MCF-7 cells and effects of exposure during fetal life on subsequent territorial behavior in male mice" *Toxicology Letters* **77**, 343–350 (1995).
41. A. M. Soto, K. C. Connencschein, K. L. Chung *et al.*, "The E-SCREEN assay as a tool to identify estrogens: an update on estrogenic environmental pollutants" *Environmental Health Perspectives* **103**, Suppl. 7, 113–122 (1995).

42. R. D. Jones, C. J. Brenneke, H. E. Hoss *et al.*, "An electroretinogram protocol for toxicological screening in the canine model" *Toxicology Letters* **70**, 223–224 (1994).
43. J. A. Venkate, S. Shami, K. Davis *et al.*, "Relative genotoxic activities of pesticides evaluated by a modified SOS microplate assay" *Environmental and Molecular Mutagenesis* **25**, 67–76 (1995).
44. R. McConnell, M. Keifer and L. Rosenstock, "Elevated quantatitve vibrotactile threshold among workers" *Amer. J. Industrial Medicine* **25**, 325–334 (1994).
45. P. J. Landirgan, D. G. Graham and R. D. Thomas, "Environmental neurotoxic illness: research for prevention" *Environmental Health Perspectives* **102**, Suppl. 2, 117–120 (1994).

8. ON CHLORINATED FATTY ACIDS AS ENVIRONMENTAL POLLUTANTS

HELENA BJÖRN *, PETER SUNDIN ** and CLAS WESÉN †

INTRODUCTION

For some decades chlorinated pollutants have attracted considerable attention. One reason is the environmental harm caused by persistent, organochlorine pollutants such as PCBs, DDT, DDE and chlorinated dioxins. However, usually less than 5% of the organochlorine load (extractable, organically bound chlorine; EOCl) can be accounted for by known pollutants[1–5]. The major proportion (90–99%) of the EOCl has remained unknown until the early nineties (Fig. 1). It was then discovered that most of the unidentified EOCl in fish from polluted waters and a substantial portion of the unknown EOCl in fish from remote waters, in bivalves and in sediment, consisted of chlorine in ester bound chlorinated fatty acids[6–15]. This has been confirmed by mass spectrometric studies[10,13–17].

Chlorinated fatty acids do not fit in among the "traditional" environmental, organohalogen pollutants in that they are not persistent. Preceding the identification of

Figure 1 A simplified representation of the fraction (black circle sectors) of the extractable organically bound chlorine, not accounted for by well-known environmental pollutants, in fish, bivalves and aquatic bottom sediments.

* Chemical Ecology and Ecotoxicology, Department of Ecology, Lund University, Ecology Building, SE-223 62 Lund (Sweden).

** Department of Environmental Assessment, Swedish University of Agricultural Sciences, P.O. Box 7050, SE-750 07 Uppsala (Sweden).

† Department of Technical Analytical Chemistry, Center of Chemistry and Chemical Engineering, Lund University, P.O. Box 124, SE-221 00 Lund (Sweden).

persistent organohalogens, sulphuric acid is often employed to oxidise the sample matrix (i.e. to remove the lipids). Sulphuric acid however, destroys the fatty acids which may explain why the chlorinated fatty acids have remained unnoticed for so long[9].

In chlorinated fatty acids, chlorine atoms are substituted for hydrogen atoms at certain positions in the carbon chain. A chlorine atom is relatively large, which gives the chlorinated fatty acid a bulkiness compared to the unchlorinated one[18]. As reviewed by Mu *et al.*[19], the major chlorinated fatty acids in fish, exposed to a local source such as effluents from a pulp bleachery that uses chlorine gas, are dichlorinated myristic, palmitic and stearic acids and tetrachlorinated myristic and stearic acids[9,10,13,14]. Dichlorinated species with one double bond have also been identified[14]. Most chlorinated fatty acids are present in different isomeric forms. The formation of some of the isomers is most likely related to industrial processes such as pulp cooking procedures. In fish from remote and unpolluted waters, however, the major contributors seem to be chlorohydroxy fatty acids[15,20].

The presence of chlorinated fatty acids in fish from remote waters is intriguing. Can this be explained by long-range transport of chlorinated fatty acids originating from chlorine bleached pulp mill effluents? Or is it caused by traditional, chlorinated pollutants (which are global), being partly dechlorinated and the chlorinated fragments then introduced to the fatty acid molecules during the fatty acid synthesis? Or could it be that almost the entire organochlorine load originates from chloride ions, and thus is nothing that has been introduced by man? Those questions will be discussed in this chapter.

DISTRIBUTION IN THE ENVIRONMENT

Chlorinated fatty acids are present in both triacylglycerols (storage lipids) and phospholipids (membrane lipids) of different fish species such as eel, salmon, pike and herring (Fig. 2)[12,15,20,21], and—possibly—in acyl sterols of a lean fish like pike[20]. Migratory fish like salmon and eel store triacylglycerols in the muscles and adipose tissue. These lipid depots also contain the majority of the chlorinated fatty acids[4,15]. However, the concentration of chlorinated fatty acids seems to be higher in phospholipids than in triacylglycerols[15,20,21].

Other species that have been found to contain chlorinated fatty acids are bivalves[11], lobster[17], and jellyfish[22]. So far, though, only a few marine mammals have been investigated. Kawano *et al.*[23] found that a large proportion of the EOCl in dolphin, whale and seal from the Pacific ocean was associated with compounds of a lipid nature.

Similarly, *brominated* fatty acids constitute a large proportion of the organically bound bromine in marine fish and mammals[23–25]. Brominated fatty acids have also been found in phospholipids isolated from marine sponges[26,27].

It has also been demonstrated that sedimented material both from the Gulf of Bothnia, Sweden, and the Oslo Fjord, Norway, contained chlorinated fatty acids that most likely originated from pulp mill bleach effluents[7].

The load of persistent, chlorinated pollutants is related not only to exposure, but also to age and fat content (e.g. Larsson *et al.*[28]). However, no such relationships have yet been found for chlorinated fatty acids.

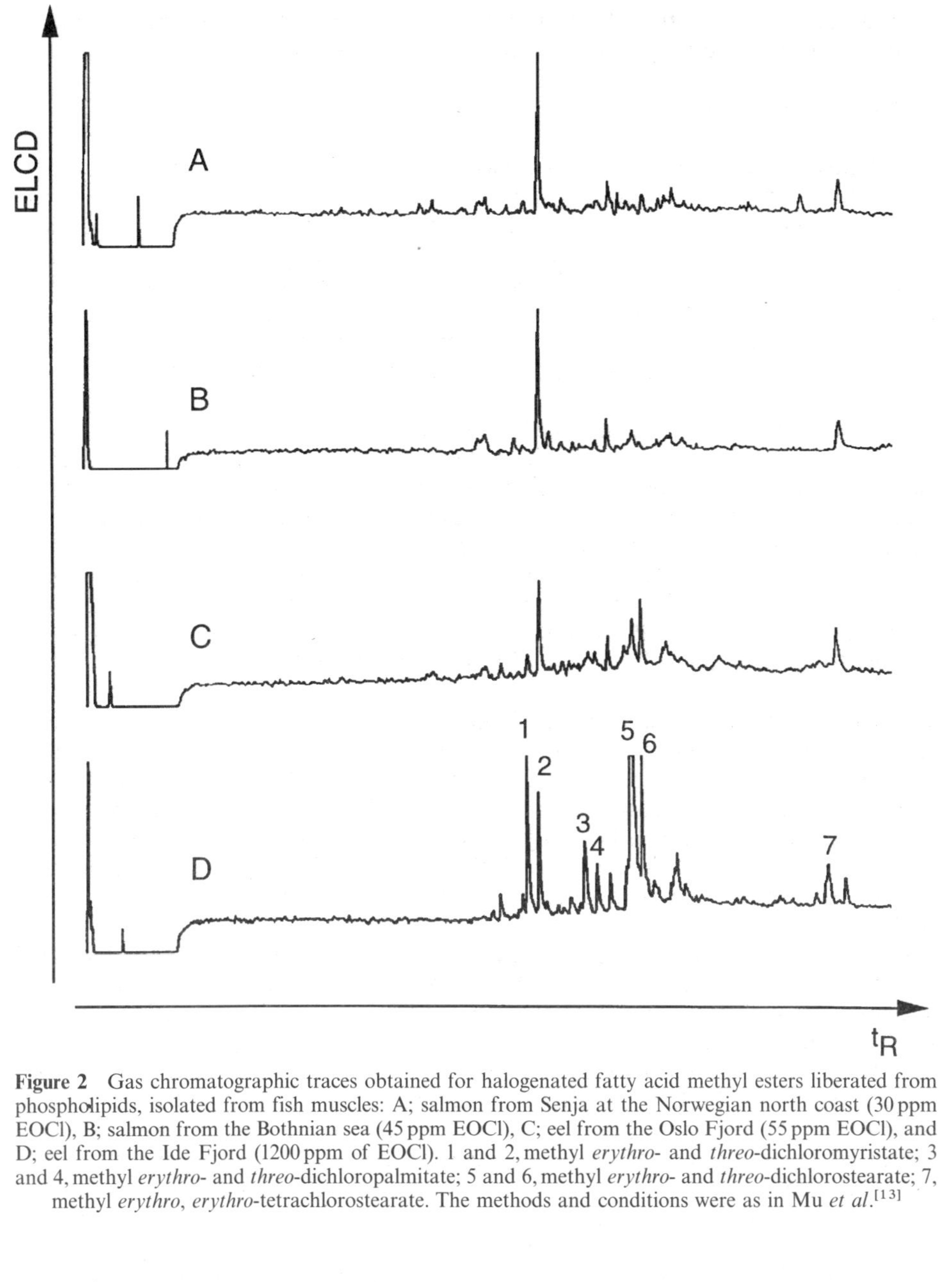

Figure 2 Gas chromatographic traces obtained for halogenated fatty acid methyl esters liberated from phospholipids, isolated from fish muscles: A; salmon from Senja at the Norwegian north coast (30 ppm EOCl), B; salmon from the Bothnian sea (45 ppm EOCl), C; eel from the Oslo Fjord (55 ppm EOCl), and D; eel from the Ide Fjord (1200 ppm of EOCl). 1 and 2, methyl *erythro-* and *threo*-dichloromyristate; 3 and 4, methyl *erythro-* and *threo*-dichloropalmitate; 5 and 6, methyl *erythro-* and *threo*-dichlorostearate; 7, methyl *erythro*, *erythro*-tetrachlorostearate. The methods and conditions were as in Mu *et al.*[13]

ORIGIN

The origin of chlorinated fatty acids is far from clear. Many possible sources exist, both anthropogenic and natural. In addition to direct chlorination in anthropogenic processes it is also important to consider the indirect or secondary formation from chlorinated precursors of anthropogenic origin, where the actual mechanism is natural (biotic or abiotic). A truly natural production, however, might involve endogenic, enzymatic chlorination of unsaturated fatty acids.

1. *Anthropogenic Sources*

a. Chlorination of Unsaturated Fatty Acids. Chlorinated fatty acids can inadvertently be produced in pulp bleaching with chlorine[29,30]. In 1984, the annual world production of chlorine bleached pulp was roughly 50 million tonnes, which would have consumed about 3.25 million tonnes of chlorine if bleached solely by Cl_2[31]. As a result, between 20 and 580 ppm thereof would have ended up in chlorinated fatty acids[29,30,32], corresponding to a production of 300–9500 tonnes of dichlorostearic acid. Thus, chlorine bleached pulp mill effluents have probably been the dominating, direct anthropogenic source, and possibly still are where elemental chlorine is still in use.

In some western countries, chlorine has been used to bleach flour in order to modify the cake-baking characteristics. Chlorinated fatty acids have been found both in bleached flour and in flour-containing food[33–35]. Sanitary practices where foodstuff, such as poultry, are washed in chlorinated water to counteract bacterial contamination, lead to the formation of chlorinated lipids[36]. Other possible sources to chlorinated fatty acids might be chlorination of drinking water and sewage water[37,38]. Also, a large number of minor sources may exist, like swimming pools, dishwashers, and different house-hold cleaning-agents, where disinfection with OCl^- is likely to result in the chlorination of unsaturated fatty acids.

b. Formation from Anthropogenic Precursors. Another possible anthropogenic source of chlorinated fatty acids is chlorinated paraffins, which are widely used as plasticizers in lubricants, paints and fire retardants[39]. Murphy and Perry[40–42] showed that various microorganisms can oxidise 1-chloroalkanes and incorporate the chlorinated fatty acids formed directly into cellular phospholipids. In agreement with these results, Omori *et al.*[43] showed that certain microorganisms can degrade chlorinated paraffins co-metabolically to a variety of chlorinated long-chain fatty acids.

Rainbow trout have been found to metabolically degrade assimilated chlorinated paraffins into smaller molecules, which have been suggested to become incorporated into lipids through other biochemical pathways[44]. Jernelöv[45] proposed that certain small, organochlorine compounds could be endogenically incorporated during fatty acid biosynthesis. In an experiment where fish were exposed to vinyl chloride (not being part of the EOCl), the EOCl concentration in the fish increased rapidly. Afterwards, the fish were placed in clean water and the concentration of EOCl decreased, but still remained at a level higher than would have been expected if the vinyl chloride in the fish tissues was in equilibrium with that of the surrounding water. This indicates that the vinyl chloride was metabolised to organochlorine products that were deposited in fat tissue.

Unsaturated fatty acids and organochlorine pollutants can interact under the influence of UV-radiation and form chlorinated products, among which dichlorostearic acid has been identified[46]. Reactions of this kind can be of importance in aquatic surface-layers and crops sprayed with pesticides, as well as in leaf surface waxes, which can receive pesticides from precipitation.

There are also implications that organochlorine compounds can be produced in reactions between naturally occurring organic substrates and chloride ions affected by strong oxidants of anthropogenic origin[47].

2. *Natural Sources*

a. Biotic Chlorination. For a time, many scientists considered all organohalogens as unnatural, post-industrial compounds. However, in a large number of studies from the last half century, evidence has accumulated showing a considerable natural production of organohalogens (e.g. Clutterbuck *et al.*[48] and several reviews[47,49–51]). Numerous halometabolites have been attributed to different groups of organisms, e.g. bacteria, fungi, marine algae, sponges and higher plants. The halogenated compounds range from simple alkanes, over phenols to complex antibiotics. Many of the widespread, naturally produced organohalogens are, however, halogenated humic acids[47], which are not likely to be incorporated in fatty acid synthesis. As discussed above the mere presence of halogenated, organic compounds may lead to formation of halogenated fatty acids.

Many organisms have enzymes (haloperoxidases) that can chlorinate organic compounds[49]. Halogenation can occur both inside and outside organisms. In both cases, haloperoxidases use inorganic halogen ions to produce organohalogens. For example chlorohydroxy fatty acids in jellyfish can be formed from unsaturated fatty acids[22], a process most likely governed by chloroperoxidases[47,49].

b. Abiotic Chlorination. The importance and mechanisms of abiotic, natural halogenation are still unclear. There are several suggestions that UV-radiation or ozone and sea spray aerosol play key roles. For example Keene *et al.*[52] have proposed that ozone together with chloride ions might produce elemental chlorine, which readily adds to organic molecules. Vogt *et al.*[53] suggested an autocatalytic mechanism for the release of reactive halogen from sea-salt aerosol—a process that could occur in remote waters.

UPTAKE AND BIOCHEMICAL PROPERTIES

It is not yet known how chlorinated fatty acids become incorporated into lipids. The studies done on uptake by and distribution within organisms, suggest that chlorinated fatty acids can take part in the ordinary fatty acid biochemical pathways, and thus become incorporated into cellular lipids in a similar manner as unchlorinated species.

Ewald and Sundin[18] suggested that chlorine atoms not only create a bulkiness in the chlorinated fatty acids, but also alter the conformation of the molecules (Fig. 3). Chlorination of an unsaturated fatty acid normally occurs over the double-bond. The chlorine atoms in the resulting dichlorinated fatty acids cause a restriction of the rotation around the –CHCl–CHCl– bond, which gives the carbon chain a low-energy conformation resembling that of a monounsaturated fatty acid.

In an uptake experiment with perch, radiolabelled dichlorostearic acid was compared to stearic and oleic acids. No significant discrimination occurred towards the chlorinated fatty acid[54]. Experiments with rats have shown that chlorinated fatty acids and chlorinated triacylglycerols can be assimilated from food and can also be transferred via placenta and milk to the offspring[55–57]. Other tracer studies with rats have shown that dichlorostearic acid was absorbed and distributed throughout the body much in the same way as normal lipids[58,59].

In a study of migrating salmon in Alaska[13], the chlorinated fatty acids in triacylglycerols from muscle were released to the same extent as the unchlorinated ones during the 400 km spawning migration. However, the total amount of chlorinated

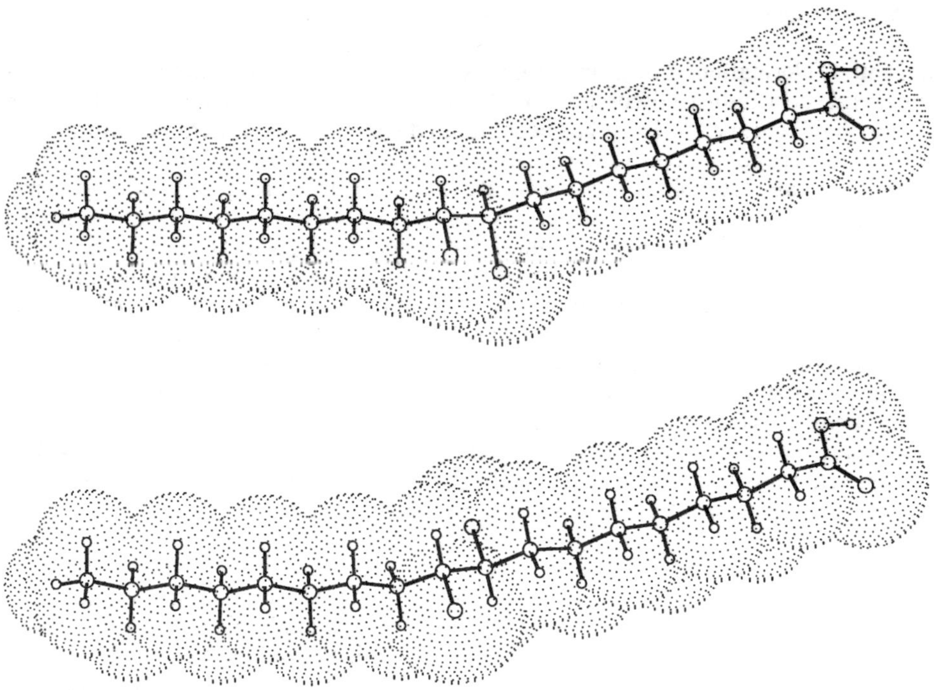

Figure 3 Space-filling projections showing energy-minimised configurations of two diastereomers of 9,10-dichlorostearic acid, illustrating the bulkiness of the –CHCl–CHCl– moiety (top: *threo*, formed by chlorination of a *cis* double bond; bottom: *erythro*, formed from a *trans* double bond).

fatty acids increased in the roe and in muscle phospholipids. The finding indicated a low turn-over rate of chlorinated fatty acids in phospholipids, because the total mass of muscle phospholipids decreased concurrently. The increased amount of chlorinated fatty acids in the roe was, however, due to the growth of the developing roe. Chlorinated fatty acids were also analysed in a stationary fish (grayling) both from a lake where the migrating salmon spawned and from a lake without migrating salmon. The lipids of grayling in the salmon spawning lake contained higher amounts of chlorinated fatty acids than lipids of grayling in the other lake. However, the chlorinated fatty acids in grayling had other identities than those in the migrating salmon. This suggests that the chlorinated fatty acids were partly metabolised during their transfer in the food web. Thus, a certain recalcitrance of chlorinated fatty acids is implied and the migratory salmon can be looked upon as a carrier of chlorinated fatty acids to a remote area.

Those results are supported by a recent laboratory study of the uptake and transport of dichlorostearic acid in an experimental food-chain, in which the chlorinated fatty acid was transported to organisms of higher trophic levels (H. Björn, unpublished results).

The metabolism of chlorinated fatty acids is not clear. The studies performed on rats show that, in relation to oleic acid, a relatively higher percentage of the administered dichlorostearic acid is taken up in heart tissue, while oleic acid dominates in other organs[58,59]. Di-, tetra- and hexachlorostearic acids are taken up in the animals to a lesser extent than the unsaturated analogues[58] and a large proportion of the chlorinated fatty acids are eventually degraded to water-soluble compounds and excreted with the urine[58,59]. The test compounds were radioactively labelled (^3H and ^{36}Cl, respectively) and the excreted material was measured in terms of water-soluble radioactivity.

Fatty acids with even carbon numbers are the most abundant in nature. A fatty acid is catabolised through β-oxidation, where the molecule is shortened stepwise by an acetyl-group. Dichloromyristic acid (of 14 carbon atoms chain length) was found as the main metabolite when rats were fed dichlorostearic acid (18 carbon atoms)[57]. In fish with low concentrations of EOCl, the dichloromyristic acid dominates (Fig. 2(A) and 2(B))[12,13,21]. There have also been indications that the relative composition of chlorinated fatty acids in biota changes with the distance to polluted areas. In fish sampled near pulp mill effluents, the long-chain (16 to 18 carbons) chlorinated fatty acids dominated. Further away from the polluted area, the short-chained chlorinated fatty acids dominated[12]. This either indicates that the short-chained chlorinated fatty acids are dechlorinated before further degradation or that dichloromyristic acid has a higher biological stability than chlorinated fatty acids having the chlorine atoms positioned further away from the carboxylic acid group (which normally is the case in chlorinated fatty acids with longer chain lengths). A reason for this resistance to β-oxidation could be that cleavage of an acetyl moiety from dichloromyristic acid is sterically hindered by the chlorine atoms (Fig. 4).

Jones *et al.*[60] discussed if halogenated fatty acids might be β-oxidised by peroxisomes. Their assumption was based on the finding by Mohamed *et al.*[61], who showed that no β-oxidation of 9,10-dibromopalmitic acid occurred in mitochondria, but that the coenzyme A-thioester was formed. These reasonings imply another possible cause to the apparent resistance to β-oxidation. Given that the initial β-oxidation of long-chain, chlorinated fatty acids is managed by peroxisomes and that further β-oxidation is halted once the chlorinated acids reach the mitochondria, the partly degraded, chlorinated fatty acids ought to be ready for incorporation in different acylglycerols once the coenzyme A-thioesters have been formed.

It should be noticed that different diastereomeric forms of chlorinated fatty acids have been found in fish from the vicinities of pulp mills and that the different diastereomers are β-oxidised in a similar manner.

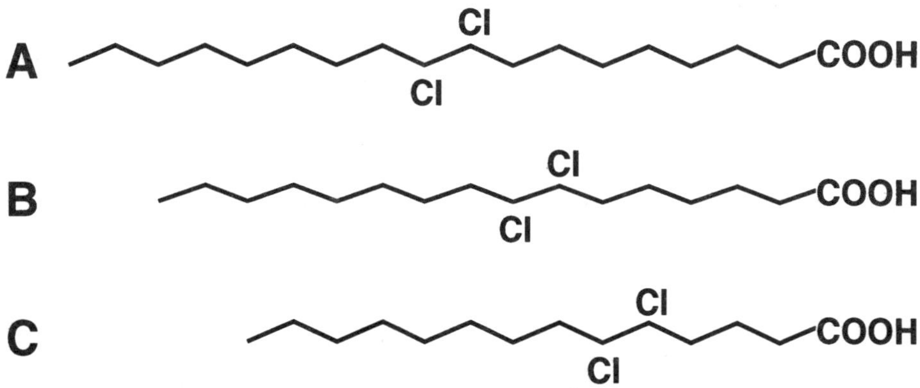

Figure 4 Chain shortening by β-oxidation that puts the chlorine atoms closer to the functional group. (A) 9,10-dichlorostearic acid, (B) 7,8-dichloropalmitic acid, and (C) 5,6 dichloromyristic acid.

TOXICITY AND PHYSIOLOGICAL EFFECTS

None of the above would have any ecotoxicological relevance if chlorinated fatty acids did not have adverse consequences on organisms. However, several different effects have been observed, many of which are related to reproduction processes.

The acute toxicity of dichlorostearic acid is relatively high to fish[29], but low to rat[56]. Common ways of assessing the physiological impact of traditional pollutants can be to investigate the detoxifying hepatic enzyme system, cytochrome P_{450}, and the activity of the hepatic enzyme EROD (7-ethoxyresurofin O-deethylase). Neither of these systems are activated by chlorinated fatty acids in fish lipids[4,62], suggesting that chlorinated fatty acids are not recognised by organisms as xenobiotics.

When rats were fed a diet rich in chlorinated fatty acids, the uptake in adult rats and transfer to suckling rats decreased with increasing degree of chlorination of the fatty acids[55,56,58]. The rats suffered sublethal effects, such as decreased growth and increased weight of the liver, kidneys and heart. In another experiment, female mice fed dichlorostearic acid showed a higher mortality after 17 months[63].

Cherr *et al.*[64] tested compounds found in bleached kraft mill effluents for effects on the fertilisation rate and sperm motility of sea urchin sperm cells. Among 12 tested compounds, including chlorophenols and chlorinated resin acids, dichlorostearic acid was the most toxic compound with regard to fertilisation rate (EC_{50} 0.057 mg/l). The reproduction of fish can also be affected by chlorinated fatty acids. The influence of different compounds on the steroid production in testes and ovaries of goldfish have been studied in sensitive *in vitro*-systems[65]. Fatty acids with a high chlorine content caused a significant reduction of the arachidonic acid stimulated testosterone production (Fig. 5). When zebrafish were fed a diet rich in chlorinated fatty acids, both the amount of roe laid decreased as well as the hatching frequency of the roe[4]. In the same study, a similar experiment with blue mussels resulted in physiological stress and an increased mortality. Both the zebrafish and the blue mussels bioaccumulated EOCl, dominated by chlorinated fatty acids, from their food.

Wilson *et al.*[66] showed in an experiment with rats that *brominated* fatty acids altered the ratio between the natural, essential fatty acids in the same way as DDT[67]. The changed ratio of essential fatty acids was proposed to cause disturbances in the production of arachidonic acid[66], which is a prostaglandin precursor. Gribble[51], on the other hand, has pointed out that monochlorinated prostaglandins have higher anti-tumour capacity than their unchlorinated analogues. Additionally, dichlorostearic acid shows anti-mutagenic properties in the Ames test[68].

After exposing mammalian cells to mono- and dichlorostearic acids, the ATP leakage increased, which possibly resulted from an impact on the membrane function[18]. The ratio between saturated and unsaturated fatty acids in phospholipids is an important factor determining membrane fluidity and permeability, and a change in this ratio or incorporation of bulky fatty acids may cause increased leakage of cell components. Additionally, because the melting point of a chlorinated fatty acid is higher than that of the corresponding, unsaturated analogue[69], an animal can be expected to have reduced tolerance to low temperatures if chlorinated fatty acids are substituted for unsaturated species in the membrane lipid synthesis. Thus, because dichlorostearic acid seems to be assimilated much in the same way as unchlorinated analogues[54,55,58,59], membrane-related, physiological malfunctions may result from exposure to chlorinated fatty acids.

Thus, both negative and positive effects have been found for chlorinated fatty acids and probable derivatives thereof (chlorinated prostaglandins). Because

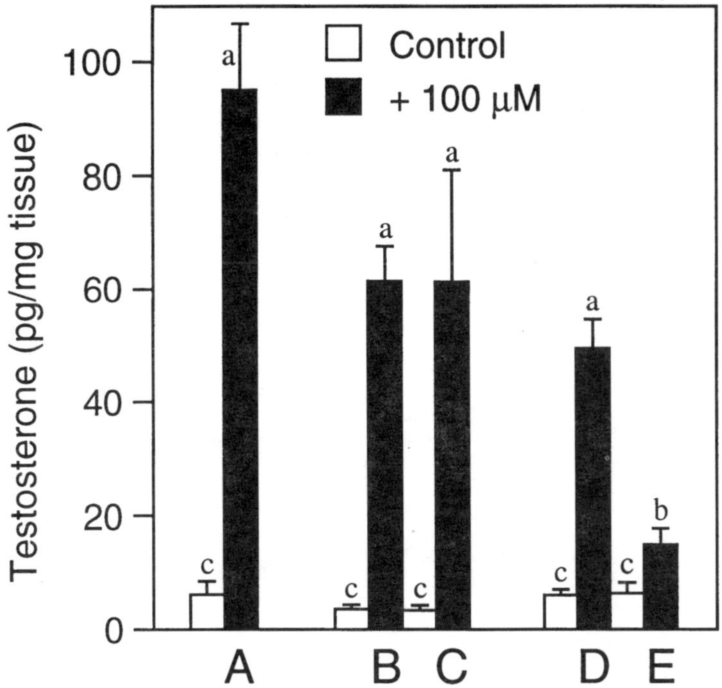

Figure 5 Effects of free fatty acids, released from salmon and eel lipids, on arachidonic acid (100 μM) stimulated testosterone production by goldfish testis pieces incubated for 18 h at 18°C. Values represent mean ± SEM of four replicate incubations. (A) no fatty acids added, (B) salmon from Senja in northern Norway (30 ppm EOCl), (C) salmon from the northern Baltic Sea (40 ppm EOCl), (D) eel from the Oslo Fjord (55 ppm EOCl), and (E) eel from the Ide Fjord (1200 ppm EOCl). Columns marked by different lower case letters represent means differing significantly on a total level of $p < 0.05$ (one way Anova, followed by Duncans multiple range test). For details about the fish lipids, see Fig. 2 and Håkansson *et al.*[4]. (Dr Glen Van Der Kraak, University of Guelph, Canada, is gratefully acknowledged for providing these results and giving us permission to use this figure.)

chloroperoxidases are part of the defence mechanisms of the body[49], certain chlorinated fatty acids may be valuable to the individual. It may be so that a balance exists, implying that low concentrations of chlorinated fatty acids are positive to a living organism, while high concentrations are detrimental.

CONCLUSIONS

The reviewed evidence points towards a picture where chlorinated fatty acids are not only a post-industrial part of the environment. However, the increased concentrations of certain di- and tetrachlorinated fatty acids due to anthropogenic activities, implicates a risk to altered membrane function in exposed organisms. This possibly illuminates part of the question why well-known environmental pollutants fail to fully explain all the physiological disturbances that have been observed during the last decades in different organisms[70]. Therefore, to gain a complete picture, more research is needed, in particular to investigate the origin, effects and properties of chlorohydroxy fatty acids.

References

1. G. Lunde, J. Gether and E. Steinnes, "Determination of volatility and chemical persistence of lipid-soluble halogenated organic substances in marine organisms" *Ambio* **5**, 180–182 (1976).
2. A. Södergren, B. E. Bengtsson, P. Jonsson, S. Lagergren, Å. Larsson, M. Olsson and L. Renberg, "Summary of results from the Swedish project Environment/Cellulose" *Wat. Sci. Technol.* **20**, 49–60 (1988).
3. K. Martinsen, A. Kringstad and G. E. Carlberg, "Methods for determination of sum parameters and characterization of organochlorine compounds in spent bleach liquors from pulp mills and water, sediment and biological samples from receiving waters" *Wat. Sci. Technol.* **20**, 13–24 (1988).
4. H. Håkansson, P. Sundin, T. Andersson, B. Brunström, L. Dencker, M. Engwall, G. Ewald, M. Gilek, G. Holm, S. Honkasalo, J. Idestam-Almquist, P. Jonsson, N. Kautsky, G. Lundberg, A. Lund Kvernheim, K. Martinsen, L. Norrgren, M. Pesonen, M. Rundgren, M. Stålberg, M. Tarkpea and C. Wesén, "*In vivo* and *in vitro* toxicity of fractionated fish lipids, with particular regard to their content of chlorinated organic compounds" *Pharmacol. Toxicol.* **69**, 459–471 (1991).
5. W. H. Newsome, P. Andrews, H. B. S. Conacher, R. R. Rao and A. Chatt, "Total organochlorine content of fish from the Great lakes" *JAOAC Int.* **76**, 703–706 (1993).
6. C. Wesén, G. E. Carlberg and K. Martinsen, "On the identity of chlorinated organic substances in aquatic organisms and sediments" *Ambio* **19**, 36–38 (1990).
7. M. Remberger, P-Å. Hynning and A. H. Neilson, "Gas chromatographic analysis and gas chromatographic-mass spectrometric identification of components in the cyclohexane-extractable fraction from contaminated sediment samples" *J. Chromatogr.* **508**, 159–178 (1990).
8. C. Wesén, K. Martinsen, G. Carlberg and H. Mu, "Chlorinated fatty acids are major chloroorganic compounds in fish exposed to pulp bleach liquors" **In**: (A. Södergren, ed.), *Environmental Fate and Effects of Bleached Pulp Mill Effluents* (Swedish Environmental Protection Agency Report 4031, 1992) pp. 207–219.
9. C. Wesén, H. Mu, A. Lund-Kvernheim and P. Larsson, "Identification of chlorinated fatty acids in fish lipids by partitioning studies and by gas chromatography with Hall electrolytic conductivity detection" *J. Chromatogr.* **625**, 257–269 (1992).
10. C. Wesén, H. Mu, P. Sundin, P. Frøyen, J. Skramstad and G. Odham, "Gas chromatographic-mass spectrometric identification of chlorinated octadecanoic acids in eel lipids" *J. Mass Spectrom.* **30**, 959–968 (1995).
11. C. Wesén, H. Mu, P. Sundin, O. Ringstad and G. Odham, "Occurrence of halogenated fatty acids in bivalve lipids" **In**: (A. Grimvall and E.W.B. de Leer, eds.), *Naturally-Produced Organohalogens* (Kluwer Acad. Publ., Dordrecht, The Netherlands, 1995) pp. 307–316.
12. C. Wesén, "Chlorinated fatty acids in fish lipids" *Dissertation* (Lund University, Sweden, 1995).
13. H. Mu, C. Wesén, T. Novak, P. Sundin, J. Skramstad and G. Odham, "Enrichment of chlorinated fatty acids in fish lipids prior to analysis by capillary gas chromatography with electrolytic conductivity detection and mass spectrometry" *J. Chromatogr.* **731**, 225–236 (1996).
14. H. Mu, C. Wesén, P. Sundin and E. Nilsson, "Gas chromatographic and mass spectrometric identification of tetrachloroalkanoic and dichloroalkenoic acids in eel lipids" *J. Mass Spectrom.* **31**, 517–526 (1996).
15. H. Mu, G. Ewald, C. Wesén, P. Sundin and E. Nilsson, "Mobilisation of chlorinated fatty acids during spawning migration of sockeye salmon, and impact on a stationary fish" *Manuscript* (1997).
16. P. Sundin, C. Wesén, H. Mu and G. Odham, "GC-MS detection of dichlorooctadecanoic acid in eel lipids" **In**: (A. Södergren, ed.), *Environmental Fate and Effects of Bleached Pulp Mill Effluents* (Swedish Environmental Protection Agency Report 4031, 1992) pp. 195–198.
17. J. E. Milley, R. K. Boyd, J. M. Curtis, C. Musial and J. F. Uthe, "Dichloromyristic acid, a major component of organochlorine load in lobster digestive gland" *Environ. Sci. Technol.* **31**, 535–541 (1997).
18. G. Ewald and P. Sundin, "Release of ATP from ELD cells after exposure to stearic, monochlorostearic, dichlorostearic and oleic acid" *Pharmacol. Toxicol.* **73**, 159–162 (1993).
19. H. Mu, C. Wesén and P. Sundin, "Analysis of halogenated fatty acids: 1. Formation and occurrence in lipids" *Trends Anal. Chem.* **16**, 266–274 (1997).
20. G. Vereskuns, "Chlorinated fatty acids in lipids of pike (*Esox lucius*) from Latvian lakes and from the Baltic" *Honour's thesis* (Dept Biology, Univ. of Lund, Sweden, 1996).
21. H. Björn, P. Sundin, C. Wesén, H. Mu, K. Martinsen, A.L. Kvernheim, J. Skramstad and G. Odham, "Chlorinated fatty acids in membrane lipids of fish" (Submitted, 1997).
22. R. H. White and L. P. Hager, "Occurrence of fatty acid chlorohydrins in jellyfish lipids" *Biochem.* **16**, 4944–4948 (1977).

23. M. Kawano, H. Yoshioka, Y. Tejima and R. Tatsukawa, "Occurrence of organic halogens in marine mammals" **In**: (A. Grimvall and E. W. B. de Leer, eds.), *Naturally-Produced Organohalogens* (Kluwer Acad. Publ., Dordrecht, The Netherlands, 1995) pp. 333–337.
24. G. Lunde, "Analysis of arsenic and bromine in marine and terrestrial oils" *J. Am. Oil Chem. Soc.* **9**, 155–157 (1972).
25. I. J. Tinsley and R. R. Lowry, "Bromine content of lipids of marine organisms" *J. Am. Oil Chem. Soc.* **57**, 31–33 (1980).
26. W. Lam, S. Hahn, E. Ayanoglu and C. Djerassi, "Phospholipid studies of marine organisms. 22. Structure and biosynthesis of novel brominated fatty acid from a hymeniacidonid sponge" *J. Org. Chem.* **54**, 3428–3432 (1989).
27. N. M. Carballeira and F. Shalabi, "Novel brominated phospholipid fatty acids from the caribbean sponge *Petrosia* sp" *J. Natural Products* **56**, 739–746 (1993).
28. P. Larsson, C. Backe, G. Bremle, A. Eklöv and L. Okla, "Persistent pollutants in a salmon population (*Salmo salar*) of the southern Baltic sea" *Can. J. Fish. Aquat. Sci.* **53**, 62–69 (1996).
29. J. M. Leach and A. N. Thakore, "Compounds toxic to fish in pulp mill waste streams" *Prog. Wat. Technol.* **9**, 787–798 (1977).
30. R. H. Voss and A. Rapsomatiotis, "An improved solvent extraction based procedure for the gas chromatographic analysis of resin and fatty acids in pulp mill effluents" *J. Chromatogr.* **346**, 205–214 (1985).
31. A. Kringstad and K. Lindström, "Spent liquors from pulp bleaching" *Environ. Sci. Technol.* **18**, 236a–248a (1984).
32. D. B. Easty, L. G. Borchart and B. A. Wabers, "Removal of wood-derived toxics from pulping and bleaching wastes" *US EPA Report 600/2-78-013*, Cincinnatti, Ohio, USA (1978).
33. C. Komo-Suwelack, E. Schulte, and L. Acker, "On the chemistry and analysis of flour bleaching 1. The treatment of flours with chlorine and its analytical detection" *Z. Lebensm. Unters. Forsch.* **196**, 169–172 (1983).
34. D. L. Heikes, "Mass spectral identification and gas chromatographic determination of chlorinated bleaching adducts in flour-containing food items" *J. Agric. Food Chem.* **40**, 489–491 (1992).
35. D. L. Heikes, "Procedure for supercritical fluid extraction and gas chromatographic determination of chlorinated fatty acid bleaching adducts in flour and flour-containing food items utilizing acid hydrolysis–methylation and Florisil column cleanup techniques" *J. Agric. Food Chem.* **41**, 2034–2037 (1993).
36. H. M. Cunningham and G. A. Lawrence "Volumetric method for determining traces of organic chlorine in lipids" *JAOAC* **62**, 482–484 (1979).
37. T. M. Gibson, J. Haley, M. Righton and C. D. Watts, "Chlorination of fatty acids during water treatment disinfection: reactivity and product identification" *Environ. Technol. Letters* **7**, 365–372 (1986).
38. S. Onodera, S. Iriko and S. Suxuki, "Chemical changes of organic compounds in chlorinated water. XVIII. Kinetics and mechanisms for chlorohydrin formation during reactions of aqueous chlorine with lipid components" *J. Toxicol. Environ. Health* (Japan) **38**, 247–257 (1992).
39. K. L. Houghton, "Chlorinated paraffins" **In**: *Encyclopedia of Chemical Technology*, 4th edn, **vol. 6**, (J. Wiley & Sons, Chichester, 1991) pp. 78–87.
40. G. L. Murphy and J. J. Perry, "Incorporation of chlorinated alkanes into fatty acids of hydrocarbon-utilizing mycobacteria" *J. Bacteriol.* **156**, 1158–1164 (1983).
41. G. L. Murphy and J. J. Perry, "Assimilation of chlorinated alkanes into fatty acids by hydrocarbon-utilizing fungi" *J. Bacteriol.* **160**, 1171–1174 (1984).
42. G. L. Murphy and J. J. Perry, "Chlorinated fatty acid distribution in *Mycobacterium convolutum* phospholipids after growth on 1-chlorohexadecane" *Appl. Environ. Microbiol.* **53**, 10–13 (1987).
43. T. Omori, T. Kimura and T. Kodama, "Bacterial cometabolic degradation of chlorinated paraffins" *Appl. Microbiol. Biotechnol.* **25**, 533–557 (1987).
44. J. R. Madeley and D. N. Birtley, "Chlorinated paraffins and the environment. 2. Aquatic and avian toxicology" *Environ. Sci. Technol.* **10**, 1215–1221 (1980).
45. A. Jernelöv, "Hypotes kan förklara organiskt klor i fisk" *Kem. Tidskr.* **4**, 119–122 (1989) (In Swedish).
46. W. J. Schwack, "Photoinduced additions of pesticides to biomolecules. 2. Model reactions of DDT and methoxychlor with methyl oleate" *J. Agric. Food Chem.* **36**, 645–648 (1988).
47. A. Grimvall, "Evidence of naturally produced and man-made organohalogens in water and sediments" **In**: (A. Grimvall and E.W.B de Leer, eds.), *Naturally-Produced Organohalogens* (Kluwer Acad. Publ., Dordrecht, The Netherlands, 1995) pp. 3–20.
48. P. W. Clutterbuck, S. L. Mukhopadhyay, A. L. Oxford and H. Raistrick, "Studies in the biochemistry of micro-organisms. (A) A survey of chlorine metabolism by moulds. (B) Caldariomycin. $C_5H_8O_2Cl_2$, a metabolic product of *Caldariomyces fumago* Woronichin" *Biochem. J.* **34**, 664–677 (1940).
49. S. L. Neidleman and J. Geigert, *Biohalogenation: principles, basic roles and applications*, (Ellis Horwood Ltd, Chichester, England, 1986)

50. K. C. Engvild, "Chlorine-containing natural compounds in higher plants" *Phytochemistry* **25**, 781–791 (1986).
51. G. W. Gribble, "The diversity of natural organochlorines in living organisms" *Pure Appl. Chem.* **68**, 1699–1712 (1996).
52. W. C. Keene, W. A. P. Pzenny, D. J. Jacob, R. A. Duce, J. N. Galloway, J. J. Schultz-Tokos, H. Sievering and J. F. Boatman, "The geochemical cycling of reactive chlorine through the marine troposphere" *Global Biogeochem. Cycles* **4**, 407–430 (1990).
53. R. Vogt, P. J. Crutzen and R. Sander, "A mechanism for halogen release from sea-salt aerosol in the remote marine boundary layer" *Nature* **383**, 327–330 (1996).
54. G. Ewald, P. Sundin, J. Skramstad and P. Frøyen, "Distribution of ^{14}C from ingested, radiolabelled dichlorostearic, stearic and oleic acids in body and in lipids of perch, *Perca fluviatilis*" *Aquat. Toxicol.* **36**, 129–139 (1996).
55. H. M. Cunningham and G. A. Lawrence, "Placental and mammary transfer of chlorinated fatty acids in rats" *Food Cosmet. Toxicol.* **15**, 183–186 (1977).
56. H. M. Cunningham, "Toxicology of compounds resulting from the use of chlorine in food processing" **In**: (R. L. Jolly, W. A. Brungs, R. B. Cumming and V. A. Jacobs, eds.) *Water Chlorination. Environmental Impact and Health Effects*, **Volume 3**, (Ann Arbor Science Publishers Inc., Michigan, 1980) pp. 995–1005.
57. H. B. S. Conacher, R. K. Chadha, J. F. Lawrence, S. M. Charbonneau and F. Bryce, "Mammary transfer and metabolism in the rat of halogenated fatty acids of halogenated olive oil" *Lipids* **19**, 637 (1984).
58. H. M Cunningham and G. A. Lawrence, "Absorption and metabolism of chlorinated fatty acids and triglycerides in rats" *Food Cosmet. Toxicol.* **15**, 101–103 (1977).
59. H. M. Cunningham and G. A. Lawrence, "Absorption and distribution studies on chlorinated oleic acid and extracts of chlorinated lipid and protein fractions of flour in rats" *Food Cosmet. Toxicol.* **15**, 105–108 (1977).
60. B. A. Jones, I. J. Tinsley, G. Wilson and R. R. Lowry, "Toxicology of brominated fatty acids: Metabolite concentration and heart and liver changes" *Lipids* **18**, 327–334 (1983).
61. H. F. Mohamed, T. L. Andreone and R. L. Dryer, "Mitochondrial metabolism of (D,L)-*threo*-9, 10-dibromopalmitic acid" *Lipids* **15**, 255–262 (1980).
62. A. Goksøyr and H. E. Larsen, "Effekter av klorerte fettsyrer og fettextrakter på cytokrom P_{450}-systemet hos laks" *Report* (Laboratory for Marine Molecular Biology, University of Bergen, Norway 1993) (In Norwegian).
63. R. A. Neal, "Known and projected toxicology of chlorinated by-products" **In**: (R. L. Jolly, W. A. Brungs, R. B. Cumming and V. A. Jacobs, eds.): *Water Chlorination. Environmental Impact and Health Effects*, **Volume 3**, (Ann Arbor Science Publishers Inc., Michigan, 1980) pp. 1007–1017.
64. G. N. Cherr, J. M. Shenker, C. Lundmark and K. O. Turner, "Toxic effects of selected bleach kraft mill effluent constituents on sea urchin sperm cell" *Environ. Toxicol. Chem.* **6**, 561–569 (1987).
65. G. Van Der Kraak, "The influence of calcium ionofore and activators of protein kinase C on steroid production by preovulatory ovarian follicles of the goldfish" *Biol. Reproduct.* **42**, 231–238 (1990).
66. G. R. Wilson, I. J. Tinsley and R. R. Lowry, "Fatty acid composition of liver lipids in rats fed brominated fatty acids" *Lipids* **18**, 661–663 (1983).
67. I. J. Tinsley and R. R. Lowry, "An interaction of DDT in the metabolism of essential fatty acids" *Lipids* **7**, 182–185 (1971).
68. G. Vereskuns, C. Wesén, K. Skog and M. Jägerstad, "Inhibitory effect of 9,10-dichlorooctadecanoic acid on the mutagenic activity of MeIQx, 2-AF and B[a]P in the Ames/Salmonella test" (Submitted, 1997).
69. F. D. Gunstone, "Halogenation" **In**: (F. D. Gunstone, J. L. Harwood and F. B. Padley, eds.), *The Lipid Handbook* (Chapman and Hall, London, 1986) pp. 465–466.
70. B. Brunström, D. Broman, L. Dencker, C. Näf, E. Vejlens and Y. Zebühr, "Extracts from settling particulate matter collected in the Stockholm archipelago waters: Embryolethality, immunotoxicology and EROD-inducing potency of fractions containing aliphatics/monoaromatics, diaromatics or polyaromatics" *Environ. Toxicol. Chem.* **11**, 1441–1449 (1992).

9. MOLECULAR AND GENETIC TOXICOLOGY OF ARSENIC

TOBY G. ROSSMAN *

EXPOSURE OF HUMANS TO ARSENIC COMPOUNDS

Arsenic is a common environmental contaminant of our air, water and food. Sources of arsenic contamination are both natural and man-made. It is found especially in ores that contain copper or lead, and as a result of volcanoes. Leaching of arsenic from rocks and soil into hot spring water has been reported in many places around the world. Arsenic contaminates the drinking water supply in some areas of the United States[1]. In the Western hemisphere, high arsenic content is associated with areas formed during the pre-Andes pre-Rocky Mountains period, which explains why problems with arsenic in the drinking water are seen in Argentina, Chile (Antofagasta), Oregon and Alaska[2]. Major man-made sources of arsenic in the environment include mainly nickel and copper smelting, burning of arsenic-containing coal, biocide use and glass manufacturing. Arsenic compounds are used in wood preservative (chromated copper arsenate), insecticides, herbicides (as weed killers for railroad and telephone posts), desiccants to facilitate mechanical cotton harvest, algaecides, in glass manufacturing, nonferrous alloys and sheep dips. The burning of wood treated with chromated copper arsenate caused severe arsenic poisoning in Wisconsin resulting from inhalation of the combustion fumes[3]. Roxarsone (3-nitro-4-hydroxy phenylarsonic acid) or arsenilic acid is used to fatten swine or poultry, and it has been argued that arsenic is an essential element[4,5]. Use of arsenical pesticides may increase arsenic accumulation in various vegetables, grains, fruits, meats and tobacco products, even if such usage has not occurred recently, since arsenic compounds can accumulate in the soil. The environmental biochemistry of arsenic has been reviewed[6].

A new use for arsenic compounds has recently occurred as a result of the discovery that crystals made of gallium arsenide are better superconductors than silicon and are now being used in semiconductors, integrated circuits, diodes, infrared detectors, and laser technology. Airborne particles of gallium arsenide (GaAs) have the potential of dissolution in the lung or stomach[7,8]. The very toxic gas arsine (AsH_3) is used to make gallium arsenide. Arsine and As(III) halogenides are more toxic than other arsenic compounds. These compounds enter erythrocytes where they cause hemolysis leading to jaundice and red urine. The resulting arsenic acid also damages the kidneys. A recent review of arsine toxicity has been published[9]. The remainder of this review will be devoted to the less toxic but carcinogenic arsenic species.

The major routes of arsenic uptake by humans are food and water via the GI tract and inhalation of polluted air by the lung[10]. Approximately 95% of soluble trivalent arsenic compounds are absorbed from the GI tract[1]. Seafood is especially high in

* The Nelson Institute of Environmental Medicine and the Kaplan Comprehensive Cancer Center, New York University Medical Center, 550 First Ave., New York, NY 10016 (USA).

arsenic content, but the form of arsenic in seafood is the less toxic arsenobetaine which is excreted in the urine without metabolism[11]. After absorption through the lungs or the GI tract, most arsenic is transported in the blood (where it can be bound to hemoglobin in erythrocytes) to other parts of the body and distributed to various organs. Arsenic in humans tends to accumulate in hair and nails, followed by skin and lungs. Approximately 70% of arsenic is excreted, mainly in urine, and arsenic intoxication is usually assessed by determining urinary arsenic content. The half life of arsenic in humans is about 10 hours[1].

ARSENIC TOXICOLOGY

Acute symptoms of arsenic poisoning in humans who have ingested AsO_3 are characterized by profound gastrointestinal inflammation and can include constriction of the throat, gastric pain, vomiting, diarrhoea, dehydration, leg cramps, irregular pulse, shock, stupor, paralysis and coma[1,12,13]. In survivors, exfoliative dermatitis, peripheral neuritis, cardiac abnormalities, and reversible anemia and leukopenia may develop. Chronic arsenic poisoning is characterized by numbness, tingling and sometimes hyperkeratosis, pain and temperature sensations in the extremities. Other symptoms may include peripheral vascular disorders (e.g. Raynaud's syndrome), anemia, leukopenia, electrocardiogram abnormalities, and paralysis and atrophy of hand muscles[1,12,13]. This review will focus on cellular and molecular aspects of arsenic toxicity and carcinogenicity. A number of excellent reviews which deal with other aspects of arsenic toxicity exist[1,3,10,12-16].

It is important to consider the speciation of arsenic because many organic and inorganic forms co-exist in our environment and the toxicology of arsenic is complicated by its ability to convert between oxidation states and organometalloid forms. It has been known for some time that arsenite is more toxic than arsenate[17,18]. This may be due in part to different rates of cellular uptake. At equimolar concentration, arsenite uptake into BALB/3T3 cells is 4-fold higher than is uptake of arsenate[19]. Greater uptake of arsenite compared to arsenate is also seen using rabbit erythrocytes or rat liver slices[20,21]. In rat hepatocytes, arsenite readily enters the cell, but arsenate does not[22]. It was suggested that arsenite, having a pK_a of 9.23 (the lowest pK_a), is uncharged at physiological pH and can therefore pass through the cell membrane faster than can arsenate which has a pK_a of 2.20 and is negatively charged[19,22]. It is argued that the concentration gradient would be maintained by the rapid methylation of arsenite once inside the cell, an argument which can apply only to those cells which methylate arsenite (see below). In other cells, extensive binding to intracellular components could play a similar role. However, arsenate uptake (but not arsenite uptake) is blocked by phosphate ion, and by energy poisons suggesting possible competition for a common transporter[23], perhaps a Na/Pi cotransporter[24]. No specific transporters involved in transporting arsenic compounds into mammalian cells have been identified. If arsenate cannot enter hepatocytes, it must enter other cells and be reduced by them to arsenite, which is then extruded and carried to the liver for methylation. Evidence for this view is presented below.

Arsenate (As^{+5}) is similar in structure to inorganic phosphate and can take its place in many reactions. It is known to inhibit metabolic reactions in mitochondrial oxidative phosphorylation by substituting for inorganic phosphate with subsequent formation of an unstable arsenate ester that spontaneously decomposes. Arsenate incubated with rat liver mitochondria generate an arsenate ester[25]. This "arsenolysis reaction" has the effect of uncoupling ATP synthesis in oxidative phosphorylation and also in glycolysis in the reaction catalyzed by glyceraldehyde 3-phosphate dehydrogenase[3,26]. Exposure of rodents to arsenate results in hepatic mitochondrial damage[27].

In contrast to arsenate, arsenite (As^{+3}) tends to bind to vicinal thiol groups. During World War II a great deal of research was conducted on the arsenite–thiol interaction due to the risk of the chemical warfare agent Lewisite (chlorvinyldichloroarsine). It was found that some enzymes reacted with Lewisite by forming a ring involving two thiol groups. This ring was not reversible by most monothiol substances, prompting research into dithiols and the discovery of British Anti-Lewisite. A search for natural dithiol compounds lead to the discovery of lipoic acid, a cofactor of pyruvic oxidase, now called pyruvate dehydrogenase (PDH) multienzyme complex[3,15]. PDH multienzyme complex is an important mitochondrial enzyme system which controls the supply of C_2 fragments to the mitochondria via production of acetylCoA. Like Lewisite, arsenite exerts toxic effects by reacting with vicinal thiols in the cell, whether on lipoic acid (which has 2 thiol groups on adjacent carbon atoms) or on proteins where two cysteine residues come into close enough proximity to be "vicinal". Besides PDH multienzyme complex, thiolase, glutathione reductase, and DNA ligase II may also be important targets for arsenite in mammalian cells[10,26,28]. Arsenite also binds specifically to glucocorticoid receptors (which have vicinal thiol groups) and prevents steroids from binding[29], and inhibits two steps in ubiquitin-dependent proteolysis, a process that degrades abnormal proteins and controls levels of others[30], which suggests the involvement of vicinal thiol groups in that process. Compounds which contain vicinal thiol groups can also be used in chelation therapy for treating arsenic intoxication[31].

Metabolism of Arsenic Compounds

The biomethylation of arsenic was first recognized 1839 by Gmelin who reported on the garlic-like odor of a fungal product, trimethylarsine[32]. Total arsenic is often elevated in greenhouse air due to volatilization of methylarsines produced by soil bacteria and fungi and lack of air exchange[15]. The metabolism of arsenic compounds in mammals has recently been reviewed[3,14,33–35]. Reduction of arsenate to arsenite is necessary before methylation can occur. While it is not known which cells carry out the reduction *in vivo*, reduction has been documented in a number of cell lines *in vitro*[19,24,36]. The reduction of arsenate to arsenite requires glutathione (GSH)[19,34,37,38]. In *E. coli*, the product of the *arsC* gene is an arsenate reductase[39], so it is reasonable to expect that mammalian cells also have this enzyme. Arsenate reductase activity has been measured in liver cytosol of some mammalian species[14], but the enzyme responsible has not been characterized. The reduction of arsenate by GSH can also take place non-enzymatically[40–42] and can proceed further to produce the arsenite-GSH conjugate $As(GS)_3$:

$$As(V)O_4^{3-} + 2GSH \rightarrow As(III)O_3^{3-} + GSSG + H_2O,$$

$$As(III)O_3^{3-} + 3GSH \rightarrow As(GS)_3 + 3OH^-.$$

So far, there is no evidence that $As(GS)_3$ is formed in mammalian cells. Instead, at least in erythrocytes, arsenite appears to form a ternary complex with GSH and a macromolecule, possible hemoglobin[20]. In cells other than erythrocytes, proteins other than hemoglobin would be expected to play a similar role. In arsenite-treated

Chinese hamster V79 cells, all intracellular arsenite appears to be bound to macromolecules, probably protein[36].

Methylation occurs primarily in the liver with much smaller amounts in kidney and lung[21,37,43–45]. Arsenite is methylated by enzymatic transfer of the methyl group from S-adenosyl methionine (SAM) to monomethylarsonic acid [MMAA(V)], which is reduced (probably by GSH) to monomethylarsonous acid [MMAA (III)], which is likewise methylated enzymatically to dimethylarsinic acid [DMAA(V)][14,33,40]:

$$AsO_3^{3-} + SAM \rightarrow AsO_3CH_3^{2-} \quad \text{(methylarsonic acid)},$$

$$AsO_3CH_3^{2-} + GSH \rightarrow AsO_2CH_3^{2-} \quad \text{(methylarsonous acid)},$$

$$AsO_2CH_3^{2-} + SAM \rightarrow AsO_2(CH_3)_2^{1-} \quad \text{(dimethylarsinic acid)}.$$

GSH appears to be required as its depletion has been shown to block arsenic methylation[46]. Glutathione is protective against arsenite toxicity[15,47]. Approximately 50% of excreted arsenic in human urine is dimethylarsinate and 25% is monomethylated, the remainder being inorganic[48]. Recently, evidence of a possible human genetic polymorphisms in the control of methyltransferase activity was presented[49]. Humans may be more sensitive to arsenic than other species (such as rats) in part because arsenic methylation in humans is less efficient than it is in other species[35,50]. The enzymes arsenite methyltransferase and MMAA methyltransferase have recently been purified from rabbit liver[51]. Both reactions require SAM, have similar pH optima, and appear on the same band after gel electrophoresis. Thus, both activities appear to reside on the same 60 kDa protein. Although other thiols can substitute for GSH in the *in vitro* reaction, the thiol utilized *in vivo* is likely to be GSH. Crude liver extracts also appear to contain inhibitors of the methylation reaction[51].

The methylation of arsenic compounds has been considered a detoxification mechanism since the methylated metabolites are less reactive with tissue constituents than are arsenite or arsenate, show increased LD_{50}'s and are rapidly excreted in urine[35,48,52,53]. However, when DMAA was administered orally to mice, it caused oxidative damage and DNA strand breaks in the lung. The strand breaks are caused by dimethylarsine, a further metabolite[54] via the dimethylarsenic peroxy radical $(CH_3)_2AsOO^{\bullet}$. Dimethylarsine was also mutagenic in bacteria[55]. Since arsenite can induce lipid peroxidation in various rat tissues[56], it is possible that this effect is also due to dimethylarsenic peroxy radical. However, there is very little human exposure to DMAA, and certainly not in the concentrations used in some of these studies[53]. It is also unlikely that exposure to inorganic arsenic would result in high tissue accumulation of dimethylarsine or its peroxy radical. When DMAA was ingested by human volunteers, most was excreted unchanged[48,53], but approximately 5% in rodents or 4% in humans was converted to trimethylarsine oxide $AsO(CH_3)_3$. Fungi form trimethylarsine oxide and then reduce it to trimethylarsine, $As(CH_3)_3$, the volatile compound with the garlic-like odor noted by Gmelin.

No arsenic methylation was found to occur in rat erythrocytes, brain, or intestine[37], and at most very small amounts occur in kidney and lung[21] thus confirming the view that methylation *in vivo* takes place predominantly, if not only, in liver[45]. In cells other than liver, cellular protection against arsenic toxicity is probably via active extrusion of arsenic from the cell, thus lowering its intracellular concentration to subtoxic levels. In bacteria, *ars* operons have been identified that express energy-dependent pumps to extrude arsenic and antimony compounds[57,58]. Like arsenic, antimony is a member of the VB group of compounds and is chemically similar to arsenic. The plasmid-encoded *arsA* and *arsB* gene products in *E. coli* form a membrane bound ATP-coupled pump with structural and functional similarities to mammalian P-glycoprotein, the multidrug resistance (*mdr*) gene product. The ArsA protein is a 63 kDa peripheral membrane protein that hydrolyzes ATP upon stimulation by As(III) or Sb(III). The ArsB protein is a 45 kDa integral membrane protein that spans the *E. coli* inner membrane 12 times. It is the membrane anchor for the ArsA protein and probably forms the anion-conducting pathway. The ArsC protein is an arsenate reductase. A chromosomal *ars* operon with strong homology to the plasmid operon has been identified in *E. coli*[59,60] and similar *ars* operons have also been found in other microorganisms[61] as well as in *Leishmania* strains resistant to pentostam, an antimony-containing drug used in the treatment of leishmaniasis[62]. It has been proposed that arsenate-resistance in *Leishmania* consists of three steps: (1) reduction of arsenate to arsenite. (2) conjugation of arsenite with glutathione to form $As(GS)_3$, (3) extrusion by a glutathione-linked pump that is perhaps similar to the multidrug resistance-associated protein (MRP)[63,64].

Mammalian cells also appear to protect themselves from arsenic compounds by using efflux pumps. SA7, an arsenite-resistant subline of Chinese hamster ovary (CHO) cells cross-resistant to As(V), Zn(II), Fe(II), Co(II), Hg(II) and Sn(II), accumulated less arsenic compared to its wild-type parental cell line and had elevated levels of GSH and GSH transferase[65,66], suggesting that the substrate of the pump is a GSH conjugate. Recently, we found evidence for an up-regulated efflux pump in arsenite-resistant Chinese hamster V79 sublines As/R7 and As/R27[36]. These cells show faster efflux of [^{73}As]arsenite, but do not share the phenotype of SA7 cells, in that they are cross-resistant to antimonite but not to Fe(II), and do not show elevated levels of GSH or GSH S-transferase (Rossman, Wang, and Goncharova, unpublished data). An arsenite-hypersensitive variant As/S5 exhibited increased [^{73}As]arsenite accumulation compared with the parental line[36]. Efflux could be inhibited by the protonophore carbonyl cyanide *m*-chlorophenylhydrazine, suggesting energy-dependence. Inhibitors of glutathione S-transferase also decreased arsenite eflux, suggesting the involvement of an arsenite-glutathione complex such as $As(GS)_3$. Biliary excretion of arsenate also requires GSH[67]. Analysis of the extrusion products from V79 cells given arsenate or arsenite failed to show the appearance of $As(GS)_3$. Rather, all label in the product of the transport reaction appeared to be arsenite in both cases. Similar results were seen with BALB/3T3 cells[19]. This indicates that intracellular reduction of arsenate to arsenite occurs, and that arsenite is transported as an unconjugated species, although it is possible that a GSH conjugate is transported and immediately dissociates.

Although treatment of cells with arsenite causes at least a temporary depletion of GSH, so far there is no evidence that $As(GSH)_3$ exists in mammalian cells. In rat erythrocytes treated with arsenite, there is evidence for the formation of a mixed As(III) complex with protein and GSH[38]. This is consistent with the high molecular weight complexes seen in other systems (see below).

No methylated arsenic products were seen in arsenite-exposed BALB/3T3 cells (a mouse fibroblast line) or their medium[19], and the same is true for V79 cells (a Chinese hamster fibroblast line)[36]. However, DMAA appeared in the medium of arsenite-treated rat hepatocytes[22], supporting the idea that the liver is the main site of arsenic methylation. In V79 cells exposed to labeled arsenite, all intracellular ^{73}As was associated with high molecular weight material, possibly protein. Similar results were seen using rabbit erythrocytes[20]. It was previously suggested that some arsenite detoxification might also be achieved by protein binding, as in the marmoset monkey which does not methylate arsenic[68]; nor does the chimpanzee[69] or the guinea pig (although a polymorphism exists in that species)[70]. A number of other New World animals also fail to methylate arsenic, and the intriguing evolutionary implications of this finding are discussed by Aposhian[14]. Three arsenite-binding proteins have been identified in rabbit liver, but it is not clear if these act to detoxify[71].

The Carcinogenicity and Genetic Toxicology of Arsenic Compounds

Chronic arsenic exposure has become of more concern than acute exposure mainly because of its carcinogenic effects[72,73]. Epidemiological studies clearly indicate that inorganic arsenic compounds are human carcinogens, but they are not tumorigenic in most animal bioassays. Because there is no good animal model, arsenic compounds are the only compounds that IARC considers to have sufficient evidence for human carcinogenicity, but inadequate evidence for animal carcinogenicity[74]. Many cases of skin cancer have been reported among people exposed to arsenic through medical treatment with inorganic trivalent arsenic compounds. In some instances, skin cancers have occurred in combination with other cancers, such as liver angiosarcoma, intestinal, and urinary bladder cancers and meningioma. Occupational exposure to inorganic arsenic, especially in mining and copper smelting, has quite consistently been associated with an increased risk of cancer[72]. The strongest epidemiological association between arsenic ingestion and an internal cancer is for the bladder[75], followed by kidney and liver[76]. The increase in cancer risk observed in epidemiological studies is attributed mainly to the presence of inorganic trivalent arsenic[12,72].

Unlike many other carcinogens, arsenite does not behave as a mutagen at single gene loci in *E. coli*, or in mammalian cells[77,78]. The inability of arsenite to induce the SOS system in *E. coli* is consistent with its lack reaction with DNA[79]. Chinese hamster G12 cells[80] can detect mutation at the *gpt* locus, as well as increased methylation (leading to loss of expression) as is caused by carcinogenic nickel compounds[81]. Since arsenite fails to induce *gpt*$^-$ variants in these cells[82], it causes neither mutation nor hypermethylation changes at that locus (but see section: Effects on Gene Expression below).

Although arsenite is not mutagenic, it does effect the mutagenicity of other carcinogens, probably via effects on DNA repair. The first evidence that arsenic compounds may inhibit DNA appeared in 1969[83]. Since that time, arsenite has been shown to enhance the mutagenicity of UV in *E. coli*[84] and to enhance the mutagenicity of UV, N-methyl-N-nitrosourea, and methylmethane sulfonate[85–89] and the clastogenicity of diepoxybutane[90], UV and X-rays[91] in mammalian cells. Arsenite's effect on UV-induced clastogenesis is maximal during the late G1 to early S phase of the cell cycle[92]. Arsenite inhibits the completion of DNA excision repair[87,93,94], probably via effects on DNA ligase II, which is especially sensitive to arsenite in the cell[28]. Arsenite alters the mutational spectrum (but not the strand bias) of UV-irradiated Chinese hamster ovary cells[89]. Arsenite also potentiates killing by UV in excision-proficient human fibroblasts by inhibiting excision of pyrimidine dimers, but has no effect on UV killing in cells from Xeroderma pigmentosum patients who lack nucleotide excision repair[95]. Recent results using the single cell comet assay have confirmed that arsenite prevents the completion of DNA repair in human white blood cells and in human SV40-transformed fibroblasts, consistent with an inhibition of DNA ligase and/or repair synthesis[96].

These comutagenic actions of arsenite suggest that important DNA repair enzymes may contain vicinal thiol groups that are targets for arsenite's action. DNA ligases have been shown to contain essential thiol groups[97], but whether they occur in the correct geometry to bond with arsenite is not known. Vicinal thiols are a common feature of proteins which are involved with DNA, such as the zinc fingers of transcription factors. A number of other DNA repair proteins that might be targets for arsenite inhibition is discussed in Wienke and Yager[90]. Evidence also points to effects of arsenite on postreplication repair of UV-induced damage[98], a process whose enzymology is not well characterized.

A lack of mutagenicity by arsenite was also seen at two loci in Syrian hamster embryo cells, where arsenite caused cell transformation and cytogenetic damage[99,100]. Similar results were found using 10T1/2 mouse embryo cells, where arsenite induced morphological transformation and increased steady-state levels of *c-myc* transcripts but did not induce mutations[101]. Arsenite also induces neoplastic transformation in BALB/3T3 mouse embryo cells which gave rise to sarcomas after subcutaneous injection into nude mice[102], increased transformation by bovine papillomavirus[103], and caused anchorage-independent growth but no focus formation or immortality in diploid human fibroblasts[101]. Arsenite also transforms human osteosarcoma (HOS) cells to anchorage-independence (Rossman and Hu, unpublished).

It must not be concluded, however, that arsenite is strictly a non-genotoxic carcinogen. Although there is a lack of mutagenicity at single gene loci, there is ample demonstration from a number of laboratories of sister chromatid exchanges (SCE) and chromosome aberrations induced by arsenite[91,104–106]. Arsenite-induced DNA-protein crosslinks have been reported in human cells[107]. Treatment of CHO cells in the G2 phase of the cell cycle with arsenite lead to poorly condensed chromosomes and chromatid breaks. Upon further growth, aneuploidy and micronuclei were seen[108,109]. Micronuclei are also found in the bone marrow of mice treated with arsenite[110]. Arsenite induced gene amplification at the *dhfr* locus in human and rodent cells, but failed to cause amplification of SV40 sequences in SV40-transformed human keratinocytes or Chinese hamster cells[82,99,111]. This suggests that arsenite treatment does not result in signaling typical of DNA-damaging agents (which induce SV40 amplification in these systems), but rather appears to feed into checkpoint pathways common to those involving p53, whose disruption lead to cellular gene amplification[112]. Chromosome aberration and endoreduplication were induced by arsenite, but not arsenate, at concentrations 10^{-6} M and higher, whereas both arsenite and arsenate cause SCE at concentrations as low as 10^{-8} M, but not in a dose dependent manner[113,114].

As mentioned earlier, when DMAA is given to animals, lung-specific DNA damage is seen. This damage is caused by the DMAA peroxy radical[54]. DNA strand breaks and DNA-protein crosslinks are also induced by DMAA in cultured cells[115,116]. Although doubts have been expressed that humans exposed to inorganic arsenic could accumulate sufficient DMAA (and its peroxy radical) for genotoxic effects to ensue, nevertheless the possibility exists that some of arsenite's genotoxic effects might be caused either by DMAA peroxy radical or by oxygen free radicals resulting from GSH or lipoic acid depletion. Toxicity and genotoxicity of a number of metals does involve oxygen radicals[117–119]. The addition of superoxide dismutase to the culture medium was able to block arsenite-induced SCE in human lymphocytes[120], suggesting that oxygen free radicals mediate the SCE induction by arsenite. Vitamin E (α-tocopherol, an antioxidant) protects human fibroblasts from arsenite toxicity[121]. Further support for an oxygen radical involvement in arsenite genotoxicity is the finding that an X-ray sensitive CHO cell variant, XRS-5, is hypersensitive to killing and micronucleus induction by arsenite[109]. These cells have normal GSH levels but are deficient in catalase activity. Addition of catalase to the medium blocked micronucleus induction by arsenite, suggesting a H_2O_2 intermediate in that process. Arsenite-induced SCE and micronuclei were also blocked by squalene, a component of shark liver oil used in Asian folk medicine, which is structurally similar to β-carotene, an antioxidant[122].

Genetic toxicology endpoints have been used as biomarkers of arsenic exposure. SCE is one of the most sensitive endpoints, although its mechanism is not well understood. In a recent study of subjects in Argentina who had more than 0.13 mg/l (0.13 ppm) arsenic in their drinking water for a period of at least 20 years, a significant elevation in lymphocyte SCE is seen[2]. Other health effects in these individuals are hyperkeratosis, melanosis, actinic keratosis and basal cell carcinoma. Delay of cell cycle progression was seen in the lymphocytes of Mexicans drinking water with a high arsenic content[123]. A 1.8 fold increase in micronuclei in exfoliated bladder cells of individuals drinking arsenic-contaminated well water in Nevada was reported[124]. The bladder cell micronucleus assay has been suggested to be the most appropriate biological marker of arsenic genotoxicity[125]. Recently, micronuclei were studied in bladder cells of arsenic-exposed individuals utilizing fluorescent *in situ* hybridization (FISH) with a centromeric probe[126]. There is evidence for interindividual variation in aneuploidy and cell-cycle delay in response to arsenic exposure[49].

The mechanisms by which arsenic compounds cause human cancers are not known. No studies have been carried out identifying changes in oncogenes or tumor suppressor genes in arsenic-induced cancers. Arsenite did not give positive results in a bioassay testing for promotional activity[127], but DMAA did[128]. In Ha-ras transgenic mice, arsenite increased the numbers of papillomas induced by the tumor promoter phorbol ester[129], so arsenite could be considered a "co-promoter" in that system (the authors suggest the term "tumor enhancer"). The genotoxic effects of arsenite would be more likely to result in loss of tumor suppressor functions (e.g. by deletion) than in mutation of oncogenes. Finally, it is also likely that effects on gene expression play an important role in arsenic carcinogenesis.

Effects on Gene Expression

The exposure of cells to temperatures higher than those they are usually accustomed to results in the expression of the so-called heat shock proteins (HSPs). Hsp genes are also activated by stimuli seemingly unrelated to heat, including arsenite. It has been suggested that all of the inducers of HSPs are agents which can cause the denaturation of cellular proteins (in the case of arsenite, by reacting with vicinal thiols)[130]. Arsenite induces HSP27, which is sufficient to confer the thermoresistant phenotype[131], as well as some stress proteins that might be tissue-specific[132]. Arsenite-induced thermoresistance has been demonstrated in V79 cells[133,134]. All of the "classical" heat shock proteins are induced by arsenite, as are a number of novel proteins[135,136]. Heat and arsenite induce cross-tolerance to each other in hepatoma cells[137] but not in V79 cells, where heat failed to induce arsenite tolerance[134].

One of the arsenite-inducible proteins (32 kDa) has been identified as heme oxygenase (HO)[138], a protein that is also inducible by UVA, cadmium chloride, and hydrogen peroxide, but not very well by heat shock. HO is thought to protect against oxidant damage. HO cleaves the heme tetrapyrrole ring in cellular hemoproteins involved in redox reactions, thus causing cellular redox potential to shift towards reduction. The enzymatic product of HO, biliverdin, is then converted to bilirubin which is also an antioxidant[139]. Induction of HO is responsive to cellular levels of GSH[140], and its induction by arsenite was blocked by antioxidants[141]. An arsenite-resistant variant of a human lung adenocarcinoma cell line has elevated levels of heme oxygenase[142] and arsenite-resistance in these cells could be blocked by tin-protoporphyrin, an inhibitor of heme oxygenase. Wild type cells, but not the arsenite-resistant variant, showed increased oxidant content after exposure to arsenite for 24 hours, thus supporting the role of oxidants in arsenite toxicity and the protective role of heme oxygenase.

Metallothionein, a small metal-binding protein, can also be considered a stress protein[143]. It was induced in the liver of rats fed arsenite[144] and in V79 cells exposed to arsenite[145]. A cDNA subtraction library was made between RNA from HeLa cells growing for 24 hours in the presence and absence of 5 μM sodium arsenite. This lead to the identification of metallothionein II and ferritin H chain as arsenite-inducible genes in human cells[146]. V79 cells carrying metallothionein I expression vectors show increased resistance to arsenite[147]. A number of investigators have suggested that a major role of MT is to protect against electrophilic agents such as free radicals and reactive metabolites[147–152]. Metallothionein is capable of scavenging hydroxyl radicals and superoxide anions *in vitro*[153,154]. Many of the physical and chemical agents which induce metallothionein also give rise to oxidative stress[148,155]. The fact that arsenite both induces metallothionein and is protected by it support a role for oxygen free radicals in arsenite toxicity. In the case of V79 cells, which do not appear to methylate arsenic[36] and thus cannot generate DMAA peroxy radical, the most likely involvement of oxygen would be via depletion of GSH after arsenite treatment, or perhaps via reaction with lipoic acid which is also an antioxidant[156].

Arsenite has also been shown to increase expression of the multidrug resistance gene *mdr* which codes for P-glycoprotein[157]. Expression of *c-fos* and *c-myc* (but not of *erbB* or *c-H-ras*) were also induced by arsenite treatment[158,159]. Treatment of primary human keratinocytes in culture with arsenite induced a unique cytokine profile which included transforming growth factor α and granulocyte-macrophage colony stimulating factor. These same cytokines were seen in arsenite-treated Ha-ras transgenic mice, a model for skin cancer[129]. Arsenite induces Phase 2 enzymes (the so-called "Electrophile Counterattack") in the liver[160]. Induction of these enzymes appears be mediated by enhancer elements that contain AP-1-like sites. AP-1 is composed of the Jun and Fos proteins, both of which are induced by oxidants[161]. (Induction of *c-fos* by arsenite was mentioned above). The DNA-binding activity of Fos and Jun are also controlled by oxidation/reduction[161]. Arsenite can also enhance the mitogenic effect of sub-optimal serum concentrations on quiescent C3H10T$^{1/2}$ cells[162], possibly via induction of c-Fos and/or HSP70.

Some of the effects of arsenite on gene expression might be mediated by changes in protein phosphorylation. Arsenite has been shown to mimic tumor necrosis factor in inhibiting specific patterns of protein phosphorylation, probably via inactivation of protein phosphatase 2A(PP2A)[163]. The induction of HSP28 is associated with such hyperphosphorylation of the protein[164], which can be caused by arsenite treatment[113]. The p34 subunit of RPA, a DNA binding protein needed for DNA repair, is also modulated by PP2A[165], which may help to explain some of the effects of arsenite on DNA repair. The tumor suppressor p53 is also modulated by reversible phosphorylation[166], which may help to explain the effects of arsenite on gene amplification. There is also evidence that the perturbation of keratinocyte differentiation by arsenite may take place via inhibition of a protein kinase[167].

Another way to alter gene expression is to alter the patterns of cytosine methylation of CpG sequences in genes. Recently it was found that treatment of cells with arsenate and arsenite, but not DMAA, could produce hypermethylation in the p53 gene promoter[168]. Arsenite does not appear to cause gene silencing of a transgenic *E. coli gpt* gene in Chinese hamster G12 cells, in contrast to nickel compound which do[81,82]. However, there are probably multiple pathways by which genes can become hypermethylated, and sequence context may be important. Finally, the degradation rates of proteins might be decreased by arsenite via inhibition of ubiquitin-dependent proteolysis[30], resulting in up-regulation of these proteins.

INDUCIBLE TOLERANCE TO ARSENITE

Both prokaryotic and some eukaryotic cells develop tolerance when exposed to arsenic compounds. I will use the word “tolerance” to refer to a temporary resistance. In *E. coli*, the *ars* operon is up-regulated by prior exposure to arsenic or antimony compounds[57,58]. Both wild type Chinese hamster V79 cells and their arsenite-resistant variants exhibit the phenomenon of arsenite- or antimonite-inducible arsenite tolerance[169]. The inducible tolerance requires *de novo* mRNA and protein synthesis, and differs from the heat shock response[134]. Although, as mentioned above, arsenite induces metallothionein which is somewhat protective against arsenite, the amount of protection afforded by metallothionein is much less than that afforded by induction of arsenite tolerance by arsenite[145,147]. Preliminary results suggest that the arsenite-inducible arsenite tolerance mechanism is not mediated by up-regulation of an arsenite efflux pump (Rossman, Mukhopadhyay, and Rosen, unpublished data). An arsenite-hypersensitive V79 subline, As/R27D, fails to show arsenite-inducible arsenite tolerance, as do a number of different human cell lines (including normal diploid fibroblasts from three individuals, meduloblastoma cells, SV40-transformed keratinocytes, and HeLa cells)[170]. The greater sensitivity (10-fold) to arsenite of human fibroblasts compared with Chinese hamster ovary cells was first pointed out by Lee and coworkers[171], who suggested that the explanation lies in the fact that CHO cells contain more antioxidants than do human fibroblasts. We have found that all human cells tested are more sensitive to arsenite than are rodent cells. Human keratinocytes were especially sensitive. In general, human cells resemble arsenic hypersensitive Chinese hamster As/R27D cells, which have lost an inducible protective mechanism found in wild type Chinese hamster cells. We suggest that the lack of an arsenite-inducible tolerance mechanism may play a role in this greater sensitivity[170]. This difference between human and rodent cells may underlie the different carcinogenic effects of arsenic compounds in humans and rodents.

Acknowledgments

I would like to thank Drs. H. V. Aposhian and M. J. Mass for allowing me to see their recent articles prior to publication. I thank Drs. E. T. Snow, B. P. Rosen and M. Vahter for their helpful suggestions with a number of issues discussed in this paper. Finally, I thank Ms. E. Cordisco for her patient help in preparing this manuscript. This work was supported by United States Public Health Service Grant CA57352 and is part of New York University's Nelson Institute of Environmental Medicine Center programs supported by Grants ES 00260 from the National Institute of Environmental Health Sciences.

References

1. A. Leonard, "Arsenic" **In**: *Metals and their Compounds in the Environment* (E. Merian, ed.) (Verlagsgesellschaft, New York, 1991). pp. 751–774.
2. D. Lerda, "Sister-chromatid exchange (SCE) among individuals chronically exposed to arsenic in drinking water" *Mutat. Res.* **312**, 111–120 (1994).
3. H. V. Aposhian, "Biochemical toxicology of arsenic" *Rev. Biochem. Toxicol.* **10**, 265–299 (1989).
4. A. Schmidt, M. Anke, B. Groppel and H. Kronemann, "Effects of As-deficiency on skeletal muscle, myocardium and liver. A histochemical and ultrastructural study" *Exper. Pathol.* **25**, 195–197 (1984).
5. E. O. Uthus and F. H. Nielsen, "Determination of the possible requirement and reference dose levels for arsenic in humans" *Scand. J. Work Environ. Health* **19**, 137–138 (1993).
6. S. Tamaki and W. T. Frankenberger Jr., "Environmental biochemistry of arsenic" *Rev. Environ. Contamin. and Toxicol.* **124**, 79–110 (1992).
7. D. Carter and W. TY. Bellamy, "Toxicology of the group III–V intermetallic semiconductor, gallium arsenide" **In**: *Biological Monitoring of Toxic Metals* (T. W. Clarkson, L. Friberg, G. F. Nordberg and P. R. Sager, eds.) (Plenum Press, New York, 1988) pp. 455–468.
8. S. J. S. Flora, "Alterations in some hepatic biochemical variables following repeated gallium arsenide administration in rats" *Internat. Hepatology Commun.* **5**, 97–103 (1996).
9. W. T. Klimecki and D. E. Carter, "Arsine toxicity: chemical implications and mechanistic implications" *J. Toxicol. Environ. Hlth.* **46**, 399–409 (1995).
10. H. V. Aposhian and M. M. Asposhian, "New developments in arsenic toxicity" *J. Amer. Coll. Toxicol.* **8**, 1273–1281 (1989).
11. M. Vahter, E. Marafante and L. Dencker, "Metabolism of arsenobetaine in mice, rats and rabbits" *Sci. Total Environ.* **30**, 197–211 (1983).
12. P. J. Landrigan, "Arsenic" **In**: *Environmental and Occupational Medicine Second Edition* (W. N. Rom, ed.) (Little Brown and Co., Boston, MA, 1992) pp. 773–779.
13. G. G. Garcia-Vargas and M. E. Cebrián, "Health Effects of Arsenic" **In**: *Toxicology of Metals* (L. W. Wang, ed.) (CRC Press, Boca Raton, 1996) pp. 423–438.
14. H. V. Aposhian, "Enzymatic methylation of arsenic species and other new approaches to arsenic toxicity" *Ann. Rev. Pharmacol. and Toxicol.* **37**, 397–419 (1997).
15. V. Bencko, "Arsenic" **In**: *Advances in Modern Environmental Toxicology Vol XI: Genotoxic and Carcinogenic Metals: Environmental and Occupational Occurrence and Exposure* (L. Fishbein, A. Furst, and M. A. Mehlman, eds.) (Princeton Scientific Publishing Co., Inc. Princeton NJ, 1987) pp. 1–30.
16. K. S. Squibb and B. A. Fowler, "The toxicity of arsenic and its compounds" **In**: *Biological and Environmental Effects of Arsenic* (B. A. Fowler, ed.) (Elsevier, Amsterdam, 1983) p. 233.
17. J. J. Gordon and J. H. Quastel, "Effects of organic arsenicals on enzyme systems" *Biochem. J.* **42**, 337–350 (1948).
18. S. A. Peoples, "Arsenic toxicity in cattle" *Ann. N.Y. Acad. Sci.* **111**, 644–649 (1964).
19. F. Bertolero, G. Pozzi, E. Sabbioni and U. Saffiotti, "Cellular uptake and metabolic reduction of pentavalent to trivalent arsenic as determinants of cytotoxicity and morphological transformation" *Carcinogenesis* **8**, 803–808 (1987).
20. M. Delnomdedieu, M. M. Basti, M. Styblo, J. D. Otvos and D. J. Thomas, "Complexation of arsenic species in rabbit erythrocytes" *Chem. Res. Toxicol.* **7**, 621–627 (1994).
21. B. Georis, A. Cardenas, J. P. Buchet and R. Lauwerys, "Inorganic arsenic methylation by rat tissue slices" *Toxicology* **63**, 73–84 (1990).
22. S. A. Lerman, T. W. Clarkson and R. J. Gerson, "Arsenic uptake and metabolism by liver cells is dependent on arsenic oxidation state" *Chem.-Biol. Interact.* **45**, 401–406 (1983).
23. R. N. Huang and T. C. Lee, "Cellular uptake of trivalent arsenite and pentavalent arsenate in KB cell cultures in phosphate-free medium" *Toxicol. Appl. Pharmacol.* **136**, 243–249 (1996).
24. H. Murer, D. Markovich and J. Biber, "Renal and small intestinal sodium-dependent symporters of phosphate and sulphate" *J. Exper. Biol.* **196**, 167–181 (1994).
25. T.-L. Chan, B. R. Thomas and C. L. Wadkins, "Formation and isolation of an arsenylated component of rat liver mitochondria" *J. Biol. Chem.* **244**, 2883–2890 (1968).
26. K. A. Rein, B. Borreback and J. Bremer, "Arsenite inhibits oxidation in isolated rat liver mitochondria" *Biochem. Biophys. Acta.* **574**, 487–494 (1979).
27. B. A. Fowler, "International conference on environmental arsenic: an overview" *Environ. Health Perspect.* **19**, 239–242 (1977).
28. J.-H. Li and T. G. Rossman, "Inhibition of DNA ligase activity by arsenite: A possible mechanism of its comutagenesis" *Mol. Toxicol.* **2**, 1–9 (1989).
29. S. Lopez, Y. Miyashita and S. S. Simons, "Structurally based, selective interaction of arsenite with steroid receptors" *J. Biol. Chem.* **265**, 16039–16042 (1990).

30. N. S. Klemperer and C. M. Pickartt "Arsenite inhibits two steps in the ubiquitin-dependent proteolytic pathway" *J. Biol. Chem.* **264**, 19245–19252 (1989).
31. M. M. Jones, "Chemistry of chelation: chelating agent antagonists for toxic metals" **In**: *Toxicology of Metals* (R. A. Goyer and M. G. Cherian, eds.)(Springer-Verlag, New York, 1995) **Vol. 115**, Chapter 13, pp. 279–304.
32. J. S. Fruton and S. Simmonds, "Special aspects of amino acid metabolism" *General Biochemistry* (John Wiley & Sons, 1953).
33. D. J. Thompson, "A chemical hypothesis for arsenic methylation in mammals" *Chemico-Biol. Interact* **88**(2–3), 89–114 (1993).
34. M. Styblo, M. Delnomdedieu and D. J. Thomas, "Biological mechanisms and toxicological consequences of the methylation of arsenic" **In**: *Toxicology of Metals Biochemical Aspects* (R. A. Goyer and G. Cherian, eds.) (Springer-Verlag, New York, 1995) pp. 407–433.
35. M. Vahter, "Metabolism of arsenic" **In**: *Biological and Environmental Effects of Arsenic* (B. A. Fowler, ed.)(Elsevier, Amsterdam, 1983) pp. 171–198.
36. Z. Wang, S. Dey, B. P. Rosen and T. G. Rossman, "Efflux mediated resistance to arsenicals in arsenic resistant and -hypersensitive Chinese hamster cells" *Toxicol. Appl. Pharmacol.* **137**, 112–119 (1996).
37. J. P. Buchet and R. Lauwerys, "Study of inorganic arsenic methylation by rat liver *in vitro*: Relevance for the interpretation of observations in man" *Arch. Toxicol.* **57**, 125–129 (1985).
38. S. L. Winski and D. E. Carter, "Interactions of rat red blood cell sulfhydryls with arsenate and arsenite" *J. Toxicol. & Environ. Health* **46**, 379–397 (1995).
39. K. L. Oden, T. B. Gladysheva and B. P. Rosen, "Arsenate reduction mediated by the plasmid-encoded ArsC protein is coupled to glutathione" *Mol. Microbiol.* **12**, 301–306 (1994).
40. N. Scott, K. M. Hatlelid, N. E. MacKenzie and D. E. Carter, "Reactions of arsenic (III) and arsenic (V) species with glutathione" *Chem. Res. Toxicol.* **6**, 102–106 (1993).
41. M. Delnomdedieu, M. M. Basti, J. D. Otvos and D. J. Thomas, "Reduction and binding of arsenate and dimethylarsinate by glutathione: a magnetic resonance study" *Chemico-Biol. Interact.* **90**, 139–155 (1994).
42. M. Delnomdedieu, M. M. Basti, J. D. Otvos and D. J. Thomas, "Transfer of arsenite to dithiols: a model of interaction" *Chem. Res. Toxicol.* **6**, 598–602 (1993).
43. M. Vahter and E. Marafante, "Effects of low dietary intake of methionine, choline, or proteins on the biotransformation of arsenite in the rabbit" *Toxicol. Lett.* **37**, 41–46 (1987).
44. E. Marafante and M. Vahter, "The effect of methyltransferase inhibition on the metabolism of [^{74}As] arsenite in mice and rabbits" *Chem. Biol. Interact.* **50**, 49–57 (1984).
45. E. Marafante, M. Vahter and J. Envoll, "The role of the methylation in the detoxication of arsenate in the rabbit" *Chem. Biol. Interact.* **56**, 225–238 (1985).
46. J. P. Buchet and R. R. Lauwreys, "Study of factors influencing the *in vivo* methylation of inorganic arsenic in rats" *Toxicol. Appl. Pharmacol.* **91**, 65–74 (1987).
47. H. Huang, C. F. Huang, D. R. Wu, C. M. Jinn and K. Y. Jan, "Glutathione as a cellular defence against arsenite toxicity in cultured Chinese hamster ovary cells" *Toxicology* **79**, 195–204 (1993).
48. J. P. Buchet, R. Lauwerys and H. Roels, "Comparison of the urinary excretion of arsenic metabolites after a single dose of sodium arsenite, monomethylarsonate or dimethylarsinate in man" *Int. Arch. Occup. Environ. Health* **48**, 71–79 (1981).
49. L. Vega, M. E. Gonseblatt and P. Ostrosky-Wegman, "Aneugenic effect of sodium arsenite on human lymphocytes *in vitro*: an individual susceptibility effect detected" *Mutat. Res.* **334**, 365–373 (1995).
50. H. V. Aposhian, "Arsenic toxicology: does methylation of arsenic species have an evolutionary significance?" *Metal Ions Biol. Med.* **4**, 399–401 (1996).
51. R. Zakharyan, Y. Wu, G. M. Bogdan and H. V. Aposhian, "Enzymatic methylation of arsenic compounds: assay, partial purification and properties of arsenite methyltransferase and monomethylarsonic acid methyltransferase of rabbit liver" *Chem. Res. Toxicol.* **8**, 1029–1038 (1995).
52. G. K. H. Tam, S. M. Charbonneau, F. Bryce, C. Pomroy and E. Sandi, "Metabolism of inorganic arsenic [^{74}As] in humans following oral ingestion" *Toxicol. Appl. Pharmacol.* **50**, 319–322 (1979).
53. E. Marafante, M. Vahter, H. Norin, J. Envall, M. Sandstrom, A. Christakopoulos and R. Ryhage, "Biotransformation of dimethylarsinic acid in mouse, hamster and man" *J. Appl. Toxicol.* **7**, 111–117 (1987).
54. K. Yamanaka and S. Okada, "Induction of lung-specific DNA damage by metabolically methylated arsenics via the production of free radicals" *Environ. Health Perspect.* **102**(3), 37–40 (1994).
55. K. Yamanaka, M. Hoshino, M. Okamoto, R. Sawamura, A. Hasegawa and S. Okada, "Induction of DNA damage by dimethylarsine, a metabolite of inorganic arsenic, is for the major part likely due to its peroxyl radical" *Biochem. Biophys. Res. Commun.* **168**, 58–64 (1990).
56. O. Ramos, L. Carrizales, L. Yanez, J. Mejia, L. Batres, D. Ortiz and F. Diaz-Barriga, "Arsenic increased lipid peroxidation in rat tissues by a mechanism independent of glutathione levels" *Environ. Health Perspect.* **103**, 85–88 (1995).

57. B. P. Rosen, "Resistance mechansims to arsenicals and antimonials" *Rev. Clinical Basic Pharmacol.* **6**, 251–263 (1995).
58. S. Dey and B. P. Rosen, "ATP-coupled pumps for heavy metals and metalloids" **In**: *Handbook of Biological Phys.* **Vol. II**: *Transport processes in membranes* (W.N. Konings, H.R. Kabach, J.S. Lolkema, eds.) (Elsevier, Amsterdam, 1996) pp. 1–19.
59. A. Carlin, W. Shi, S. Dey and B. P. Rosen, "The ars operon of *Escherichia coli* confers arsenical and antimonial resistance" *J. Bacteriol.* **177**, 981–986 (1995).
60. C. Diorio, J. Cai, J. Marmor, R. Shinder and M. S. DuBow, "An *Escherichia coli* chromosomal *ars* operon homolog is functional in arsenic detoxification and is conserved in gram-negative bacteria" *J. Bacteriol.* **177**, 2050–2056 (1995).
61. G. Ji and S. Silver, "Regulation and expression of the arsenic resistance operon from *Staphlococcus aureus* plasmid pI258" *J. Bacteriol.* **174**, 3684–3694 (1992).
62. S. Dey, B. Papadopoulou, G. Roy, K. Grondin, D. Dou, B. P. Rosen and M. Ouellette, "High level arsenite resistance in *Leishmania* is due to an active extrusion system" *Molec. Biol. Parasitol.* **67**, 49–57 (1994).
63. I. Leier, G. Jedlitschky, U. Buchholz, S. P. C. Cole, R. G. Deeley and D. Keppler, "The *MRP* gene encodes an ATP-dependent export pump for leukotriene C_4 and structurally related conjugates" *J. Biol. Chem.* **269**, 27807–27810 (1994).
64. M. Muller, C. Meijer, G. J. R. Zaman, P. Borst, R. J. Scheper, N. H. Mulder, E. G. E. de Vries and P. L. M. Jansen, "Overexpression of the gene encoding the multidrug resistance-associated protein results in increased ATP-dependent glutathione S-conjugate transport" *Proc. Natl. Acad. Sci. USA* **91**, 13033–13037 (1994).
65. T. C. Lee, M. L. Wei, W. J. Chang, I. C. Ho, J. F. Lo, K. Y. Jan and H. Huang, "Elevation of glutathione S-transferase activity in arsenic-resistant Chinese hamster ovary cells" *In vitro Cell. Devel. Biol.* **25**, 442–448 (1989).
66. J. F. Lo, H. F. Wang, M. F. Tam and T. C. Lee, "Glutathione S-transferase in an arsenic-resistant Chinese hamster ovary cell line" *Biochem. J.* **288**, 977–981 (1992).
67. A. Gyurasics, F. Varga and Z. Gregus, "Glutathione-dependent biliary excretion of arsenic" *Biochem. Pharmacol.* **42**, 465–468 (1991).
68. M. Vahter and E. Marafante, "Reduction and binding of arsenate in marmoset monkeys" *Arch. Toxicol.* **57**, 119–124 (1985).
69. M. Vahter, R. Couch, B. Nermell and R. Nilsson, "Lack of methylation of inorganic arsenic in the chimpanzee" *Toxicol. Appl. Pharmacol.* **133**, 262–268 (1995).
70. S. M. Healy, R. A. Zakharyan and H. V. Aposhian, "Enzymatic methylation of arsenic compounds: *In vitro* and *in vivo* deficiency of the methylation of arsenite and monomethylarsonic acid in the Guinea pig" *Mutat. Res.* **386**, 229–239 (1997).
71. G. M. Bogdon, A. Sampayo-Reyes and H. V. Aposhian, "Arsenic binding proteins of mammalian systems: I. Isolation of three arsenite-binding proteins of rabbit liver" *Toxicology* **93**, 175–193 (1994).
72. IARC (International Agency for Research on Cancer), "IARC Monographs on the Evaluation of Carcinogenic Risk of Chemicals to Man" *Some Metals and Metallic Compounds* (World Health Organization, Lyon, France, 1980) **Vol. 23**, pp. 39–141.
73. Z. Wang, and T. G. Rossman, "The carcinogenicity of arsenic" **In**: *Toxicology of Metals* (Louis W. Chang, ed.) (CRC Press, Boca Raton, 1996) pp. 219–227.
74. J. Wilbourne, L. Haroun, E. Heseltine, J. Kaldor, D. Partensky and V. Vainio, "Response of experimental animals to human carcinogens: an analysis based upon the IARC Monographs programme" *Carcinogenesis* **7**, 1853–1863 (1986).
75. A. H. Smith, C. Hopenhayn-Rich, M. N. Bates, H. M. Goeden, I. H. Hertz-Picciotto, I. Duggan, R. Wood, M. Kosnett and M. T. Smith, "Cancer risk from arsenic in drinking water" *Environ. Health Perspect.* **97**, 259–267 (1992).
76. M. N. Bates, A. H. Smith and C. Hopenhayn-Rich, "Arsenic ingestion and internal cancers: a review" *Amer. J. Epidemiol.* **135**, 462–476 (1992).
77. T. Rossman, S. Meyn and W. Troll, "Effects of sodium arsenite on the survival of UV-irradiated *E. coli*: Inhibition of a *recA*-dependent function" *Mutat. Res.* **30**, 157–162 (1975).
78. T. G. Rossman, D. Stone, M. Molina and W. Troll, "Absence of arsenite mutagenicity in E.coli and Chinese hamster cells" *Environ. Mutagen.* **2**, 371–379 (1980).
79. T. G. Rossman, M. Molina and L. W. Meyer, "The genetic toxicology of metal compounds: I. Induction of λ prophage in *E. coli* WP2s (λ)" *Environ. Mutagen.* **6**, 59–69 (1984).
80. C. B. Klein and T. G. Rossman, "Transgenic chinese hamster V79 cell lines which exhibit variable levels of gpt mutagenesis" *Environ. Mol. Mutagen.* **16**, 1–12 (1990).
81. Y.-W. Lee, C. B. Klein, B. Kargacin, K. Salnikow, J. Kitahara, K. Dowjat, A. Zhitkovich, N. T. Christie and M. Costa, "Carcinogenic nickel silences gene expression by chromatin condensation and DNA methylation: A new model for epigenetic carcinogens" *Mol. Cell Biol.* **15**, 2547–2557 (1995).

82. J.-H. Li, *Ph.D. Thesis* (New York University, NY, 1989).
83. E. G. Jung and B. Trachsel, "Molekularbiologische Untersechunger zur arsencarcinogenese," *Arch. Klin. Exp. Dermatol.* **237**, 819–826 (1970).
84. T. G. Rossman, "Enhancement of UV-mutagenesis by low concentrations of arsenite in *E. coli*" *Mutat. Res.* **91**, 207–211 (1981).
85. T.-C. Lee, K. C. Lee, Y. J. Tseng, R. Y. Huang and K. Y. Jan, "Sodium arsenite potentiates the clastogenicity and mutagenicity of DNA crosslinking agents" *Environ. Mol. Mutagen.* **8**, 119–128 (1986).
86. T.-C. Lee, R. Y. Huang and K. Y. Jan, "Sodium arsenite enhances the cytotoxicity clastogenicity and 6-thioguanine-resistant mutagenicity of ultraviolet light in Chinese hamster ovary cells" *Mutat. Res.* **148**, 83–89 (1985).
87. J.-H. Li and T. G. Rossman, "Mechanism of comutagenesis of sodium arsenite with N-methyl-N nitrosourea" *Biol. Trace Elem. Res.* **21**, 373–381 (1989).
88. J.-H. Li and T. G. Rossman, "Comutagenesis of sodium arsenite with ultraviolet radiation in Chinese hamster V79 cells" *Biol. Metals* **4**, 197–200 (1991).
89. J.-L. Yang, M.-F. Chen, C.-W. Wu and T.-C. Lee, "Posttreatment with sodium arsenite alters the mutation spectrum induced by ultraviolet light irradiation in Chinese hamster overy cells" *Environ. Mol. Mutagen.* **20**, 156–164 (1992).
90. J. K. Wiencke and J. W. Yager, "Specificity of arsenite in potentiating cytogenetic damage induced by the DNA crosslinking agent diepoxybutane" *Environ. Mol. Mutagen.* **19**, 195–200 (1992).
91. A. N. Jha, M. Noditi, R. Nilsson and A. T. Natarajan, "Genotoxic effects of sodium arsenite on human cells" *Mutat. Res.* **284**, 215–221 (1992).
92. H. Huang, C. F. Huang, J. S. Huang, T. C. Wang and K. Y. Jan, "The transition from late G1 to early S phase is most vulnerable to the coclastogenic effect of ultraviolet radiation plus arsenite" *Intern. J. Radiat. Biol.* **61**, 57–62 (1992).
93. R. D. Snyder, G. F. Davis and P. Lachmann, "Inhibition by metals of X-ray and ultraviolet-induced DNA repair in human cells" *Biol. Trace Element Res.* **21**, 389–398 (1989).
94. S. F. Lee-Chen, C. T. Yu, D. R. Wu and K. Y. Jan, "Differential effects of luminol, nickel, and arsenite on the rejoining of ultraviolet light and alkylation-induced DNA breaks" *Environ. Molec. Mutagenesis* **23**, 116–120 (1994).
95. T. Okui and Y. Fujiwara, "Inhibition of human excision DNA repair by inorganic arsenic and the comutagenic effect in V79 Chinese hamster cells" *Mutat. Res.* **172**, 69–76 (1986).
96. A. Hartmann and G. Speit, "Effect of arsenic and cadmium on the persistence of mutagen-induced DNA lesions in human cells" *Environ. Mol. Mutagen.* **27**, 98–104 (1996).
97. S. Söderhall and T. Lindahl, "DNA ligase of eukaryotes" *FEBS Lett.* **67**, 1–8 (1987).
98. S. F. Lee-Chen, C. T. Yu and K. Y. Jan, "Effect of arsenite on the DNA repair of UV-irradiated Chinese hamster ovary cells" *Mutagenesis*, **7**, 51–55 (1992).
99. T. C. Lee, N. Tanaka, W. P. Lamb, T. M. Gilmer and J. C. Barrett, "Induction of gene amplification by arsenic" *Science* **241**, 79–81 (1989).
100. J. C. Barrett, P. W. Lamb, T. C. Wang and T. C. Lee, "Mechanisms of arsenic-induced cell transformation" *Biol. Trace Element Res.* **21**, 421–429 (1989).
101. J. R. Landolph, "Molecular mechanisms of transformation of C3H/10T1/2 Cl 8 mouse embryo cells and diploid human fibroblasts by carcinogenic metal compounds" *Environ. Health Perspect.* **102**(3), 119–125 (1994).
102. U. Saffiotti and F. Bertolero, "Neoplastic transformation of BALB/3T3 cells by metals and the quest for induction of a metastatic phenotype" *Biol. Trace Element Res.* **21**, 475–482 (1989).
103. L. A. Kowalski, S. S. Tsang and A. J. Davison, "Arsenic and chromium enhance transformation of bovine papillomavirus DNA-transfected C3H/10T1/2 cells" *Cancer Letters* **103**, 65–69 (1996).
104. G. R. Paton and A. C. Allison, "Chromosome damage in human cell cultures induced by metal salts" *Mutat. Res.* **16**, 332–336 (1972).
105. M. L. Larramendy, N. C. Popescu and J. DiPaolo, "Induction by inorganic metal salts of sister chromatid exchanges and chromosome aberrations in human and Syrian hamster strains" *Environ. Mutagen.* **3**, 597–606 (1981).
106. T.-C. Lee, M. Oshimura and J. C. Barrett, "Comparision of arsenic-induced cell transformation, cytotoxicity, mutation and cytogenetic effects in Syrian hamster embryo cells in culture" *Carcinogenesis* **6**, 1421–1426 (1985).
107. J.-T. Dong and X.-M. Luo, "Arsenic-induced DNA-strand breaks associated with DNA-protein crosslinks in human fetal lung fibroblasts" *Mutat. Res.* **302**, 97–102 (1993).
108. J.-R. Gurr, Y.-C. Lin, I.-C. Ho, K.-J. Jan and T.-C. Lee, "Induction of chromatid breaks and tetraploidy in Chinese hamster ovary cells by treatment with sodium arsenite during the G2 phase" *Mutat. Res.* **319**, 135–142. (1993).

109. T. S. Wang and H. Huang, "Active oxygen species are involved in the induction of micronuclei by arsenite in XRS-5 cells" *Mutagenesis* **9**, 253–257 (1994).
110. H. Tinwell, S. C. Stephens and J. Ashby, "Arsenite as the probable active species in the human carcinogenicity of arsenic: mouse micronucleus assays on Na and K arsenite, orpiment, and Fowler's solution" *Environ. Health Perspect.* **95**, 205–210 (1991).
111. T. G. Rossman and D. Wolosin, "Differential susceptibility to carcinogen-induced amplification of SV40 and *dhfr* sequences in SV40-transformed human keratinocytes" *Mol. Carcinogen.* **6**, 203–213 (1992).
112. L. R. Livingstone, A. White, J. Sprouse, E. Livanos, T. Jacks and T. D. Tlsty, "Altered cell cycle arrest and gene amplification potential accompany loss of wild-type p53" *Cell* **70**, 923–935 (1992).
113. R. N. Huang, I. C. Ho, L. H. Yih and T. C. Lee, "Sodium arsenite induces chromosome endoreduplication and inhibits protein phosphatase activity in human fibroblasts" *Environ. & Molec. Mutagen.* **25**, 188–196 (1995).
114. T. S. Kochhar, W. Howard, S. Hoffman and L. Brammer-Carleton, "Effect of trivalent and pentavalent arsenic in causing chromosome alterations in cultured CHO cells" *Toxicol. Letters* **84**, 37–42 (1996).
115. K. Yamanaka, M. Tezuka, K. Kato, A. Hasegawa and S. Okada, "Crosslink formation between DNA and nuclear proteins by *in vivo* and *in vitro* exposure of cells to dimethylarsinic acid" *Biochem. Biophys. Res. Commun.* **191**, 1184–1191 (1993).
116. M. Tezuka, K.-I. Hanioka, K. Yamanaka and S. Okada, "Gene damage induced in human alveolar type II (L-1232) cells by exposure to dimethylarsinic acid" *Biochem. Biophys. Res. Commun.* **191**, 1178–1183 (1993).
117. C. B. Klein, K. Frenkel and M. Costa, "The role of oxidative processes in metal carcinogenesis" *Chem. Res. Toxicol.* **4**, 592–604 (1991).
118. E. T. Snow, "Metal carcinogenesis: mechanistic considerations" *Pharmacol. Ther.* **53**, 31–65 (1992).
119. T. G. Rossman, "Metal mutagenesis" **In**: *Toxicology of Metals-Biochemical Aspects* (R. A. Goyer and M. G. Cherian, eds.) (Springer-Verlag, New York, 1994) pp. 374–405.
120. I. Nordenson and L. Beckman, "Is the genotoxic effect of arsenic mediated by oxygen free radicals?" *Hum. Hered.* **41**, 71–73 (1991).
121. T. C. Lee and I. C. Ho, "Differential cytotoxic effects of arsenic on human and animal cells" *Environ. Health Perspect.* **3**, 101–105 (1994).
122. S.-R. Fan, I-C. Ho, F. L.-F. Yeoh, C.-J. Lin and T.-C. Lee, "Squalene inhibits sodium arsenite-induced sister chromatid exchanges and micronuclei in Chinese hamster ovary-K1 cells" *Mutat. Res.* **368**, 165–169 (1996).
123. P. Ostrosky-Wegman, M. E. Gonseblatt, R. Montero, L. Vega, H. Barba, J. Espinosa, A. Palao, C. Cortinas, G. García-Vargas, L. M. del Razo and M. Cebrián, "Lymphocyte proleration kinetics and genotoxic findings in a pilot study on individuals chronically exposed to arsenic in Mexico" *Mutat. Res.* **250**, 477–482 (1991).
124. M. L. Warner, L. E. Moore, M. T. Smith, D. A. Kalman, E. Fanning and A. H. Smith, "Increased micronuclei in exfoliated bladder cells of individuals who chronically ingest arsenic-contamined water in Nevada" *Cancer Epidemiol. Biomarkers and Prevent.* **3**, 583–590 (1994).
125. A. H. Smith, C. Hopenhayn-Rich, M. Warner, M. L. Biggs, L. Moore and M. T. Smith, "Rationale for selecting exfoliated bladder cell micronuclei as potential biomarkers for arsenic genotoxicity" *J. Toxicol. Environ. Health* **40**(2–3), 223–224 (1993).
126. L. E. Moore, M. L. Warner, A. H. Smith, D. Kalman and M. T. Smith, "Use of the fluorescent micronucleus assay to detect the genotoxic effects of radiation and arsenic exposure in exfoliated human epithelial cells" *Environ. Mol. Mutagen.* **27**, 176–184 (1996).
127. J. E. Milner, "The effect of ingested arsenic on methylcholanthrene-induced skin tumors in mice" *Arch. Environ. Health* **18**, 7–11 (1996).
128. S. Yamamoto, Y. Konishi, T. Matsuda, T. Murai, M. Shibata, I. Matsui-Yuasa, S. Otani, K. Kuroda, G. Endo and S. Fukushima, "Cancer induction by an organic arsenic compound, dimethylarsinic acid (cacodylic acid), in F344/DuCrj rats after pretreatment with five carcinogens" *Cancer Res.* **55**, 1271–1276 (1995).
129. M. I. Luster, J. L. Wilmer, D. R. Germolec, J. Spalding, T. Yoshida, K. Gaido, P. O. Simeonova, F. G. Burleson and A. Bruccoleri, "Role of keratinocyte-derived cytokines in chemical toxicity" *Toxicol. Letters* **82–83**, 471–476 (1995).
130. J. Ananthan, A. L. Goldberg and R. Voellmy, "Abnormal proteins serve as eukaryotic stress signals and trigger the activation of heat shock genes" *Science* **232**, 522–524 (1986).
131. M. A. Deaton, P. D. Bowman, G. P. Jones and M. C. Powanda, "Stress protein synthesis in human keratinocytes treated with sodium arsenite, phenyldichloroarsine and nitrogen mustard" *Fund. Appl. Toxicol.* **14**, 471–476 (1990).

132. J. L. Pipkin, J. F. Anson, W. G. Hinson, E. R. Burns, D. A. Casciano and D. M. Sheehan, "Cell cycle-specific effects of sodium arsenite and hyperthermic exposure on incorporation of radioactive leucine and phosphate by stress proteins from mouse lymphoma cell nuclei" *Biochim. et Biophys. Acta* **927**, 334–344 (1987).
133. T. Hatayama, E. Kano, Y. Taniguchi and K. Nitta, "Role of heat-shock proteins in the induction of thermotolerance in Chinese hamster V79 cells by heat and chemical agents" *Int. J. Hyperthermia* **1**, 61–74 (1991).
134. Z. Wang, G. Hou and T. G. Rossman, "Induction of arsenite tolerance and thermotolerance by arsenite occur by different mechanisms" *Environ. Health Perspect.* **102**, 97–100 (1994).
135. Y.-L. Kim, J. Shuman, M. Sette and A. Przybyla, "Arsenite induces stress proteins in cultured rat myoblasts" *J. Cell Biol.* **96**, 393–400 (1983).
136. K. Ohtsuka, A. Masuda, A. Nakai and T. Nagata, "A novel 40-kDa protein induced by heat shock and other stresses in mammalian and avian cells" *Biochem. Biophys. Res. Commun.* **166**, 642–647 (1990).
137. J. van Rijn, J. van den Berg, F. A. Wiegant and R. van Wijk, "Sensitization to X-rays by sodium arsenite or heat in normal cells and in cells with an induced tolerance for heat and arsenite" *Radiat. and Environ. Biophys.* **34**, 169–175 (1995).
138. S. M. Keyse and R. M. Tyrrell, "Heme oxygenase is the major 32-kDa stress protein induced in human skin fibroblasts by UVA radiation, hydrogen peroxide, and sodium arsenite" *Proc. Natl. Acad. Sci. USA* **85**, 99–103 (1989).
139. R. Y. Stocker, Y. Yamamoto, A. F. McDonagh, A. N. Glazer and B. N. Ames, "Bilirubin is an antioxidant of possible physiological importance" *Science* **235**, 1043–1046 (1987).
140. D. Lautier, P. Luscher and R. M. Tyrrell, "Endogenous glutathione levels modulate both constitutive and UVA radiation/hydrogen peroxide inducible expression of the human heme oxygenase gene" *Carcinogenesis* **13**, 227–232 (1992).
141. T.-C. Lee and I.-C. Ho, "Modulation of cellular antioxidant defense activities by sodium arsenite in human fibroblasts" *Arch. Toxicol.* **69**, 498–504 (1995).
142. T.-C. Lee and I.-C. Ho, "Expression of heme oxygenase in arsenic-resistant human lung adenocarcinoma cells" *Cancer Res.* **54**, 1660–1664 (1994).
143. P. L. Goering and B. R. Fisher, "Metals and Stress Proteins" **In**: *Toxicology of Metals* (R. A. Goyer and M. G. Cherian, eds.) (Springer-Verlag, New York, 1995) **Vol. 115**, Chapter 11, pp. 229–266.
144. A. Albores, J. Koropatnick, M. G. Cherian and A. J. Zelazowski, "Arsenic induces and enhances rat hepatic metallothionein production *in vivo*" *Chem.-Biol. Interactions* **85**, 127–140 (1992).
145. Z. Wang, *Ph.D. Thesis* (New York University, NY, 1994).
146. A. Guzzo, C. Karatzios, C. Diorio and M. S. DuBow, "Metallothionein-II and ferritin H mRNA levels are increased in arsenite-exposed HeLa cells" *Biochem. Biophys. Res. Commun.* **205**, 590–595 (1994).
147. E. I. Goncharova and T. G. Rossman, "The antimutagenic effects of metallothionein may involve free radical scavenging" **In**: *Genetic Response to Metals* (B. Sakar, ed.) (Marcel Dekker, Inc., New York, 1995) pp. 87–100.
148. J. W. Bauman, J. Liu, Y. P. Liu and C. D. Klassen, "Increase in metallothionein produced by chemicals that induce oxidative stress" *Toxicol. Appl. Pharmacol.* **110**, 347–354 (1991).
149. L. S. Chubatsu and R. Meneghini, "Metallothionein protects DNA from oxidative damage" *Biochem. J.* **291**, 193–198 (1993).
150. E. I. Goncharova and T. G. Rossman, "A role for metallothionein and zinc in spontaneous mutagenesis" *Cancer Res.* **54**, 5318–5323 (1994).
151. M. Sato and I. Bremner, "Oxygen free radicals and metallothionein" *Free Rad. Biol. Med.* **14**, 325–337 (1993).
152. M. A. Schwarz, J. S. Lazo, J. C. Yalowish, I. Reynolds, V. E. Kagan, V. Tyurin, Y.-M. Kin, S. C. Watkins and B. R. Pitt, "Cytoplasmic metallothionein overexpression protects NIH 3T3 cells from *tert*-Butyl hydroperoxide toxicity" *J. Biol. Chem.* **269**, 15238–15243 (1994).
153. P. T. Thornalley and M. Vasak, "Possible role for metallothionein in protection against radiation induced oxidative stress. Kinetics and mechanism of its reaction with superoxide and hydroxyl radicals" *Biochem. Biophys. Acta* **827**, 36–66 (1985).
154. J. Abel and N. de Ruiter, "Inhibition of hydroxyl radical-generated DNA degradation by metallothionein" *Toxicol. Lett.* **47**, 191–196 (1989).
155. A. Basu and J. S. Lazo, "A hypothesis regarding the protective role of metallothioneins against the toxicity of DNA interactive anticancer drugs" *Toxicol. Lett.* **50**, 123–135 (1990).
156. L. Packer, E. H. Wiott and H. J. Tritschler, "Alpha-Lipoic acid as a biological antioxidant" *Free Radical Biol. and Med.* **19**, 227–250 (1995).

157. K. V. Chin, S. Tanaka, G. Darlington, I. Pastan and M. M. Gottesman, "Heat shock and arsenite increase expression of the multidrug resistance (MDR1) gene in human renal carcinoma cells" *J. Biol. Chem.* **265**, 221–226 (1990).
158. R. M. Gubits, "*c-fos* mRNA levels are increased by the cellular stressors, heat shock and sodium arsenite" *Oncogene*, **3**, 163–168 (1988).
159. J.-H. Li, P. C. Billing and A. R. Kennedy, "Induction of oncogene expression by sodium arsenite in C3H/l0Tl/2 cells: inhibition of c-myc expression by the Bowman-Birk protease inhibitor" *Cancer J.* **5**, 354–358 (1992).
160. T. Prestera, Y. Zhang, S. R. Spencer, C. A. Wilczak and P. Talalay, "The electrophile counterattack response: protection against neoplasia and toxicity" *Advances in Enzyme Regulation* **33**, 281–296 (1993).
161. I. Kullik and G. Story, "Transcriptional regulators of the oxidative stress response in prokaryotes and eukaryotes" *Redox Report* **1**, 23–29 (1994).
162. R. van Wijk, M. Welters, J. E. M. Souren, H. Ovelgonne and F. A. C. Wiegant, "Serum-stimulated cell cycle progression and stress protein synthesis in C3H10T1/2 fibroblasts treated with sodium arsenite" *J. Cell. Physiol.* **155**, 265–272 (1993).
163. G. R. Guy, J. Cairns, S. B. Ng and Y. H. Tan, "Inactivation of a redox-sensitive protein phosphatase during the early events of tumor necrosis factor/interleukin-1 signal transduction" *J. Biol. Chem.* **268**, 2141–2148 (1993).
164. I. Vietor and J. Vilcek, "Pathways of heat shock protein 28 phosphorylation by TNF in human fibroblasts" *Lymphokine and Cytokine Res.* **13**, 315–323 (1994).
165. R. R. Ariza, S. M. Keyse, J. G. Moggs and R. D. Wood, "Reversible protein phosphorylation modulates nucleotide excision repair of damaged DNA by human cell extracts" *Nucl. Acid. Res.* **24**, 433–440 (1996).
166. W. Zhang, C. McClain, J. P. Gau, X. Y. Guo and A. B. Deisseroth, "Hyperphosphorylation of p53 by okadaic acid attenuates its transcriptional activation function" *Cancer Res.* **54**, 4448–4453 (1994).
167. D. J. Kachinskas, M. A. Philips, Q. Qin, J. D. Stokes and R. H. Rice, "Arsenate perturbation of human keratinocyte differentiation" *Cell Growth and Different.* **5**, 1235–1241 (1994).
168. M. J. Mass and L. Wang, "Arsenic alters cytosine methylation patterns of the tumor suppressor gene p 53 in human lung cells: a model for a mechanism of carcinogenesis" *Mutat. Res.* **386**, 263–277 (1997).
169. Z. Wang and T. G. Rossman, "Stable and inducible arsenite resistance in Chinese hamster cells" *Toxicol. Appl. Pharmacol.* **118**, 80–86 (1993).
170. T. G. Rossman, E. I. Goncharova, T. Rajah and Z. Wang, "Human cells lack the inducible tolerance to arsenite seen in Chinese hamster cells" *Mutat. Res.* **386**, 307–314 (1997).
171. T.-C. Lee, J. L. Ho and K. Y. Jan, "Differential cytotoxicity of sodium arsenite in human fibroblasts and Chinese hamster ovary cells" *Toxicology* **56**, 289–299 (1989).

10. URBAN AIR POLLUTION AND HEALTH

I. J. BEVERLAND*

INTRODUCTION

Recently there has been increasing scientific evidence of association between urban air quality and human health effects[e.g. 1–3]. In many urban areas vehicle pollution is now regarded as one of the most important factors affecting air quality[4,5] and developments involving large amounts of road traffic must be considered carefully by public health officials. Similarly there is evidence on the association between industrial air pollution and the health of the inhabitants of urban areas. This chapter outlines some of the important aspects of these problems.

Despite recent attention, the problem of poor air quality has been present in densely populated areas for many years (Fig. 1). For example, urban smogs resulted in 4000 and 1000 premature deaths in London during the winters of 1952 and 1956, respectively. These smogs were caused by the burning of coal in domestic premises

Figure 1 Urban air pollution in Edinburgh before the implementation of smoke control areas. The formation of smog was likely under meteorological conditions which inhibited atmospheric dispersion.

* Department of Public Health Sciences, University of Edinburgh, Medical School, Teviot Place, Edinburgh EH8 9AG (UK).

Figure 2 Air pollution problems from domestic fuel combustion in Edinburgh have largely been eliminated. However the rapid growth of road traffic is resulting in deteriorating air quality. Problems can be particularly severe when dispersion is limited in deep street canyons.

and this form of pollution has been dramatically reduced by the introduction of the Clean Air Acts of 1956 and 1968. These Acts permitted action on a local scale by providing the means to set up smoke control areas within towns and cities.

In the 1970s and 1980s attention shifted towards wider scale pollution issues such as the long-range transport of acidic material, stratospheric ozone depletion and global climate change. Many research resources have been deployed to examine these areas and it is important that some of the expertise gained is now used in the study of deteriorating urban air quality because of current and projected traffic levels[5] (Fig. 2).

POLLUTANTS AND HEALTH EFFECTS

The risk to human health from air pollution is dependent on a number of important factors including: the concentration of the pollutant, the dose–response relationship for a range of susceptible groups, the number of people exposed to the pollutant, and the duration of exposure. In the European region four main classifications of air pollution phenomena have been identified[6]:

A. Episodic Pollution

(i) Summer-type smog episodes, characterised by increased levels of ozone (O_3) and nitrogen dioxide (NO_2).

(ii) Winter-type smog episodes characterised by increased levels of sulphur dioxide (SO_2) and suspended particulate material (SPM).

B. Long-term Pollution

(iii) Long-term exposure in urban areas to pollutants emitted into the atmosphere including SO_2, NO_2, SPM, carbon monoxide (CO) and a range of hydrocarbons.

(iv) Long-term multimedia exposure to persistent pollutants (this could include exposure to certain chemical forms of heavy metals such as lead and arsenic).

DESCRIPTION OF SOME OF THE MAIN POLLUTANTS

The descriptions that follow are intended to provide only a brief review of the sources, kinetics and health effects of some of the main urban air pollutants. More comprehensive accounts can be found in references [6] and [7].

1. *Sulphur Dioxide* (SO_2)

Atmospheric sulphur dioxide is formed mainly from the combustion of fossil fuels, especially coal. Industrial processes, e.g. metal ore smelting, can be important local sources of SO_2. In contrast vehicles tend not to be significant sources because the sulphur content of petrol is small (approx. 0.04% by mass). Diesel contains larger amounts of sulphur (0.2% by mass) resulting in some SO_2 emissions from these vehicles.

Concentrations of SO_2 in urban areas tend to reflect the extent to which coal and oil are burnt and changes in fuel usage have greatly reduced emissions over the last 30 years. For domestic heating, coal has been replaced in most UK towns and cities by natural gas (which contains negligible amounts of sulphur) and electrical forms of heating. A small number of urban areas have not implemented smoke control areas and have high concentrations of SO_2 from domestic coal fires in the local atmosphere. For example, in 1993, annual mean concentrations of 8 and 25 ppb (23 and 72 $\mu g\,m^{-3}$) were observed in Edinburgh and Belfast, respectively[8], illustrating the effect of smoke control areas in the former city.

As a result of the high solubility of SO_2 in water only small amounts of the gas penetrate into the alveoli of the lungs and correspondingly the effects of this pollutant tend to be observed in the upper airways. It is well established that high concentrations of SO_2 cause bronchoconstriction in normal and asthmatic subjects[9]. Both normal and asthmatic subjects vary widely in their response to SO_2[6]. Most normal subjects show a response to concentrations in excess of 4000 ppb (11.4 $mg\,m^{-3}$) but very little response when concentrations fall below 2000 ppb

(5.7 mg m^{-3}). A small number of asthmatic patients may develop symptoms at concentrations as low as 200 ppb (572 μg m^{-3}).

2. *Nitrogen dioxide* (NO_2)

Nitrogen oxides are released into the atmosphere mainly as nitric oxide (NO) from fuel combustion. NO reacts readily with ozone to form nitrogen dioxide (NO_2). Road transport is responsible for approximately 50% of total UK NO_2 emissions, while in urban areas this source becomes ever more important[10,11]. The natural background level in remote rural areas in the UK is between about 1 and 4 ppb (2 and 8 μg m^{-3}). European urban areas generally experience average NO_2 concentrations of between 10 and 48 ppb (20 and 90 μg m^{-3}) with hourly peaks of several hundreds of ppb in bad conditions.

Of the oxides of nitrogen only NO_2 is probably of such toxicity to be implicated in health effects at ambient levels. Exposure to high concentrations (e.g. in an occupational context) results in airway constriction and at extreme exposures pulmonary oedema and other inflammatory damage may result. However, it is thought that at peak ambient levels experienced in cities that only asthmatics are likely to experience any change in airway resistance when levels of NO_2 exceed 300 ppb (560 μg m^{-3})[4,12]. An important complicating factor in epidemiological studies is that indoor exposure to NO_2 can be very high (in excess of 1000 ppb (1880 μg m^{-3})) because of the use of unventilated gas cookers. For example, Hasselblad[13] showed an association between respiratory infection in young children and this form of cooking.

3. *Ozone* (O_3)

Trophospheric O_3 is a strong oxidising agent formed by the action of sunlight on NO_2. O_3 producing and scavenging processes can be characterised by the following reactions:

$$NO_2 + h\nu \rightarrow NO + O^{\bullet},$$
$$O^{\bullet} + O_2 \rightarrow O_3,$$
$$O_3 + NO \rightarrow NO_2 + O_2,$$

where $h\nu$ is ultraviolet radiation and $O^{\bullet}$ is an oxygen radical. The presence of hydroxyl radicals and volatile organic compounds in the atmosphere shifts the reaction equilibrium towards high concentrations of O_3. Other compounds are formed during these photochemical reactions include nitric acid, hydrogen peroxide, peroxyacyl nitrates (PAN) and aldehydes[10].

O_3 concentrations are generally lower in city centres than those in the suburbs and rural areas because of the scavenging reaction with NO. Indeed the highest levels are usually found in polluted air masses which have travelled considerable distances from their source regions. Maximum hourly concentrations can exceed 200 ppb (400 μg m^{-3}) compared to background levels of about 15 ppb (30 μg m^{-3}).

As a powerful oxidant O_3 will react with many biological tissues. Acute O_3 exposure at levels above 350 ppb (700 μg m^{-3}) can result in transient reductions in lung function and inflammation of the lower airways[6]. Other studies have suggested that respiratory effects and eye irritation could take place at concentrations above 50 ppb (100 μg m^{-3})[14–17]. Fewer studies have examined the effect of chronic exposure to elevated O_3 concentrations. However there is some evidence that there is a decline in pulmonary function associated with chronic exposure to O_3 concentrations above 45 ppb (90 $\mu g\,m^{-3}$)[18] although it is not conclusive that this association is causal.

4. *Hydrocarbons*

A wide range of hydrocarbons are present in the urban atmosphere and the reader is referred to the PORG Report[10] which gives a detailed listing of the compounds measured in the UK. Hydrocarbons come mainly from the evaporation and use of petrol and diesel fuels, from natural gas leakage and from evaporation of industrial solvents. Benzene is used as an octane booster in unleaded petrol, but is also present in leaded petrol and the main contribution is created in the exhaust during normal engine operation. Benzene concentrations in ambient air are generally between 1 and 50 ppb (3–160 $\mu g\,m^{-3}$). Higher ambient levels are found in urban areas and concentrations of several hundred ppb have been reported in studies near petrol stations, storage tanks and other industrial installations[19].

Two of the hydrocarbons monitored are of particular health significance; benzene is a known human carcinogen and 1,3-butadiene is a probable human carcinogen. However epidemiological studies of population exposure in urban areas tend to suggest that the health risks associated with these compounds in ambient urban atmospheres are immeasurably small. Evidence of the harmfulness of many of the hydrocarbons and metal compounds (following section) comes from studies of human exposure in the workplace and from experiments with laboratory animals. Generally greater credence is given to the former since it is difficult to extrapolate from results in animals to effects on man. Having said this, there are also problems in extrapolating from high occupational exposures to possible effects at much lower concentrations found in ambient air. These problems are illustrated in the derivation of the UK standard for ambient benzene concentrations. EPAQS[20] show that the health risks associated with the genotoxic nature of benzene are difficult to quantify because of the long latency periods involved, the relative rarity of the resulting health outcomes and differences between occupational groups which have been studied and the general population. Nonetheless given the importance of setting a standard as part of the overall strategy for improvement of air quality EPAQS[20] describe an approach which uses available occupational health data and extrapolation through carefully considered safety factors.

5. *Metals*

At sufficiently high atmospheric concentrations metals can be damaging to human health[5]. Effects can be direct through inhalation, or indirect through deposition into the human ecosystem and subsequent ingestion or absorption. The effects of metal pollution vary widely and include damage to the respiratory and cardiovascular systems, skin diseases and effects on the nervous system. The International Agency for Research on Cancer (IARC) has identified certain forms of some metals which are known to be carcinogenic, including compounds containing arsenic, nickel and chromium.

The main anthropogenic sources of metals present in the atmosphere are fuel combustion, metallurgical industries, vehicle emissions and waste incineration. With the exception of mercury, metals tend to be volatilised at high temperature and then recondense on particulate matter in the atmosphere. Atmospheric mercury remains principally in elemental vapour form. The speciation of metals in particulate matter is complex and poorly understood[e.g. 7]. For example, they may occur as elemental metals, in inorganic form such as oxides or chlorides, or in organic form such as alkyl compounds[5]. Many of the analytical techniques to not differentiate between these different forms which can limit the validity of subsequent risk assessments.

6. *Particulate Matter* (*PM*)

Until relatively recently particulate air pollution was measured by drawing air through a filter and measuring the darkness of the resulting stain. However, with increasing contribution from vehicular sources, particulates are now less black than the coal derived material of previous decades. Therefore new methods of measurement have been adopted which measure particulate levels by the mass of particles sampled through a size selective orifice. To date, in the UK, sampling inlets with a collection efficiency of 50% for particles less than 10 µm aerodynamic diameter have been used (termed PM_{10}). The mean concentrations in UK urban areas are generally in the range 20–30 $\mu g\, m^{-3}$ (with the exception of higher concentrations in Belfast for the reasons mentioned above).

The PM_{10} fraction consists of particles that can enter the respiratory system although only the sub-micron particles can enter deeply into the lungs. Lawther *et al.*[21] showed that urban pollution consisted mainly of these small particles—for the air in central London half the PM_{10} mass comprises particles of less than 1 µm in diameter and half the total number of particles are less than 0.1 µm.

The chemical composition of particles in the urban atmosphere varies considerably at different locations and times. However it has been shown that the fine fraction (below about 2.5 µm diameter) is about half carbon and half inorganic material (mainly ammonium sulphate and ammonium nitrate). These particles are derived mainly from fuel combustion especially by diesel vehicle engines. The sub-micron particles can remain suspended for weeks in the atmosphere and readily penetrate into buildings.

Many epidemiological studies (discussed in more detail in the following section) have consistently shown an association between particulate air pollution and not only exacerbations of illness in people with respiratory disease but also rises in cardiovascular morbidity/mortality. Animal studies have shown that sub-micron particles can cause a marked airway inflammatory response[22,23]. Seaton *et al.*[1] suggest that ultra-fine particles provoke alveolar inflammation, with release of mediators capable of causing exacerbations of lung disease and of increasing blood coagulability, thus possibly explaining the observed increases in cardiovascular deaths associated with urban pollution episodes.

EPIDEMIOLOGICAL STUDIES

This section uses particulate material as an example but the concepts discussed apply to most of the pollutants described above.

The increase in interest in urban air quality has been motivated largely by recent epidemiological studies which have indicated that exposure to airborne pollutants may cause adverse health effects in the general population. For example, research into the health effects of PM escalated dramatically in the 1980s and many publications have compared air quality and health outcomes in American cities, including Philadelphia, Detroit, Cincinnati, Minneapolis, Seattle, and St. Louis[2,3].

These studies together with similar research in Europe provide a reasonably consistent link between elevated levels of PM and increased risks of respiratory- and cardiovascular-related mortality/morbidity. Individuals with pre-existing respiratory conditions appear to be most susceptible however there is also some evidence for effects within the general population. At this point it is useful to distinguish between acute effects (e.g. pneumonia or acute respiratory or cardiovascular distress) from the way in which long-term exposure to pollutants may initiate or promote development of chronic disease. Such chronic exposure may lead to increased or earlier hospitalisation or premature death; irrespective of whether a preceding air pollution episode had a role in triggering these acute effects[24].

ACUTE EFFECT STUDIES

Most of the studies of acute effects of air pollution are either time series studies of a defined population or longitudinal studies of a panel of subjects. In both of these the study group is followed over a period of time during which health effects, pollution levels and potential confounding factors are monitored. Then the same day or previous day(s) pollution patterns are examined in relation to health outcomes in the study group. The studies need large population groups to give reliable outcomes and subjects are not monitored on an individual basis. Instead the health outcome data is obtained from routine health records on deaths, hospital admissions and emergency room visits.

In contrast to the population studies described above, less severe acute effects (e.g. asthma symptoms or lung function changes) are based on detailed studies of groups or panels of individuals.

In most circumstances pollution levels only explain a small percentage of daily variation in measurements of ill health, therefore very careful statistical analysis is required. For example, there are long-term fluctuations such as seasonal trends, weather influences and communicable disease epidemics which must be accounted for before the effect of pollution is quantified.

Several notable reviews of acute effect mortality studies include those mentioned in references [25–28]. One of the difficulties in interpretation of these publications is that the measure of particulate pollution is not always the same and some 'translation' between total suspended particulates and PM_{10} or $PM_{2.5}$ is often necessary. However, all of the reviews/meta-analyses show strong agreement between the magnitude of the effect in different studies, suggesting an approximate increase of 1% in premature acute deaths from an increase of $10\,\mu g\,m^{-3}$ in PM_{10}.

Lipfert[29] and Quenel *et al.*[30] provide comprehensive reviews of the studies for hospital admissions related to air pollution and there are a range of publications on emergency room visits[e.g. 31–38]. These are discussed in detail by COMEAP[24] together with publications which focus on restricted activity days, asthma and respiratory symptoms.

One of the approaches to improved understanding of the degree of causation in the association between air quality and health is to conduct multi-centre studies to determine geographical differences. Four recent studies of this nature have been conducted in Europe and the USA. The multi-centre APHEA study by Katsouyanni (Municipal Institute of Medical Research, Athens) and colleagues used co-ordinated and standardised time-series analysis techniques to assess short-term effects of air pollution in 15 cities covering a large part of Europe[39]. Elliot (London School of Hygiene and Tropical Medicine) and colleagues assessed the association between small area variations in air pollution and health (the SAVIAH study) in 4 European cities[40]. Brunekreef (Wageningen University, Netherlands) and colleagues studied pollution effects on asthmatic children in Europe (the PEACE study)[41]. Similarly the national human exposure assessment survey (NHEXAS) by Sexton (University of Minnesota) and colleagues is a multi-centre study aimed at monitoring human exposure to a range of environmental pollutants[42].

STUDIES RELATED TO LONG-TERM EXPOSURES

It is important to examine the role of exposure to air pollutants in the development of chronic diseases or in increasing (annual) death rates in communities[24]. The biologically relevant indices of exposure for these long term effects may be different to those for the acute effect studies described above. Cumulative lifetime exposure, based on annual average concentrations, may be appropriate. However it is often difficult to assess this because of poor historical data on pollution levels and population movement. It is also possible that long-term effects are linked to episodic exposures to pollution within particular age groups, e.g. very young or very old age groups. These aspects of air pollution are poorly understood and very difficult to address adequately in epidemiological studies.

Confounding variables are very important in studies related to long-term exposure to air pollutants. Many factors can affect the development of chronic disease including socio-economic conditions, smoking habits, availability of medical services and exposure to a range of acute illnesses. Often these factors are poorly documented and difficult to take into account even in the most detailed cohort studies.

However, despite the problems outlined above there is a growing body of evidence on the long-term effects of air pollution. Initially this evidence was circumstantial and subject to confounding problems, e.g. Evans *et al.*[43] describe how cross sectional studies have been used to show that mortality is higher in more polluted locations in the UK and USA. Earlier work by Daly[44] and Gardner[45] showed associations between mortality and morbidity and coal consumption rates.

More definitive evidence has resulted from detailed studies which have examined mortality in closely observed groups, with appropriate adjustments for smoking habits and other characteristics, before examining the effect of air pollution at the subject's residence. The best known study of this type is the 'Six Cities' study in the USA[3]. Over 8000 adults in six towns with different pollution climates were followed for 14 to 16 years. Mortality rates, adjusted for confounding factors (including sex, age, smoking habits, occupation and body mass index) were shown to be associated with the pollution level in the town of residence (especially particulate pollution). The ratio of mortality rate in the most and least polluted cities ($PM_{2.5}$ concentrations: $30\,\mu g\,m^{-3}$ and $11\,\mu g\,m^{-3}$, respectively) was 1.3 with the other cities occupying intermediate positions based on the observed pollution level (Fig. 3). While the relationship in Fig. 3 is very pronounced COMEAP[24] point out some limitations of this study, including the possibility that contrasts in social and some lifestyle characteristics may not have been adequately represented.

Similar increases in mortality were noted in polluted areas in a large cohort study conducted by the American Cancer Society, primarily related to smoking[46]. These cohort studies provide more reliable evidence regarding the association between mortality and the long-term pollution climate. However much uncertainty surrounds the assessment of exposure in the earlier years of the population and quantitative risk assessment is even more difficult than in the acute effect studies. The possibility

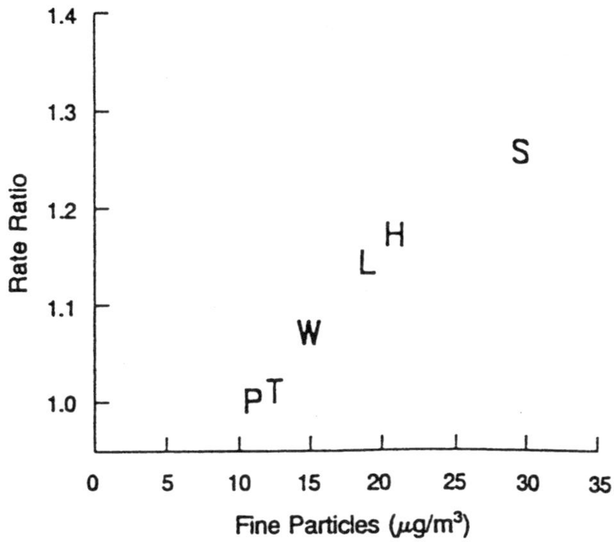

Figure 3 Some results from the 'Six Cities' study in the USA. Mortality rates, adjusted for confounding factors, were shown to be associated with the long-term mean pollution level in the town of residence. Mortality rates are illustrated as ratios for the more polluted cities to the least polluted city. P denotes Portage, Wisconsin; T Topeka, Kansas; W Watertown, Massachusetts; L St Louis; H Harriman, Tennessee; and S Steubenville, Ohio. (Reprinted with permission from Dockery *et al.* (1993))[3].

of air pollution resulting in increased cancer rates is particularly problematic. For example, there has been considerable interest in the carcinogenic potential of benzo(a)pyrene (BaP) in particulate matter but the confounding effects of smoking have tended to be overwhelming[47].

Similarly, it has been difficult to quantify the relationship between morbidity and pollution climate although associations have been noted in the UK and US[7]. Collectively the studies in this area indicate a consistent adverse effect from exposure to air pollution and bronchitic symptoms, e.g. Schwartz[48] found a 7% increase in chronic bronchitis for each 10 $\mu g\,m^{-3}$ increase in the annual mean of total suspended particulates.

SYNERGISTIC EFFECT OF A 'COCKTAIL' OF AIR POLLUTANTS

Some animal experiments have suggested that certain pollutants can combine in a synergistic fashion to increase damage or inflammation to the lung. Such effects have been suggested for ozone and acidic aerosols, and oxidant gases and particulates[49]. However very few toxicological data on mixtures of pollutants are available. The Advisory Group on the Medical Aspects of Air Pollution Episodes (AGMAAPE) considered the evidence for health effects from mixtures of pollutants found in winter and summer episodes in the UK[49,50]. The conclusion from AGMAAPE was that chamber studies on human subjects suggested no clear evidence of synergistic interaction between pollutants, though the studies were limited in scope. Ozone was found to have the greatest effect on lung function but overall effects were generally no greater than expected by summing the effect of each individual pollutant. AGMAAPE found

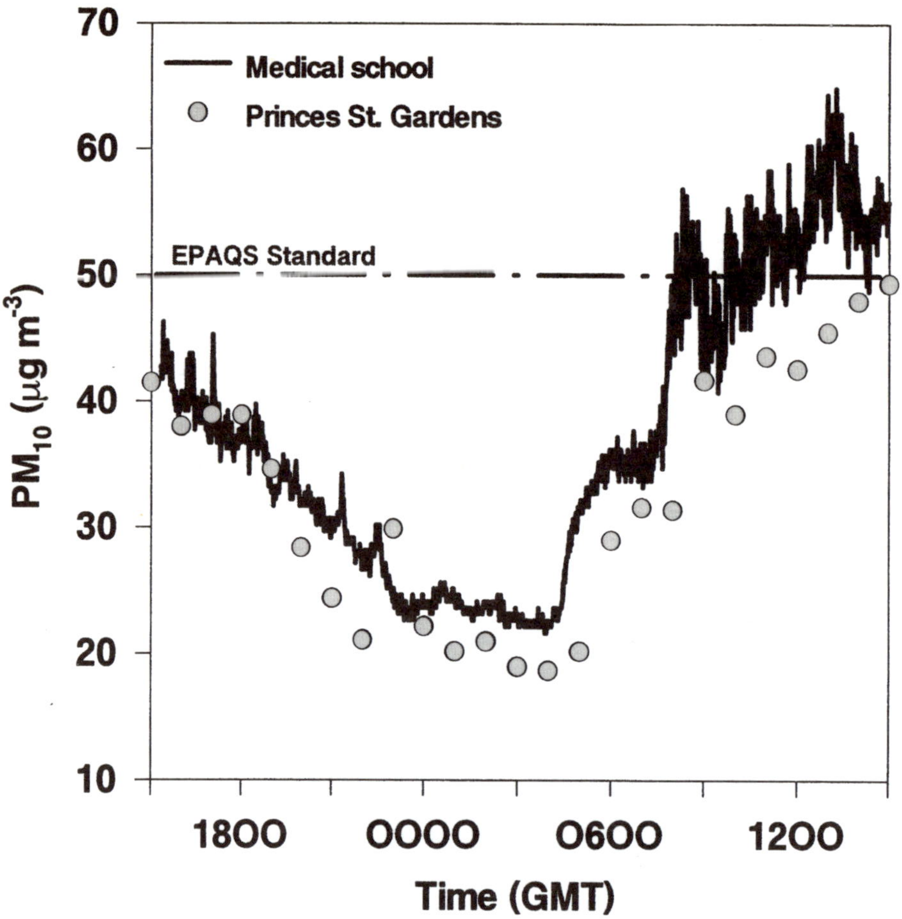

Figure 5 Diurnal patterns in PM_{10} concentrations at two separate locations within central Edinburgh. Measurements were made over a 24 hour period in March 1996.

temporal variations in PM_{10} levels introduce many uncertainties into the extrapolation of average exposure from point measurements. During the first practical session a walk through park land and areas of high traffic density in central Edinburgh indicated that transitory spatial variations are marked with very high levels of PM_{10} immediately adjacent to traffic queuing at intersections (Fig. 4). In the second practical session pronounced diurnal patterns in PM_{10} concentrations were noted at two separate locations within the city (Fig. 5). Epidemiological studies in progress will use a combination of personal exposure measurements and pollution modelling techniques to allow for these variations.

Indeed, by studying the determinants of personal exposure it should be possible to improve estimates of exposures in epidemiological studies that might not make use of personal monitoring, e.g. where the large populations or historical data are being studied. If the important influences are known then information about these variables can be collected to derive a proxy index of exposure[54].

An area of recent interest has been the development of techniques for measuring DNA adducts in people exposed to genotoxic carcinogens[55]. Adducts form when such compounds or their metabolites interact with DNA. If the adducts are not repaired it is possible they may have a role in initiating carcinogenesis[54]. Adducts can also form with haemoglobin and other proteins although these are not thought to be connected with carcinogenic processes. It is possible that adduct formation could be used to provide a proxy measure of exposure and, in some cases, health outcome in epidemiological studies. This could be particularly useful in investigations of carcinogens with long latency periods. Adduct formation might also be used to identify susceptible groups in the population allowing detailed epidemiological studies to be focused on subjects most at risk. More research is required in this important field.

ENHANCED EXPOSURE OF ROAD USERS

The elevated exposure of traffic wardens mentioned above is likely to affect other road users. Children in pushchairs may experience higher exposures because their breathing zone is at approximately the same height as vehicle exhausts. Similarly, cyclists may suffer adverse effects because of heavier breathing while exercising close to the source of pollution. However, there is some evidence that motorists may suffer from higher pollution levels than cyclists in the same area[56]. A range of studies have demonstrated higher levels of nitrogen oxides, benzene, and carbon dioxide inside cars compared to outside[e.g. 57–59]. From some of these studies it is evident that levels inside cars are dependent on traffic density and driving speed; and that urban commuters experience concentrations of up to eight times higher than background levels. Consequently it is not surprising that other surveys have found elevated levels of carbon monoxide in the blood of commuting motorists compared to cyclists[56] Acute exposures of motorists to hydrocarbons may occur while refuelling[19].

GEOGRAPHICAL ASPECTS OF AIR POLLUTION EPIDEMIOLOGY

It is well known that disease patterns do not have uniform geographical distributions and public health scientists have used this to investigate possible environmental causal factors. These patterns reflect regional differences in environments and socio-economic factors[60]. For example, Williams *et al.*[61] suggested that 'toxic factors' (distinct from the non-specific influences in social class) in Scottish urban environments influenced the national patterns of bronchitis and lung cancer.

Lloyd *et al.*[62] reviewed an extensive study of high rates of lung cancer in a small industrial town in the East of Scotland (Armadale, East Lothian). It is useful to examine this earlier work since it provides additional insight into some of the ways the exposure assessment problems discussed above could be tackled. At the time of the high incidence of lung cancer the local industry in this former mining area had declined to a small steel foundry in the south-east of the town and a small brickwork to the south of the town. With a population of approximately 7000 Armadale had similar socio-economic characteristics to neighbouring communities. However, routinely published mortality data for the period 1969–1973 showed the highest standardised mortality rates (SMRs) for total mortality and lung cancer in all of Scotland and high SMRs for other major diseases including ischaemic heart disease, cerebrovascular disease, bronchitis and other cancers. These high rates appeared to concur with changes in the metallurgical processes in the steel foundry that would have altered the nature of the air pollution within the town.

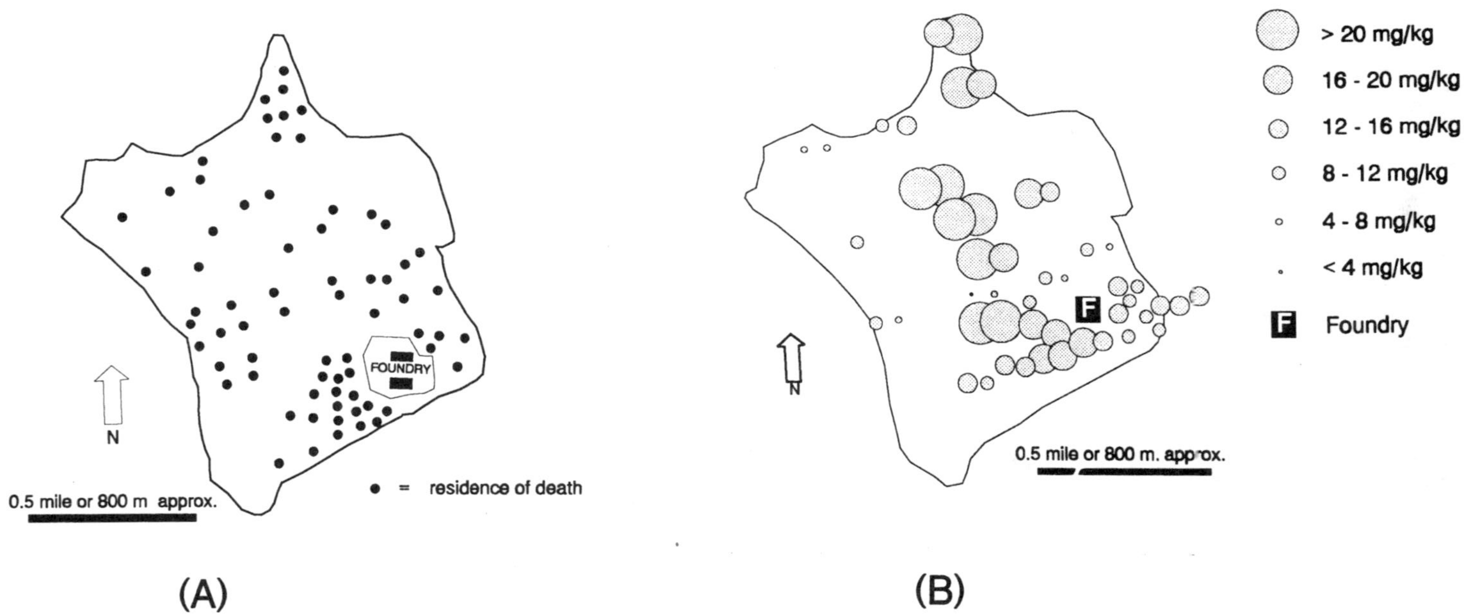

Figure 6 (A) Geography of lung cancer in Armadale, Scotland; 1968–1976; dots are residences of deaths; the ground slopes from the foundry northwards down to a valley. (B) The pattern of arsenic pollution in soil cores at various sites in Armadale. (Reprinted with permission from Lloyd (1995).)

To obtain information on detailed patterns of pollution, a high density of sampling sites is required. With many pollution monitoring devices the cost of replication is prohibitive and there may be problems with vandalism or instrument faults during extended measurements. Lloyd[60] describes how some of these difficulties can be overcome by using low-technology samplers (LTS). The initial approach was to divide up the town into 400 m^2 grid squares and to give each square an index of environmental pollution by counting the numbers and varieties of lichens at one site within the square[63]. Then indigenous and transplanted mosses and lichens, soil cores, grasses and farbric were used as LTS[64]. The LTS were subsequently analysed for a range of metals including iron, manganese, lead, zinc, chromium, nickel, cobalt and arsenic.

Residential locations of deaths within the town were obtained and used to examine if the geographical distribution of the lung cancer would provide a clue about its possible causation[60]. A cluster of deaths in the residential zone directly to the south-west of the foundry (Fig. 6(A)) was shown to be statistically significant using spatial analytical techniques[65,66]. Meteorological data had shown this area as probably the most vulnerable to air pollution from the foundry. The small cluster in the north of the town was not statistically significant and its biological significance was unclear. There was no significant clustering in either area before or after the period of investigation (1969–1973).

The lichen index study indicated that most pollution was near the foundry especially along an axis running south-west to north-east corresponding to the common wind directions. This was reinforced by the results from the low technology samplers—soil core studies showed heavy metal contamination was highest to the south-west of the foundry and further high concentrations to the north (Fig. 6(B)). The authors suggested that under unfavourable atmospheric conditions pollution moved first south-west from the foundry to the area with the large cluster of lung cancer deaths and then north down a gentle slope into the small valley at the north of the town (i.e. the area with the small cluster). This flow pattern was confirmed using a polystyrene model of Armadale's topography in a wind tunnel.

An alternative approach is to look for clues about the adverse effect of environmental pollution on reproductive processes. Lloyd[60] examined one obstetric parameter, the sex ratio of births (male : female) on a time-series basis covering the period with the lung cancer epidemic and mapped the distribution of sex ratios within the town. In Armadale the sex ratios of births were high during the 1960s with an exceptionally high value of 2.2 in 1967. Within the town the highest ratio (of 340 with 22 births) was in 1967 in the zone with the large cluster of lung cancer.

The data from these studies could suggest a short-latency carcinogenic mechanism, e.g. from a tumour promoter released into the atmosphere during changes in the metallurgical processes of the local foundry. The pattern of mortality agreed with that of pollution after the influence of geography on air flow patterns had been assessed. The LTSs provided valuable data on the distribution of pollutants and demonstrated that the zones of maximum pollution coincided with clusters of lung cancer. In particular the soil cores were useful in determining the historical pollution load within different zones. The low cost of the LTS is a great advantage in this type of study and the detailed spatial information obtained can be used to pinpoint optimal sites for conventional sampling devices. Similarly the preliminary data from these experiments indicate that obstetric epidemiological parameters may have a sentinel role to play in studies of environmental toxicity[60].

This earlier work provides some insight into techniques that might be used in environmental epidemiological investigations of air pollution associated ill health. Current approaches in the investigation of disease risks in small geographical areas are summarised by Elliot[67]. The present challenge is for specialists in public health, environmental chemistry and geographical information systems to provide simple low cost approaches to some of the exposure assessment problems which beset the current generation of epidemiological investigations.

CURRENT RESEARCH THEMES

A number of research programmes on the general themes discussed in the preceding sections are currently underway at the Universities of Edinburgh and Aberdeen and these are described here to illustrate some of the issues raised earlier.

1. *A Time Series and Cohort Study to Investigate the Relationship between Urban Pollution and Cardiorespiratory Ill-Health*

This study is investigating the temporal relationships between concentrations of outdoor pollutants, and morbidity/mortality from defined cardiorespiratory diseases using data collected over a much longer time period than most previously published work. Urban pollution in the city of Edinburgh is being related to cardiovascular and respiratory mortality (approximately 2000 cases per annum) and emergency hospital admissions (approximately 10,000 cases per annum). Seasonal, meteorological, pollen and fungal spore effects are taken into account. Simple modelling techniques are being used to apply differential exposure estimates across the city. Indicators derived from the national census are being used to resolve confounding caused by socio-economic variables. Possible susceptible groups are being studied in a cohort of 1600 people who have had measurements of fibrinogen and other risk factors on recruitment. Thus we aim to determine the influence of a number of cardiorespiratory risk factors on the subsequent relationship between outdoor pollution concentrations and cardiorespiratory ill-health.

2. *Air Pollution and Cardiovascular Disease: An Investigation of the Relationship Between Particulate Air Pollution and Blood Coagulation Factors*

Particulate air pollution has been shown to be associated with increased death rates from heart attack and stroke in older people. Seaton *et al.*[1] suggested that this may be a consequence of pulmonary inflammation leading to release of mediators of fibrinogen production and consequent increased coagulability of the blood. This might explain both the observed short-term and the possible longer term effects of pollution on cardiovascular mortality. We propose investigating this association, using time-series analysis of the inter-relationships of exposures to particulate pollution and clotting factors in the blood. We shall adapt the techniques that we have used to show a relationship between ambient temperature and clotting factors, studying 50 subjects aged 60+ years in both Belfast and Edinburgh over 2 years. We shall compare measures of blood clotting factors to routinely recorded air pollution data, but also make estimates of individuals' exposures to particulate pollution from activities diaries and measurements of the exposures of a sample using portable dust samplers. The study is intended to explain the link between particulate pollution

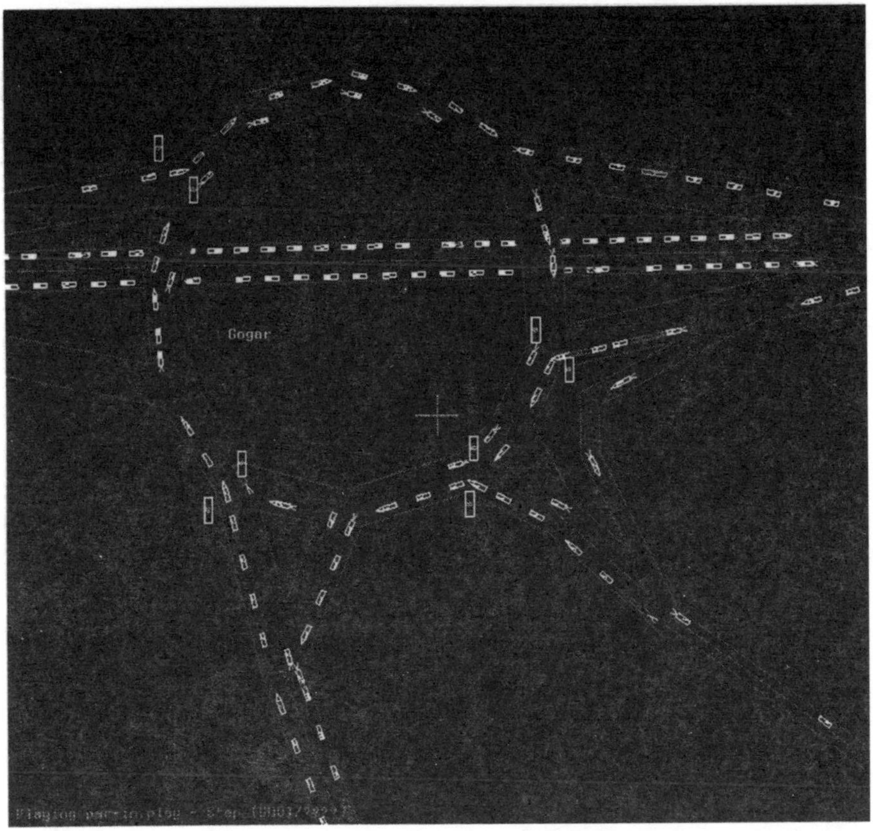

Figure 7 PARAMICS traffic simulation computer model. Individual vehicles, are modelled giving more realistic congestion and flow patterns (up to 250,000 vehicles simultaneously). Supercomputing power enables PARAMICS to study large-scale traffic patterns at high speeds.

and cardiovascular disease and may indicate the levels of exposure to particles at which such effects might be expected to occur.

3. *Assessing Health Implications of Urban Air Pollution Using Validated Traffic Simulation and Pollutant Dispersion Models*

In relation to the misclassification of exposure problems described above it will be important to develop air pollution measurement and modelling techniques to provide highly resolved temporal and spatial pollution concentration data necessary for detailed health effects studies. In proposed research in Edinburgh detailed measurements of pollutant levels will be made on 2 spatial scales (street canyon scale and city scale). These measurements will be used to assess human exposure to airborne pollutants and to validate a coupled traffic simulation and pollutant dispersion model. The validated model will be evaluated as a tool to predict and ultimately prevent episodes of traffic congestion and pollution.

Significant progress has been made at the Edinburgh Parallel Computing Centre (EPCC) and an associated company (Quadstone Ltd) on the development of a computer model (PARAMICS) which simulates the movement of large numbers of vehicles at a microscopic scale[68]. While most traffic simulators treat traffic as a fluid, microscopic traffic simulation models individual vehicles, giving more realistic congestion and flow patterns (up to 250,000 vehicles at individual vehicle scale). Supercomputing power enables PARAMICS to study large-scale traffic patterns at high speeds[69,70]. In the project we intend to integrate the traffic simulation model with existing advection-diffusion models[71,72] in order to predict pollutant concentration fields necessary for estimating human exposure. The integrated model will be validated at both street canyon scale and at city scale by appropriate measurements. An important advantage of a validated model would be the ability to predict the occurrence of traffic congestion and associated air quality problems to enable pro-active air quality management (Fig. 7).

Exposure estimates from the detailed measurements and coupled models will be used as additional input data to the time series study described above. The limiting factor in the ongoing study is the resolution of the exposure data which is based on relatively crude extrapolations from fixed sampling points. By adding the more accurate and better resolved temporal and spatial exposure data from the proposed work, a much more reliable measure of any causal association will be achieved. Assessment of the research outcome will through testing whether the project provides exposure assessment and modelling methods that account for a higher proportion of the variance in adverse health effects than the methods currently in use.

CONCLUSIONS

- The nature of pollution in urban areas has changed over the last 50 years following rapid reduction in domestic fuel combustion but similarly rapid growth in road traffic.
- The main pollutants of health concern are particulate matter, nitrogen dioxide, ozone and a range of hydrocarbons. The effects of heavy metals and other toxic materials may be important in localised situations.
- The application of environmental epidemiology is central to the quantification of the health effects of urban air pollution.
- Current epidemiological research shows associations between acute and chronic health effects and air pollution, although there are fewer studies and less confidence for the latter.
- There is some evidence of consistency in the magnitude of the health effects associated with the main pollutants in both the USA and Europe.
- The main problems in epidemiological studies are confounding effects and bias. Confounders include social, economic and lifestyle factors and the effects of pollutant mixtures and climatic factors. Misclassification bias can result from poor knowledge of personal exposures in the population being studied.
- The problems of susceptible groups and groups with high exposures within the general population need further investigation.
- Misclassification bias is closely related to spatial variations in pollution levels. Two possible approaches to gaining further knowledge of these variations are suggested. In the first, the deployment of a network of low technology sampling devices would provide a simple and reliable indication of spatial variations at low temporal resolution. Judicious application of GIS technology should allow these pollution variations to be related to spatial variations in health outcomes. In the second method, the development and validation of

computer models of traffic movement and pollution measurement would allow both retrospective and prospective estimates of exposure to pollution.
This chapter has concentrated on health effects and pollution issues in Europe and north America. It is important to remember that these problems are very severe and worsening rapidly in developing countries[e.g. 73,74]. It is hoped that the techniques of risk assessment developed in the research programmes and ideas described here can be applied to the management of risk in the areas of greatest concern.

References

1. A. Seaton, W. MacNee, K. Donaldson and D. Godden, "Particulate air pollution and acute health effects" *Lancet* **345**, 176–178 (1995).
2. T. Reichhardt, "Weighing the health risks of airborne particulates. Recent epidemiological studies link an increase in mortality and sickness to particulates, but do the data justify a tighter air quality standard" *Environmental Science and Technology* **29**, 360–364 (1995).
3. D. W. Dockery, C. A. Pope III, X. Xu, J. D. Spengler, J. H. Ware, M. E. Fay, B. G. Ferris, Jr. and F. E. Speizer, "An association between air pollution and mortality in six US cities" *NEJM* **329**(24), 1753–1759 (1993).
4. QUARG, "Diesel vehicle emissions and urban air quality" *Second Report of the Quality of Urban Air Review Group* (*QUARG*) (Prepared at the request of the Department of the Environment. Copies from: University of Birmingham, Birmingham, UK, 1993).
5. QUARG, "Urban air quality in the United Kingdom" *First Report of the Quality of Urban Air Review Group* (*QUARG*) (Prepared at the request of the Department of the Environment. DOE, PO Box 135, Bradford, UK, 1993).
6. WHO, *Concern for Europe's Tomorrow. Health and the Environment in the WHO European Region* (Wissenschaftliche Verlagsgesellschaft mbH, Stuttgart, 1995).
7. WHO, *Air Quality Guidelines for Europe* (World Health Organisation, Regional Publications, Copenhagen, 1987).
8. J. S. Bower, G. F. J. Broughton, P. G. Willis and H. Clark, *Air Pollution in the UK: 1993/94* (AEA Technology, Harwell, UK, 1995).
9. AGMAAPE, "Sulphur dioxide, acid aerosols and particulates" *Second report of the Advisory Group on the Medical Aspects of Air Pollution Episodes* (HMSO, London, 1992).
10. PORG, "Ozone in the United Kingdom" *Third Report of the Photochemical Oxidants Review Group* (*PORG*) (Prepared at the request of the Department of the Environment. London, UK, 1995).
11. INDITE, "Impact of nitrogen deposition in terrestrial ecosystems" *First Report* (HMSO, London, 1993).
12. WHO, *Environmental Health Criteria 4: oxides of nitrogen* (World Health Organisation, WHO, Geneva, 1977).
13. V. Hasselblad, D. M. Eddy and D. J. Kotchmar, "Synthesis of environmental evidence: nitrogen dioxide epidemiology studies" *J. Air Waste Manag.* **42**, 662–671 (1992).
14. P. L. Kinney, J. H. Ware, J. D. Spengler, D. W. Dockery, F. E. Speizer and B. G. Ferris, "Short-term pulmonary function change in association with ozone levels" *Amer. Rev. of Respiratory Diseases* **139**, 56–61 (1989).
15. M. Krzyanowski, J. J. Quackenboss and M. D. Lebowitz, "Relation of peak expiratory flow rates and symptoms to ambient ozone" *Arch. Envir. Health* **47**, 107–115 (1992).
16. G. Hoek, P. Fischer, B. Brunekreef, E. Lebret, P. Hofschreuder and M. G. Mennen, "Acute effects of ambient ozone on pulmonary function of children in the Netherlands" *Amer. Rev. of Respiratory Diseases* **147**, 111–117 (1993).
17. D. M. Spektor, M. Lippman, G. D. Thurston, P. J. Lioy, J. Stecko, G. OConner, E. Garshick, F. E. Speizer and C. Hayes, "Effects of ambient ozone on respiratory function in healthy adults exercising outdoors" *Amer. Rev. of Respiratory Diseases* **138**, 821–828 (1988).
18. J. Schwartz, "Lung function and chronic exposure to air pollution: a cross sectional analysis of NHANES II" *Environmental Research* **50**, 309–321 (1989).
19. L. Fishbein, "An overview of environmental and toxicological aspects of aromatic hydrocarbons. 1. Benzene" *Science of the Total Environment* **40**, 189–218 (1984).
20. EPAQS, *Benzene* (The Expert Panel on Air Quality Standards, HMSO, London, 1994).
21. P. J. Lawther, J. Ellison and R. E. Waller, "Some aspects of aerosol pollution" *Proc. Roy. Soc. A* **307**, 223–234 (1968).

22. J. Ferin, G. Oberdorster and D. P. Penny, "Pulmonary retention of ultra-fine and fine particles in rats" *Am. J. Respir. Cell Mol. Biol.* **6**, 535–542 (1992).
23. J. Ferin, G. Oberdorster and D.P. Penny, "Increased pulmonary toxicity of ultra-fine particles? 1. particle clearance, translocation, morphology" *J. Aerosol Sci.* **21**, 381–384 (1990).
24. COMEAP (Committee on the Medical Effects of Air Pollutants) *Non-Biological Particles and Health* (HMSO, London, 1995).
25. D. V. Bates "Health indices of the adverse effects of air pollutants: the question of coherence" *Environ. Res.* **59**, 336–349 (1992).
26. D. W. Dockery and C. A. Pope, "Acute respiratory effects of particulate air pollution" *Ann. Rev. Public Health* **15**, 107–132 (1994).
27. J. Schwartz, "Air pollution and daily mortality: a review and meta analysis" *Environ. Res.* **64**, 36–52 (1994).
28. C. A. Pope, D. W. Dockery and J. Schwartz, "Review of the epidemiological evidence of health effects of air pollution" *Inhalation Toxicology* **7**, 1–18 (1995).
29. F. W. Lipfert, "A critical review of studies of the association between demands for hospital services and air pollution" *Environ. Health Perspectives* **101**, Supplement 2, 229–268 (1993).
30. P. Quenel, P. Medina, I. Pirard, Y. Momas, B. LeMoullec and W. Festy, "Health service based morbidity indicators as a measure of health effects of an air pollution episode" *Eur. J. Public Health* **4**, 201–206 (1994).
31. D. V. Bates and R. Sizto, "Relationship between air pollution levels and hospital admissions in Southern Ontario" *Can. J. of Public Health* **74**, 117–122 (1983).
32. F. W. Lipfert and T. Hammerstrom, "Temporal patterns in air pollution and hospital admissions" *Environ. Res.* **59**, 374–399 (1992).
33. G. D. Thurston, K. Ito, C. G. Hayes, D. V. Bates and M. A. Lippmann, "Respiratory hospital admissions and summertime haze air pollution in Toronto, Ontario: consideration of the role of atmospheric aerosols" *Environ. Res.* **65**, 271–290 (1994).
34. C. A. Pope, "Respiratory hospital admissions associated with PM10 pollution in Utah, Salt Lake and Cache Valleys" *Arch. Environ. Health* **46**, 90–97 (1991).
35. S. H. Lamm, T. A. Hall, A. Engel, F. H. Reuter and D. L. White, "PM10 particulates: are they the major determinant of pediatric respiratory admissions in Utah County, Utah" *Ann. Occup. Hyg.* **38**, Supplement 1, 969–972 (1994).
36. J. Schwartz, "Air pollution and hospital admissions for the elderly in Birmingham, Alabama" *Am. J. Epidemiol.* **139**, 589–598 (1994).
37. J. Schwartz, "PM10, ozone and hospital admissions for the elderly in Minneapolis—St. Paul, Minnesota" *Arch. Environ. Health* **49**, 366–374 (1994).
38. S. Walters, R. K. Griffiths and J. G. Ayres, "Temporal association between hospital admissions for asthma in Birmingham and ambient levels of sulphur dioxide and smoke" *Thorax* **49**, 133–140 (1994).
39. J. Sunyer, D. Zmirou, C. Spix, *et al.*, "Short-term effects of air pollution on health: a European approach using epidemiologic time-series data" *Epidemiology* **6**, No. 4 Supplement, S38(1996).
40. B. Kriz, M. Bobak, M. Martuzzi, D. Briggs, E. Livesley, E. Lebret, P. Fischer, B. Wojtyniak, P. Gorynski and P. Elliot, "Respiratory health in the SVIAH study" *Epidemiology* **6**, No. 4 Supplement, S31(1996).
41. B. Brunekreef, G. Hoek, W. Roemer, *et al.*, "PEACE study. (1) Background and design" *Epidemiology* **6**, No. 4 Supplement, S65(1996).
42. K. Sexton, M. A. Callahan and D. Kleffman, "Exposure surveillance in the United States: a national human exposure assessment survey" *Epidemiology* **6**, No. 4 Supplement, S29(1996).
43. J. S. Evans, T. Josteson and P. L. Kinney, "Cross sectional mortality studies and air pollution risk assessment" *Environ. Int.* **10**, 55–83 (1984).
44. C. Daly, "Air pollution and causes of death" *Br. J. Prev. Soc. Med.* **13**, 14–27 (1959).
45. M. J. Gardner, M. D. Crawford and J. N. Morris, "Patterns of mortality in middle and early old age in the county boroughs of England and Wales" *Br. J. Prev. Soc. Med.* **23**, 133–140 (1969).
46. C. A. Pope, M. J. Thun, M. M. Namboodiri, D. W. Dockery, J. S. Evans, F. E. Speizer and C. W. Heath, "Particulate air pollution as a predictor of mortality in a prospective study of US adults" *Am. J. Respir. Crit. Care Med.* **151**, 669–674 (1995).
47. R. Cederlof, R. Doll, B. Fowler, B. Friberg, N. Nelson and V. Vonk, "Air pollution and cancer risk assessment methodology and epidemiological evidence" *Environ. Health Perspect.* **22**, 1–12 (1978).
48. J. Schwartz, "Particulate air pollution and chronic respiratory disease" *Environ. Res.* **62**, 7–13 (1993).
49. Department of Health, "Advisory Group on the Medical Aspects of Air Pollution Episodes (AGMAAPE)" *Fourth Report: Health effects of exposure to mixtures of air pollutants* (HMSO, London, 1995).
50. Department of Health, "Advisory Group on the Medical Aspects of Air Pollution Episodes (AGMAAPE)" *Second Report: sulphur dioxide, acid aerosols and particulates* (HMSO, London, 1992).

51. M. Watt, D. Godden, J. Cherrie and A. Seaton, "Individual exposure to particulate air pollution and its relevance to thresholds for health effects: a study of traffic wardens" *Occup. Envir. Med.* **790**, 792 (1995).
52. H. Ozkaynak, J. Xue, J. Spengler, L. Wallace, E. Pellizzari and P. Jenkins, "Personal exposure to airborne particles and metals: results from the particle team study in Riverside, California" *J. Exposure Analysis & Environ. Epidemiology* **6**, 57–78 (1996).
53. M. Brauer, P. Koutrakis and J. D. Spengler, "Personal Exposures to Acidic Aerosols and Bases" *Envir. Science and Technology* **23**, 1408–1412 (1989).
54. D. Coggon, "Assessment of exposure to environmental pollutants" *Occup. and Envir. Med.* **52**, 562–564 (1995).
55. K. Hemminki, A. Dipple, D. E. G. Shuker, F. F. Kadlubar, D. Segerback and H. Bartsch, *DNA Adducts: identification and biological significance* (International Agency for Research on Cancer, Lyon, IARC Sci. Publ. No. 125, (1994).
56. A. Rowell and M. Fergusson, *Bikes not fumes—The emission and health benefits of a modal shift from motor vehicles to cycling* (Cyclists' Touring Club, Godalming, Surrey, 1991).
57. W. A. M. DenTonkelaar and W. Rudolf, "Air quality inside motor cars" **In**: *Indoor Air Quality* (K. Aurand *et al.*, ed.) (Gustav Fischer Verlag, Stuttgart, 1982) pp. 219–233.
58. A. J. Hickman, "Personal exposures to carbon monoxide and oxides of nitrogen" *Research Report 206* (Transport and Road Research Laboratory, Crowthorne, Berkshire, UK" 1990).
59. C. C. Chan, "Commuter exposures to volatile organic compounds" *PhD Dissertation* (Harvard School of Public Health, Boston, Massachusetts, USA, 1990).
60. O. L. Lloyd, "The exploration of the possible relationship between deaths, births and air pollution in Scottish towns" **In**: *The Added Value of Geographical Information Systems in Public and Environmental Health* (M. J. C. deLepper *et al.* ed.) (Kluwer Academic Publishers, Netherlands, 1995) pp. 167–180.
61. F. L. R. Williams, O. L. Lloyd and W. G. Berry, "Mortality from non-malignant respiratory diseases in Scotland between 1959 and 1983" *Ambio* **16**, 206–210 (1987).
62. O. L. Lloyd, R. Barclay, M. M. Lloyd and Armadale Group Practice, "Lung cancer and other health problems in a Scottish industrial town" *Ambio* **14**, 322–328 (1985).
63. F. A. Yule and O. L. Lloyd, "An index of atmospheric pollution survey in Armadale, central Scotland" *Water, Air and Soil Pollution* **22**, 27–45 (1984).
64. F. A. Y. Gailey and O. L. Lloyd, "Methodological investigations into low technology monitoring of atmospheric metal pollution. 1. The effects of sampler size on metal concentrations. 2. The effects of length of exposure on metal concentrations. 3. The degree of consistency of the metal concentrations" *Environ. Pollution* **B12**, 41–59, 61–74, 85–109 (1986).
65. A. B. Lawson and F. L. R. Williams, "Armadale: a case study in environmental epidemiology" *J. Roy. Statist. Soc.* **A157**, 285–298 (1994).
66. O. L. Lloyd, "Mortality in a small industrial town: problems of analysis and interpretation" **In**: *Current Approaches to Occupational Health* (A. W. Gardner, ed.) (John Wright, Bristol, UK, 1981) pp. 283–210.
67. P. Elliott, "Investigation of disease risks in small areas" *Occup. Envir. Med.* **52**, 785–789 (1995).
68. G. Duncan, "PARAMICS: Wide area microscopic simulation of ATT and traffic management" *Proceedings of ISATA 1995*. ISTA-95ATS044 (1995).
69. G. Cameron, B. J. N. Wylie and D. McArthur, "PARAMICS—Moving vehicles on the connection machine" *Proceedings of Supercomputing 1994* (Anonymous IEEE Computer Society Press, 1994) 291–300.
70. G. Duncan, "PARAMICS: wide area microscopic simulation of ATT and traffic management" *Proceedings of International Symposium on Automotive Technology and Automation*, Anonymous ISTA-95ATS044 (1995).
71. A. A. Hassan and J. M. Crowther, "Modelling of pollutant dispersal in a street canyon" *Paper presented at the Institute of Physics Conference on Air Pollution* (University of Hertfordshire, England, July 1996).
72. A. A. Hassan and J. M. Crowther, "A simple model of pollutant concentrations in a street canyon" *Paper presented at the Institute of Physics Conference on Air Pollution* (University of Hertfordshire, England, July 1996).
73. WHO, "Air Pollution in the World's Megacities" *Environment* **36**(2), 4–13–25–37 (1994).
74. K. R. Smith, "Fuel combustion, air pollution exposure, and health: The situation in developing countries" *Ann. Rev. Energ. Env.* **18**, 529–566 (1993).

11. THE EFFECT OF TOXIC SUBSTANCES ON THE DEVELOPMENT OF DISEASES IN AQUATIC ORGANISMS

BRIAN AUSTIN *

INTRODUCTION

Although there is much evidence to point to the presence of pollutants/toxicants in the aquatic environment, the link to incidences of 'disease' is often unclear. The situation is not helped by confusion among individuals about the precise meaning of the term disease. Therefore, mortalities in animals and plants resulting from a chronic discharge of pollutants do not necessarily conjure up the image of a disease situation. Conversely, an infection caused by bacteria or viruses would fulfill most individuals' notion of a disease. However, it would be pertinent to enquire whether or not long-term damage to an organism resulting from chronic discharges of noxious material would be regarded as disease. A complication is that the damage might well occur long after the pollutant has been removed from the vicinity of the aquatic organism. Therefore, proof of cause and effect may be notoriously difficult to acquire. In addition, it should be emphasised that where a choice exists motile organisms are likely to move away from polluted areas.

Experimental data have clearly demonstrated that potentially toxic materials may accumulate in aquatic organisms. Yet, it is largely conjecture that such compounds are responsible for disease situations. Moreover, reports describing alleged associations between pollution and disease in the aquatic environment often fail to establish the presence of toxic materials in the diseased hosts. However, it is reasonable to argue that the toxicant may well lead to the development of a disease situation, long after its effective removal from the host. From the literature, it is abundantly clear that there are strong proponents and equally vocal opponents to the view that the presence of toxic compounds in the aquatic environment is directly responsible for the development of disease outbreaks. Often, these views reflect national concerns. The evidence will be presented below.

AQUATIC POLLUTION

It is common knowledge that waterways, including seas, estuaries, rivers and lakes, receive pollutants/toxicants, naturally, such as from the collapse of algal blooms, and as a result of human activity. The pollutants, many of which are derived from land-based sources[1], encompass environmental oestrogens, organisms from ship-ballast waters, plant nutrients and plastics[2], pulp mill effluents[3], heavy metals (including organotin from anti-fouling paints[4]), hydrocarbons[5], pesticides[6], and bacteria[7] and viruses from sewage, which may be introduced accidentally, such as

* Department of Biological Sciences, Heriot-Watt University, Riccarton, Edinburgh EH14 4AS, Scotland (UK).

a result of discharges from ships [a topical example would be the Braer oil spill[8]] or deliberately, e.g. during the Gulf War[9]. Aquaculture facilities have been associated with the introduction of pollutants, i.e. bacteria[10] and organic and inorganic compounds[11], into receiving waters. For example, freshwater prawn farms have been reported to contribute suspended solids, total nitrogen and phosphorus and pigments into the aquatic environment[11]. Fish facilities lose nitrate/nitrite and ammonia in the effluent[11]. Nevertheless, whatever the perceived pollutant, there is a clear need for good quantitative data, and effective monitoring procedures to determine the extent and longevity of the pollution[12].

THE PRESENCE OF TOXIC SUBSTANCES IN AQUATIC ORGANISMS

There is an increasing literature concerning the presence of pollutants/toxicants in aquatic organisms. However, many data have been obtained from laboratory-based experiments, rather than studies of whole organisms in the aquatic environment. Therefore, there is justifiable concern about the relevance of the data to events in the natural environment. Nevertheless, there is evidence that foreign chemicals, e.g. heavy metals and pesticides, can accumulate in aquatic organisms, although there appear to be quantitative differences between species and the precise habitat. The gist of the relevant information is summarised below:

1. *Heavy Metals*

It is well established that vertebrates and invertebrates are capable of accumulating heavy metals from the aquatic environment[13–16]. For example, cadmium, copper, lead and zinc have been detected by atomic absorption spectrophotometry in the tissues (gill, muscle, vertebrae and viscera) of rabbitfish (*Siganus oramin*) from the polluted waters around Hong Kong[17]. Mercury has been found in fish and shellfish from India[18]. In this example, emphasis was placed on effluent which received discharge from a chlor-alkali plant. Certainly, some seasonality was apparent in the concentration of mercury, with highest levels occurring during the monsoon period. It is interesting that the concentration of mercury in *Chanos chanos*, *Liza macrolepis*, *Penaeus indicus* and *Crassostrea madrasensis* was considered insufficient to cause any adverse effects on human health[18]. Surely, it would be relevant to enquire about possible deleterious effects on aquatic species in view of other studies which have clearly tried to establish a link between polluted (aquatic) environments and disease.

Sixteen elements, namely arsenic, cadmium, cobalt, caesium, copper, iron, magnesium, manganese, molybdenum, nickel, rubidium, selenium, silver, strontium and zinc, were found in the liver and ovary of cod (*Gadus morhua*) caught from off the coast of Newfoundland, Canada[14]. Boron levels were considered high for muscle from fish caught at one particular location. Interestingly, the levels for cadmium, lead and mercury were lower than amounts reported for cod, which were obtained from the north east Atlantic, the North Sea and the Baltic sea[14]. Ironically the levels for cadmium, lead and mercury were considered to be below the maximum limit permitted in foods[14]. This must cast some doubt about whether these quantities of heavy metals are actually harmful to the fish. In addition to this study, cadmium has been found to be accumulated in the marine sponge, *Halichondria paniceae*[16] and American lobster (*Homarus americanus*)[19]. Also, organotins, namely tributyltin and triphenyltin, from antifouling paints, have been recorded in Japanese gastropods[4].

Experiments have been conducted to investigate the uptake and retention/loss of heavy metals, i.e. cadmium, lead and mercury, from aquatic organisms. Of these, lead was regarded as the least toxic, whereas mercury accumulated in greatest amounts, and was highly toxic to clams but less so to tilapia and mussels[20]. Moreover, differences were recorded in the precise site that the toxicants accumulated. Thus, there was affinity for gill, liver and muscle by lead, cadmium and mercury respectively.

Studies have sought to determine the possible accumulation of heavy metals, i.e. lead, in fish and their parasites. In particular, it was found that much higher concentrations of lead occurred in parasites than the muscle, liver and intestines of fish. For example from an examination of five fish, 4–16 µg of lead/g wet weight were found in the intestinal parasite *Acanthocephalus lucii* compared to 0.1–0.3 µg/g from the host, i.e. perch (*Perca fluviatilis*). Here, the highest concentration occurred in the intestine rather than liver or muscle[21]. Similarly, much higher concentrations of lead were found in the parasites *Pomphorynchus laevis* (54 µg/g), and *Paratenuisentis ambiguus*/*Anguilicola crassus* compared to the hosts, i.e. chub (*Leuciscus cephalus*)[22] and eel (*Anguilla anguilla*)[23] respectively.

2. *Pesticides*

As a result of an examination of crustaceans, fish and molluscs living in a vicinity of a sewage outfall, chlordane, dieldrin, hexachlorobenzene and SIGMA-DDT were found in tissues[6]. Generally, livers of elasmobranch and tetraodontiform fish contained higher levels of organochlorine pesticides than muscle. Indeed, the levels in muscle were extremely low. Furthermore, DDT and PCB have been recovered from four species of fish in the Vaike Vain Strait, western Estonia[24].

3. *Hydrocarbons*

Echinoderms, namely *Amphiura filiformis* and *Ophiothrix fragilis*, are very sensitive to oil pollution[5]. The sub-lethal stress from the presence of hydrocarbons has led to a decline in the number of symbiotic bacteria in the echinoderms. Yet, there is an absence of sound data pointing to the presence/accumulation of hydrocarbons in aquatic organisms.

EFFECT OF TOXIC SUBSTANCES ON AQUATIC ORGANISMS

There has been widespread publicity concerning the devastating effect of high concentrations of pollutants on aquatic life-forms. Television has repeatedly shown pictures of oil-coated birds in the aftermath of large-scale accidents involving oil tankers, e.g. the Exxon Valdez in Alaska and the Amoco Cadix off Brittany, France. In addition, toxins, such as released by the collapse of algal blooms, may have a devastating effect on fish, leading to massive kills over wide geographical areas. Such acute exposure to pollutants has an immediate and observable outcome. In many cases, the aquatic organism will die quickly or be seriously damaged with little chance of recovery. In this situation, the cause of the damage and its effect on aquatic organisms is comparatively easy to prove. In addition, experimental data have determined the acute toxicity of organophosphorus pesticides, notably diazinon and malathion[25] leading to death in fish (*Cyprinus carpio* and *Barilius vagra*). In contrast, chronic discharges may lead to long term damage of the aquatic organism[26]. Here, the damage, which may eventually become manifested in the animal or plant, may not be directly attributed to the original pollutant. Thus, the organism may be initially weakened/immunosuppressed[27], allowing subsequent colonisation by opportunistic bacterial or fungal pathogens, which lead to the ultimate demise of the organism. There is some evidence that pesticides, namely atrazine and lindane, may well adversely effect the immune system of fish[28]. Also, exposure to heavy metals[29] and unknown components of sewage sludge[30,31] have been implicated with adversely effecting the immune response of fish. Clearly, a situation leading to impairment of a principle host defence system, i.e. the immune response, will have long term repercussions for the well-being of the individuals. However, to re-iterate previous concerns, the proof is notoriously difficult to obtain.

Experiments have shown that injection of sublethal concentrations of heavy metals, i.e. cadmium and copper, into sea bass (*Dicentrarchus labrax*) led to inhibition of phagocytosis and the production of reactive oxygen intermediates, with respect to the bacterial fish pathogen, *Aeromonas salmonicida*[32]. The half-inhibition values for cadmium and copper were reported as 1 mg/kg and 250 μg/kg, respectively. Similarly, cadmium, copper and mercury have been found to be toxic for invertebrates, e.g. the gammarid amphipod, *Parhalella natalensis*[33]. Sub-lethal concentrations of mercury have been determined to effect food transformation in *Heteropneustes fossilis*[34]. In particular, the presence of mercury reduced the rates of feeding, absorption, conversion and metabolism.

The effects of ammonia, copper and phenol has been investigated on the ultrastructure of rainbow trout (*Oncorhynchus mykiss*) gills, using scanning electron microscopy[35]. In particular, ammonia caused the development of circular depressions, and pitting of the epithelium (this is especially relevant in discussions concerning the development of ulcers in fish from polluted marine environments). Exposure to copper led to fusion of the gill lamellae, and swelling to the tips of the filaments and epithelium (gill disease may also be associated with pollution). Phenol destroyed the epithelial layers as far as the cartilage[35].

Hydrocarbons have been associated with damage to DNA, i.e. breakages in the double-stranded DNA, and enzyme induction in dab (*Limanda limanda*) and starfish (*Asterias rubens*)[36]. Furthermore, the presence of polycyclic aromatic hydrocarbons in underlying sediments has been correlated with chlorophyll-deficiency mutations in mangroves in Puerto Rico[37].

Expose to the antifoulant, tributyltin (TBT), has been determined to reduce the growth rate of veliger larval development of the bivalve *Scrobicularia plana*[38]. It was considered that the presence of TBT in the aquatic environment may well have contributed to the reduction of clam populations in North America[38].

Sewage contaminated sites, which often contain elevated numbers of enteric bacteria, have been reported to exert an effect on fish in terms of enhanced antibody levels to the bacteria[39]. Thus, the presence of the bacteria led to a measurable immune response by the fish. In this case, there was a clear association between the presence of pollution and the host response. Furthermore in a series of long term experiments, conducted over three months, it was demonstrated that brown trout (*Salmo trutta*) developed histological changes in the liver following exposure to unknown components of biologically-treated domestic sewage[40]. Exposure to sewage sludge has also been implicated with effects on growth and protein synthesis in dab (*Limanda limanda*)[41]. In this study, it was determined that protein growth rate was higher in controls than in fish exposed to sewage sludge. Essentially, it was argued that changes in the balance between protein synthesis and protein growth exert an important influence on the general response of fish to such environmental stress.

DISEASES OF AQUATIC ORGANISMS

There is no universally accepted definition of disease. A compromise definition has been proposed, as follows:

> "...a disease is the sum of the abnormal phenomena displayed by a group of living organisms in association with a specified common characteristic or set of characteristics by which they differ from the norm of their species in such a way as to place them at a biological disadvantage..."[42].

An interpretation is that the term disease includes most unhealthy events which occur in an individual of any species. Such unhealthy events would be caused by biotic, including micro-organisms, and abiotic factors, such as pollutants. The precise reasons for an outbreak of disease will reflect the presence of:

- the host
- the disease-causing situation, such as the presence of a micro-organism (=pathogen) or a pollutant, e.g. carcinogen.

With host–pathogen interactions, the development of disease will often follow a decline in the resistance of the host, which may occur during reproductive cycles or as a result of environmental stress such as overcrowding or poor hygiene, and an increase in numbers and virulence of the pathogen[43]. Causal agents of disease include viruses, parasites, fungi and bacteria[44] (Table I). Regarding the aquatic

Table I Examples of pathogens and parasites of aquatic animals

Pathogen/parasite	Disease	Host range
Bacteria		
Aerococcus viridans subsp. *homari*	Gaffkemia	Lobsters
Aeromonas salmonicida	Furunculosis	Salmonids
Leucothrix mucor	Shell disease	Lobsters, shrimps
Vibrio spp. (*V. anguillarum*, *V. damsela*, *V. ordalii*, *V. salmonicida*)	Vibriosis	Many marine species
Fungi		
Exophiala salmonis	—	Salmonids
Fusarium solani	Gangrene	Lobsters
Ichthyophonus hoferi		Many fish species
Saprolegnia spp	Saprolegniasis	Many freshwater fish species
Protozoans		
Bonamia	Bonamiasis	Oysters
Hexamita nelsoni	Hexamitiasis	Oysters
Marteilia	Marteiliasis	Oysters
?	Proliferative kidney disease	Salmonids
Viruses		
Atlantic cod adenovirus		Atlantic cod
Baculovirus	Baculovirus disease	Shrimp
Infectious pancreatic necrosis virus	Infectious pancreatic necrosis	Many fish species
Herpes virus	Herpes	Many marine animal species
Lymphocystis virus	Lymphocystis	Many marine fish species
Viral haemorrhagic septicaemia virus	Viral haemorrhagic septicaemia	Many fish species

environment, emphasis has been placed on diseases of farmed (=aquaculture) animal species, i.e. fish and invertebrates, notably clams, lobsters, oysters and shrimps.

THE ROLE OF TOXIC OCCURRENCES IN DISEASE OF AQUATIC ORGANISMS

There is an steadily increasing literature regarding the possible relationship between aquatic pollution and the incidence of fish diseases[45–53]. Yet, the link between disease and pollution is often tenuous. The gist of many arguments centres on the presence of greater numbers of diseased fish in polluted compared with supposedly clean environments. The definition of polluted and non-polluted environments can be misleading. Is the control, i.e. "clean" site, really unpolluted? The criteria for determining the relative levels of pollution remain obscure. Therefore, there must be concern about comparisons between allegedly polluted and unpolluted sites. Moreover, it is unclear why motile organisms remain in polluted sites. Many studies have emphasised the North Sea[52,54–56], and the reader might be forgiven for wondering if politics have influenced the posturing between nations as to the relevance/influence of pollution on diseases of aquatic organisms. As such, the outcome of studies have apparently reflected national concerns. Yet, it remains to be proven that the alleged pollutants were definitely responsible for the increased incidence of disease. It would be pertinent to enquire if diseased fish caught in polluted sites might have recently migrated from cleaner areas. The converse argument could also apply regarding fish caught in non-polluted areas.

Diseases of fish, which have been linked to pollution, include:

- ulceration[57]
- fin/tail rot[57]
- gill disease[35]
- neoplasia[50]
- epidermal papilloma[54]
- liver damage[50,58,59]
- parasitic infiltration[60].

Of these, ulceration, fin/tail rot and gill disease may be linked to bacterial involvement, possibly after damage to the host. For example, the presence of ammonia has been associated with the development of circular depressions, and pitting of the epithelium[35]. Maybe, such damage could lead to bacterial colonisation, and thus ulceration. Neoplasia and hyperplasia may reflect viral activity, as well as the presence of carcinogens. Liver damage could well suggest the accumulation of toxic compounds, such as heavy metals. However, the presence of parasites is more difficult to reconcile, insofar as a polluted environment could be expected to lead to a reduction in parasitic activity.

1. *Contaminated Diet*

During 1993–1994, heavy mortalities of tropical reef fish were recorded in Florida, USA. Disease signs included head lesions, ulcers, fin and tail rot, and the presence of a mucus layer on the body surface. Bacteria were recovered; and amoebae, *Brooklynella hostilis*, *Uronema marinum* and turbellarians were present as parasites[61].

Yet, it was not considered that these micro-organisms were the primary cause(s) of the mortalities. In fact, it was reasoned that the microbes might have been secondary invaders of already damaged/compromised hosts. Instead, the author considered that the initiator of the unhealthy condition might well have been the presence of toxic materials, notably toxins from macroalgae (*Caulerpa* spp.) or benthic dinoflagellates (*Gambierdiscus toxicus*), which comprised the principal food. Yet again, the definitive proof was lacking.

2. *Heavy Metals*

The presence of 30–60 μg of copper/ml and 10 μg of iron/ml in sea water led to the exacerbation of vibriosis[62,63]. The debilitating effect, which reflected both concentration and the time of exposure to the heavy metal[64], may be assessed by coagulation of the mucus layer of the gills, which inhibits oxygen transport leading to respiratory stress[65]. Higher levels (100–250 μg/ml) of copper in water led to the increased susceptibility of Japanese eels to infection by *Edwardsiella tarda*[66]. In this example, it seemed likely that increased susceptibility to disease could be attributed to a reduction in lymphocytes and granulocytes in the blood, leading to reduced phagocytosis[67]. Another example centred on the interaction of cadmium with parasitism in three-spined stickleback (*Gasterosteus aculeatus*)[68].

During 1986–1988, a large scale survey of 5942 fish sought to examine the health status of common dab (*Limanda limanda*) caught in Dutch coastal waters, of which two locations were dump sites, which received acids of titanium dioxide[69]. Certainly, there has been repeated inferences to an association between pollution with titanium dioxide and fish diseases[54,55,70]. In the study by Vethaak and van der Meer[69], the diseases included epidermal hyperplasia/papilloma, lymphocystis, liver nodules and infections with the parasite, *Glugea* spp. A higher incidence of epidermal hyperplasia was recorded in fish caught from the titanium dioxide dump sites than at other locations. Unfortunately, chemical analyses to determine levels of pollutants in the fish were not carried out. Interestingly, there was no association between the incidence of disease and heavy metal dumping.

3. *Pesticides*

In 1993, 35 fish species were surveyed in the Vaike Vain Strait between the islands of Saaremaa and Muhu, western Estonia[24]. Including evidence obtained from local fisherman, it would appear that there have been increasing reports of diseases, such as "cauliflower disease", lymphocystis and ulceration. Could the increasing incidence of disease reflect the quantities of DDT and PCB in the fish[24]? It is tempting to make the connection between tissue levels of pollutants and disease. However, the definitive evidence remains to be obtained.

4. *Nitrites*

The presence of nitrites at 6 mg/l of water have been reported to increase the susceptibility of channel catfish to infection by *Aeromonas hydrophila*[71].

5. *Hydrocarbons*

There is evidence that the presence of hydrocarbons impairs mucus, leads to defective immune systems in fish, and induces hyperplasia[60,72,73]. The presence of neoplasia in soft-shell clams (*Mya arenaria*) has been found in oil contaminated sites[72]. Furthermore, an association has been made between hydrocarbons and increased incidences of parasites, such as *Trypanosoma* spp. in fish[60].

Levels of 100 μg of hydrocarbons/g were associated with increased mortalities in juvenile winter flounder (*Pseudopleuronectes americanus*)[74]. Here, earlier mortalities occurred in juvenile rather than adult fish.

Evidence has been presented for a relationship between pollution with hydrocarbons and an increased prevalence of infection of the Eastern oyster (*Crassostrea virginica*) by the protozoan parasite *Perkinsus marinus*[75]. In this example, the sediments were contaminated with 2.42 mg/g of aromatic hydrocarbons, including fluoranthene, phenanthene and pyrene. However, other pollutants, such as PCB and chlorinated pesticides, were not detected. It was concluded that there was a direct relationship between the concentration of the pollutants and the prevalence of infection. Furthermore, it was reasoned that the pollutants may well reduce resistance of the host to disease[75].

Liver hypertrophy in winter flounder has been associated with hydrocarbons, using experimental oil–sediment mixtures[76].

6. *Stress*

The role of "stress" in disease processes is unclear. Indeed, it would be pertinent to enquire what is meant by the term, "stress". Notwithstanding, there is an increasing literature describing stress-mediated conditions in aquatic animals. A common theme is that the presence of a pathogen or parasite in a disease process is blamed on stress. Yet, the nature of the stressor is rarely considered. For example, *Bacillus* sp. was recovered from a wide range of septicaemic fish, including *Chrysichthys nigrodigitatus*, *Clarias gariepinus*, *Cyprinus carpio* and *Heterobanchus bidorsalis*, in Nigeria. Stress, attributed to unnamed environmental factors, was associated with the disease[77]. In other examples, the stressors have been identified. Thus, Mellergaard and Nielsen[78] attributed oxygen-deficiency as the likely reason for outbreaks of the viral diseases, lymphocystis and epidermal papilloma, in the common dab from southern Kattegat, Denmark during 1984–1993. The evidence centered around the observation that following severe oxygen depletion in 1986–1988, the incidence of lymphocystis and epidermal papilloma in the common dab increased and peaked at 14.7% and 3.3%, respectively, in 1989. Indeed, there was a negative correlation between the incidences of these diseases and oxygen levels[78].

Furthermore, stress has been linked to bacterial disease in channel catfish (*Ictalurus punctatus*)[79].

7. *Sewage*

A role for sewage and diseases of aquatic organisms has been repeatedly stated. Publications have described the incidence of fish parasites and the association with sewage sludge dumping[80]. Interestingly in this example, gastro-intestinal metazoan parasites in flatfish, i.e. the long rough dab (*Hippoglossoides platessoides*) and the common dab (*Limanda limanda*), were examined at sewage dump sites in Scottish coastal waters. However, an association between sewage sludge dumping and parasites was not proven.

A survey carried out in 1988 examined the health status of 9608 flounder collected from 16 sites in the Dutch Wadden Sea[57]. The underlying aim of the study was to equate stress factors, associated with fresh water drainage sluices, with disease.

Overall, the conclusion was reached that there was a higher incidence of skin ulcers and fin rot among fish caught near drainage sluices than elsewhere.

Pollution by domestic sewage, i.e. leakage from a septic tank, was attributed to a new skin disease in rainbow trout (otherwise infected with enteric redmouth disease for which there might also be a link with sewage sludge[7] in Scotland during 1992[81]. The disease was characterised by the presence of extensive skin lesions and muscle necrosis to a depth of approximately 1 mm over the entire flank of the fish from the operculum to tail. From diseased animals, two apparently new pathogens, i.e. *Serratia plymuthica* and *Pseudomonas pseudoalcaligenes*, were recovered. The skin lesions—but not enteric redmouth disease—declined in significance shortly after the leaking septic tank was repaired.

Eutrophic waters, particularly those associated with faecal pollution and high levels of organic material, have been blamed for diseases caused by other enteric bacteria, including *Citrobacter freundii*[43], *Edwardsiella tarda*[82], *Providencia rettgeri*[83] and *Serratia marcescens*[84]. Indeed, poultry faeces, which was used to fertilise fish ponds, was blamed for mass mortality in silver carp (*Hypophthalmichthys molitrix*) in Israel[83].

8. *Contaminated Diets*

Improperly stored (damp) food may permit the development of fungal and bacterial blooms, which in turn may lead to the accumulation of toxins, e.g. mycotoxins. Unpublished data (B. Austin) have pointed to several mass mortalities in rainbow trout populations, which have been blamed on such contaminated food. In addition, a suggestion has been made that the causal agent of streptococcicosis may be transmitted in diets containing contaminated fish meal, i.e. derived from diseased fish[85]. Similarly, contaminated food may be responsible for transmission of mycobacteriosis[86].

Feeding Chinook salmon (*Oncorhynchus tshawytscha*) with contaminated crude fish offal was blamed for an infection by a new fish pathogen, i.e. *Rhodococcus* sp. This organism led to pronounced exophthalmia and eye loss[87]. Minced trash fish, used as food, was implicated as a possible cause of an outbreak of botulism in rainbow trout[88].

9. *Evidence from Histology*

Parasitic lesions and neoplasia in dab livers, as revealed by histopathology, have been linked to pollution of the North Sea[89]. Furthermore, histopathology techniques have been used successfully to study the impact of oil pollution on mussels[90].

10. *Parasites*

Associations have been made between pollution and parasites[45,91–93]. Various modes of action have been proposed. For example, interactions between heavy metals (cadmium) and parasites have been discussed[68]. Möller[92] considered that pollutants may act on free-living stages of the parasite, or on ectoparasites attached to the host. It is conjecture whether or not parasites, which accumulate/sequester heavy metals (such as from the diet), actually protect the host from the toxic effects at sub-lethal exposures.

CONCLUSIONS

There are specific examples, e.g. concerning bacteria in sewage, of a clear relationship between pollution and disease. However, scepticism must be expressed over generalised statements concerning the role of polluted seas, e.g. the North Sea, and disease. Yet, definitive experiments to silence potential sceptics remain to be done.

References

1. K. Collins, "Towards a cleaner marine environment: European Community priorities for protection" *Mar. Poll. Bull.* **29**, 641–643 (1994).
2. E. D. Goldberg, "Emerging problems in the coastal zone for the twenty-first century" *Mar. Poll. Bull.* **31**, 152–158 (1995).
3. K. J. Lehtinen, M. Notini and L. Landler, "Tissue damage and parasite frequency in flounders, *Platichys flesus* chronically exposed to bleached kraft pulp mill effluents" *Ann. Zool. Fennici.* **21**, 23–28 (1984).
4. T. Horiguchi, H. Shiraishi, M. Shimizu, S. Yamazaki and M. Morita, "Imposex in Japanese gastropods (Neogastropoda and Mesogastropoda): effects of tributyltin and triphenyltin from antifouling paints" *Mar. Poll. Bull.* **31**, 402–405 (1995).
5. L. C. Newton and J. D. McKenzie, "Echinoderms and oil pollution: a potential stress assay using bacterial symbionts" *Mar. Poll. Bull.* **31**, 453–456 (1995).
6. A. G. Miskiewicz and P. J. Gibbs, "Organochlorine pesticides and hexachlorobenzene in tissues of fish and invertebrates caught near a sewage outfall" *Environ. Poll.* **84**, 269–277 (1994).
7. D. J. Dudley, M. N. Guentzel, M. J. Ibarra, B. E. Moore and B. P. Sagik, "Enumeration of potentially pathogenic bacteria from sewage sludges" *Appl. Environ. Microbiol.* **39**, 118–126 (1980).
8. W. R. Turrell, "Modelling the Braer oil spill: a retrospective view" *Mar. Poll. Bull.* **28**, 211–218 (1994).
9. M. I. Evans, P. Symens and C. W. T. Pilcher, "Short-term damage to coastal bird populations in Saudi Arabia and Kuwait following the 1991 Gulf War marine pollution" *Mar. Poll. Bull.* **27**, 157–161 (1993).
10. B. Austin, "Taxonomy of bacteria isolated from a coastal, marine fish-rearing unit" *J. Appl. Bacteriol.* **53**, 253–268 (1982).
11. D. A. Ziemann, W. A. Walsh, E. G. Saphore and K. Fulton-Bennett, "A survey of water quality characteristics of effluent from Hawaiian aquaculture facilities" *J. World Aquacult. Soc.* **23**, 180–191 (1992).
12. A. C. Ibe and G. Kullenberg, "Quality assurance–quality control (QA–QC) regime in marine pollution monitoring programmes: The GIPME perspective" *Mar. Poll. Bull.* **31**, 209–213 (1995).
13. K. Hylland, C. Haux and C. Hogstrand, "Hepatic metallothionein and heavy metals in dab *Limanda limanda* from the German Bight" *Mar. Ecol. Prog.* **Ser. 91**, 89–96 (1992).
14. J. Hellou, W. G. Warren, J. F. Payne, S. Belkhode and P. Lobel, "Heavy metals and other metals in three tissues of cod *Gadus morhua* from the northwest Atlantic" *Mar. Poll. Bull.* **24**, 452–458 (1992).
15. P. Vas and J. D. M. Gordon, "Trace metals in deep-sea sharks from the Rockall Trough" *Mar. Poll. Bull.* **26**, 400–402 (1993).
16. T. M. E. Olesen and J. M. Weeks, "Accumulation of Cd by the marine sponge *Halichondria panicea* Pall: effects upon filtration rate and its relevance for biomonitoring" *Bull. Environ. Contam. Toxicol.* **52**, 722–728 (1994).
17. K. M. Chan, "Concentrations of copper, zinc, cadmium and lead in rabbitfish (*Siganus oramin*) collected in Victoria Harbour, Hong Kong" *Mar. Poll. Bull.* **31**, 277–280 (1995).
18. K. O. Joseph and J. P. Srivastava, "Mercury in finfishes and shellfishes inhabiting Ennore estuary" *Fish. Technol.* **30**, 15–118 (1993).
19. C. L. Chou and J. F. Uthe, "Cadmium in American lobster *Homarus americanus* from the area of Belledune Harbour New Brunswick Canada 1980–1992 results" *Can. Tech. Rep. Fish Aquat. Sci.* 1–25 (1993).
20. T. W. Kureishy and C. D'Silva, "Uptake and loss of mercury, cadmium and lead in marine organisms" *Indian J. Exp. Biol.* **31**, 373–379 (1993).
21. B. Sures, H. Taraschewski and E. Jackwerth, "Comparative study of lead accumulation in different organs of perch (*Perca fluviatilis*) and its intestinal parasite *Acanthocephalus lucii*" *Bull. Environ. Contam. Toxicol.* **52**, 269–273 (1994).

22. B. Sures, H. Taraschewski and E. Jackwerth, "Lead accumulation in *Pomphorynchus laevis* and its host" *J. Parasitol.* **80**, 355–357 (1994).
23. B. Sures, H. Taraschewski and E. Jackwerth, "Lead content of *Paratenuisentis ambigus* (Acanthocephala), *Anguillicola crassus* (Nematodes) and their host *Anguilla anguilla*" *Dis. Aquat. Org.* **19**, 105–107 (1994).
24. H.-R. Voigt, "Fish surveys in the Vaike Vain Strait between the islands of Saaremaa and Muhu, western Estonia" *Proc. Estonian Acad. Sci. Ecol.* **4**, 128–135 (1994).
25. M. K. Alam and O. E. Maughan, "Acute toxicity of selected organophosphorus pesticides to *Cyprinus carpio* and *Barilius vagra*" *J. Environ. Sci. Health, Part B Pestic Fs. Contam. Agric. Wastes* **28**, 81–89 (1993).
26. F. L. Mayer, D. L. Woodward and W. J. Adams, "Chronic toxicity of pydraul 50E to lake trout" *Bull. Environ. Contam. Toxicol.* **51**, 289–295 (1993).
27. J. G. Vos, H. Van Loveren, P. W. Wester and A. D. Vethaak, "Toxic effects of environmental chemicals on the immune system" *Trends Pharmacol. Sci.* **10**, 289–293 (1989).
28. M. Cossarini-Dunier, "Effect of the pesticides atrazine and lindane and of manganese ions on cellular immunity of carp, *Cyprinus carpio*" *J. Fish. Biol.* **31** (Suppl. A), 67–73 (1987).
29. J. G. O'Neill, "The humoral immune response of *Salmo trutta* L. and *Cyprinus carpio* L. exposed to heavy metals" *J. Fish. Biol.* **19**, 297–306 (1981).
30. C. J. Secombes, T. C. Fletcher, J. A. O'Flynn, M. J. Costello, R. Stagg and D. F. Houlihan, "Immunocompetence as a measure of the biological effects of sewage sludge pollution in fish" *Comp. Biochem. Physiol.* **100C**, 133–136 (1991).
31. C. J. Secombes, T. C. Fletcher, A. White, M. J. Costello, R. Stagg and D. F. Houlihan, "Effects of sewage sludge on immune response in the dab, *Limanda limanda* (L.)" *Aquat. Toxicol.* **23**, 217–230 (1992).
32. N. Bennani, A. Schmid-Alliana and M. Lafaurie, "Immunotoxic effects of copper and cadmium in the sea bass *Dicentrarchus labrax*" *Immunopharmacol. Immunotoxicol.* **18**, 129–144 (1996).
33. U. G. Bhat and K. Vamsee, "Toxicity of heavy metals Cu, Cd and Hg to the gammarid amphipod *Parhalella natalensis* (Stebbing)" *Sci. Total Environ. Suppl.* 2, 887–987 (1993).
34. R. James, K. Sampath, V. J. Pattu and G. Devakiamma, "Utilization of *Eichhornia crassipes* for the reduction of mercury toxicity on food transformation in *Heteropneustes fossilis*" *J. Aquacult. Trop.* **7**, 189–196 (1992).
35. R. S. Kirk and J. W. Lewis, "An evaluation of pollutant induced changes in the gills of rainbow trout using scanning electron microscopy" *Environ. Technol.* **14**, 577–585 (1993).
36. J. M. Everaarts, H. M. Sleiderink, P. J. Den Besten, R. S. Halbrook and L. R. Shugart, "Molecular responses as indicators of marine pollution: DNA damage and enzyme induction in *Limanda limanda* and *Asterias rubens*" *Environ. Health Perspectives* **102**, 37–43 (1994).
37. E. J. Klekowski, J. D. Corredor, J. M. Morelli and C. Del Castillo, "Petroleum pollution and mutation in mangroves" *Mar. Poll. Bull.* **28**, 166–169 (1994).
38. J. M. Ruiz, G. W. Bryan and P. E. Gibb, "Effects of tributyltin (TBT) exposure on the veliger larvae development of the bivalve *Scrobicularia plana* (da Costa)" *J. Exp. Mar. Biol. Ecol.* **186**, 53–63 (1995).
39. R. A. Robohm, C. Brown and R. A. Murchelano, "Comparison of antibodies in marine fish from clean and polluted waters of the New York Bight: relative levels against 36 bacteria" *Appl. Environ. Microbiol.* **38**, 248–257 (1979).
40. F. Bucher and R. Hofer, "The effects of treated domestic sewage on three organs gills kidney and liver of brown trout *Salmo trutta*" *Water Res.* **27**, 255–261 (1993).
41. D. F. Houlihan, M. J. Costello, C. J. Secombes, R. Stagg and J. Brechin, "Effects of sewage sludge exposure on growth, feeding and protein synthesis of dab (*Limanda limanda* L.)" *Mar. Environ. Res.* 331–353 (1994).
42. E. J. M. Campbell, J. G. Scadding and R. S. Roberts, "The concept of disease" *Brit. Med. J.* **2**, 757–762 (1979).
43. B. Austin and D. A. Austin, *Bacterial Fish Pathogens, Disease in Farmed and Wild Fish*, 2nd edn. (Ellis Horwood, Chichester, 1993).
44. B. Austin, *Marine Microbiology* (Cambridge University Press, Cambridge, 1988).
45. R. M. Overstreet and H. D. Howse, "Some parasites and diseases of estuarine fish in polluted habitats of Mississippi" *Ann. N. Y. Acad. Sci.* **298**, 427–462 (1977).
46. C. J. Sindermann, "Pollution-induced diseases and abnormalities of fish and shellfish: a review" *Fish. Bull.* **76**, 716–749 (1979).
47. V. Dethlefsen, "Observations on fish diseases in the German Bight and their possible relation to pollution" *Rapp proce.—verb. run. Cons. Int. Explor. Mer.* **179**, 110–117 (1980).
48. V. Dethlefsen, "Studies of pathology in relation to pollution" **In**: *Report of the ICES Workshop on the Use of Pathology in Studies of the Effects of Contaminants* (J. Thulin, ed.) (ICES, CM 1986/E.40 ref. F) pp. 22–35 (1986).

49. P. Nounou, R. Martoja and L. Orcel, "Ulcerations of marine fishes and mammals caught in French coastal waters" *Rapp. Sci. Tech.—Centre Nat. Explor. Oceans* **43**, 1–99 (1980).
50. D. C. Malins, B. B. McCain, J. T. Landahl, M. S. Myers, M. M. Krahn, D. W. Brown, S. L. Chan and W. T. Roubal "Neoplastic and other diseases in fish in relation to toxic chemicals: an overview" *Aquat. Toxicol.* **11**, 43–67 (1980).
51. D. Bucke, S. W. Feist, M. G. Norton and M. S. Rolfe, "A histopathological report of some epidermal anomalies of Dover sole, *Solea solea* L., and other flatfish species in coastal waters off South-East England" *J. Fish Biol.* **14**, 565–578 (1983).
52. A. D. Vethaak and T. ap Rheinallt, "Fish disease as a monitor for marine pollution: the case of the North Sea" *Rev. Fish. Biol. Fish.* **2**, 1–32 (1982).
53. K. Gopal and S. P. Pathak, "Possible causes of outbreaks of fish diseases and mortality in polluted waters" *J. Adv. Zool.* **14**, 53–60 (1993).
54. V. Dethlefsen and B. Watermann, "Epidermal papilloma of North Sea dab, *Limanda limanda*: histology, epidemiology and relation to dumping from TiO_2 industry" *ICES Spec Meeting on Diseases of Commercially Important Marine Fish and Shellfish* **8**, 1–30 (1980).
55. V. Dethlefsen, B. Watermann and M. Hoppenheit, "Diseases of North Sea dab (*Limanda limanda* L.) in relation to biological and chemical parameters" *Arch. Fishereiwissenschaft* **37**, 101–237 (1987).
56. A. H. McVicar, D. W. Bruno and C. O. Fraser, "Fish diseases in the North Sea in relation to sewage sludge dumping" *Mar. Poll. Bull.* **19**, 169–173 (1988).
57. A. D. Vethaak, "Diseases of flounder (*Platichthys flesus* L.) in the Dutch Wadden Sea, and their relation to stress factors" *Netherlands J. Sea Res.* **29**, 257–272 (1992).
58. D. C. Malins, B. B. McCain, M. S. Myers, D. W. Brown, M. M. Krahn, W. T. Roubal, M. H. Schiewe, J. T. Landahl and S. L. Chan, "Field and laboratory studies of the aetiology of liver neoplasms in marine fish from Puget Sound" *Environ. Health Persp.* **71**, 5–16 (1987).
59. N. Peters, A. Köhler and H. Kranz, "Liver pathology in fishes from the lower Elbe as a consequence of pollution" *Dis. Aquat. Org.* **2**, 87–97 (1987).
60. R. A. Khan, "Crude oil and parasites of fish" *Parasitol. Today* **3**, 99–100 (1987).
61. J. H. Landsberg, "Tropical reef-fish disease outbreaks and mass mortalities in Florida, USA: What is the role of dietary biological toxins?" *Dis. Aquat. Org.* **22**, 83–100 (1995).
62. M. C. Rødsaether, J. Olafsen, J. Raa, K. Myhre and J. B. Steen, "Copper as an initiating factor of vibriosis (*Vibrio anguillarum*) in eel (*Anguilla anguilla*)" *J. Fish. Biol.* **10**, 17–21 (1977).
63. T. Nakai, T. Kanno, E. R. Cruz and K. Muroga, "The effects of iron compounds on the virulence of *Vibrio anguillarum* in Japanese eels and ayu" *Fish. Pathol.* **22**, 185–189 (1987).
64. K. J. Baker, M. D. Knittel and J. L. Fryer, "Susceptibility of chinook salmon, *Oncorhynchus tshawytscha* (Walbaum), and rainbow trout, *Salmo gairdneri* Richardson, to infection with *Vibrio anguillarum* following sublethal copper exposure" *J. Fish. Dis.* **6**, 267–275 (1983).
65. B. A. Westfall, "Coagulation film anoxia in fishes" *Ecol.* **26**, 283–287 (1945).
66. K. Mushiake, K. Muroga and T. Nakai, "Increased susceptibility of Japanese eel *Anguilla japonica* to *Edwardsiella tarda* and *Pseudomonas anguilliseptica* following exposure to copper" *Bull. Jap. Soc. Sci. Fish.* **50**, 1797–1801 (1984).
67. K. Mushiake, T. Nakai and K. Muroga, "Lowered phagocytosis in the blood of eels exposed to copper" *Fish. Pathol.* **20**, 49–53 (1985).
68. D. Pascoe and P. Cram, "The effect of parasitism on the toxicity of cadmium to the three-spined stickleback, *Gasterosteus aculeatus* L" *J. Fish. Biol.* **10**, 467–472 (1977).
69. D. Vethaak and J. van der Meer, "Fish disease monitoring in the Dutch part of the North Sea in relation to the dumping of waste from titanium dioxide production" *Chem. Ecol.* **5**, 149–170 (1991).
70. K.-J. Lethinen, "Effects on fish exposed to effluent from a titanium dioxide industry and tested with rotary-flow technique" *Ambio* **9**, 31–33 (1980).
71. L. A. Hanson and J. M. Grizzle, "Nitrite-induced predisposition of channel catfish to bacterial diseases" *Prog. Fish. Cult.* **47**, 98–101 (1985).
72. P. P. Yevich and C. A. Barszcz, "Neoplasia in soft-shell clams (*Mya arenaria*) collected from oil-impacted sites" *Ann. N. Y. Acad. Sci.* **298**, 409–426 (1977).
73. W. E. Haensly, J. M. Neff, J. R. Sharp, A. C. Morris, M. F. Bedgood and P. D. Beom, "Histopathology of *Pleuronectes platessa* L. from Aber Wrac'h and Aber Benoit, Brittany, France: long term effects of the Amoco Cadiz crude oil spill" *J. Fish. Dis.* **5**, 365–391 (1982).
74. R. A. Khan, "Influence of concurent exposure to crude oil and infection with *Trypanosoma murmanensis* (Protozoa: Mastigophora) on mortality in winter flounder, *Pseudopleuronectes americanus*" *Can. J. Zool.* **69**, 876–880 (1991).
75. F.-L. E. Chu and R. C. Hale, "Relationship between pollution and susceptibility to infectious disease in the Eastern oyster, *Crassostrea virginica*" *Mar. Environ. Res.* **38**, 243–256 (1994).
76. G. L. Fletcher, M. J. King, J. W. Kicenuik and R. F. Addison, "Liver hypertrophy in winter flounder following exposure to experimentally oiled sediments" *Comp. Biochem. Physiol. C.* **73**, 457–462 (1982).

77. G. A. Oladosu, O. A. Ayinla and M. O. Ajiboye, "Isolation and pathogenicity of a *Bacillus* sp. associated with a septicaemic condition in some tropical freshwater fish species" *J. Appl. Ichthyol.* **10**, 69–72 (1994).
78. S. Mellergaard and E. Nielsen, "Impact of oxygen deficiency on the disease status of common dab *Limanda limanda*" *Dis. Aquat. Org.* **22**, 101–114 (1995).
79. G. R. Walters and J. A. Plumb, "Environmental stress and bacterial infections in channel catfish, *Ictalurus punctatus* Refinesque" *J. Fish. Biol.* **17**, 177–185 (1980).
80. R. Siddall, A. W. Pike and A. H. McVicar, "Parasites of flatfish in relation to sewage dumping" *J. Fish. Biol.* **45**, 193–209 (1994).
81. B. Austin and M. Stobie, "Recovery of *Serratia plymuthica* and presumptive *Pseudomonas pseudoalcaligenes* from skin lesions in rainbow trout, *Oncorhynchus mykiss* (Walbaum), otherwise infected with enteric redmouth" *J. Fish. Dis.* **15**, 541–543 (1992).
82. F. P. Meyer and G. L. Bullock, "*Edwardsiella tarda*, a new pathogen of channel catfish (*Ictalurus punctatus*)" *Appl. Microbiol.* **25**, 155–156 (1973).
83. Y. Bejerano, S. Sarig, M. T. Horne and R. J. Roberts, "Mass mortalities in silver carp *Hypophthalmichthys molitrix* (Valenciennes) associated with bacterial infection following handling " *J. Fish. Dis.* **2**, 49–56 (1979).
84. A. M. Baya, A. E. Toranzo, B. Lupiani, Y. Santos and F. M. Hetrick "*Serratia marcescens*: a potential pathogen for fish" *J. Fish. Dis.* **15**, 15–26 (1992).
85. T. Minami, "*Streptococcus* sp., pathogenic to cultured yellowtail, isolated from fishes for diets" *Fish. Pathol.* **14**, 15–19 (1979).
86. M. P. Dulin, "A review of tuberculosis (mycobacteriosis) in fish" *Vet. Med/Small Animal Clinician* 735–737 (1979).
87. R. Claveau, "Néphrite granulomateuse à *Rhodococcus* spp. dans un élevage de saumons de l'Atlantique *Salmo salar*)" *Méd. Vétérin. Québec.* **21**, 160–161 (1991).
88. H. H. Huss, A. Pedersen and D. C. Cann, "The incidence of *Clostridium botulinum* in Danish trout farms. 1. Distribution in fish and their environment" *J. Fd. Technol.* **9**, 445–450 (1974).
89. D. Bucke and S. W. Feist, "Histopathological changes in the livers of dab *Limanda limanda* L." *J. Fish. Dis.* **16**, 281–296 (1993).
90. P. P. Yevich and C. A. Barszcz, "Histopathology as a monitor for marine pollution. Reports of the histopathologic examination of the animals collected for the U.S. 1976 Mussel Watch Program" *Rapp. Proc. verb. Réun.* **182**, 96–102 (1983).
91. M. C. Das and A. K. Shrivastava, "Fish mortality in Naini Tal Lake (India) due to pollution and parasitism" *Hydrobiol. J.* **20**, 60–64 (1984).
92. H. Möller, "Pollution and parasitism in the aquatic environment" *J. Parasitol.* **17**, 353–361 (1987).
93. R. A. Khan and J. Thulin, "Influence of pollution on parasites of aquatic animals" *Adv. Parasitol.* **30**, 201–238 (1991).

12. TOXICOLOGY IN THE WORKING ENVIRONMENT

ANDREW WATTERSON *

INTRODUCTION

This chapter defines occupational toxicology and examines the dimensions of toxicological hazards and risks posed in workplaces. It also looks at the range of substances used in occupational settings and what is and is not known about the hazards such substances present to workers. The details of toxicological methodologies and testing are dealt with elsewhere in the volume and this chapter simply highlights the key tests used. The major problems and limits of occupational toxicology are, however, identified. The problems relating to attempts to establish safe thresholds or occupational exposure levels for workers are particularly important and these are set in scientific and also social context. Different strategies for establishing safe levels of exposure are described and discussed.

The role of occupational toxicology in assisting occupational hygienists, occupational health and safety practitioners and managers in removing or reducing the risks to workers from hazardous substances is outlined. Central to this role is the development of a 'healthy and safe workplace'. The concepts of clean production and toxics reduction are two key approaches underpinned by toxicology directly relevant to 'healthy and safe workplaces'.

The chapter then looks at occupational carcinogens followed by sections on how toxicology has revealed the hazards and risks to workers of substances which damage a variety of systems and organs. The sections cover liver toxicity, kidney toxicity, reproductive toxicity, neurotoxicity, immunotoxicity, respiratory sensitisers and allergens, skin toxicity, cardiovascular toxicity, and blood toxicity.

The chapter ends with conclusions about the way forward for occupational toxicology and further reading.

DEFINITIONS

Toxicology has been described as:

"...a hybrid science built on advances in biochemistry, physics, pathology, physical chemistry, pharmacology and public health"[1].

Occupational toxicology has been defined as:

"the study of adverse effects of agents that may be encountered by workers during the course of their employment"[2].

* Professor of Occupational and Environmental Health and Director, Centre for Occupational and Environmental Health, De Montfort University, Scraptoft Campus, Scraptoft, Leicester LE7 9SU (UK).

Toxicology applied to the working environment entails study, through toxicological methods, of any substances, metals or other materials which may affect people at work. Working environments may cover those in industry, commerce, education, the health sector, the leisure sector, transport. The term occupational toxicology and not industrial toxicology is therefore most accurate.

Occupational toxicology and the risk assessments, control standards and auditing of those standards have provided many of the early warnings about wider environmental problems through illnesses detected in production workers and other users of toxic substances. The working environment is often the frontline and source of wider environmental pollution. Workers are the 'canaries' for populations outside the workplace.

TOXICS USE AND EXPOSURE

There are around 10 million known chemical substances. Within Western Europe there are approximately 120,000 existing substances on European Union lists and in many industrialised countries there will be 60,000–70,000 substances used in workplaces. Around 1,000 new chemicals are introduced each year. Working out the uptake, distribution, storage, metabolism, excretion and effect of even single substances on this list in a vast range of working conditions is a mammoth task. Working out interactions between these substances in the workplace is still beyond our capacity as advanced industrial societies. Trying to assess long term low level effects of workers' exposures to such substances is not possible even with sophisticated toxicological and epidemiological techniques.

Some consideration is needed of how exposures outside the workplace may affect workers within and *vice-versa*. The inter-relationship between toxic substances inside and outside the workplace has been increasingly recognised in recent decades. The capacity of toxicology to predict the likely human effects of workplace exposures to toxic substances may be better than for some areas of environmental exposure but it is limited. The limits partly relate to the wide range of conditions—temperature, humidity, ventilation, noise, vibration, physical and mental stress and exertion, length of exposure—and the wide range of other substances and processes to which employees may be exposed in the workplace. Toxicity may also be affected by age, gender, health status and ethnicity. Early toxicity studies in some instances set exposure standards based on limited human data often involving young military personnel: across the world it is not possible to replicate such a cohort as typical of the workforce.

DATA ON OCCUPATIONAL DISEASES IN THE UK

Occupational and occupationally-related diseases affect large numbers of the working population. In 1990, a survey conducted for the UK Health and Safety Commission suggested that 2 million people with illness believe those illnesses were caused or made worse by work. It is also estimated that 10% of general-practitioner (family doctor) consultations are work-related.

The toll of occupational illnesses resulting from exposures to toxic substances in the UK workplace is large. Occupational cancers have attracted most public attention and it is conservatively estimated that between 3% and 6% of all cancers in the country are caused each year by workplace exposures and 3–6% of cancers are viewed to have workplace exposures as contributory factors. UK trade unions have estimated 12% or more of cancer deaths could be occupationally related. The lower occupationally related cancer death rate per year in the UK is 5,000 and the higher, given by some occupational health commentators, is 12,000. HSC estimated that there were at least 2,000 premature deaths each year from occupational disease, 8,000 deaths each year where work contributed to mortality, and at least 80,000 new cases of work-related disease each year. More than 500,000 people were estimated to be suffering continuing damage to health from work.

Some substances present major hazards to a wide range of the workforce and indeed to the wider public. Asbestos-related deaths in the 1980s were estimated to run at 2,000 a year with as many as 50,000 deaths likely to occur between 1982 and 2012 at the threshold levels set for asbestos in the early 1980s. Now epidemiologists have revised their risk estimates upwards and consider that annual mortality from asbestos may run at 6,000 deaths a year. Inquest verdicts in the last five years have begun to record deaths from mesothelioma among such groups as teachers due to asbestos exposure in schools. Other occupational groups such as maintenance and gas workers are now considered at higher risk by epidemiologists than before. However, studies from pathology and toxicology revealed the asbestos threat as early as the 1910s and 1920s.

Occupationally related respiratory diseases also present serious threats to health. The UK project—SWORD—which is investigating occupational related respiratory diseases suggests that there may be three times the number of such diseases occurring than are reported. Occupational asthma is a growing problem with between 2% and 5% of all asthma cases estimated to be work-related. This would give approximately 1,500 new occupationally related cases each year but only 500 cases each year are reported in the UK. An estimated 85,000 people suffered from work-related asthma in 1990 and more recent research indicates that such figures are the tip of an iceberg[3]. Occupationally related dermatitis causes 850,000 lost working days each year and it is estimated that there were 85,000 people with work-related dermatitis in 1990, yet few received compensation. Accurate figures on the immunological, neurological and cardiovascular diseases caused by workplace exposures to toxic substances are simply not available.

HOW TOXIC SUBSTANCES AFFECT WORKING

The routes of entry into the body of substances found in the workplace are through inhalation, ingestion and skin absorption. Substances in the workplace may occur as dusts, fumes, mists, vapours and gases. Much of the testing done on industrial and commercial chemicals relates to ingestion through LD and other tests rather than Lethal Concentration tests (LC) or skin penetration and absorption indicators: yet ingestion is often the least likely source of occupational exposure. Skin and lung exposures are usually far more important. There is therefore a need to weigh the significance of routes of entry in terms of a particular occupation and also to consider the total dose of a substance a worker may be exposed to by a combination of routes of entry. Much of this research is still relatively speculative as indeed have been studies in the past which can accurately assess exposures from multiple routes of entry or even the effectiveness of Personal Protective Equipment (PPE) especially gloves in reducing contamination in the workplace.

Toxic substances may have local or systemic or effects both: the former often being easier to identify than the latter in the workplace. Likewise the acute effects of exposure to a substance may be relatively easily recognised in the workplace whereas the chronic effects may not.

THE PROBLEMS WITH OCCUPATIONAL TOXICOLOGY

"Human beings have always been intuitive toxicologists, relying on their senses of sight, taste and smell to detect harmful or unsafe food, water and air. As we have come to recognize that our senses are not adequate to assess the dangers inherent in exposure to a chemical substance, we have created the sciences of toxicology and risk assessment to perform this function. Yet despite this great effort to overcome the limitations of intuitive toxicology, it has become evident that even our best scientific methods still depend heavily on extrapolations and judgements in order to infer human health risks from animal data"[4]. Toxicology has informed occupational risk assessments of a variety of substances. The latest emphasis among many toxicologists has been to claim a detailed understanding of the mechanisms of toxicity of particular chemicals and to eschew quantitative risk assessments used by agencies like the USA Environmental Protection Agency and the 'Qstar' system as too speculative and reliant on statistical methods with flawed inputs. The Q* is the quantitative assessment of a chemical's oncogenicity based on the slope of the dose response curve from animal tests producing a positive oncogenic response. The slope describes the change in tumour incidence over the change in dose.

The assumptions are that mechanistic assessments do not fall into the same traps as exclusively statistical methods and that somehow individual risk assessments so derived will allow us to assess the many thousands of substances to which workers are exposed to in a wide range of occupational activities and conditions. This is fallacious and the risk assessments with the greater margins for error—along the lines of the EPA—are the ones which offer the greatest protection to workers. Indeed Kraus and her colleagues, in a large study of US and Scandinavian toxicologists found that 'controversies over chemical risks may be fueled as much by limitations of risk assessment and disagreements among experts as by public misconceptions'. Of course, toxicology risk assessments may rely on combining both mechanistic and quantitative methods.

The convention for setting 'thresholds' in many countries related to 'safe' standards set for most of the working population for the most of the time over an 8 hour working day and a 40 hour working week. Some formulae, based on the toxicity data, were provided for calculating safe limits over a longer working week and also for mixtures. The validity and value of the thresholds require careful scrutiny[5] and there may be debates about the toxicity data generated, the toxicity models developed, the specific mechanisms of toxicity at work and the significance of the data gaps which exist on a particular substance[4].

Toxicologists typically attempt to establish data for the following:

- No Observable Effect Levels (NOELs). This is the 'highest exposure level at which no morphological, physiological or functional modification of any kind is detectable under the test conditions'
- No Observable Adverse Effect Levels (NOAELs) 'the highest dose level at which no biological adverse effects occur'
- Lowest Observable Adverse Effect Level (LOAELs). This operates to set exposure levels when data are lacking to set the NOAEL.

Typically safety factors of 10 are used so exposure standards are 10 times higher on the basis of differences between test species and 10 times higher between animals and humans producing a somewhat arbitrary figure of a safety factor 100 times higher than the data indicate. This is the traditional response of toxicology to testing substances. However, there are in occupational toxicology two very well established philosophies for the setting of threshold levels not one. One philosophy leads to the setting of thresholds on the basis of no significant indicators of likely adverse effects. This approach uses known adverse effect levels and works down measuring with biochemical, pre-clinical and other tests viewed as relevant likely predictors of ill-health. The other more precautionary philosophy works upwards from levels which are known to have no effects on humans or animals and exposure limits are set just below the lowest exposure level which causes a statistically significant deviation from the norm in the test organism[6].

Soviet and Eastern European standards in the 1970s and 1980s based on a precautionary approach to toxicology tended to be much tighter than those in the West and always set more explicit safeguards for women and young children. Unfortunately, the implementation and enforcement of these standards often left much to be desired and there was little development of occupational epidemiology to check the outcomes of worker exposure to toxic substances.

The precautionary toxicology principle led Sweden, for instance, during the 1970s and 1980s to adopt occupational standards for solvents, such as trichloroethylene, which were often five or more times tighter than those in the UK and USA. What is also significant is that the UK and USA standards have in some instances but not all been tightened and thresholds lowered closer to those in Sweden following additional research. Table I graphically illustrates this point. Hence the Swedish model has much to recommend it in practice.

Data and resources only exist to set a relatively small number of thresholds, occupational exposure levels, maximum exposure levels, ceilings or maximum acceptable concentrations of toxic substance for workplace chemical exposure. Such thresholds cover just several hundred of perhaps 60 or 70,000 substances relatively widely used in the workplace and well over 100,000 substances with which workers may come into contact.

In the USA, the Occupational Health and Safety Agency (OSHA) has established permissible exposure levels (PELs) for some substances at workplaces. Yet recently one study identified 120 rodent carcinogens which had no OSHA PEL. These substances ranged from sodium nitrite to which an estimated 928,000 workers exposed in that country were exposed through chlorinated paraffins with 573,000 workers down to such substances as carbazole with estimated 70 plus workers and benzoyl hydrazine with just 20 workers exposed[7].

In addition to the lack of threshold standards for many substances, there are other concerns about the thresholds that have been established. "Occupational exposure limits are the highest legally permitted levels of human exposure to toxic substances. Environmental exposure of the public to the same air pollutants is regulated much

Table I Threshold limits on trichloroethylene

		Threshold limit value	Short term exposure limit
1981	Sweden	20 ppm (110 mg/m^3)	50 ppm (250 mg/m^3)
1980	UK	100 ppm (535 mg/m^3)	150 ppm (800 mg/m^3)
1996	UK	100 ppm (535 mg/m^3)	150 ppm (802 mg/m^3)

more strictly. Workers exposure to toxic substances should be no greater in the workplace than in other regulated settings of human exposure to toxic substances"[8].

The precise toll of exposure to toxic substances in the workplace is unknown but there is general agreement that under-reporting of occupational diseases caused by or related to toxics exposures is significant. In terms of occupational respiratory disease, occupational or occupationally-related cancer, occupational dermatitis and occupational neurological damage, indications are that the figure is high.

The picture is further confounded by an absence of complete toxicity data on many of the substances used in the workplace as Table II shows.

Indeed it could be argued that through a lack of resources, time and toxicologists we are going faster and faster in terms of identifying, developing and sometimes introducing new molecules and compounds into the workplace and yet running slower and slower in terms of assessing, controlling, reducing or removing the risks associated with occupational exposures.

Toxicology is the principal and first discipline used to assess the hazards presented by substances in the workplace. The 'safety net' used to ensure that those substances 'cleared' by toxicology have been adequately tested and assessed is provided by epidemiology: the study of the distribution of diseases and the correlation and likely causes of patterns of diseases. Occupational epidemiology attempts to track groups of workers exposed to 'cleared' or 'approved' substances. Occupational epidemiology, however, may find difficulties in identifying causation and it may be of limited value for instance because the data about the exposure of workers to toxic substances may be poor, the population studied may be too small or incomplete, the follow up period for tracking diseases which could be occupationally caused may be too short, studies may not fully take account of 'healthy worker effects' on morbidity and mortality. Hence such studies may produce 'negative' results and substances may be cleared when in fact the studies are not able to produce any conclusive results at all which would either clear or indict a chemical. It is important that toxicologists and those using toxicology to make risk assessments are aware of the dangers of negative epidemiology in the tracking of substances in the workplace.

A number of international industry-based toxicology agencies have stressed the need for epidemiological studies to monitor chemicals to ensure that the toxicology provided to clear the substance in the first place through the regulatory system has proved adequate. Sadly, in a number of instances such follow-up has not occurred[9]. Sometimes, too, it is clear that non-scientific factors have impinged upon the testing processes and a small number of labs in the US have in the past been found to have fraudulently falsified the results of regulatory toxicity tests. There have also been other cases of scientific fraud reported in labs. The application of Good Laboratory Practice standards and the monitoring and auditing of results provide a framework

Table II Percentage of chemicals with toxicity data

Category	Complete health hazard assessment data available	No toxicity information available
Pesticides/inerts	10%	38%
Cosmetics	2%	56%
Drugs	18%	25%
Food additives	5%	46%
Chemicals in commerce	0%	76–82%

(*Source*: NAS Toxicity Testing 1984)[2].

within which such abuses should be reduced although it is unlikely they will be removed entirely.

Epidemiology has identified occupational diseases, as have a small number of clinical case studies, which toxicology either did not detect or, more usually, where toxicologists did identify potential hazards to workers from lab tests but the results of their research were ignored or suppressed. Examples of epidemiological findings which have revealed occupational diseases relate to studies of polyvinyl chloride workers exposed to vinyl chloride monomer who contracted angiosarcoma, wood workers who contracted nasal cancer, gas workers and dye workers who contracted bladder cancer. In most if not all of these cases, toxicologists had provided data about potential hazards.

OTHER SIGNIFICANT INFLUENCES ON OCCUPATIONAL TOXICOLOGY

Social, political and economic influences may affect either the research questions posed by toxicologists, the results that they generate and the use of those results. For instance the value of LD_{50} tests has been questioned by some toxicologists in certain circumstances as the cheapest and most easily used statistical method but not the one which provides the best indicator of toxicity. In other studies on formaldehyde toxicity, some regulators have used toxicology and later epidemiological studies to conclude that this chemical was a potential human carcinogen and others have used the same test to conclude that the chemical presented no workplace carcinogenicity risk.

THE LIMITS OF TOXICOLOGY AND THE ROLE OF HEALTH-BASED EXPOSURE LIMITS

These were well understood in the 1930s and 1940s in terms of the limited knowledge base of the subject, the complexities of occupational exposures, the difficulty of extrapolating acute, high level exposures in animals to chronic long term low level effects in people and even the diet and gender issues raised by chemical exposures[10].

The traditional view of toxicology is that all substances are potentially toxic but this may cloud work on occupational toxicology. The classical toxicological philosophy is based on Paracelus who argued that the dose determined toxicity. For instance, excessive and rapid intake by humans of water can be fatal and adverse effects may be enhanced linked to drug abuse. However, this has led some to argue that everything is 'hazardous' and for some to suggest that concerns about occupational and indeed environmental hazards from toxic substances are much exaggerated. Yet research indicates that for many chemicals extensively used in the workplace, exposure at the permitted limits over thirty or forty years may still lead to more than 10% of the workforce dying from an occupational disease[8].

These data have led groups like the Santa Clara Centre for Occupational, Safety and Health in California to argue for the establishment of what have been termed health-based exposure limits (HBELs). These limits would basically eliminate occupational disease risks and adopt the NOEL approach and standard risk evaluation techniques used by the USA Environmental Protection Agency by setting standards or operating guidelines which set standards at levels which are believed to present little or no risk of disease. The EPA system uses Integrated Risk Information

Systems (IRIS) and Health Effects Assessment Summary Tables (HEST). The methods involved are critically geared to determining safe exposure limits on the basis of systemic and carcinogenic studies of chronic and subchronic exposures: they are not predictors of risk which is how current OEL and TLV systems often operate.

An illustration of how such a guideline would work is provided by benzene. The current lowest occupational exposure limit (OEL) set for benzene is 0.5 ppm or 1.6 mg/m^3 which presents a risk of developing cancer for workers of 1 in 500 (or 2,000 per million workers exposed). To reduce a worker's risk of cancer from benzene exposures to one in a million would require a limit of 0.00021 ppm or 0.00063 mg/m^3. This is 2000 times lower than the current OEL.

Many HBELs may be set too low for measurements of exposure to occur but the reason for using HBELs are twofold. Firstly they starkly reveal the enormous gap between existing limits for toxic substances and levels which would provide more effective protection for workers. Secondly, as toxicology guidelines, they emphasise the need for occupational hygienists using such data to introduce the principle of lowering exposure as a matter of course[8].

Working out dose response relationships in an occupational setting with varied conditions, multiple exposures and employees with widely different health status and susceptibility is far more complex than some toxicology texts suggest. Nor are all issues resolved: for instance the question of threshold for carcinogens or the question of idiosyncratic responses of individuals to certain substances or mixtures remain unanswered. Indeed, this is precisely why the threshold standards for workplaces based on toxicology tests refer or did refer only to standards which protect most of the workforce, most of the time, in controlled conditions, from irreversible damage. The toxic torts cases going through the courts worldwide indicate quite clearly that the balance of probabilities standards of proof in a number of instances have not been met in toxicology testing whereas the proofs beyond reasonable doubt that chemicals are hazardous have been met.

It should be noted, however, that not all substances for instance cause cancer in the workplace; not all substances are occupational teratogens and the hazards and the risks that flow from toxic substances in the workplace vary significantly from industry to industry and in some instances from company to company and workplace to workplace, and country to country. Historically great emphasis was placed upon acute toxicity testing and the Lethal Dose 50 tests (LD_{50}) which, because of cheapness and statistical ease, dominated and still does dominate much occupational toxicology. A list of substances with low acute toxicity in the WHO classifications could then be interpreted as low hazard but further investigation frequently revealed that the long term low exposure to such substances would produce a list of potential human carcinogens, neurotoxins and immunotoxic substances and the risk assessments for such chemicals would rise significantly.

There are now legal requirements in most European countries to test new substances in terms of toxicity and such tests often cover the gamut of acute, sub chronic, chronic, neurological, reproductive, neurotoxic and immunotoxic tests. European Union law also require controls over the classification and labelling of dangerous substances—in the CHIP regulations. Even with these controls there are still problems with the range and depth of toxicity testing, the categorisation and labelling of substances which in some instances does not reflect best standards worldwide. Labels on substances to provide information for users about toxicity have also especially in the UK context, not necessarily flagged up international assessments of potential chronic damage from long term exposures. The legal requirements should set high control standards or exposures and ensure occupational hygiene standards reflect the latest and safest toxicology findings. Data sheets provided with toxic substances should also, as happens with the USA Environmental Protection Agency data sheets on pesticides, flag up where significant data gaps exist on substances which are available commercially.

Greatest public concern appears to have been expressed about carcinogenicity but other areas of toxicity have been neglected until quite recently. These include immunotoxicity, gerontogens (those agents which affect older people in particular ways), asthmagens (those agents which may cause asthma), neurotoxicity especially chronic toxicity and this partly relates to the problem of measuring both exposure and effect of low doses.

There may be significant problems in identifying the specific effects of many toxic substances for a variety of reasons. These problems exist for a whole range of organs. The reasons include the following factors:

1. There may be no differences in the type of damage done to the body either from toxic substances or a range of drugs or other factors.
2. There are major problems with differentiating non-occupational and occupational causes of diseases.
3. Diseases linked to toxicity of substances may be very common: for instance lung or heart diseases which may be caused by occupational exposures may not be identified as such. Only the unusual nature of some diseases has led to recognition of occupational causes. Vinyl chloride monomer as a cause of angiosarcoma in humans for instance or cancer of the scrotum from exposure to soot are examples of the latter: lung cancers from exposure to asbestos and bronchitis from welding fumes are examples of the former. This highlights the importance of epidemiological studies which may confirm, through human data, the toxicology research undertaken before substances are cleared for commercial use.
4. Some toxic substances may lack clear clinical or pathological signs or symptoms especially in the early stages of exposure.
5. Some agents in tissues can rarely be monitored even at high levels.
6. Exposures may be at low levels over long periods of time in the workplace and the cause-effect relationship may be hard to establish. This has led to serious under-estimates of occupational diseases due to toxic substances.
7. Many substances have never been tested for toxic effects.

OCCUPATIONAL HEALTH PRACTICE LINKED TO TOXICOLOGICAL RESEARCH

Toxicology should inform good health and safety policies and procedures in the workplace. Occupational health practice relies on toxicology and occupational hygiene which are directly linked to selection of substances for use in the workplace and the engineering and other controls needed to ensure that substances are used with lowest or least risks and exposures.

The key principle of occupational health practice is to remove the hazard at source if this at all possible and hence no risk will apply to the users. If this cannot be achieved, a scientific response should be determined by the techniques and knowledge available not by cost or time considerations. Economics and law, however, do play a part in determining the extent and thoroughness of testing and product

Table III Hierarchy of health and safety steps to deal with toxic substances

HEALTHY AND SAFE WORKPLACES
1. remove hazard at source (Clean Production)
↓
2. if 1 cannot be achieved substitute with a less hazardous substance or a safe subsistence (Toxics reduction)
↓
3. if 1 and 2 cannot be achieved, isolate and enclose the hazard
↓
4. if this cannot be done, local exhaust ventilation and other engineering controls may be used
↓
5. if this cannot be achieved, personal protective equipment should be used as a last result with time limits built in to engineer out or remove the hazard at source
↓
6. training, information, supervision and instruction are used to support 2 to 5
HEALTHY AND SAFE WORKERS

selection for use in the workplace. In law—common law, statute law and case law—a lesser requirement than technically possible often applies. To do what is reasonably practicable brings into play cost and technology. If the costs are high, the technical difficulty of bringing in a solution to toxic exposures is high and the hazards and risk from those hazards are low or moderately low, the courts would view it as not reasonably practical to remove a hazard. Here it is the law and not the toxicologist or health and safety practitioner which determines what is reasonably practicable. In the best circumstances, step 1 should apply.

In terms of saving money from compensation cases, avoiding adverse publicity from occupational diseases cases and damage to market image, ensuring uninterrupted production, reducing insurance costs and replacement costs of employees adversely affected by toxics exposures, some Scandinavian and North American employers have begun to use toxicology as a means to achieve 1. through what are termed "toxics reduction" and "cleaner production". For industry, cleaner if not clean production is viewed as the way forward for both environmental and work environment reasons. For all workplaces where toxicology will have a role, the key approach is now to work towards toxics reduction if not toxics removal[11]. The implementation of good practice is again underpinned by good toxicology which recognises its limits, its gaps in knowledge and the debates about the validity of its understanding of the mechanisms of toxicity in the context of moving towards practical public health precautionary policies on toxics use in our workplaces.

TOXICS REDUCTION

The term is a relatively new one; it may also be called 'pollution prevention'. Toxics reduction is a policy and practice which reduces or removes known or suspect toxic substances from the workplace rather than relying on complicated risk assessments and then risk reduction programmes which may leave the toxic substances in the workplace or the environment. In the 19th century, some toxic metals and certain forms of phosphorous were removed from the workplace: toxics reduction in early action. The UK occupational physician, Legge, advocated toxics reduction in the 1920s and 1930s. The practice was very simple and the obstacles familiar. Governments were unwilling to set standards and enforce standards, employers were reluctant to phase out materials unless compelled to do so because of the time and effort as well as resources involved in such measures.

The process of toxics reduction may be straightforward or complex. An occupational hygienist was called in to a factory to advise on extraction of toxic solvent fumes used to wash swarf (metal particles) from a finished item. The solution was simple: water performed the washing as effectively as the solvent: instant toxics reduction occurred saving the company the cost of new extraction and the cost of future purchases of the solvent. Occupational Hygiene standards immediately improved in the factory.

Sometimes toxics reduction policies have not been fully thought through or are too fragmentary. For instance, when unleaded petrol was introduced without catalytic converters fitted to cars, increasing benzene levels in the air were recorded—an increased occupational risk for car park attendants, commercial drivers, ferry operators and urban workers. There are dangers with reducing or removing toxic substances in one part of production process may lead to problems either earlier or later in the chain. This may then lead to greater risks for some workers against others, or workers as opposed to consumers, or vice versa. As the example of trichloroethylene cited later illustrates or orchard workers asked to spray fungicides at great risk to themselves in an effort to cut residue levels on fruit in warehouses. This latter example lends itself to toxics reduction and clean production through phasing out the use of fungicides and improving storage methods and retail turnover.

Toxics use reduction techniques may involve process recycling, effective operations and maintenance, production unit redesign or modification, production unit modernisation and input substitution or product reformulation.

Within such a philosophy and practice of good occupational health, toxicology should inform best practice. For instance, toxicological research may reveal that exposures to substances causes adverse effects: this should not in the first instance be used to screen out either employees or potential employees who may have 'a particular susceptibility' to the substance; rather it should lead to the removal of the substance itself. Sufficient data now exist to indicate that often such substances will damage many and not a few employees if they work long enough with and/or are exposed to high enough levels of such substances. This would be the case with a number of respiratory sensitisers and possibly with occupational carcinogens too.

CARCINOGENS

'Cancer' is not one homogenous disease: cancers may differ in aetiology and impact and also survival rates depending on a range of factors. "Carcinogenesis is a complex process involving sequential genetic events which result in altered functions of the genes that control normal cellular growth, with subsequent clonal growth of the resulting 'preneoplastic' or neoplastic cells. Chemical carcinogens may act by inducing mutations and/or by altering gene and cellular growth control"[2]. Genetic toxicology is defined as "the identification of agents that interact with nucleic acids to alter hereditary material of living organisms: agents that produce such alterations are termed 'genotoxic'". The agents so identified can then be assessed for carcinogenicity based on evidence of mutagenicity in test results[2]. Details of these tests are provided elsewhere in this volume. This branch of toxicology underpins much of the work done to identify occupational carcinogens in recent decades and also has a key role in investigations of workplace reproductive hazards and has been further developed by *in vitro* tests as well as *in vivo* ones.

Carcinogens present a major challenge to health and safety practitioners because, unlike many other substances, there may not be a dose response relationship below which no adverse effect can be detected. No solid data prove that there are thresholds for carcinogens and there is much ignorance about predicting low dose curves as indicators of human carcinogenicity on the basis of high dose data from lab tests. On this basis, occupational toxicology should adopt the precautionary principle and all occupational carcinogens should be treated as if no safe thresholds exist for exposure of workers.

Concern about occupational carcinogens has dominated occupational health for many decades. Percival Pott identified cancer of the scrotum in chimney sweeps in the late eighteenth century through clinical case studies but toxicology and epidemiology now dominate the identification and assessment of workplace carcinogens. Studies of occupational carcinogens have in some instances drawn attention and funds away from other serious workplace hazards described below. This partly reflects the seriousness of cancers which often prove fatal and partly the relatively poor success rates of treatments. If there is no cure, prevention becomes the only strategy and the UK Health and Safety Executive in the 1970s produced a guidance note which outlined a precautionary policy based on the recognition that many substances had not been fully tested or tested at all for carcinogenicity and, by the time a substance was positively identified, it would often be too late to prevent worker exposure and death. Several countries—especially Denmark and Sweden—proceeded to use toxicological data and the lack of data—to prohibit or control chemicals where there were indications of potential carcinogenicity.

The role that workplace exposures to toxic substances play in the aetiology of occupational cancer can be complex. The extent to which genotoxic and epigenetic carcinogens can always and fully be separated is not clear. Toxics in the workplace may also be co-carcinogens or promoters. This makes anything other than a nil exposure to carcinogens in the workplace a strategy of risk for exposed workers. (Tables IV and V)

LIVER TOXICITY

The liver is the organ which has a primary part to play in the body's metabolism by removing or neutralising toxic substances in the body. Damage to hepatocytes, which may be temporary or permanent, causes jaundice. The liver may be affected by many substances which do not relate to workplace exposures—alcohol and prescribed and non-prescribed drugs are obvious examples—and these substances may affect how well the body detoxifies toxic occupational substances.

Those substances which act directly on the liver, and there are over 100 chemicals in this category, include organic solvents such as carbon tetrachloride, chloroform, trichloroethylene and tetrachloroethane. Other substances may act indirectly by interfering with the metabolic pathways: these include drugs which pharmaceutical workers may encounter, aflatoxins which agricultural and food workers may encounter and MDA used in plastics.

Other workers may also be exposed to infectious agents which damage the liver. These include various hepatitis viruses which may affect health and emergency workers; farm workers and public health workers may be exposed to leptospirosis and

Table IV Selected International Agency for Research on Cancer carcinogens which may occur in the workplaces

1. *Human carcinogens*				
Aflatoxins	Arsenic	Asbestos	Benzene	Benzidine
Beryllium	BisCME Bis(chloromethyl)-ether	Cadmium	Chromium	Diethylstilboestrol
Ethylene oxide	Hepatitis B	Hepatitis C	Mustard gas	Oestrogens
[various]	Radon	Solar radiation	Vinyl chloride	
2. *Mixtures*				
Coal tar pitches	Coal tars	Mineral oils	Shale oils	Tobacco products
Wood dust				
3. *Probable human carcinogens*				
Acrylamide	Acrylonitrile	Benzo(a)pyrene	Captafol	Epichlorhydrin
Ethylene dibromide	Formaldehyde	MOCA	Silica	Styrene
Tetrachloroethylene	Radiation (UV A,B,C)	Vinyl bromide	Vinyl fluoride	
4. *Mixtures*				
Creosote	Diesel engine exhaust	Non-arsenical	Insecticides	Polychlorinated biphenyls (PCBs)
5. *Possible human carcinogens*				
Acetaldehyde	Acetamide	Amitrole	Atrazine	Bracken fern
Carbon black extracts	Carbon tetrachloride	Ceramic fibres	Chlordane	Chlordecone
Chloroform	Chlorophenols	Chlorophenoxy herbicides	Cobalt	DDT
Dibenzopyrene	Dichlormethane	Dichlorvos	Dimethylformamide	Ethylene thiourea
Furan	Griseofulvin	Heptachlor	Hexachlorobenzene	Hexachlorocyclohexane
Hydrazine	Lead	Lead compounds (inorganic)	Magenta (with CI Basic Red)	Nickel (metallic)
Pentachlorophenol	Potassium bromate	Progestins	Propylene oxide	Propylene oxide
Rockwool	Saccharin	Slagwool	Styrene	Toluene diisocyanate (TDI)
Vinyl acetate				
6. *Mixtures*				
Bitumens	Chlorinated paraffins	Marine diesel fuels	Petrol engine exhaust	Heavy residual fuel oils
PBBs welding fumes				
7. *Lung carcinogens include*:				
Asbestos	Arsenic	Beryllium	Cadmium	Chromium
Diesel exhaust	Nickel	Radon	Silica	

In the USA it is estimated that approximately 9–10,000 males and 900–1,900 females develop lung cancer because of past exposure to occupational carcinogens[12].

(*Sources*: Various IARC and NIOSH reports)[13].

Table V Some reported cancers caused by or associated with certain occupations and industries

Substance	Site(s) of cancer	Industry/occupation
2,4-D	Soft tissue sarcomas	Pesticide manufacture, agriculture
Arsenic	Lung, skin	Pesticides, other
Asbestos	Pleura, peritoneum, lung, larynx, GIT, kidney, leukaemia	Construction, shipbuilding, insulation, maintenance, textiles
Benzene	Leukaemia	Boot and shoe manufacture, chemicals, printers, other
Benzidine	Bladder	Chemicals
Chromium	Lung	Metals, welders
Coal tars and pitches	Skin, lung	Chemicals, steel, gas production, pitch workers, coke oven operators
Mineral oils	Skin	Engineering
Nickel	Lung, nose	Metals, smelters, engineering
Polycyclic aromatic hydrocarbons	Lung, bladder, skin, kidneys	Aluminium production, coal and coke production, iron and steel foundries
Vinyl chloride monomer	Liver	Plastics
Wood	Nose	Furniture making

(*Sources*: Fischman, Cadman and Desmond in Le Dou[17], pp. 182–208; Vainio in Stacey[2], p. 159).

Table VI Substances linked to occupational liver disease

Substance	Effect	Occupational group
Arsenic	Cirrhosis, angiosarcoma, hepatocellular carcinoma	Pesticides, wood, vinters, smelters
Beryllium	Granulomatous disease	Ceramics
Carbon tetrachloride (CTC)	Acute hepatocellular damage cirrhosis	Dry cleaning, solvent workers, cleaners
Chlordecone (DBCP)	Hepatotoxicity	Pesticides
Chloroform	Acute hepatocellular industry	Fridge workers, cleaners
Dimethylformamide (DMF)	Acute hepatocellular injury	Solvents, chemicals wallpaper manufacturers
Dioxins	Porphyria cutanea tarda	Pesticides
Ethyl alcohol	Cirrhosis/fibrosis	Bartenders, wine producers, distillers
Halothane	Acute hepatocellular injury	Anaesthesia
Hydrazine	Steatosis	Rocket fuels
Methylene dianiline	Cholestasis	MDA manufacturers
2-nitropropane	Acute hepatocellular	Painters injury
Phosphorous	Acute hepatocellular injury	Arms industry
Polychlorinated biphenyls (PCBs)	Sub-acute liver injury	Electrical
Tetrachloroethane	Acute/sub-acute hepatocellular injury	Aircraft manufacturers
Trichloroethylene	Acute hepatocellular injury	Solvents
Trinitrotoluene	Acute hepatocellular injury	Arms industry
Vinyl chloride	Angiosarcoma	Plastics

(*Sources*: R. J. Harrison in LaDou[17]; Pransky in Levy and Wegman[1]).

coxiella burnetti as may animal care workers and slaughterhouse workers; and health care workers dealing with children may be exposed to cytomegalovirus[1]. (Table VI)

Epidemiological studies at the end of the 1980s showed laundry workers, dry cleaners, petrol station staff, asphalt workers and bar staff all had organic solvent exposures and higher than expected liver cancer rates.

KIDNEY TOXICITY

As with other organs, a variety of substances can damage the kidneys including drugs. Occupational exposures may also result in kidney damage and in the USA it is estimated that up to 4 million workers may be exposed to nephritic chemicals[2]. It may be possible to identify occupational damage to the organ from the specific site of that damage in the kidney. Indirect damage to the kidney through chemical effects causing cardiac or other damage may be much harder to identify and would require epidemiological data. Also 'the kidneys have great reserve capacity and can function adequately despite progressive loss of nephrons, work-related kidney disease is typically not diagnosed until considerable dysfunction has occurred'[1].

Biochemical damage to the kidney occurs with heavy metals and antibiotics, organic solvents, iron, zinc and ethylene glycol and immunological damage may occur through exposure to mercury. In addition renal toxicity has resulted from exposure to widely used industrial solvents such as toluene, carbon tetrachloride in the past and trichloroethylene; plastic manufacture involving acrylonitrile and styrene and pesticides such as bipyridls like paraquat and diquat[2]. Trichloroethylene illustrates the dilemmas facing those using toxicological assessments where the risk may vary from workplace to the environment. Trichloroethylene was once phased out of many workplaces but is now being re-introduced in some parts of the UK as a more environmentally friendly substitute for CFCs. However, the textile industry has produced good case studies of the effective substitution of trichloroethylene by water: a generally agreed low toxicity liquid! (Table VII.)

REPRODUCTIVE HAZARDS INCLUDING TERATOGENS

Adverse reproductive outcomes may be the result of genetic and environmental exposures, diseases such as chicken pox and mumps, and the effects of prescribed and illegal drug usage, alcohol use and smoking. Such outcomes may include a wide range of effects ranging from teratogenicity to declining fertility.

Table VII Substances reported to have damaged the kidney in the workplace

Substance	Effect	Occupation
Cadmium	Nephrotoxicity	Welding, engineering
Carbon disulphide	Nephrotoxicity	Chemicals, paints, textiles
Carbon tetrachloride	Nephrotoxicity	Chemicals and other
Coke oven emissions	Renal cancer	Steel works and other
Ethylene dichloride	Nephrotoxic	Resins, rubber, insecticides
Lead	Nephrotoxicity	Chemicals and other, paint, batteries
Mercury (inorganic)	Nephrotoxicity	Chemicals, paints
Methanol	Nephrotoxicity	Chemicals and other
Pentachlorophenol	Nephrotoxic	Pesticide
Phenol	Nephrotoxic	Chemicals, various
Toluene	Nephrotoxic	Painters
Trichloroethylene	Nephrotoxic	Solvent, cleaning

(*Sources*: Frumkin and Millius in Levy and Wegman[1], 2nd ed., Kaysen in LaDou[17], pp. 259–266).

Some adverse reproductive outcomes may be related to occupational hazards which include chemical and other exposures like radiation, extremes of heat and cold and also to exposure to infection, viruses and zoonoses in the workplace. Effects of adverse occupational exposures include:

(a) for males, declining sperm counts, infertility, impotence.

(b) for females, menstrual problems, infertility, chromosomal abnormalities in the foetus, birth defects, spontaneous abortions, low birth weight, neonatal deaths, infant deaths.

About 3% of babies are born with a congenital abnormality which requires medical treatment. 20% of these abnormalities are linked to gene mutations and 5% with chromosomal aberrations. Much remains unknown about the causes of the other abnormalities. One recent text identified 1200 teratogens based on lab data out of a list of 2200 agents[14].

Genetic toxicology through tests for aberrant chromosomes is important in the identification of those workplace substances which may cause infertility, spontaneous abortions and neoplasia[2]. However, in the USA in the early 1990s of the 70,000 widely used commercial chemicals, only 2800 had been tested for teratogenicity of which 62% only were probably not teratogenic in the test[2]. An OECD survey in 1990 found that only 367 of 948 organic chemicals and 148 of 390 inorganic chemicals had adequate data to assess the reproductive hazards of these substances: researchers concluded that 'even for chemicals with high production volumes and likely human exposure, there is apt to be little available data for assessing the potential risk to reproduction and development'[15].

Workplace exposures may have varied effects on women depending on the time of the pregnancy with insults in the first trimester being critical as this is the time of most significant foetal development.

In some occupations, adverse reproductive effects due to exposure to a variety of agents have been identified in a number of studies. Occupations listed include various hospital workers, paper and print workers, chemical workers, agro-chemical workers, plastics workers and metal workers. (Table VIII)

In humans, ionising radiation has been reported to cause infertility, spontaneous abortions, malformations and childhood leukaemias; heat caused malformations and noise infertility and malformations[1]. Viruses have been linked to malformations, hepatitis B to neonatal chronic hepatitis, herpes to malformations, rubella to malformations and toxoplasmosis to malformations. All these may occur as occupational exposures. Indeed some occupations will run or have run relatively high risks of exposure and hence the greater likelihood of adverse effects. Hospital workers, for instance, are exposed to many of these reproductive hazards including stress and lifting and carrying, staff in operating theatres have had adverse reproductive effects linked to exposure to anaesthetic gases.

NEUROTOXICITY

A wide range of substances in the workplace may cause central and peripheral neurological damage in humans. Damage may affect the brain and spinal cord—the central nervous system; or the peripheral nervous system. Central nervous system damage relates to brain function. Peripheral nervous system damage may relate to peripheral neuropathy 'initially manifesting itself as intermittent numbness and

Table VIII Possible factors influencing reproductive outcomes based on experimental data

Factor	Effect	Workplace
Acrylonitrile	Teratogen	Chemicals, textiles
Anaesthetic gases	Miscarriages	Health workers
Arsenic	Fetotoxic, teratogen, transplacental carcinogen	Agriculture, wood preserving
Benomyl	Teratogen	Agriculture
Benzene	Fetotoxic, teratogen, menstrual disorders	Chemicals
Benzo(a)pyrene	Transplacental carcinogen	Fuel/fumes
Cadmium	Spontaneous abortions, impaired, implantation, teratogen. Male and female damage	Engineering, chemicals, batteries, paints, smelting, mining
Carbaryl	Teratogen, sperm damage	Pesticides
Carbon disulphide	Menstrual disorders	Chemicals,textiles
Chlorodifluromethane	Teratogen	
Chloroprene	Teratogen	Chemicals
Chromium	Teratogen	Chemicals, engineering
DBCP	Decreased fertility, sperm damage	Agriculture
DDT	Decreased fertility, fetotoxic, hormonal alterations,	Agriculture
Endrin	Teratogen	Pesticides
Ethylene oxide	Fetotoxic	Chemicals and other
Formaldehyde	Menstrual disorders	Chemicals, health workers, textiles
Glycol ethers	Teratogen	Chemicals
Hydrogen sulphide	Spontaneous abortions	Textiles, sewers, paper mills, petrochemical plants, iron smelters and coke oven plants, food processing and tanneries
Kepone	Reduced reproductive capacity in offspring	Pesticides
Lead	Decreased fertility, fetotoxic, impaired implantation, teratogen, sperm damage, hormonal alterations	Various
Manganese	Decreased fertility, impaired implantation	Various
Mercury	Fetotoxic, teratogen, menstrual disorders	Chemicals, pesticides
Methyl ethyl ketone	Teratogen	Chemicals, engineering
Methyl formamide	Teratogen	Chemicals
Oestrogens	Spontaneous abortions	Pharmaceutical
Organophosphates	Reduced fertility	Agriculture, chemicals
PBBs	Fetotoxic	Chemicals
PCBs	Decreased fertility, fetotoxic, hormonal alterations	Engineering
Phthalates	Teratogen	Chemicals
Selenium	Fetotoxic	
Styrene	Menstrual disorders	Fibre glass, plastics, shipbuilding
Thallium	Fetotoxic	
Thiram	Teratogen	Pesticides
Triethyl lead	Spontaneous abortions	
Vinyl chloride	Fetotoxic, teratogen, transplacental carcinogen	
Xylene	Fetotoxic, menstrual disorders	Chemicals, paints, pesticides
Carbon disulphide	Sperm damage	Chemicals

(*Sources*: Rudolph and Sonquits Forest in LaDou[17]; Letz in LaDou[17]).

tingling in the hands and feet; motor weakness in the feet or hands may develop somewhat later and progress to the development of an ataxic gait or an inability to grasp heavy objects[1].

Neurological damage may be measured by a variety of techniques including:

1. neurophysiological tests including nerve conduction tests, sensation and co-ordination tests.
2. checking levels of acetylcholinesterase in blood and serum. Field kits exist to do this testing where carbamates and organophosphates may be used in agriculture and exposed workers may be hours or days away from clinic-based biological health monitoring.
3. neuropsychological tests which include standard test batteries which measure intellectual functioning such as indicators of attention, concentration and orientation, language, verbal and visual recall, motor skills and conceptual and executive functions.
4. neuropsychiatric tests which attempt to identify such sequelae as dementia, delirium, depression, mania, delusions and hallucinations as well as stress disorders[14].

There may, however, be difficulties in identifying and measuring the effect of workplace exposures. This partly relates to the lack of data on the neurological effects of a large number of substances to which workers may be exposed in various settings. There may also be confounding factors such as exposure to neurotoxic substances—solvents for instance in painting and hobby activities such as boat building and model making—in non-occupational settings. Data on the incidence and prevalence of work-caused and work-related neurological diseases are therefore very incomplete and the size of this particular problem is significantly under-estimated.

There is considerable debate about the link between chemical exposures and brain tumours in humans. There are some epidemiological studies which have linked gliomas with certain types of agro-chemical and other industrial chemical exposure. Occupational groups associated with brain cancer excesses include metal workers (aluminium and lead and machinists); oil and petroleum workers; professional groups like dentists, medical staff, vets and chemists; and rubber and PVC workers[2]. Other types of neurological disease associated with occupational exposure which in the past have been discounted by health professionals and are now being more widely reported include mental illnesses linked to solvent, lead and carbon monoxide exposure; confusional states linked to solvents again and to certain metals like manganese and the organotins, insecticides; and finally organophosphate pesticides linked to psychotic effects[2]. These links perhaps indicate the need for regulators of workplace hazards to pay greater attention to both toxicological and epidemiological research findings than has hitherto been the case.

Alcohol and other drugs, legal and otherwise, may result in neurotoxic effects too. The problems of neuropsychiatric and neuropsychological toxicity provide additional problems of diagnosis and measurement. Different countries may also recognise different occupational neurological diseases: Sweden, for instance, compensates painters exposed to a range of solvents whereas the UK does not recognise such a disease. Acute effects of short term high level exposures may be simply identified but the longer term effects of such exposures and the effects of long term low level exposure may prove much harder to measure.

Inhalation and ingestion provide the major route for metal toxicity and skin absorption is a key route for solvent exposure with inhalation. (Tabe IX)

Table IX Substances found in the workplace reported to have caused neurological damage

Substance	Effects	Occupation
Acrylamide	Acute hepatocellular injury encepha lopathy, peripheral neuropathy	
Arsenic	Peripheral neuropathy	Metal production, pesticides
Benzene		
Carbamates	Cholinesterase effects and central nervous system effects	Pesticide manufacture, agriculture
Carbon disulphide	Encephalopathy, axonal neuropathy, acute psychosis	Rayon and cellophane industries, soil fumigant and laquers
Carbon monoxide	Encephalopathy, peripheral neuropathy, Parkinsonism	Transport fumes, blast furnace operations, space heaters and oil and gas burners
2, 4-D	Peripheral neuropathy	Pesticide manufacture, agriculture
Ethylene oxide	Encephalopathy, peripheral neuropathy	
Lead	Encephalopathy and peripheral neuropathy	General
Manganese	Encephalopathy, ataxia, later Parkinson disease-like Symptoms occur, acute psychosis	Engineering, aircraft industry, steel, aluminium, magnesium and cast iron production
Methyl *n*-butyl Ketone (MBK)	Peripheral neuropathy, fatigue	Textiles
Methyl chloride	Ataxic gait	Chemicals, foaming agent, solvent
Methyl ethyl ketone	Neuropathy	Solvents, painting, laquer, printing, glues, resins
Mercury	Tremor, weakness, peripheral neuropathy is uncommon; chronic exposure leads to ataxia, mental impairment	Chemicals, pharmaceuticals, dentistry, plastics, paper, various
Nickel	Headache	Engineering
Nitrous oxide	Encephalopathy, Peripheral neuropathy	Health sector, dentistry
Organophosphates	Acute and chronic effects neuropathy, delayed neuropathy signs include inhibition of acetylcholinesterase	Agriculture
Perchloroethylene	Neuropsychiatric effects, sensory and motor cranial neuropathies	Chemicals, dry cleaning, engineering
Styrene	Neuropsychological effects	Plastics, fibreglass
Thallium	Encephalopathy, ataxia (high doses)	
Tin: organic	Encephalopathy	
Toluene	Abnormal cognitive functioning, optic atrophy, ataxic gait, neuropsychiatric effects, sensor and motor cranial neuropathies	Engineering, labs, sports goods, paints and glues
Trichloroethylene	Neuropathy, encephalopathy	Solvent, adhesives, paint removers, dry cleaners, degreaser in engineering
Vinyl chloride	Encephalopathy, neuropsychological effects, neuropathy	
Xylene	Neuropsychological effects	Solvents, chemical production, paints, laquers and adhesives

(*Sources*: Bleecker in Bleecker[16], pp. 207–233; Estrin and Parry in LaDou[17]; and Baker and Schottenfield in Levy and Wegman[1], p. 521).

Table X Substances known or associated with visual damage in the workplace

Substance	Effect	Occupation/industry
Triethylamine	Visual disturbances	Foundry industry
Lead	Optical neuropathy	Metals, paints
Carbon monoxide	Disc swelling	Transport, indoor exposures
Organic solvents	Loss of colour vision	Chemicals, painters, micro-electronics, printers, ink manufacturers, other
Styrene	Loss of colour vision	Plastics, fibreglass, chemicals
Carbon disulphide	Abnormal pupillary reactions	Rayon industry, cellophane
Organophosphates	Blurred vision and retrobulbar optic neuropathy, fundus abnormalities	Agriculture and chemical production
Magnesium	Abnormal pupillary reactions	Metals, engineering
Toluene	Abnormal pupillary reactions, retro-chiasmal disease	Paints, solvents, rubber industry, other
Mercury	Cranial nerve palsies	Chemicals
Methyl chloride	Diplopia	Chemicals
Trichloroethylene	Disc swelling	Solvents, paints, degreaser in engineering, dry cleaning

(*Sources*: Mergler in Bleecker[16], p. 172; and Striph and Miller in Bleecker[16], pp. 172–186).

Some substances in the workplace may also prove neurotoxic to the visual system and in the mid 1980s over 30 chemicals in the US were identified as having adverse effects on this system. Effects of such exposures can include visual disturbances such as blurred vision, double vision, changes in colour perception, dimmed vision, visual hallucinations, and occasional partial or complete loss of vision[16]. (Table X)

IMMUNOTOXICITY

In 1700 the founder of occupational medicine, Ramazzini, noted that repeated exposures to flour dust led to respiratory diseases. He did not understand the mechanisms of the disease which later were revealed to be immune hypersensitivity[16]. Immunotoxicity was a relatively neglected subject in occupational toxicology until the 1980s and 1990s although a few studies of chemical workers in the 1970s did identify some adverse immunological effects. This is distinct from the larger number of studies of animals which have been carried out and it does raise the question of the validity of extrapolating animal data to humans. Again this reveals the value of a precautionary policy in dealing with occupational hazards because toxicological fields may have either not existed or have had incomplete data or unconfirmed toxicity tests and mechanisms with which to assess risk to workers of such hazards.

The immune system acts to protect the body from tumours and pathogens through interactions of "white blood cells and blood serum molecules. Three distinct immunity mechanisms work in concert to recognize and eliminate bacteria, fungi, viruses and cancer cells"[18]. Primary and secondary responses also occur some over years.

Table XI Substances reported to have caused immune system effects in the workplace

Substance	Effect	Industry/occupation
Asbestos	Immunosuppression	Building, general
Benzene	Immunosuppression	Chemicals, general
Cotton, jute, sisal	Hypersensitivity (byssinosis)	Textiles, rope makers
Cyclophosphamide	Carcinogenicity	Health, lab workers
Diisocyanates	Hypersensitivity	Printing, paints, other
Epoxy resins	Autoimmunity	Garages, engineering, chemicals
Nickel	Hypersensitivity	Metals Engineering
Trichloroethylene	Autoimmunity	Chemicals, solvents

(*Sources*: German in LaDou[17], pp. 149; Stacey[2], pp. 64; Rosenstock and Cullen[21]).

Macrophages, neutrophils and natural killer cells take out foreign or cancerous cells. This is the response of the non-specific immunity mechanism. B Cells and antibody and complement proteins bind to antigens and hence neutralise such foreign cells. This is called humoral immunity. The white blood cells, the T cells, can also bind to the antigens and become involved in a process whereby macrophages destroy the foreign organisms. A final mechanism involving cytotoxic T-cells can destroy viruses[18].

Between the 1970s and 1990s research papers have for instance documented the immunotoxicity of a large number of pesticide including those in the organochlorine, organophosphate, carbamate and metals groups. Epidemiological studies provide further evidence that agricultural workers exposed to pesticides may experience higher rates of diseases which can be associated with damage to the immune system and a number of studies have demonstrated suppression of human T-cell counts and functions after exposure to pesticides. Farmers and farm workers, because of the healthy worker effect and often lower rates of smoking, generally show lower rates of heart disease and lung cancer than the wider population. However, several cancers which occur in immuno-deficient patients also appear elevated for farm workers[18].

Adverse effects from occupational immunotoxic substances could range from colds through to cancers and allergic skin and respiratory problems[2]. More specific causes of occupational asthma due to immune responses are listed in the next section of this chapter. (Table XI.)

RESPIRATORY SENSITISERS, ALLERGENS AND CARCINOGENS

Again, this is a complex field with exposures resulting from home and work environments and work impacts affected by non-work exposures and *vice versa*. Smoking may affect lung function and produce carcinogens but this appears to have been an excuse for inaction in some workplaces against specific occupational carcinogens and respiratory sensitisers. In some instances, too, many decades elapsed before toxicological and other scientific literature indicating potential human hazards was acted upon in terms of prescribing occupational diseases. Occupational bronchitis in coal miners and respiratory damage from welding fumes are two such examples.

Fitness, age and lung size may also influence, to a greater or lesser extent, the health effects of occupational respiratory exposures. It is important to note, however, that exposures determine effects. Although there may be differences in individual susceptibility to substances which are inhaled, the problem is created in the first instance by the inhalation and not the individual susceptibility. This point is central to removal and control strategies which may flow from toxicological research. Some strategies for dealing with respiratory sensitisers have led to policies which screen out either likely susceptible new employees or those employees adversely affected in the workplace by respiratory sensitisers. The first response to toxicological data which indicate sensitisation should be to remove the substance and only further down the line should screening of workers be considered.

Occupational asthma presents a major problem for workers. Two main types of occupational asthma have been identified[3]:

1. Allergic occupational asthma which occurs where a worker who may have used a substance for years or days without apparent ill-effects suddenly develops a violent reaction to a chemical. This is sensitisation affecting the immune system and once sensitised, a person is sensitised for life, and will react to levels of the chemical many times below which other harmful effects occur. It should be noted that exposure to chemicals and not individual sensitisation to them is of course the cause of occupational asthma and therein lies the solution. The individual sensitisation is a manifestation of the effect.

2. Non-allergic occupational asthma occurs after exposure to workplace irritants and produces chest tightness similar to symptoms caused by 1. This differs from allergic occupational asthma because no immune system effects occur and later very low exposures to the irritant may not trigger asthmatic attacks. Lung damage done by irritant chemicals can make people more vulnerable to potential sensitisers. Factors influencing both the cause and severity of occupational asthma may include exposure to an irritant or sensitiser or both, the dose, the length of exposure and the effect. The effect relates to the degree of disability caused and the biological or non-biological response; the immuno or non-immunological response. Since 1985, the term 'reactive airways dysfunction syndrome' (RADS) has been coined to describe the respiratory damage done by toxic inhalations.

Most physicians who deal with respiratory diseases accept that thousands of substances in the workplace will, through inhalation, cause occupational asthma. Even in 1980, at least 2000 agents were reported to cause asthma in the working environment. Asthma is the most commonly reported occupational lung disease in the UK. Diisocyanates are by far the major cause of recorded occupational asthma cases in most countries. No one knows exactly what the overall prevalence of occupational asthma is in the British population. Estimates in Japan are that 15% of asthma cases in adult male workers have occupation causes and 2% of all cases in the USA. In the UK, estimates indicate that between 5% and 25% of those exposed to respiratory sensitisers will be sensitised and, in certain instances, some sensitisers will cause almost 100% sensitisation in exposed workers. The 1990 Labour Force survey in the UK, analysed by the HSE, etimates that 20,000 people believe they have occupational asthma and 50,000 more believe their asthma is made worse by work[3].

Table XII Reported respiratory effects of certain workplace substances

Substance	Effect	Industry
Amoebae	Humidifier fever	Offices, shops, factories
Animals	Asthma	Laboratories, vets, agriculture
Barley (mouldy)	Extrinsic allergic alveolitis 'malt worker's lung'	Distilleries
Coal	Pneumoconiosis	Mining
Colophony	Asthma	Soldering
Cotton, flax, hemp, sisal	Byssinosis	Textiles
Drugs	Asthma	Pharmaceutical industry
Enzymes	Asthma	Pharmaceutical industry
Formalin/ formaldehyde	Asthma	Scientific, technical, textiles
Grains and other crops	Asthma	Agriculture, food
Hay (mouldy)	Extrinsic allergic alveolitis 'farmers lung'	Agriculture
Isocyanates	Asthma	Printing, manufacture
Silica and quartz	Silicosis	Mining, sand blasting, foundry work, ceramics, stone masons
Sugar cane (mouldy)	Extrinsic allergic alveolitis 'bagassosis'	Plantation workers
Woods	Asthma	Timber and furniture trades
Asbestos	Asbestosis and mesothelioma	Textiles, construction and maintenance, engineering
Hops	Occupational asthma	Brewing
Glutaraldehyde	Occupational asthma	Health and lab workers
Metals (especially platinum, nickel, chromium, cobalt, vanadium and tungsten carbide)	Occupational asthma	Metal and engineering workers

(*Sources*: Parkes[27]; Christiani and Wegman in Levy and Wegman[1]; Sheppard *et al.* in LaDou[17]).

A range of substances may damage lung function and some may be carcinogens in the respiratory system. (Table XII)

OCCUPATIONAL TOXICITY TO THE SKIN

The skin is a major route of entry for substances into the body. Those substances which are percutaneously absorbed and may then affect other organs and systems in the body include the aromatic nitro and amino compounds, hydrocarbons, metals, pesticides and pharmaceutical products. Some major examples of chemicals which are absorbed through the skin are phenol which causes blood disorders; aromatic amines, paraquat and petrol linked to renal toxicity; and MBK and other solvents, lindane and organophosphate pesticides causing neurological damage; and aromatic amines and carbon tetrachloride causing gastro-intestinal disorders[18].

The epidermis provides most of the skin's resistance to chemical unlike the dermis and the subcutaneous fat. Occupational agents may cause skin cancer—sunlight and coal tar for instance act synergistically to cause an elevated risk of skin cancer in roofers—or non life threatening but nevertheless painful and socially disabling diseases. These would include chloracne, and primary irritant and allergic contact dermatitis.

Substances known to cause contact dermatitis include cement, fibre glass, ethylene oxide and hydrofluoric acid. Hairdressers may be at risk from repeated exposures to irritant chemicals. Common airborne irritants in a work setting include both acids and alkalis, aluminium, cement, cleaning agents, epoxy resins, formaldehyde, plant particles, plastics particles, solvents, stone particles, wood dust and wool[19]. (Table XIII)

Table XIII Substances known to be absorbed through or damaging to the skin in the workplace

Substance	Type of skin damage/ penetration	Industry/occupation
Aliphatic hydrocarbons	Percutaneous absorption	Chemicals, engineering, etc.
Animals	For example orf (sheep)	Farmers, vets and others
Arsenic	Skin cancer	Agriculture, lead workers, dyers, copper smelters, brass makers, chemicals, textiles, painters, pesticide users
Benzidine	Percutaneous absorption	Chemicals
Chemical agents (including soaps, acids, alkalis, oils)	Skin disorders	Chemical and a wide range of other industries and occupations
Chromium	Allergic contact dermatitis	Metal and engineering workers
Citrus fruit	Skin disorders	Agriculture and horticulture
Creosote	Skin cancer	Timber and brick workers
Dioxin	Chloracne	Chemicals, agriculture as pesticide contaminant
Epoxy resins	Allergic contact dermatitis	Various including painters, assembly workers in car, aircraft industries, construction workers, painters
Food (apples, fish, potatoes, beans, eggs, carrots, flour, meats, milk, peaches, spices)	Allergic contact urticaria	Agriculture and food processors and handlers
Formaldehyde	Allergic contact dermatitis	Model and mould makers, resin workers, embalmers, textile and insulation workers, wood workers
Glass dust	Skin disorders	Glass industry and trades
Glutaraldehyde	Allergic contact dermatitis	Health workers
Metals (many)	Percutaneous absorption	Engineering and chemicals
Nickel	Allergic contact dermatitis	Metals, engineering
Organic solvents	De-fatters	General
PCBs and PBBs	Chloracne	Electrical transformers and other
Penicillin	Allergic contact dermatitis	Health workers
Pesticides (many)	Percutaneous absorption	Agriculture and agro-chemical manufacture
Polycyclic aromatic hydrocarbons	Skin cancer	Soot and carbon black workers, coal tar, pitch users and oil refining and distillation workers
Thiuram	Allergic contact dermatitis	Rubber workers
Xylene	Contact urticaria	Solvents

(*Sources*: Fisher in Adams[19]; Zugerman in Adams[19], pp. 127–157; 113–126; Nethercott in LaDou[17], pp. 209–220).

CARDIOVASCULAR TOXIC AGENTS

These have been a much neglected group of substances to which workers may be exposed with adverse effects. Research on their effects is far from complete. Again, a range of influences in terms of age, blood pressure, diet, fitness, smoking, weight may influence the impact of occupational exposures. Blood pressure which may cause cardiovascular damage can result from exposure to substances which damage the kidneys. These would include occupational exposure to metals such as cadmium, gold, iron, lead; solvents such as carbon tetrachloride and toluene, pesticides such as paraquat, pentachlorophenol and certain chlorinated hydrocarbon insecticides. Cadmium may also cause emphysema which will create some cardiovascular problems.

The acute toxicity of high levels of carbon monoxide has often been described but the toxicity of chronic low level exposures to substances may be considerable. Methylene chloride produces carbon monoxide as a metabolite and so may have a 'knock on' effect. Epidemiological and toxicological studies have indicated the cardiovascular toxicity of the following substances. (Table XIV)

TOXICITY TO THE BLOOD

"The hematologic system is a primary endpoint of effect for a variety of occupational health problems. It is also a conduit of unwanted material to other organ systems and often an early indicator of important effects in other tissues"[1]. Adverse effects may include one or several diseases such as leukaemia, aplastic anaemia, coagulation, metabolism, denaturation of haemoglobin and hemolysis.

Workplace chemicals were found to present hazards to the blood in factory workers even in the 1800s. (Table XV)

Table XIV Substances linked with cardiovascular toxicity

Substance	Effects
Arsine	Cardiac arrhthymia, peripheral arterial occlusive disease
Arsenic	Myocardial injury
Antimony	Hypertension
Cadmium	Hypertension
Carbon disulphide	Atherosclerosis
Carbon monoxide	Atherosclerosis
Chlorofluorocarbons	Cardiac arrhthymia
Cobalt	Myocardial injury
Lead	Myocardial injury hypertension
Organic nitrates	Nonatheromatous ischemic heart disease
Organophosphate pesticides	Cardiac arrhythmia
Trichloroethylene	Cardiac arrhythmia
1,1,1-trichloroethane	Cardiac arrhythmia

(*Source*: Benowitz[17] and Theriault in Levy and Wegman[1], p. 565).

Table XV Reported adverse effects of chemicals on the blood

Chemical	Disease	Industry/occupation
Aluminium	Porphyria	Phosphorous binder
Aniline dyes	Red blood cells	Dyeing and chemical industry
Arsenic	Aplastic anaemia	Glass, paints, enamels, pesticides, tanning agents
Arsine	Red blood cells	Electronics, industrial reaction
Benzene	Aplastic anaemia and acute myeloid leukaemia	Chemicals, labs, coke byproduct plant, rubber manufacture
Carbon monoxide	Red blood cells	Various
Chemotherapy agents	Aplastic anaemia	Pharmaceutical and health workers
Copper	Red blood cells	Engineering
Ethylene glycol monomethyl or monobutyl ether	Aplastic anaemia	Cleaning agent, dyes, inks, paints and laquer production
Ethylene oxide	Lymphoma	Chemical production workers
Hexachlorobenzene	Porphyria	Pesticide
Lead	Red blood cells, porphyria	General
Lindane	Aplastic anaemia, toxic thrombocytopenia	Agriculture
Nitrates and nitrites (inorganic)	Red blood cells	Chemicals
Pentachlorophenol	Aplastic anaemia	Pesticide manufacture and use
Vinyl chloride	Porphyria	Plastics production
Radiation	Aplastic anaemia	Various

(*Sources*: Rugo and Damon in LaDou[17], pp. 155–169; Goldstein and Kipen in Levy and Wegman[1], pp. 575–590).

CONCLUSIONS

The need for more research on the toxicology of substances must be linked, in the context of regulatory toxicology and occupational health policy and practice, to the fact that there will always be insufficient resources, staff, time and money to test all substances fully. Hence research needs must operate in parallel with the international adoption of precautionary policies and standards which build in the greatest toxicity safety margins for substances in the workplace. There is still some considerable way to go before this point is reached. There should also be a full assessment of the need to develop such substances only if they bring large and clear benefits to society.

Toxicology has helped to identify major hazards to workers. Occupational toxicology has also shown that in some critical instances, it has been incapable of identifying hazards with current techniques and models or has large data gaps. This is why occupational toxicology should underpin the development of public health precautionary policies by acknowledging what it knows, what it doesn't know and what are its limits.

FURTHER READING

The classic text book on toxicology is Cassarett and Doull's toxicology: the basic science of poisons[20]. However, this is not particularly accessible and does not focus exclusively on occupational toxicology. The best one volume paperback guide to the subject comes in Neil Stacey's edited volume, Occupational Toxicology. A number of single volume text books on Occupational Health and Occupational Medicine also provide excellent chapters on various toxicological topics: these texts include Levy and Wegman, LaDou, Rosenstock and Cullen[21] listed below. Sax's Dangerous Properties of Industrial Materials[22] provide the most extensive summary toxicological data on the widest range of industrial chemicals for health and safety managers. G. D. Clayton's Patty[23] listed below provides the most comprehensive guide to relatively recent toxicological thinking and a large number of substances. Williams[24] provides some context from the 1980s for very recent developments. Geiser has produced the best work on toxics reduction[25,26]. Specific texts cover such topics as occupational respiratory diseases[27] and occupational skin diseases[19] in great detail.

The most up to date information on toxicology is, however, provided in the scientific journals such as Archives of Environmental Health, Journal of Toxicology and Applied Pharmacology, Toxicology, Scandinavian Journal of Work, Environment and Health, American Journal of Industrial Medicine and Occupational and Environmental Medicine.

References

1. B. S. Levy and D. H. Wegman (eds.), *Occupational Health: recognizing and preventing work-related disease* (Little, Brown and Company, Boston, 1995).
2. N. H. Stacey (ed.), *Occupational Toxicology* (Taylor and Francis, London, 1993).
3. R. O'Neill, *Asthma at Work* (TUC/Sheffield Occupational Health Project, Sheffield, 1995).
4. N. Krause, T. Malmfors and P. Slovic, "Intuitive Toxicology: Expert and Lay Judgements of Chemical Risk" *Risk Analysis* **12**, 215–232 (1992).
5. B. I. Castleman and G. E. Ziem, "Corporate Influence on Threshold Limit Values" *Amer. J. Industrial Medicine* **13**, 531–559 (1988).
6. World Health Organisation, *Health Hazards of the Human Environment* (WHO, Geneva, 1972).
7. Mark C. Smith, D. C. Christiani and K. T. Kelsey, *Chemical Risk Assessment and Occupational Health: Current Applications, Limitations and Future Prospects* (Auburn House, Westport, Connecticut, 1994); L. S. Gold, G. B. Garfinkel and T. H. Slone, *Setting Priorities among Possible Carcinogenic Hazards in the Workplace* (in above).
8. WHIN No. 45/46, *Health Based Chemical Exposure Guidelines*, 20–21 (1995/96).
9. A. E. Watterson, "Pesticide reproductive health hazards in humans and public health policy options: some unanswered questions and undocumented answers arising from the benomyl debate" *J. Public Health Medicine* **16**, 141–144 (1994).
10. A. Hamilton and H. L. Hardy, *Industrial Toxicology*, 2nd edition, revised and enlarged (Paul Hoeber Inc, New York, 1949).
11. R. Gottlieb (ed.), *Reducing Toxics: a new approach to policy and industrial decision-making* (Pollution Prevention Education and Research Center, Island Press, Washington, 1995).
12. K. Steenland, D. Loomis, C. Shy and N. Simonsen, "Review of Occupational lung carcinogens" *Am. J. Indust. Med.* **29**, 474–490 (1996).
13. IARC Monographs on the Evaluation of Carcinogenic Risks to Humans. List of IARC Evaluations (IARC, Lyon, 1990)
14. T. H. Shepherd (ed.), *Catalog of Teratogenic Effects*, 7th ed. (Johns Hopkins University, Baltimore, 1992).
15. M. Paul (ed.), *Occupational and Environmental Reproductive Hazards* (Williams and Wilkins, Baltimore, 1993); P. K. Working and D. R. Mattison, *Reproductive and Developmental Toxicity Testing* (in above).
16. M. L. Bleecker (ed.), *Occupational Neurology and Clinical Toxicology* (Williams and Wilkins, Baltimore, 1994).
17. J. LaDou (ed.), *Occupational Medicine* (Prentice Hall, Connecticut, International Agency for Research on Cancer, 1990).
18. R. Repetto and S. S. Baliga, *Pesticides and the Immune System: the Public Health Risks* (World Resources Institute, Washington, 1996).
19. R. M. Adams (ed.), *Occupational Skin Diseases*, 2nd ed. (W. B. Saunders, Philadelphia, 1990).
20. J. Doull and C. D. Klassen (eds.), *Casarett and Doull's Toxicology: the basic science of poisons* (McGraw, New York, 5th ed., 1996).

21. L. Rosenstock and M. Cullen (eds.), *Textbook of Clinical Occupational and Environmental Medicine* (W. B. Saunders, Philadelphia, 1994).
22. N. I. Sax, *Dangerous Properties of Industrial Materials*, 9th ed. (Van Nostrand, New York, 1995).
23. G. D. Clayton *et al., Patty's Industrial Hygiene and Toxicology*, 3rd and 4th ed. **volumes 1–3** (John Wiley & Sons, New York, 1993/94).
24. P. L. Williams and J. L. Burson (eds.), *Industrial Toxicology* (Van Nostrand, New York, 1985).
25. K. Geiser, "The Greening of Industry: Making the Transition to a Sustainable Economy" *Technology Review* 66–72 (1991).
26. K. Geiser, "Protecting Reproductive Health and the environment: Toxics Use Reduction" *Environmental Health Perspectives Supplements* **101**, 221–225 (1993).
27. W. R. Parkes (ed.), *Occupational Lung Disorders*, 3rd ed. (Butterworth-Heinemann, Oxford, Reinhold, New York, 1994).

13. FUNGAL TOXINS AS DISEASE ELICITORS

J. P. F. D'MELLO* and A. M. C. MACDONALD*

Fungi synthesise a diverse array of secondary metabolites, some of which are known to be toxic to plants, animals and humans. Certain species of *Cochliobolus*, *Alternaria*, *Pyrenophora*, *Septoria* and *Periconia* produce metabolites which have been definitively linked with specific plant diseases and are, consequently, regarded as host-specific toxins. Other fungi, including particular species of *Rynchosporium*, *Fusarium*, *Alternaria*, *Fusicoccum* and *Exserohilum* synthesise non-specific toxins, some of which may act as virulence factors in plant disease. Many plant pathogenic species of *Aspergillus*, *Penicillium*, *Fusarium* and *Alternaria* also produce metabolites with the potential to precipitate disease in animals and humans. These substances are known as mycotoxins. Following infection with these fungi, plant products such as seeds and fruit regularly become contaminated with mycotoxins which may then directly or indirectly enter the human food chain. Forages infected with particular species of *Acremonium*, *Phomopsis* and *Pithomyces* may also contain mycotoxins capable of causing disease in ruminant animals. Mycotoxins commonly occur in the spores of fungi, including those emanating from *Stachybotrys* and *Alternaria* and inhalation, therefore, represents another route of entry into the body. Although mycotoxins have been implicated in many animal and human disorders, it is their carcinogenic potential which has evoked global concern. In addition, recent findings linking mycotoxins with neurotoxic, hepatotoxic and immunosuppressive effects have provided the impetus for continued monitoring and research. Some mycotoxins are associated with profound reproductive disorders in animals but the implications for human fertility need to be addressed. Mycotoxins may be regarded as unavoidable contaminants of food and feed even when good farming and manufacturing practices have been implemented. Regulation of maximum permissible levels is effected by a combination of legislative and advisory measures and is an area of continuing activity. Recent field trials have failed to provide a consensus regarding the efficacy of fungicides to control mycotoxin production. In addition, the development of fungicide-resistant strains of *Fusarium* is causing some concern with respect to potential changes in the pattern and levels of mycotoxin production. It would appear, therefore, that attempts to exploit disease-resistant plant genotypes represent a more promising strategy for further reductions in mycotoxin contamination of primary food commodities.

INTRODUCTION

Secondary metabolism in fungi results in the synthesis of a diverse array of macromolecules, some of which are endowed with toxic properties towards other organisms. Those fungal metabolites which have been definitively implicated in plant disease are conventionally referred to as host-specific or host-selective toxins

* Department of Crop Science and Technology, The Scottish Agricultural College, West Mains Road, Edinburgh EH9 3JG (UK).

(HSTs). The notion that plant pathogenic fungi may produce molecules that are toxic to their respective host plants has a considerable history, but it is only relatively recently that these compounds have been identified and characterized. As a consequence, the list of host-selective toxins has grown significantly in recent years and collation of data relating to these substances now appears to be opportune. Other fungal metabolites are endowed with phytotoxic properties but their role in plant disease remains to be determined. These compounds are designated as non-selective toxins and are characterized by their diversity of origin and chemical structure. Fungi also produce secondary compounds with the potential to elicit disease in animals and humans. These metabolites are known as mycotoxins, while the deleterious effects they precipitate are referred to as mycotoxicoses. The global occurrence of mycotoxins in primary raw materials, animal feeds and food products is a matter of considerable concern, since a number of these contaminants are attributed with carcinogenic properties. In addition, evidence of co-occurrence and synergism among these mycotoxins is emerging which offers a different perspective to the interpretation of well-known cases of mycotoxicoses. This review will focus on recent developments on the nature and effects of HSTs, non-selective toxins and mycotoxins since standard texts[1–6] are available to provide background reading, including details of chemical structures and properties.

HOST-SPECIFIC PHYTOTOXINS

Stringent criteria must be fulfilled before fungal metabolites can formally be classified as HSTs[1,2]. The pre-requisites are: (i) quantitative correlation between toxin production and pathogenicity of the fungus; (ii) quantitative correlation between host sensitivity to the toxin and susceptibility to the fungal pathogen; (iii) similarities in gross and histological manifestations induced by the toxin and fungal pathogen; (iv) toxin release from the germinating spores at invasion sites; (v) synthesis of toxin-induced receptors and/or novel mediating proteins in the host plant.

As shown in Table I, a wide range of HSTs have been identified and characterized in terms of biological activity. The diversity of secondary metabolism in fungi is reflected in differences in chemistry and disease effects of HSTs. Thus, known HSTs include cyclic peptides, terpenoids, oligosaccharides and polyketides, while other HSTs are currently of undefined chemistry. A number of HSTs from *Alternaria* are structurally related to each other and occur in closely related forms, and in addition, one of these (AAL-toxin) is structurally analogous with the fumonisin mycotoxins, produced by *Fusarium moniliforme*. Despite the diversity of chemical structure, there appears to be some uniformity in their effects at tissue level (Table I). Thus, membrane damage and leakage of electrolytes are common effects in susceptible host plants.

Although it is conventional to associate HST production with three species of *Cochliobolus (Helminthosporium)* and four species of *Alternaria* (Table I), there is good evidence that other plant pathogenic fungi also exert their effects through the synthesis of such compounds. Thus the purification and characterisation of a host-selective necrosis toxin from *Pyrenophora tritici-repentis* has recently been reported.

1. *Cochliobolus HSTs*

Three well-established *Cochliobolus* HSTs have been characterized[7] in respect of specificity and virulence: victorin, T-toxin and HC-toxin, synthesised respectively, by

Table I Host-specific toxins (HSTs) produced by fungal phytopathogens[#]

Fungal pathogen	Disease	Host plant	HST	Effects*
Cochliobolus victoriae	Victoria blight	Oat	Victorin	Depolarization of the plasma membrane potential; leakage of electrolytes; lysis of protoplasts; extra-cellular synthesis of callose
Cochliobolus heterostrophus race T	Southern corn leaf blight	Maize	T-toxin	Disruption of mito-chondrial function
Cochliobolus carbonum race 1	Leaf spot disease	Maize	HC-toxin	Inhibition of root growth; stimulation of ion uptake; agent of 'induced susceptibility'
Alternaria alternata	Alternaria blotch of apple	Apple	AM-toxin	Detachment and vesiculation of grana lamellae; inhibition of photosynthetic CO_2 fixation; effects are tissue-specific
	Black spot of Japanese pear	Japanese pear	AK-toxin	Increased loss of electrolytes; depolarization of membrane potential; ultrastructural changes; callose production
	Black spot of strawberry	Strawberry	AF-toxin	Activation of phospholipase A_2; phospholipid degradation in plasma membrane
	Brown spot of tangerine	Tangerine; mandarin	ACT-toxin	Vein necrosis; electrolyte loss from leaves; plasma-membrane invaginations
	Brown spot of rough lemon	Rough lemon	ACR(L)-toxin	Uncoupling of oxidative phosphorylation
	Brown spot of tobacco	Tobacco	AT-toxin	Ultrastructural changes in mitochondria
	Alternaria stem canker of tomato	Tomato	A(A)L-toxin	Cellular necrosis; ultrastructural changes in mitochondria
	Herbicidal activity	Spotted knapweed	Maculosin	Necrotic lesions; destruction of foliage
Alternaria brassicae	Gray leaf spot	*Brassica* species	Destruxin B	Severe chlorosis; necrosis
Alternaria brassicicola	Black leaf spot	*Brassica* species	AB-toxin	Black necrotic lesions

Table I (*Continued*)

Fungal pathogen	Disease	Host plant	HST	Effects *
Alternaria tenuissima	Leaf spot	Pigeon pea	ATC-toxin	Necrosis
Pyrenophora tritici-repentis	Tan spot	Wheat	Ptr necrosis toxin	Necrosis; chlorosis; plasmalemma damage; electrolyte leakage
Septoria glycines	Brown spot	Soyabean	Uronic acid-rich 20 kDa polysaccharide	Chlorotic lesions; premature defoliation
Periconia circinata	Periconia root rot	Sorghum	Peritoxins	Inhibits mitosis and growth of primary roots; induces electrolyte leakage; at low concentrations, peritoxins selectively reproduce disease symptoms

Based on references reviewed in text. * In susceptible cells, cultivars or lines.

C. victoriae, C. heterostrophus race T and *C. carbonum* race 1. Victorin production by the fungus is associated with Victoria blight of oats. This HST is phytotoxic only to oat varieties that are susceptible to *C. victoriae*. In sensitive cultivars of oats, victorin is highly potent, severely inhibiting root growth at 50pg/ml. In structural terms, victorin represents a group of closely related partially cyclic pentapeptides in which victorin C appears to be the dominant form. The latter compound consists of glyoxylic acid, 5,5-dichloroleucine, *erythro-β*-hydroxyleucine, victalanine, *threo-β*-hydroxylysine and α-amino-β-chloroacrylic acid in a partially cyclic peptide. The presence of the glyoxylic residue appears to be critical for both recognition of the active site and for induction of toxicity. Additionally, if the aldehyde group of the glyoxylic moeity is reduced, the resulting product is not toxic. Recent evidence[8] suggests that the glycine decarboxylase complex is a component of the site of action of victorin. The toxin binds in a ligand-specific manner to at least two enzymes of the complex and in a genotype-specific mode to the P-protein of the complex. Victorin is an effective inhibitor of the enzyme *in vitro* and *in vivo*. In addition, structural components of victorin act as precursors or substrate for the complex. The toxin also affects another component of the photorespiratory cycle, the enzyme Rubisco.

T-toxin comprises a family of linear polyketols produced by *C. heterostrophus*, the fungus responsible for leaf blight of maize[7]. Four major components of this toxin have been described. The structural requirements for toxicity are, however, not stringent, with a variety of synthetic analogues possessing host-selective activity. All the observed physiological effects of T-toxin have been explained by a mitochondrial site of action.

HC-toxin is a cyclic tetrapeptide synthesised by *C. carbonum*, a fungal pathogen causing leaf spot disease of maize[7]. The structure is represented by cyclo(D-Pro-L-Ala-D-Ala-L-Aeo), with Aeo being 2-amino-9, 10-epoxy-8-oxodecanoic acid. Minor forms, constituting less than 5% of the HC-toxin have also been identified and characterized from culture filtrates. Although the mode and site of action of HC-toxin remains obscure, it is endowed with physiological effects in plants (Table I).

2. *Alternaria HSTs*

Definitive pathotypes in the collective species *A. alternata* are now recognised to exert their pathogenicity and virulence through the synthesis of HSTs (Table I), which may be released from spores prior to penetration, to bind with specific receptor sites within host cells and to neutralise defence responses[1,2]. These HSTs are low molecular weight compounds displaying considerable chemical diversity. However, three HSTs (AK-, AF- and ACT-toxins) produced by different pathotypes of *A. alternata* have a common 8-substituted-9,10-epoxy-9-methyldecatrienoic acid fragment, thereby conferring some similarities in pathogenicity and physiological effects. On the other hand, AM-toxin and Destruxin B are cyclic depsipeptides.

The actions of the *A. alternata* HSTs are directed at three principal sites: chloroplasts, plasma membranes and mitochondria (Table I)[1,2]. AM-toxin produced by *A. alternata* apple pathotype exists in three forms (I, II, III), the last two substances being derivatives of AM-toxin I. They act on the chloroplasts of susceptible cells. Marked tissue-specificity in the reaction to AM-toxin has been observed, with non-green tissues such as petals and white calli being insensitive whereas green calli induced from petals are sensitive to the toxin. AK-toxin I and II produced by the pathotype causing black spot disease of Japanese pear are both toxic to susceptible and harmless to resistant cultivars. AK-toxin mediates its toxicity via the plasma membranes of cells, causing diverse physiological and ultrastructural changes (Table I). Three forms of AF-toxin (I, II, III) have been characterized from the pathotype of *A. alternata* causing black spot of strawberry, with AF-toxin I also possessing activity against susceptible Japanese pear cultivars and showing the same host range as that of AK-toxin I, consistent with similarities in structure. The tangerine pathotype of *A. alternata* responsible for brown spot disease of mandarin and tangerine produces two types of HSTs designated ACT-toxin Ib and IIb. However, the latter form is also toxic to Japanese pear cultivars sensitive to AK-toxin. Structural analogy may also explain why AF- and ACT-toxins act at the plasma membrane level (Table I). Plasma membrane modification is a common feature of the action of *Alternaria* HSTs, facilitating pathogen entry and expression of necrotic symptoms in susceptible plants. This also applies to the ACR(L)-toxin produced by the pathotype causing brown spot disease of rough lemon. Specifically, the toxin induces a swelling of the mitochondria, reduction in numbers and vesiculation of cristae, with accompanying biochemical changes (Table I). Unifying features in the proposed actions of AK-, ACR- and AM-toxins have been reviewed recently and are summarized in Fig. 1.

Ultrastructural derangements have also been observed with AT- and A(A)L-toxins[1], giving substance to the notion that these and the other *Alternaria* HSTs listed in Table I may have underlying mechanisms similar to those illustrated in Fig. 1. The A(A)L-toxins produced by *A. alternata* f. sp. *lycopersici* deserve particular comment in view of structural similarities with the fumonisin mycotoxins (FB_1, FB_2 and FB_3) synthesised by *Fusarium moniliforme*[9–11]. The former are tricarballylic esters of a series of long-chain aminopolyols structurally related to sphinganine. Five major fractions have been identified in the group of A(A)L-toxins, designated TA, TB, TC, TD and TE, each existing as a pair of regioisomers. TA- and FB_1-induced cell death in plant and animal systems bear stereotypical

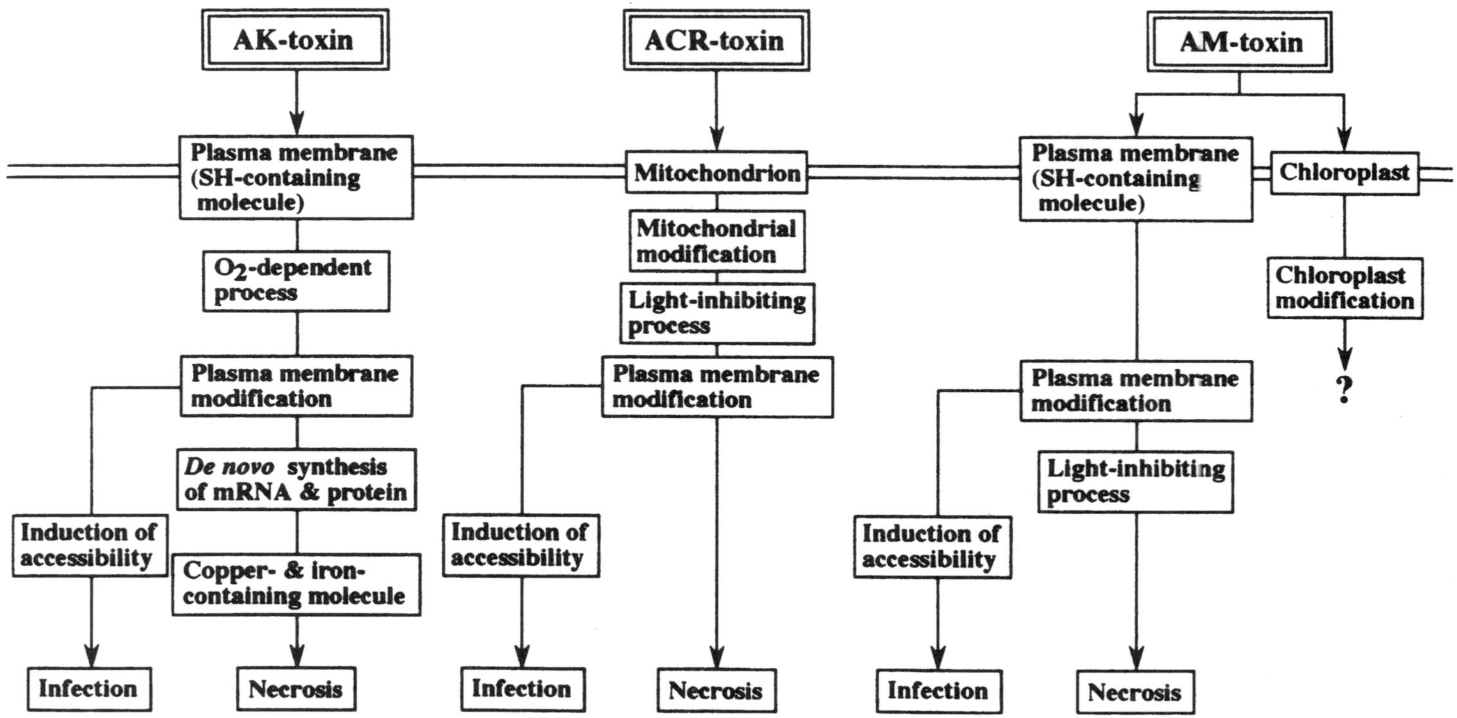

Figure 1 Summary scheme to illustrate actions of AK-, ACR- and AM-toxins in susceptible plants. Taken from Otari *et al.*, *Can. J. Bot.* **73**, S453–S458 (1995)[1].

characteristics of apoptosis. At the biochemical level, A(A)L-toxins and fumonisins have been shown to inhibit sphinganine (sphingosine) N-acyltransferase (ceramide synthase) in both animals and plants[9]. Maculosin is somewhat unique among the *A. alternata* HSTs in that although it is specific for spotted knapweed, its role is seen as a potential herbicide[12]. Its structure has been evaluated as a cyclodipeptide composed of proline and tyrosine. The bioactivity of a number of maculosin analogues is under investigation.

3. *Pyrenophora tritici-repentis HST*

Tan spot caused by *P. tritici-repentis* has emerged as a major leaf disease of wheat on a global scale. Wheat susceptibility to the pathogen is expressed by the development of tan necrosis or extensive chlorosis or a combination of these two features. Culture filtrates from the pathogen were found to contain HSTs in the form of non-dialysable and heat sensitive proteins with the capacity to induce symptoms only in the leaves of susceptible cultivars[13–15]. The HST causing necrosis has been designated Ptr necrosis toxin (Table I) and its secondary structure and amino acid composition and sequence are now documented. Sequence analysis suggests the possible presence of a membrane adhesion site and several phosphorylation sites that may confer elements of phytotoxicity.

4. *Other HSTs*

Septoria glycines Hemmi causes brown spot disease of soybeans (Table I), characterized by chlorotic lesions and premature defoliation[16]. Typical symptoms of this disease can be reproduced on cotyledons and leaves of the legume after application of the HST isolated from cultures of the fungus. The toxin has been partially characterised as a 20 kDa polysaccharide with high content of uronic acids and low levels of mannose, galactose and glucose.

Peritoxins are low molecular weight chlorinated peptides synthesised exclusively by pathotypes of the sorghum root rot fungus, *Periconia circinata*[17]. Two equally phytotoxic forms exist, designated peritoxins A and B. It is suggested that disease symptoms result from an interaction of the peritoxins with a protein receptor on or near the cell surface and disruption of the function of a signal transduction pathway.

NON-SELECTIVE TOXINS

It is neither feasible nor appropriate to review the vast literature on non-selective toxins and their phytotoxic properties. The subject is well documented[3,18] and, consequently, attention here will focus on a selection of recently characterized compounds (Table II). It should be noted that in all cases, direct association between toxin and disease aetiology remains to be established.

The three necrosis-inducing peptides (NIPs) isolated from *Rhynchosporium secalis* are typical examples of non-specific phytotoxins in that they cause necrosis in the primary leaves of both resistant and susceptible near-isogenic cultivars of barley[19]. NIPs have relative molecular masses of less than 10 kDa, two appearing to be non-glycosylated with the third being a glycopeptide. The toxic activity, however, resides in the peptide moeity.

Fusarium species are widely recognised for their production of non-selective phytotoxins arising from the diverse diseases caused by these fungi[18]. Common

Table II Non-selective toxins of fungi: collation of recent findings*

Fungal pathogen	Toxin	Effects of fungus [and toxin]
Rhynchosporium secalis	Necrosis-inducing peptides (NIPs)	Leaf scald of barley; collapse of mesophyll cells; necrotic lesions; [NIPs cause necrosis of primary leaves of both resistant and susceptible cultivars of barley]
Fusarium solani	Naphthazarin toxins	Citrus blight (wilt disease); vascular symptoms: vessel plugging and impaired hydraulic conductivity; [naphthazarins toxic to tomato, rice, beans, peas]
Fusarium solani	17 kDa peptide	Soyabean sudden death syndrome; crown necrosis; premature defoliation; pod abortion
Fusarium spp.	Enniatin	Necrosis, rot and wilt of potato, lupins and cereal plants; [enniatin induces wilting, loss of turgor and necrosis of leaves; growth of germinating wheat seedlings reduced]
	Fusaric acid	Wilt diseases of cotton, watermelon, pea, banana, tomato; [fusaric acid impairs water permeability of protoplasts; induces electrolyte loss; affects electrical membrane potential, membrane conductance and cellular ATP levels]
Fusarium moniliforme	Fumonisins (mycotoxins)	Seedling decay; stalk rot; and ear rot of maize; [fumonisin B_1 causes dose-dependent reductions in root and shoot growth in corn and tomato; effects identical to those caused by *Alternaria* A(A)L host-specific toxin]
Alternaria solani	Alternaric acid; altersolanols; tentoxin; solanopyrones; zinniol; homozinniol	Early blight disease of potato and tomato; [homozinniol phytotoxic to potato and tomato leaves]
Fusicoccum amygdali	Fusicoccin	Canker disease of almond and peach; blight of buds, blossom and fruits; leaf necrosis; cankers on twigs; [fusicoccin causes excessive transpiration and induces wilt; interacts with specific receptors in host cells]
Exserohilum turcicum	E.t. toxin	Northern leaf blight of maize; [E.t. toxin enhances lesion size, appressorial formation and ramification of germinating conidia on host leaves; virulence factor]

* See text for references.

Fusarium phytotoxins include naphthazarin toxins, enniatin, fusaric acid, lycomarasmin and even the fumonisin mycotoxins. The naphthazarin toxins of particular interest are the isomers marticin and isomarticin. Recently, the levels of isomartin in citrus trees affected by *F. solani* were compared with those in healthy trees[20]. Roots rotted by the fungus had measurably higher concentrations of the toxin than those of healthy trees. *F. solani* has also been implicated in sudden death syndrome of soybean[21]. Extracts of the fungus were found to contain 17 kDa protein which, in soybean plants caused browning of calli, necrosis on detached cotyledons and leaves and yellowing, curling and drying of attached cotyledons and leaves. Several species of *Fusarium* possess the capacity to produce the cyclohexadepsipeptide enniatin[22].

The results of virulence assays indicate that enniatin is not essential for the successful infection of potato tuber tissue by *Fusarium* species. However, it has been proposed that enniatin might still aid pathogenicity in strains of the fungus that do not produce the metabolite. The phytotoxicity of fusaric acid (5-*n*-butyl-pyridine-2-carboxylic acid) has a long history[18] and interest in association with *Fusarium* wilt diseases continues to the present time (Table II). The involvement of fusaric acid in pathogenicity has not been clearly established. However, in an elegant study[23] it was shown that sensitivity to fusaric acid, in cells from tomato cultivars resistant or susceptible to *F. oxysporum* f. sp. *lycopersici*, corresponded with differences in plant susceptibility to the fungus. *F. moniliforme* is a seed-transmitted pathogen of maize resulting in poor germination and seedling decay. It may also contribute to stalk rot, ear rot and symptomless systemic infection of maize plants (Table II). The fumonisins produced by this pathogen are classified as mycotoxins (see below), but are associated with phytotoxicity along parallel lines with their structural analogues A(A)L-toxins[10,24], the HSTs synthesised by *Alternaria alternata* (Table I). The fumonisins are long-chain polyhydroxyl alkylamines with two propane tricarboxylic acid moeities esterified to hydroxyls on adjacent carbons. Six fumonisin analogues are known to occur, of which fumonisin B_1, B_2 and B_3 (FB_1, FB_2, FB_3, respectively) cause leaf necrosis and other physiological responses (Table II) identical to those induced by the A(A)L-toxin TA, the effects being more pronounced in a susceptible tomato genotype[10].

A. solani is capable of producing a wide range of non-specific phytotoxins (Table II) and the effects of most of these have been reviewed recently[25]. However, a new metabolite, homozinniol, a novel polyketide has been isolated, chemically characterized and confirmed as phytotoxic in a leaf-spot assay against potato and tomato leaves[26].

Fusicoccum (*Phomopsis*) *amygdali* is able to infect several cultivated species of *Prunus*[27]. However, in the field the pathogen occurs chiefly on almond and peach causing a wide range of effects (Table II). Isolates of the pathogen produce a non-selective wilt toxin in the form of fusicoccin, an α-glucoside of a highly oxygenated carbotricyclic diterpene. Specific receptors in the host cells have been identified with a high affinity for the toxin, consistent with the persistence of toxic symptoms in plants infected with the fungus. Interaction with these receptors is thought to be an important initial event.

Northern leaf blight of maize is caused by the pathogen *Exserohilum turcicum* resulting in substantial yield losses. This fungus produces a phytotoxin in the form of a small peptide composed of three amino acids, glycine, serine and glutamine[28]. The compound has been designated E.t. toxin. Non-pathogenic isolates of the fungus successfully infect corn plants in the presence of the synthetic form of the toxin during inoculation. The toxin enhanced lesion size, appressorial formation and ramification of germinating conidia on host leaves. It was concluded that the toxin is an important virulence factor of the fungus (Table II).

Other non-specific toxins exist, for example, lycomarasmin and ophiobolin but there is little of any substance to add to the reviews that already exist[3].

MYCOTOXINS

Mycotoxins are those secondary metabolites of fungi which are endowed with toxic properties towards animals, including humans. The deleterious effects are referred to as mycotoxicoses and the risks arise from the presence of these substances in both the mycelium and spores of the toxigenic fungi. Mycotoxins are a diverse group of compounds produced by a wide range of fungi, normally after a phase of balanced growth. However, the synthesis of a particular mycotoxin is generally restricted to a relatively small number of fungal species and may be species or even strain specific. Of particular relevance to this review is the growing evidence that at least one HST and a number of non-specific phytotoxic metabolites of fungi may also be classified as mycotoxins.

1. *Toxigenic fungi*

The major toxigenic species of fungi and their mycotoxins are presented in Table III. It will be apparent that many fungi that produce phytotoxic compounds also synthesise mycotoxins. The various fungi and mycotoxins pose somewhat different risks to animals relative to humans and these will be highlighted in subsequent sections. Although evidence of mycotoxicoses can be traced to antiquity, the impetus for mycotoxin research did not appear until 1960, when 100,000 turkey poults in the United Kingdom died from acute necrosis of the liver and hyperplasia of the bile duct, attributed to the consumption of groundnuts infected with *Aspergillus flavus*. The research which followed this event eventually led to the identification and isolation of the responsible compounds and named aflatoxins[4,6]. This group of mycotoxins comprises aflatoxin B_1, B_2, G_1 and G_2 (AFB_1, AFB_2, AFG_1 and AFG_2 respectively). In addition, aflatoxin M_1 has been identified in the milk of dairy cows and women consuming AFB_1 from contaminated feeds or staples. The aflatoxins are a group of structurally related fluorescent heterocyclic compounds characterized by dihydrofuran or tetrahydrofuran residues fused to a substituted coumarin moeity[4,6]. The AFG molecules differ from the AFB structures in possessing a δ-lactone ring in place of a cyclopentenone ring. The presence of a double bond in the terminal furan ring of AFB_1 and AFG_1, but not in AFB_2 or AFG_2, confers

Table III Toxigenic species of fungi and their principal mycotoxins

Fungal species	Mycotoxins
Aspergillus flavus; A. parasiticus	Aflatoxins
A. flavus	Cyclopiazonic acid
A. Ochraceus; Penicillium viridicatum; P. cyclopium	Ochratoxin A
P. citrinum; P. expansum	Citrinin
P. expansum	Patulin
P. citreo-viride	Citreoviridin
Fusarium culmorum; F. graminearum	Deoxynivalenol
F. sporotrichioides; F. poae	T-2 toxin
F. sporotrichioides; F. graminearum; F. poae	Diacetoxyscirpenol
F. culmorum; F. graminearum; F. sporotrichioides	Zearalenone
F. moniliforme	Fumonisins; moniliformin; fusaric acid
Alternaria alternata	Tenuazonic acid; alternariol; alternariol methyl ether; altenuene
Acremonium coenophialum	Ergopeptine alkaloids
A. lolii	Lolitrem alkaloids
Claviceps purpurea	Ergot alkaloids
Phomopsis leptostromiformis	Phomopsins
Pithomyces chartarum	Sporidesmin A
Stachybotrys chartarum	Satratoxin H

distinct biological properties to the former two aflatoxins. It is now generally acknowledged that *A. flavus* only synthesises AFB_1 but is also capable of yielding cyclopiazonic acid[6,29], a mycotoxin recently confirmed as a co-contaminant in the batch of groundnuts which killed the turkey poults in 1960. On the other hand, *A. parasiticus* often produces all four aflatoxins. However, in both species of *Aspergillus*, there are strains which are non-aflatoxigenic. The two species develop when conditions such as temperature and humidity/water activity favour their proliferation. In the case of *A. parasiticus*, temperatures of 25–30°C are optimal for maximising aflatoxin synthesis. However, both temperature and water activity may interact in the promotion of aflatoxin synthesis[29] and the risk of contamination is, therefore, much greater in commodities emanating from the humid tropics. On the other hand, the ochratoxins, produced by several species of *Aspergillus* and *Penicillium*, have been reported to occur predominantly in temperate cereals and in the tissues of animals fed on these grains[4,5]. The ochratoxins are a family of structurally related compounds based on an isocoumarin molecule linked to L-phenylalanine. Ochratoxin A (OA) and ochratoxin B (OB) are the only forms to occur naturally in contaminated foods, and of the two, OA is more ubiquitous, often occurring with another pentaketide-derived mycotoxin, citrinin, in cereals and associated products. Citrinin is synthesised by a number of *Penicillium* species[5].

Several *Penicillium* species are also capable of synthesising patulin[4,5], a low molecular weight hemiacetal lactone with antibiotic properties. *Penicillium expansum* is of particular relevance since it is commonly associated with storage rot of apples and a wide variety of other fruits. The occurrence of patulin in apple juice has been linked with the use of mouldy fruit. Other species of *Penicillium* contaminating rice from Italy, Spain, Thailand, Burma and other countries are now recognised as producers of an open-chain nonaketide derivative known as citreoviridin[5].

The natural occurrence of the mycotoxins from *Fusarium* species is generally associated with temperate countries, since these fungi require somewhat lower temperatures for growth and mycotoxin production than the aflatoxigenic *Aspergillus* species just described. However, extensive data exists to indicate the global scale of contamination of cereal grains with a number of *Fusarium* mycotoxins[30]. *Fusarium* species synthesise a wide range of mycotoxins, of which the most important from the point of view of animal and human health are the trichothecenes, zearalenone, moniliformin and the fumonisins. The trichothecenes are subdivided into four basic groups, with types A and B representing the most important components. The type A trichothecenes include T-2 toxin, HT-2 toxin, neosolaniol and diacetoxyscirpenol (DAS), while type B trichothecenes include deoxynivalenol (DON, also known as vomitoxin), nivalenol (NIV) and fusarenon-X. All trichothecenes possess a basic tetracyclic sesquiterpene structure with a 6-membered oxygen-containing ring and an epoxide group. The synthesis of the two types of trichothecenes appears to be characteristic for a particular *Fusarium* species. Thus, for example, production of type A trichothecenes predominates in *F. sporotrichioides* and possibly also *F. poae*, whereas synthesis of type B trichothecenes occurs principally in *F. culmorum* and *F. graminearum*. A common feature of many *Fusarium* species is their ability to synthesise zearalenone (ZEN), and its co-occurrence with certain trichothecenes raises important issues regarding aditivity and/or synergism in the aetiology of mycotoxicoses in livestock and humans. ZEN (also known as F-2 toxin) is a phenolic resorcyclic lactone which also occurs as a hydroxy derivative in the form of α-zearalenol. The presence of appropriate reductases in animal tissues implies that α-zearalenol may be the active form of ZEN in animals. In respect of the co-occurrence of mycotoxins, the secondary metabolism of *F. moniliforme* may also be of significance since it is capable of producing at least three mycotoxins: the fumonisins, moniliformin and fusarin C[30]. Of these, the first two mycotoxins are of particular significance in animal and human health. The fumonisins (FB_1, FB_2 and FB_3) have already been described in this review. Moniliformin occurs as the Na or K salt of 1-hydroxycyclobut-1-ene-3,4-dione and, like the fumonisins, has been detected in maize. High proportions of *F. sambucinum* and *F. oxysporum* isolates have been reported to produce sambutoxin, a compound also found in decayed potato tubers[31].

In addition to producing HSTs, a wide range of *Alternaria* species are also capable of synthesising mycotoxins of diverse chemistry[25]. The dibenzo-α-pyrone group includes alternariol, alternariol methyl ether and altenuene. The nitrogen-containing group includes tenuazonic acid and the cyclic polypeptide tentoxin. In addition, *Alternaria* produce a number of metabolites of varied structure, including altertoxin I, an unusual partially saturated perylene.

The endophytic fungus *Acremonium coenophialum* occurs in close association with perennial tall fescue, while another related fungus, *A. lolii*, may be present in perennial ryegrass[32]. These relationships are symbiotic, enabling the host plant to withstand disease, insect and nematode infestation and drought. The fungi supply the grass with defensive secondary compounds while the plant acts as a source of essential nutrients for the fungus. These secondary metabolites are also regarded as mycotoxins since they are linked with well-defined adverse effects in cattle, sheep and horses consuming endophyte-infected grasses. Ergopeptine alkaloids, mainly ergovaline, have been associated with *A. coenophialum*-infected tall fescue toxicosis, while the indole isoprenoid lolitrem alkaloids, particularly lolitrem B have been linked with *A. lolii*-infected perennial ryegrass staggers. The ergot alkaloids are more historically associated with the sclerotia of various species of *Claviceps*, principally *C. purpurea* which may contaminate the kernels of grasses and cereals. The major alkaloids include the lysergic acid derivatives ergocristine and ergotamine, although ergosine, ergocornine and ergometrine may also occur[32].

In Australia, lupin stubble is valued as fodder for sheep, but infection with the fungus *Phomopsis leptostromiformis* is a major limiting factor due to toxicity arising from the production of phomopsins by the fungus. Mature or senescing parts of the plant, including stems, pods and seeds are particularly prone to infection[33]. Of these mycotoxins, Phomopsin A is considered to be the primary toxic agent. Phomopsin A is a hexapeptide with the sequence phe-val-ile-pro-ile-asp, modified by oxidation, chlorination and methylation.

Pithomyces chartarum is an ubiquitous saprophyte of pastures with the capacity to synthesise sporidesmin A which represents an important group of diketopiperazine derivatives arising from the formation of a cyclic anhydride between two amino acids[34]. A sulphur-containing ring imparts biological activity to the molecule.

The cellulophilic saprophyte *Stachybotrys chartarum* is associated with a wide range of substrates including straw, hay, library materials and paintings. The fungus is capable of synthesising stable macrocyclic epoxytrichothecenes such as satratoxin H[35].

2. *Natural Occurrence of Mycotoxins*

The ubiquitous distribution of the toxigenic fungi as phytopathogens and as saprophytic organisms predisposes to at least some mycotoxin contamination of raw materials and processed commodities when appropriate environmental conditions prevail[30]. Considerable data already exists to demonstrate the global scale of mycotoxin contamination of oil-seeds, nuts, fruit and animal feed. The evidence has been presented elsewhere[6,30], but there is scope for reviewing more recent data, particularly in relation to the aflatoxins, OA, patulin, the important trichothecenes and ZEN.

Not unexpectedly, aflatoxin contamination of peanuts, cottonseed and their processed derivatives is well documented. The impetus has been maintained and, in recent years, surveillance extended to other raw materials and products[36]. For example, samples of peanut butter analysed in the United Kingdom (UK) in 1986 and 1991 showed that 'crunchy' types continued to contain more aflatoxin than 'smooth' varieties. Thus, in 1991, 19% of smooth and 31% of crunchy peanut butter samples contained total aflatoxin levels in excess of 4 μg/kg. Reports in 1990 drew attention to aflatoxin contamination of tree nuts such as pistachios[36]. Data on imports into the UK between March 1990 and April 1991 and between May 1991 and April 1992 indicated that 52% and 28%, respectively, of samples exceeded the 4 μg/kg statutory limit for total aflatoxins in finished products and 38% and 25%, respectively, exceeded the 10 μg/kg limit in products destined for further processing. Elsewhere, there are similar reports of contamination of pistachio nuts, particularly small pistachio 'scalpers' in California[37] which may contain total aflatoxin concentrations of up to 149 μg/kg. In the Netherlands[38], AFB_1 levels as high as 165 μg/kg have been reported for pistachio nuts, with much lower concentrations in shells (up to 8 μg/kg). In almonds, relatively low levels of total aflatoxins were detected in a recent study, with maximum levels in an ungraded, natural batch reaching 3 μg/kg[39]. In whole dried figs, UK data[36] showed that between December 1988 and April 1992, the percentage contaminated with aflatoxins (total) at levels above 4 μg/kg fell from 26% to 16%. However, samples containing up to 427 μg/kg were found. The incidence of aflatoxins in fig paste samples above the 4 μg/kg level also fell during this period from 50% to 14%. The maximum concentration of total aflatoxins found in fig paste also declined from 165 to 15 μg/kg. In decayed figs from commercial orchards in California, total aflatoxin levels as high as 77,200 μg/kg have been detected after natural infection with *Aspergillus parasiticus*[40]. Both AFB and AFG types were detected in these samples. Most figs infected with *A. flavus*, however, were free of aflatoxins. The results of a recent survey of Egyptian foods[41] indicated highest incidence of AFB_1 contamination in nuts and seeds (82%), followed by spices (40%), herbs and medicinal plants (29%), dried vegetables (25%) and cereal grains (21%). The highest mean concentration of AFB_1 occurred in herb and medicinal plants (49 μg/kg), followed by cereals at 36 μg/kg, spices (25 μg/kg), nuts and seeds (24 μg/kg) and dried vegetables (20 μg/kg).

Surveillance of animal feeds for aflatoxin contamination has continued. In the UK, analysis conducted during the period 1987–1990 indicated that all feedingstuffs complied with legislation in force for AFB_1 levels[36]. Elsewhere, however, aflatoxin levels in particular feeds still pose serious risks to animal and human health. Thus in India, total aflatoxin levels of 3700 μg/kg were detected in a sample of groundnut cake[42]. Dairy animals consuming AFB_1-contaminated feeds are able to metabolise the mycotoxin to AFM_1 and this may be secreted in milk. Analyses of farm-gate milk in the UK between 1988 and 1989 showed low levels of AFM_1 contamination[36], but more than 50% of milk samples in Tanzania were found to contain the mycotoxin[41].

Patulin levels up to 250 mg/kg have been found in damaged or infected parts of fruit[36]. The occurrence of patulin in fruit juice has been a cause of concern in the UK and elsewhere in Europe. In recent years, there has been a considerable increase in the production of cloudy apple juices prepared by pressing the fruit and stabilising with vitamin C prior to pasteurisation of the juice. The reduction in processing steps, as compared with the procedure for production of clear juices, means that patulin losses during fining and filtration are restricted, with higher residual levels of the mycotoxin in the cloudy juices. A comparison of the patulin concentrations in the two types of juices has recently been published for samples from the UK[36] and Spain[43]. Although the UK data were derived from a relatively small number of samples, it was apparent that the incidence of patulin contamination was higher in cloudy juices, with a median value of 28 µg/kg, compared with 0–10 µg/kg for clear juices. Four cloudy samples had patulin concentrations in excess of 50 µg/kg, compared with only one of the clear juice samples. In two cloudy samples, patulin concentrations exceeded 151 µg/kg. In a more extensive study in Spain, patulin was detected in 82% of commercial samples (assumed to be clear for this comparison). However, 70% of these juices contained levels of less than 10 µg/l, although 22% had levels of 10–50 µg/l and in two samples, concentrations of 164 and 170 µg/l were recorded[43]. On a more reassuring note, patulin was absent in all 12 tested samples of apple food for children.

Some contamination of foods with OA has been reported in a survey conducted in the UK[36]. Of 104 samples of pig kidney, 12% were contaminated with 1–5 µg OA/kg, while 3% had concentrations of up to 10 µg/kg. The incidence of OA contamination within the range 1–5 µg/kg in black pudding, muesli, maize and barley was of the order of 10–13%. However, 28% and 29% of bran and oat samples, respectively, had OA concentrations in the range 1–5 µg/kg. In one sample of barley, a concentration of 45 µg/kg was detected. Mouldy flour may contain up to 2.9 mg OA/kg. A recent study in France indicated consistent contamination of cereals and oilseeds with OA, the values ranging from 0.6 to 12.8 µg/kg in positive samples[44]. Citrinin often occurs with OA in cereal and feed grains[5]. In naturally contaminated samples of barley, citrinin levels up to 790 µg/kg were detected recently.

In 1989, Scott[45] provided an exhaustive survey of the global contamination of cereal grains and animal feed with the trichothecenes. Recent data from Germany[46], Poland[47], Finland[48], The Netherlands[49], Japan[50,51], United States of America (USA)[52,53] and Canada[54] confirm the widespread distribution of these mycotoxins particularly with respect to DON, NIV and ZEN (Table IV). Of considerable significance are consistent reports of co-occurrence of the *Fusarium* mycotoxins in cereal grains. Thus in the German study[46], 20% of wheat samples were contaminated with DON and ZEN. In addition, T-2 and HT-2 toxins were detected at levels ranging from 3 to 250 and 3 to 20 µg/kg, respectively. The co-occurrence of 3-ADON with 15-ADON in Polish wheats has also been observed[47]. Furthermore, the study in The Netherlands[49] revealed the presence of DON, NIV and ZEN as co-contaminants in cereal grains. Within-country variation in DON contamination of wheat has also been observed. Highest levels in the 1991 USA harvest were seen in Missouri, North Dakota and Tennessee[53]. In the 1993 harvest, 86% of samples from Minnesota and up to 78% of samples from North and South Dakota had levels in excess of 2 mg/kg. Canadian cereal grains were also contaminated with T-2 and HT-2 toxins[54].

Contamination of maize with fumonisins has recently been reported in several countries, including Benin, Portugal, Italy, Zambia[55]; Botswana, Mozambique, South Africa, Malawi, Zimbabwe, Tanzania[56]; Argentina[57]; Costa Rica[58]; and USA[59] (Table V). Incidence rates between 82% and 100% were recorded for samples from Italy, Portugal, Zambia and Benin[55]. In most instances the predominant fumonisin is FB_1. The potential for human and animal exposure to fumonisins is an emerging issue, particularly in Africa where maize generally constitutes the staple diet of most people. In Argentina, fumonisins were detected during ear

Table IV Natural occurrence of deoxynivalenol (DON), nivalenol (NIV) and zearalenone (ZEN) in cereal grains and animal feed (mg/kg) *

Country	Cereal/feed type	DON	NIV	ZEN
Germany	Wheat	0.004–20.5	0.003–0.032	0.001–8.04
Poland	Wheat	2–40	0.01	0.01–2
Finland	Feeds and grains	0.007–0.3		0.022–0.095
	Oats	1.3–2.6		
Netherlands	Wheat	0.020–0.231	0.007–0.203	0.002–0.174
	Barley	0.004–0.152	0.030–0.145	0.004–0.009
	Oats	0.056–0.147	0.017–0.039	0.016–0.029
	Rye	0.008–0.384	0.010–0.034	0.011
Japan	Wheat	0.03–1.28	0.04–1.22	0.002–0.025
	Barley			0.010–0.658
USA	Wheat (winter), 1991	<0.1–4.9		
	Wheat (spring), 1991	<0.1–0.9		
	Wheat, 1993	<0.5–18		
	Barley, 1993	<0.5–26		
Canada	Wheat and barley	up to 0.5		up to 0.3

* Data sources presented in text. For comparison with previous data see D'Mello *et al.* (1997)[30] and Scott (1989)[45].

Table V Worldwide contamination of maize with fumonisins (μg/kg) *

Country	FB_1	FB_2	FB_3	Total
Benin	nd#–2630	nd–680	—	nd–3310
Botswana	35–255	nd–75	nd–30	35–305
Mozambique	240–295	75–110	25–50	340–395
South Africa	60–70	nd	nd	60–70
Malawi	nd–115	nd–30	nd	nd–135
Zambia	20–1420	nd–290	—	20–1710
Zimbabwe	55–1910	nd–620	nd–205	55–2735
Tanzania	nd–160	nd–60	nd	nd–225
Argentina	85–8791	nd–11300	nd–3537	85–16760
Costa Rica	1700–4780	—	—	—
Italy	10–2330	nd–520	—	10–2850
Portugal	90–3370	nd–1080	—	90–4450
USA	nd–350	—	—	—

* Data sources presented in text. # nd = not detectable.

development and was closely correlated with natural infection with *Fusarium moniliforme* and *F. proliferatum*[57]. In Costa Rica, significant regional differences were observed in contamination of maize with FB_1[58].

Mycotoxins from contaminated grains may be transmitted into beer during the brewing process[60]. The occurrence of aflatoxins in African beers has been the cause of some concern, with one study in Nigeria indicating an 80% incidence in millet beer. In those beers levels of AFB_1 ranged from 1.7 to 138 μg/l. DON survives the brewing process and has been found in high incidence in Canadian and German beers, with levels as high as 569 μg/l in the latter source. Brewing from wheat malt was associated with higher levels of DON contamination than brewing from barley malt. In Nigerian and Zambian beers, ZEN occurs at relatively high concentrations, ranging from 13 to 4600 μg/l. However, assessment of contamination may be complicated by metabolism of ZEN to α- and β-zearalenol during fermentation. FB_1 and FB_2 occur to a limited extent in beer[60].

Mycotoxins have also been detected in forages. Ergovaline concentrations in infected tall fescue seeds and forages were found in the range 2–6 mg/kg[61]. The levels of lolitrems in endophyte-infected perennial ryegrass ranged from 5–10 mg/kg[61]. In cultivars of lupins susceptible to *Phomopsis* infection, levels of phomopsin A up to 217 mg/kg were found in stubble, with much lower concentrations in resistant varieties (88 mg/kg)[33].

3. *Factors Affecting Mycotoxin Production*

The synthesis of mycotoxins is determined by a diverse array of factors, broadly divisible into biological, physical and chemical and by interactions among these factors. Many of the toxigenic species of fungi are also major plant pathogens. Thus, *Aspergillus* ear and kernel rot of maize is caused by *A. flavus*. Ear rot of maize has also been associated with a number of *Fusarium* species, including *F. graminearum* and *F. moniliforme*. A relationship between level of infection and mycotoxin production might, therefore, be expected. This assumption was confirmed in the studies of Brown *et al.*[62] who showed a direct correlation between colonisation with *A. flavus* and AFB_1 contamination of maize kernels. Genotypes of maize susceptible to infection yielded grain with high levels of AFB_1 relative to kernels from resistant varieties. Other studies confirmed that maize genotypes with low ear rot ratings had reduced AFB_1 concentrations. Similarly, in wheat, a direct quantitative assessment of the link between head blight caused by *Fusarium culmorum* and DON contamination of kernels has recently been elucidated in The Netherlands[63] using genotypes of differing resistance to the disease. A striking positive correlation was evident between the incidence of head blight after experimental inoculation at flowering and DON concentrations in the grain (Fig. 2). Work in Canada also supports the view that wheat grains from cultivars susceptible to head blight experimentally induced by *F. graminearum* or *F. culmorum* contain more DON than those from Chinese genotypes which are recognised to be resistant[64]. In natural epidemics of head blight in Minnesota, DON concentrations in wheat were positively associated with disease incidence and severity, but negatively correlated with grain yield. It should be recognised that at harvest grain is likely to be colonised by different species of moulds and mycotoxin production may be affected by fungal interactions[65]. Thus, aflatoxin formation in maize by *A. flavus* was substantially reduced by *F. moniliforme* when ears were simultaneously inoculated with both fungi[66].

A variety of interacting physical factors may affect mycotoxin production in the field and during storage of foods[29]. These include time, temperature, pH, humidity/water activity and extent of physical damage caused by insect infestation. Thus, aflatoxin contamination of groundnuts with 15% moisture is maximised after storage at 30°C. In a number of countries head blight caused by *Fusarium* species has been associated with years of high rainfall. It may be significant that the higher than average levels of DON in wheat samples from Missouri, North Dakota and Tennessee were tentatively attributed to increased rainfall in these states[52]. Laboratory studies indicate that both time and water activity may interact to affect FB_1 production in maize kernels[66]. Maximum synthesis occurred at 21 days of incubation at a water activity of 1, but on lowering water activity to 0.95, maximum production of FB_1 did not occur until 47 days had elapsed. In one study, T-2 toxin production by *F. sporotrichioides* growing in cultures at 25°C maximised at 14 days

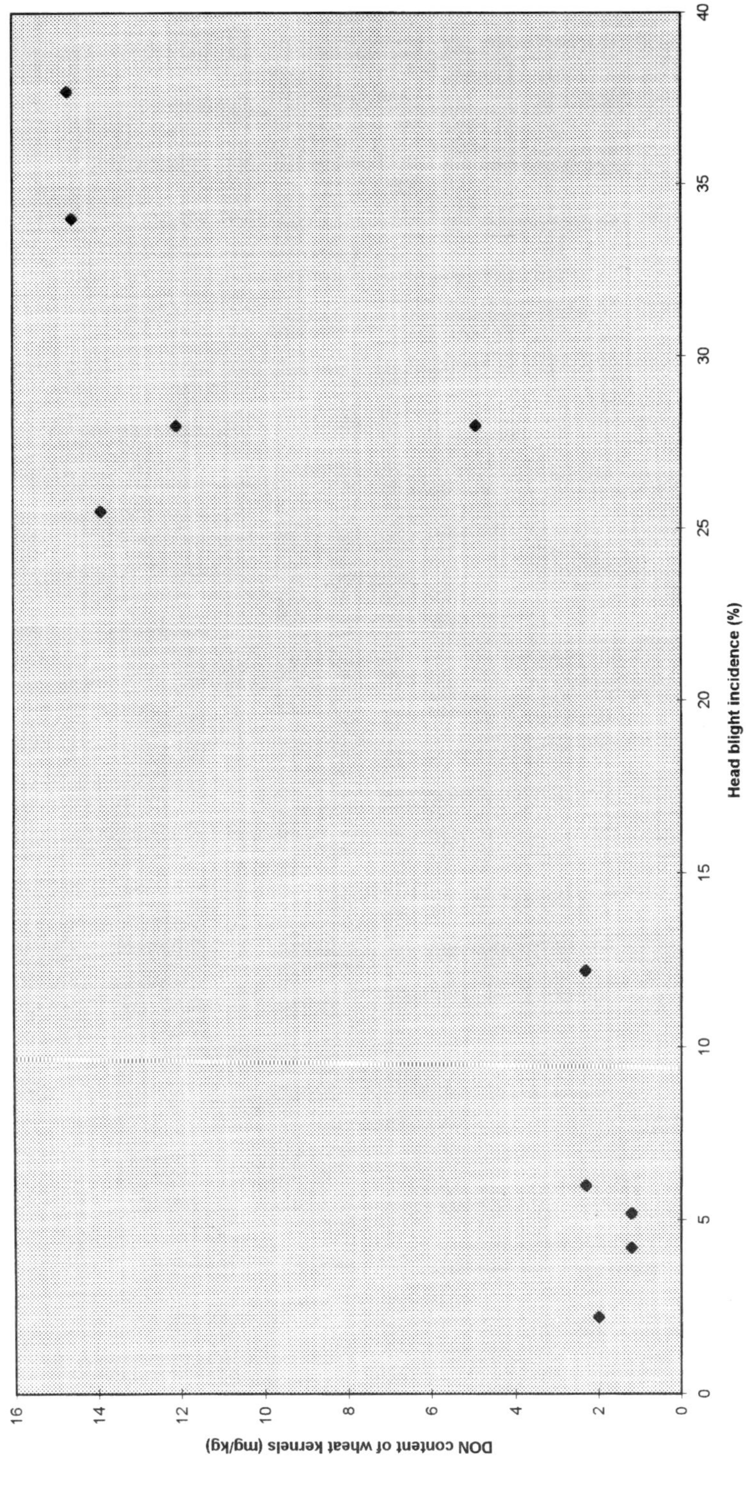

Figure 2 Correlation between *Fusarium* head blight and deoxynivalenol (DON) content of wheat kernels. From Snijders and Perkowski, 1990 (ref. [63]).

of incubation but declined markedly thereafter[67]. T-2 toxin production by this fungus is increased by agitation of broth cultures. There is some controversy concerning the effects of temperature on ZEN synthesis in laboratory cultures, with some investigators suggesting the need for thermic shock while others reporting ZEN production at constant temperatures. Of particular interest is the effect of insect invasion of grain and seeds in the field and during storage. Sommer *et al.*[68] reported values for $AFB_1 + AFG_1$ in excess of 2000 µg/kg in pistachio nuts heavily infested with navel orangeworm moths.

An important chemical factor affecting the synthesis of mycotoxins is the use of pesticides. Of particular interest here is the effect of fungicides since these are widely used to control phytopathogenic fungi. In aflatoxin-producing species of *Aspergillus*, there is conflicting evidence regarding the effects of fungicides. Thus, in one study[69] with pure cultures of *A. flavus* and *A. parasiticus*, tricyclazole was effective in reducing levels of AFB_2, AFG_1 and AFG_2, but detectable quantities of AFB_1 persisted even at the highest application of the fungicide (100 mg/l). However, in a subsequent study, sub-inhibitory levels of miconazole (0.10 µM) enhanced total aflatoxin synthesis in *A. parasiticus*[70]. Stimulatory effects on total aflatoxin production were also observed with pure cultures of this fungus after addition of fenpropimorph (250 mg/l), with an increase in the proportion of the more toxic AFB_1. As might be predicted from the foregoing account, fungicides may influence mycotoxin synthesis by *Fusàrium* species, but the effects on trichothecene and ZEN production are also somewhat variable (Table VI)[30,71]. In laboratory studies with pure cultures, dicloran, iprodione and vinclozolin were individually effective as inhibitors of DAS and ZEN synthesis in *F. graminearum* but tridemorph and carbendazim each enhanced T-2 toxin production in *F. sporotrichioides*, while 3-ADON production increased in *F. culmorum* treated with difenoconazole. Field trials with fungicides have also yielded somewhat conflicting results. Thus Boyacioglu *et al.*[72] showed that propiconazole reduced infection of wheat by an artificially applied inoculum of *F. graminearum* and DON levels also declined. However, thiabendazole had no effect on infection level, but DON contamination was markedly reduced. In a subsequent study, also with wheat inoculated with *F. graminearum*, ear blight incidence and DON concentrations in grain were unaffected by propiconazole, thiabendazole or tebuconazole applications. On the other hand, combination of tebuconazole and triadimenol in wheat inoculated with *F. culmorum* reduced ear blight, but a 16-fold increase in NIV content of grain was observed[73].

Since fungal infection of grain, nuts and fruit is often preceded by physical damage caused by insect invasion, much effort has been expended on the potential of insecticides to reduce infestation and, therefore, mycotoxin contamination. This opens up the possibility of dual-function insecticides contributing to lower overall pesticide use, an important attribute in environmental protection and management. It is now clear, however, that insecticides may exert direct effects on mycotoxin production. Thus, in pure culture studies with *Aspergillus parasiticus*, uncomplicated by insect infestation, dichlorvos, landrin, malathion and Diazinon significantly inhibited production of AFB_1 in a dose-dependent manner[74]. In several instances, AFB_1 inhibition was greater than the reduction in growth of the fungus. For example, with dichlorvos at 100 mg/l of culture broth, fungal growth declined by 29% whereas AFB_1 synthesis was reduced by 99%. At this level of application, however, Diazinon increased mycelial growth by 19% but AFB_1 production was still reduced by 23%. At 100 mg/l, the insecticide, naled, inhibited growth totally, although not at 10 mg/l at which level it precipitated a 68% reduction in AFB_1 synthesis, but

Table VI Effects of fungicides on production of *Fusarium* mycotoxins*

Fungicide	Methods	Effects
Dicloran, iprodione, vinclozolin	Added separately at levels of up to 500 μg/ml potato-dextrose broth; static culture of *F. graminearum*	Dose-related inhibition of growth and production of DAS and ZEN; total inhibition of DAS and ZEN at higher levels of each fungicide
Propiconazole, thiabendazole	Field trial with wheat; heads sprayed with *F. graminearum* inoculum 2 days after anthesis; fungicides applied separately prior to, during and after inoculation; propiconazole applied at 120 g/ha, thiabendazole at 360 g/ha	Propiconazole reduced *Fusarium* infection by 39–55%; DON levels reduced by 34–78%. Thiabendazole had no effect on infection level, but DON levels reduced by up to 83%
Propiconazole, thiabendazole, tebuconazole	Field trial with wheat inoculated with *F. graminearum* at 3 stages from beginning of flowering; fungicide rates: propiconazole, 140; thiabendazole, 280; tebuconazole, 140 g/ha	Head blight incidence ranged from 83 to 84% irrespective of fungicide (control, 87%); DON levels ranged from 12.9 to 16.7 mg/kg (control, 12.0 mg/kg)
Tebuconazole with triadimenol (Matador)	Field trial with wheat; heads inoculated with *F. culmorum*; Matador applied at 1 l/ha	Fungicide reduced head blight; 16-fold increase in NIV content of grain from fungicide-treated plants
Tridemorph	Shake-flask cultures of *F. sporotrichioides* in a defined medium; fungicide added at 6 and 36 μg/ml	At 6 μg/ml, growth enhanced but T-2 toxin and DAS production inhibited; at 36 μg/ml, growth inhibited but T-2 toxin production stimulated
Carbendazim	Cultures of *F. sporotrichioides* on potato-dextrose agar; fungicide added at 0.1 and 1.0 μg/ml	Growth unaffected by fungicide; 6-fold increase in T-2 toxin production with fungicide at 0.1 μg/ml

* Taken from D'Mello *et al.*[30].

dichlorvos was still more effective, causing a 92% inhibition of AFB_1 production[74]. AFB_2 levels were reduced below detection levels in all insecticide-treated cultures. At sub-lethal levels of insecticide, dichlorvos was more effective than naled in reducing production of AFG_1 and AFG_2. In general, AFB_1 was most resistant to inhibition by insecticides, followed by AFG_1, AFG_2 and AFB_1. In other studies[75], naled reduced AFB_1 concentrations in harvested kernels after application to a maize crop in the field which had been artificially inoculated with *A. parasiticus*. However, Bux and carbaryl were considerably more effective than naled, whereas in broth cultures naled exhibited higher efficacy. In the uninoculated maize crop, Bux and carbaryl were again highly effective in reducing natural production of AFB_1 while naled was virtually inactive. Production of OA and OB by *A. ochraceus* in yeast extract-sucrose (YES) medium and in maize kernels can be reduced in a dose-dependent manner by application of dichlorvos at levels of up to 300 mg/l[76]. Mycelial growth, measured only in the YES cultures, remained at 80% of control values at the highest level of dichlorvos addition whereas comparable data for OA and OB production were, respectively, 21% and 11% of control values. Patulin synthesis by *Penicillium urticae* is also reduced by insecticide applications in pure culture[77]. In general dose-related effects are seen with naled and dichlorvos, the former being more effective at the higher doses of application (10 and 100 mg/l).

Indeed, naled completely inhibited growth and patulin production at 100 mg/l whereas at this level of application, dichlorvos reduced growth by 50% and patulin formation by 89%. On the other hand, malathion and Sevin reduced patulin synthesis by 42.5% and 100% respectively at 100 mg/l with virtually no effect on growth of the fungus. At 100 mg/l, Diazinon increased growth of the fungus, (as it did for *A. parasiticus*), but patulin production was reduced by only 5%. Naled has also been found to be effective in reducing ZEN levels in pure cultures of *Fusarium graminearum* in liquid media or on maize kernels when the insecticide has been applied as a liquid preparation or as fumigant at concentrations of 30 and 100 µl/l[78]. Production of ZEN was completely inhibited only when naled was applied prior to inoculation of the culture media. When applied to 12-day or older cultures, naled did not inhibit ZEN synthesis but in 3–9-day cultures ZEN production was reduced by 45–92% when the insecticide was applied at 10–100 µl/l.

4. *Toxicity and Metabolism*

Mycotoxins are endowed with both acute and chronic features of toxicity. As will be demonstrated, dose and duration of exposure are important factors in the development of toxic manifestations in animals and in humans. However, it is abundantly clear that interactions among mycotoxins may occur and that the questions of synergy and additivity need to be addressed when evaluating field cases of mycotoxicology. There is now considerable evidence to indicate diverse and profound effects of mycotoxins in all classes of animals, including humans (Tables VII and VIII).

The classical assesssment of toxicity of any compound necessarily centres on the acquisition of LD_{50} data. For mycotoxins, these values of acute toxicity are subject to wide variation depending, for example, on age, sex and size of animals[4]. There are also distinct species differences in sensitivity to a particular mycotoxin (Table VII). Thus day-old ducklings are more susceptible to AFB_1 than laboratory species or adult ruminants, although even in the latter group of animals, deleterious effects have been recorded. In dairy cattle, another problem arises from the ruminal transformation of AFB_1 to a related metabolite, AFM_1 which is secreted in the milk. AFM_1 has been ascribed with both hepatotoxic and carcinogenic properties. In quantitative terms, its acute toxicity for ducklings and rats appears to be similar to that of AFB_1, but its carcinogenic potential has been estimated to be lower.

The ochratoxins act principally as nephrotoxins, possibly in conjunction with citrinin[79,80]. Field cases of porcine nephropathy, associated with the consumption of moist grain, have been reported in Denmark[5], where it is endemic. Affected kidneys are enlarged and pale, while internally, sections of renal cortex show interstitial, peritubular and periglomerular fibrosis. Tubular function may be impaired, consistent with tubular degeneration and atrophy. OA is also a potent teratogen in mammalian species and, in addition, its carcinogenicity to mice and rats has been demonstrated.

Although the trichothecenes are capable of inducing well-defined effects in non-ruminant animals[30], it is generally agreed that cattle and sheep are more tolerant. Thus, DON is widely recognised as a potent feed intake inhibitor in pigs, but recent studies with dairy cows[81] indicated that levels of up to 6 mg/kg diet failed to reduce feed intake or total milk output. Furthermore, no residues of DON or its deepoxy derivative could be detected in milk. In pigs, a diet containing 4 mg DON/kg reduced feed intake and growth by 20% and 13%, respectively[82]. At necropsy, the fundic region of the stomach of pigs fed the mycotoxin diet was more corrugated than that for control animals. Other studies[83] show that pigs fed DON at levels of up to

Table VII Deleterious effects of mycotoxins in animals *

Mycotoxins	Deleterious effects
Aflatoxin B_1	Acute toxicity: LD_{50} values, 1–17.9 mg/kg BW (laboratory animals), 0.5 mg/kg BW (ducklings); hepatic lesions; teratogenic Reduced feed efficiency, immune function and reproductive performance in ruminants
Aflatoxin M_1	Hepatotoxic
Ochratoxin A	$LD_{50} < 6$ mg/kg BW (pigs); implicated in porcine nephropathy; teratogenic; carcinogenic to mice and rats
Deoxynivalenol	LD_{50} 70 mg/kg (mice); potent feed intake inhibitor in pigs; causes erosion of gastric mucosa; affects feed choice in mink; teratogenic; immunosuppressive. Ruminants tolerant
Diacetoxyscirpenol (DAS) and T-2 toxin	LD_{50} 23 and 5 mg/kg BW, respectively (mice); mouth lesions in poultry; in sows T-2 toxin can cause infertility and abortion. Growth depression in broiler chicks caused by DAS increased by high-fat diet. Implicated in field cases of mycotoxicoses in ruminants
Zearalenone	LD_{50} range: 2–10 g/kg BW; causes vulvovaginitis in gilts and anoestrus in cycling females; feminisation in male pigs; reduced spermatogenesis; reduced milk production and hyperoestrogenism in cows; infertility in sheep
Fumonisins	Mortality in chicken embryos; onset of mortality in chickens aided by moniliformin; hepatic lesions in pigs and cattle; porcine pulmonary oedema; equine leukoencephalomalacia
Fusaric acid	Suppresses cell-mediated immunity; interactions with fumonisins and deoxynivalenol
Tenuazonic acid	LD_{50} values 125–225 mg/kg BW in mice
Ergopeptine alkaloids	Reduced growth, reproductive performance and milk production in ruminants; increased susceptibility to heat stress
Lolitrem alkaloids	Neurological effects: incoordination, staggering, shaking of head and collapse in ruminants
Phomopsins	Lupinosis in sheep: ill-thrift, liver damage, jaundice, photosensitization and death; reduced reproductive performance of ewes
Sporidesmin A	Facial eczema in sheep: liver damage, urinary lesions, photosensitization

* See text for references; BW = body weight; LD_{50} = median lethal dose.

Table VIII Mycotoxins implicated in human disease *

Mycotoxin	Disease	Food/source	Countries
Aflatoxins	Liver cancer; cirrhosis	Peanuts; maize	East and West Africa; India; Thailand; Philippines
Ochratoxins	Balkan (and possible Tunisian) endemic nephropathy	Cereal grains	Bulgaria; Romania; former Yugoslavia; Tunisia
Cyclopiazonic acid	'Kodua poisoning'	Millet	India
Citreoviridin	'Shoshin-kakke'	Rice	Japan
Fumonisins	Oesophageal cancer	Maize	South Africa
Moniliformin	'Keshan disease'	Maize	China
Ergot alkaloids	Ergotism (St. Anthony's Fire)	Rye	Europe
Satratoxin	Respiratory disorders; reduced immunocompetence; fatigue	Water-damaged indoor environment	Worldwide

* Based on references reviewed or cited in the text.

19 mg/kg diet developed gastric mucosa which was thinner and more eroded with a higher degree of folding than that of control animals. Some degree of recovery from the adverse effects on feed intake and growth is possible but this appears to be dependent on the dietary concentration of DON. *In vitro* transformation of DON by the normal gut flora of the pig has been demonstrated[84]. Flora from the caudal segments of the gut, particularly the colon, degraded DON whereas microbes from the cranial segments (duodenum and jejunum) had no such activity. DON transformation involved deepoxidation which also resulted in loss of cytotoxicity. Mink given a choice between uncontaminated and DON-containing feed displayed a preference for the former even when the DON level in the contaminated diet was as low as 0.28 mg/kg[85]. However, when no choice was offered, mink readily consumed diets containing DON at concentrations as high as 1.18 mg/kg. In laying hens, transmission of DON into egg components has been reported, but levels decline once contaminated feed has been withdrawn[86]. As regards the other trichothecenes, Perluky *et al.*[87] point to circumstantial evidence implicating DAS and T-2 toxin in field cases of mycotoxicoses. Additionally, there is evidence from experimental work to suggest specific defects in liver and kidney function in female rabbits[88]. Both mycotoxins are able to induce mouth lesions in poultry[89]. In the case of DAS, lesions were directly related to time of exposure (Fig. 3) to the mycotoxin and to its concentration in the diet (Fig. 4). Feeding a high fat diet to broiler chicks increases the growth depression caused by DAS, suggesting that such a diet facilitates lipid micellar absorption of the mycotoxin which is then able to inhibit protein synthesis at the ribosomal level[90]. T-2 toxin also induces lesions in pigs, specifically on the mucosa of the pars oesophageal region, the incidence being dose-related[91]. In addition, an interaction with DON was observed in respect of growth and feed intake, which were lowest when the diet was co-supplemented with T-2 and DON at 3.2 and 2.5 mg/kg, respectively. In sows, T-2 toxin has been shown to induce infertility[87], and after parenteral administration during the last trimester of gestation, is able to precipitate abortion within 48 hours[92].

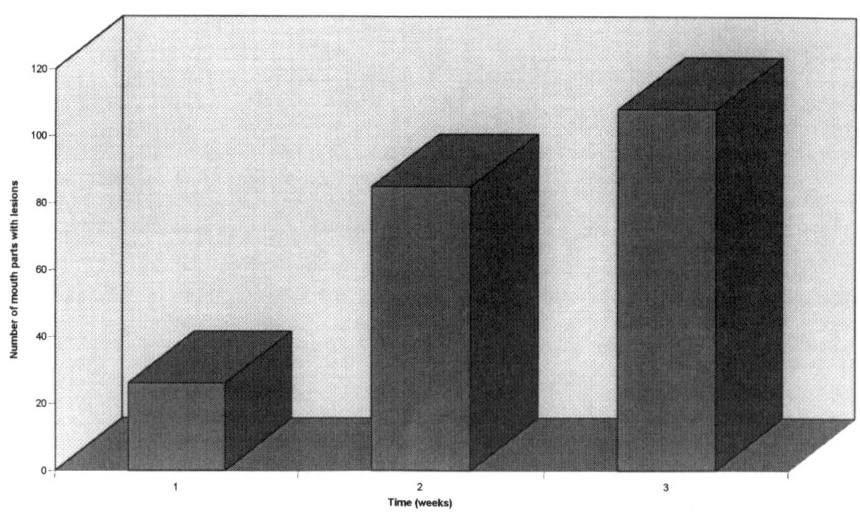

Figure 3 Effect of exposure time on the number of mouth lesions in chickens fed diacetoxyscirpenol (DAS). From Ademoyero and Hamilton, 1991 (ref. [89]).

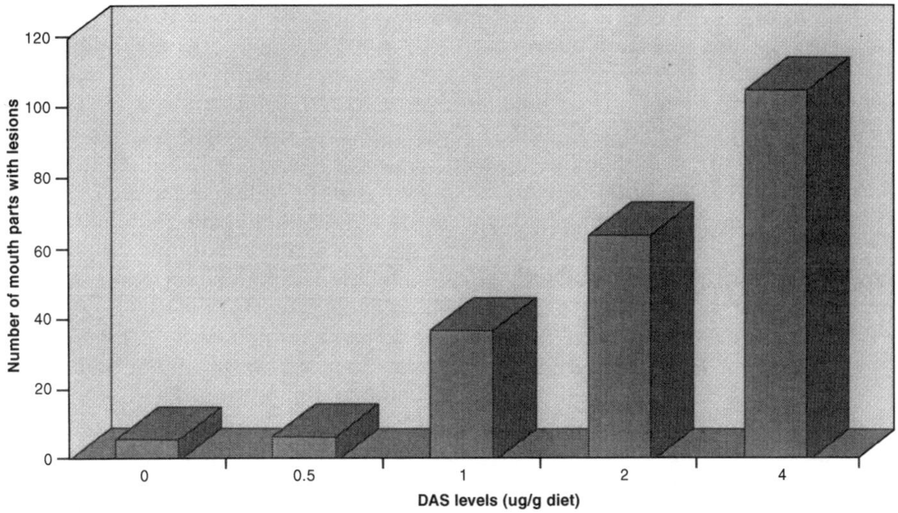

Figure 4 Effect of dietary levels of diacetoxyscirpenol (DAS) on the number of mouth lesions in chickens. From Ademoyero and Hamilton, 1991 (ref. [89]).

Acute assays indicate that ZEN is of relatively low toxicity (Table VI)[4]. However, chronic investigations demonstrate that its oestrogenic properties towards mammals are a more important feature. Thus, ZEN induces vulvovaginitis in premature gilts, anoestrus in cycling females or delayed return into oestrus post-weaning[93]. During pregnancy, ZEN reduces embryonic survival when administered above a threshold level and sometimes decreases foetal weight. ZEN may affect the uterus by decreasing LH and progesterone secretion and by altering the morphology of uterine tissues[93]. Dose level is important in eliciting these effects. Thus ZEN and NIV present in naturally contaminated maize and fed to provide, respectively, 1.8 and 6.9 mg/kg diet produced no adverse effects on sow productivity, measured in terms of conception rate, foetal numbers or foetal weights[94]. An outbreak of ZEN mycotoxicosis in pigs has been reported in Italy[95]. Although none of the sows consuming contaminated feed showed signs of hyperoestrogenism, piglets suckling these sows had swollen, reddened vulvas and some had reddened, necrotic tails. In male pigs, ZEN can depress serum testosterone, weights of testes and spermatogenesis, while inducing feminisation and suppressing libido[96]. This evidence suggests that there may be implications for humans, given the current disquiet about declining sperm counts among men in Europe. As shown in Table III, ZEN is a consistent product of *Fusarium* phytopathogens and its ubiquitous distribution in cereal grains and products is only to be expected. A more pragmatic view is that ZEN might act in concert with other agents to reduce fertility in men. In cows, infertility, reduced milk production and hyperoestrogenism have been linked with ZEN or with *Fusarium* species producing this mycotoxin. When dairy heifers were fed ZEN over three oestrous cycles, conception rates declined from 87% to 62%[97]. Additionally, ZEN from pastures in New Zealand has been implicated in the development of infertility in cattle and sheep[98]. The ovine metabolism of ZEN has recently been proposed to include synthesis of at least five metabolites including zearalanone, α-zearalenol, β-zearalenol, α-zearalanol and β-zearalanol[99]. These transformations may be significant in mammalian fertility since it is well recognised that α-zearalenol is markedly more potent than ZEN as an oestrogen and possible tumour promoter. Possible synergistic effects on fertility with T-2 toxin is possible since both often co-occur in *Fusarium*-contaminated feeds. Indeed, a report of Hungarian studies suggest that ZEN and/or T-2 toxin may cause ovarian dysfunction in cows fed an acidosis-inducing diet[95].

Feeding corn contaminated with *F. moniliforme* is often associated with the induction of leukoencephalomalacia and acute neurotoxicity in equine species and with porcine pulmonary oedema[87]. The former syndrome is an acutely fatal neurological disorder of horses and donkeys, while the latter condition in pigs is characterised by pulmonary oedema as well as pancreatic and liver lesions. Isolates of this fungus have also been found to be hepatocarcinogenic, nephrotoxic and cancer-promoting in rats. There is little doubt now that most, if not all, of these effects are attributable to the consumption of the fumonisins. In studies on the chronic toxicity of FB_1, weanling pigs fed a diet containing the mycotoxin at 100 mg/kg for 7 days followed by a diet containing 190 mg/kg for 83 days, developed nodular hyperplasia of the liver[100]. These nodules, of various diameters, were composed of solid sheets or nests of hepatocytes. In other pigs, the formation of papillary downgrowths of the stratum basale of the distal oesophageal mucosa were observed. FB_1 causes morphological and functional alterations in chicken macrophages *in vitro*, implying an immunosuppressive effect. In addition, it is becoming increasingly apparent that a part of the action of the fumonisins hinges on structural analogy. Their structural relationships with the *Alternaria* HST, A(A)L-toxin, has been described earlier in this review. However, FB_1 also bears a remarkable resemblance to sphinganine and sphingosine, intermediates in the biosynthesis and degradation of sphingolipids[9,30]. Indeed, it has been demonstrated *in vitro* and *in situ* that FB_1 blocks sphingolipid biosynthesis by specifically inhibiting sphinganine-N-acyltransferase (Fig. 5). A consequence of this inhibition is the accumulation of sphingoid bases in the sera of ponies, pigs and rats fed contaminated corn or culture material from *F. moniliforme* containing known levels of FB_1. Sphinganine also accumulates in a dose- and time-dependent manner in chick embryos after yolk sac injection of FB_1[101]. Furthermore, subcutaneous and hepatic haemorrhages were seen in these embryos. Striking effects are seen on mortality of chicken embryos[102]. Thus, eggs inoculated with 1, 10 and 100 μM FB_1 had, respectively, 50%, 70% and 100% mortality. Early embryonic changes included hydrocephalus, enlarged beaks and elongated necks. Pathological lesions were observed in the liver, heart, kidneys, lungs, intestine, testes and brain of FB_1-exposed embryos. FB_1-induced mortality also occurs in broiler chickens in a dose-dependent manner, the onset being more rapid in the dietary presence of moniliformin[103]. Limited work suggests that fumonisins, at concentrations fatal to horses and pigs, can cause mild, reversible hepatic changes and reduced lymphocyte blastogenesis in cattle after chronic feeding[104]. Fusaric acid is a common metabolite of several *Fusarium* species, co-occurring with ZEN, DON and the fumonisins. Although of minor toxicity at levels detected in nature, fusaric acid can enhance the activity of other *Fusarium* mycotoxins. Thus, a toxic interaction between fusaric acid and FB_1 has been demonstrated in the fertile chicken egg. In combination, at 5 μg each, 48% lethality was observed whereas individually, at similar low levels, the mycotoxins had virtually no effect on mortality. Fusaric acid can also cause elevations in the brain levels of serotonin in pigs and a potential synergistic interaction with co-occurring DON has been proposed in feed refusal and emesis in these animals[30].

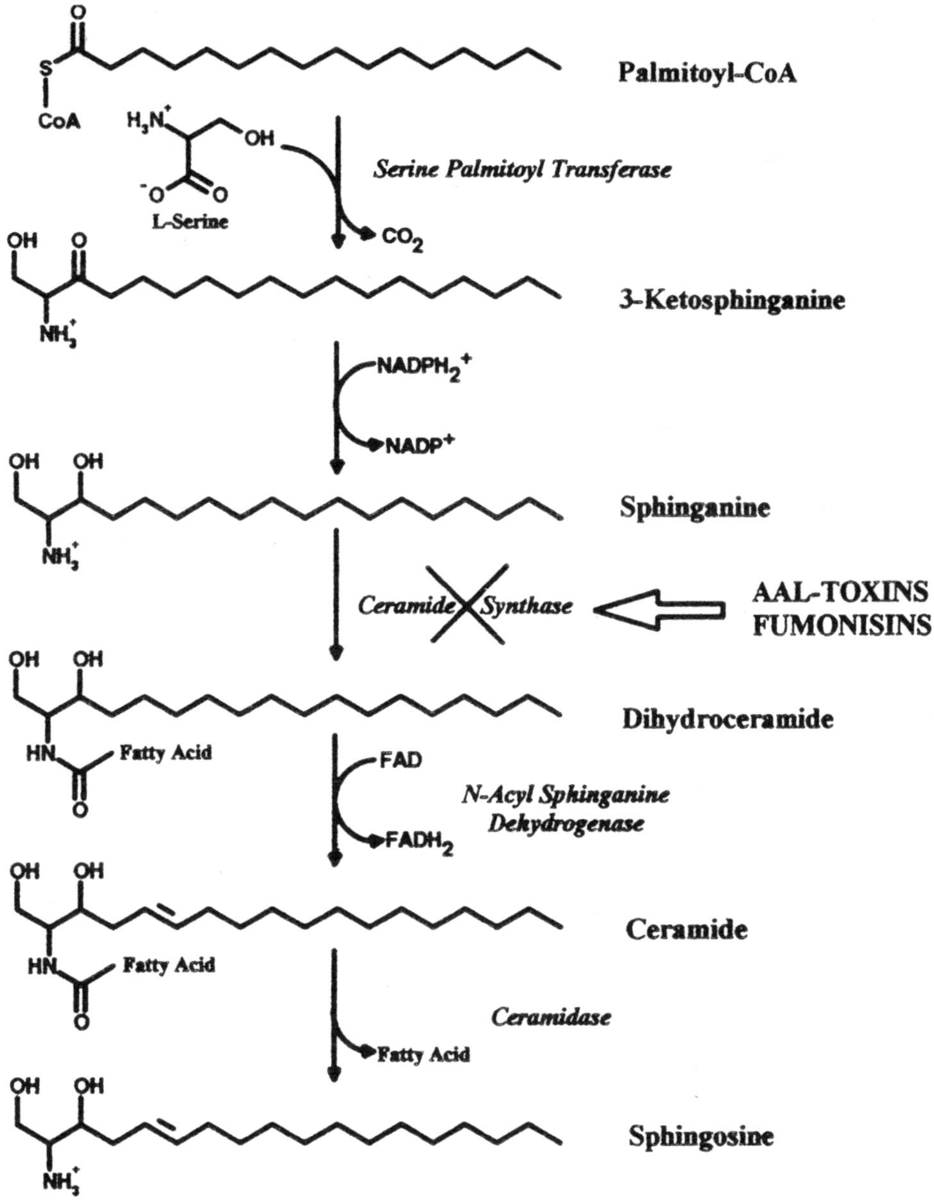

Figure 5 Proposed inhibition of sphingolipid biosynthesis by A(A)L-toxins and fumonisins. Taken from Gilchrist *et al.*, *Can. J. Bot.* **73**, S459–S467 (1995)[9]; [A(A)L = AAL].

The ergopeptine alkaloids of *Acremonium*-infected tall fescue are associated with reduced productivity in cattle and sheep[32]. These effects are mediated through impaired reproductive efficiency and milk production. In beef animals, fescue toxicosis is characterised by reduced growth and increased susceptibility to heat stress. In the case of ergovaline, concentrations as low as 50 ng/g of infected grass are sufficient to elicit at least some signs of fescue toxicosis in heat-stressed cattle.

Neurohormonal imbalances of prolactin and melatonin secretion together with altered neurotransmitter metabolism in the hypothalamus, the pituitary and pineal glands have also been observed in affected steers. In addition, the ergopeptine alkaloids are potent vasoconstrictive agents. It has been proposed that the hormonal changes in combination with restricted blood flow to internal organs may be the primary factors underlying the toxicity of the ergopeptine alkaloids[32].

Perennial ryegrass staggers in ruminants is characterized by neurological symptoms, including incoordination, staggering, shaking of the head and collapse. The quantity of lolitrem B required to precipitate these effects is only 5 μg/g of infected grass[32].

Lupinosis is a disorder which affects the liver of sheep grazing lupin stubble, resulting in ill-thrift, jaundice and death[33]. The causative agent, phomopsin A, interferes with microtubule-dependent intracellular transport mechanisms and acts as an anti-mitotic agent in the liver. Hepatic damage results in hyperammonaemia, which in turn may induce degenerative changes in the central nervous system. Affected sheep may thus appear disorientated. Other overt signs include listlessness and inappetance, with photosensitisation being apparent in some instances. In ewes, ovulation rate is depressed by administration of phomopsins leading to a marked reduction in the incidence of pregnancy and the number of lambs born[105].

Sporidesmins also cause liver damage and photosensitisation in ruminants and fallow deer[34]. The major toxin is sporidesmin A. In calves, a single oral dose of 3 mg/kg can cause 100% mortality within 3–5 days, while a dose of 0.8 mg/kg induces photosensitisation with low mortality. In the liver, sporidesmins cause inflammation of the bile ducts, progressive obliterative cholangitis, obstructive jaundice and phylloerythrin retention. The mucosa and muscular wall of the bladder may also undergo inflammatory changes.

Mycotoxins have long been implicated in human disease (Table VIII)[4–6]. One of the ancient European episodes of mycotoxicoses in humans relates to ergotism (St. Anthony's Fire) caused by the bioactive alkaloids produced in the sclerotia of *Claviceps purpurea*[4]. The alkaloids can cause constriction of peripheral blood capillaries leading to oxygen starvation and gangrene of the limbs. They also act as neurotoxins. Toxicological similarities with the ergopeptine alkaloids of endophyte grasses are worth noting. Balkan endemic nephropathy is a chronic disease occurring in rural populations of Bulgaria, Romania and the former state of Yugoslavia. In affected subjects, the kidneys are markedly reduced in size and histologically, the disease is characterized by tubular degeneration, interstitial fibrosis and glomerular defects. Tubular function is also impaired. The similarities with porcine nephropathy are striking and have led to the conclusion that OA is also the causative agent in Balkan endemic nephropathy[4,5]. A possible endemic ochratoxin-related nephropathy has also been suggested to occur in Tunisia[106]. Affected subjects were classified into those with chronic interstitial nephropathy, chronic glomerular nephropathy and chronic vascular nephropathy. Patients with chronic interstitial nephropathy had blood OA levels of 25–59 μg/l compared with 6–18 μg/l for the other groups and 0.7–7.8 μg/l for the general population.

It is generally acknowledged that had the mycotoxins only affected animal productivity they would have aroused little medical interest, despite the long and varied history of mycotoxicoses in humans (Table VIII). It was their carcinogenicity, with consequent implications for human health, which provided the impetus for world-wide research on mycotoxins. Present-day concern over mycotoxins still centres on their carcinogenic potential in humans. In tropical countries, particularly East and West Africa, India, Thailand and the Philippines, aflatoxicosis is a continuing health issue among the indigenous populations. There is now good epidemiological evdence linking chronic aflatoxin exposure with the incidence of liver cancer in these regions[4,6]. In one study, it was possible to demonstrate that men were more sensitive than women to the carcinogenic effects of aflatoxins but that in both cases there was a linear effect of dose on the development of liver cancer[107]. The epidemiological data should be interpreted with caution, as other factors such as poor diet and disease may have contributed to the incidence of liver cancer. The order of carcinogenicity is $AFB_1 > AFG_1 > AFB_2 > AFG_2$. In toxicological classification, AFB_1 has been designated as a Group 1 carcinogen (=sufficient evidence in humans for carcinogenicity), whereas AFM_1 falls in the Group 2B category (=probable human carcinogen)[6]. The aflatoxins are mainly metabolized by hepatic mixed-function oxidases and cytosolic enzymes and most of the resulting metabolites are further detoxified by conjugation with amino acids, glucuronic acid or bile salts and eliminated via the faeces and urine. Carcinogenicity of AFB_1 emanates from the formation of a reactive epoxide which then permits covalent binding to cellular components such as DNA to yield genotoxic adducts. AFB_1-DNA adduction is greater in the liver than in other organs, consistent with the epidemiological evidence specifically linking the mycotoxin with liver cancer in humans[108]. In experimental animals, levels of liver DNA adduction per unit of AFB_1 dosage generally correlates well with species susceptibility to the mycotoxin. It is important to recognise that the aflatoxins can induce acute effects in humans, and field cases continue to occur despite world-wide appreciation of the toxicological and health implications. Thus in 1974, an outbreak of liver disease occurred in India following the consumption of mouldy corn containing aflatoxins. Of 997 subjects, 97 were reported to have died during this episode[109]. Principal pathological features in the liver included destruction of centrilobular zones, thickening of central veins and cirrhosis. Epidemiological evidence has also been presented to link human oesophageal cancer in South Africa with dietary exposure to the fumonisins[30].

5. *Monitoring and Risk Assessment*

Surveillance is an essential component of the process of assessing risk from mycotoxins. Recent studies show that the risks arising from mycotoxin contamination of foods persist in Africa and Asia, with regional differences now becoming apparent. For example, in Nigeria 25% of maize-based gruels used as weaning food for children were found to be contaminated with aflatoxins, although concentrations were relatively low[110]. OA contamination was perceived not to be a problem. However, it has been suggested that levels of aflatoxin should be maintained at 'irreducible levels', this being defined as 'that concentration of a substance which cannot be eliminated from the food without involving the discarding of the food altogether, severely compromising the availability of major food supplies'. In Kenya[111], three commercial brands of maize flour were contaminated with AFB_1 and AFB_2 at levels ranging from 0.4–20 µg/kg, OA (50–1500 µ/kg) and ZEN (2.5–5.0 mg/kg). OA was the most prevalent mycotoxin in this East African study. In a rural population of southern India, aflatoxin intakes in 9 out of 12 households ranged from 0.33 to 1.5 µg/day, with individual intakes varying from 0.08 to 2.22 µg/day[112]. On a body-weight basis, children were considered to be more at risk, with those aged 1–5 years consuming 47 ng/kg/day, compared with 22 ng/kg/day for individuals over 18 years of age. In a UK survey[36], estimates of OA and patulin intakes have been estimated for different types of consumers. Based on the levels of OA found in cereals, cereal products, pig kidney and black pudding, 'extreme' consumers of all of these foods may have an intake of 9.5 ng/kg/day, with the largest individual intake arising from consumption of cereals rather than from pig products. Extreme consumers of cloudy apple juice, might have patulin intakes of 22.8 and 14.7 μg/kg body-weight per week, respectively, for children aged 2–5 years and for adults in the UK.

In order to obtain a more direct assessment of human exposure to dietary mycotoxins, surveillance is now focusing on measurements of the blood content of these substances, particularly OA. Thus a survey in Sweden showed that plasma levels of OA in most donors were below 0.3 μg/l[113]. In contrast, concentrations in the general population in Tunisia were in the range 0.7–7.8 μg/l and considerably higher in patients with chronic interstitial nephropathy[106].

Human exposure to mycotoxins can also be assessed by analysing breast milk samples. In Sierra Leone, surveillance of breast milk samples from 113 mothers showed that only 10 individuals were mycotoxin-free[114]. Eighty per cent of milk samples contained various aflatoxins and 35% were contaminated with OA. In addition, there was a high incidence of OA with the various aflatoxins in these milk samples. It was concluded that infants in Sierra Leone are exposed to OA and aflatoxins at levels, which in some cases, far exceed those permissible in animal feeds in developed economies. Significant exposure of mothers and babies to OA has also been established in Italy where intakes were calculated to exceed tolerable daily limits estimated from animal models[115].

The link between aflatoxin exposure and liver cancer has been established from epidemiological studies. However, these studies depend upon presumptive intake data rather than direct analyses of metabolites or aflatoxin-DNA adducts. Consequently, in recent years, attention has focused on measurements of the major adducts in tissues and fluids[116]. This would appear to be a promising approach for assessing exposure and risk to individuals within a population, not only with respect to aflatoxins but also in relation to other mycotoxins with carcinogenic potential.

6. *Mycotoxins and Environmental Health*

Exposure to mycotoxins may also occur through inhalation of fungal spores and contaminated dust particles. There is increasing concern about adverse health effects of fungal bio-aerosols on occupants of water-damaged or damp buildings. Under such conditions, *Stachybotrys chartarum* (*atra*) may be the predominant fungus, with the concomitant production of a macrocyclic trichothecene, satratoxin H[35]. Subjects living or working in these buildings may show a variety of symptoms, including pulmonary irritation, headaches and eye disorders as well as chronic fatigue. During the drying and milling of grain and feeding of farm animals, farmers may be exposed to air-borne mycotoxins. On the basis of a recent study in Finland[117], it was concluded that air-borne DON may occur during the handling of grain, suggesting the possibility of inhalation exposure among farmers. In commercial peanut and linseed oil extraction plants, aflatoxin concentrations of up to 250 μg/kg were found in settled and airborne dust. Mortality from total and respiratory cancer may also be higher in workers at these plants, as judged by epidemiological studies[118]. In such assessments, however, it is important to recognise that other factors, for example cigarette smoking, may confound the evidence for respiratory cancer.

7. *Regulatory and Advisory Directives*

The wide distribution, acute toxicity and carcinogenic potential of mycotoxins has led to the establishment of regulatory and advisory directives for primary foods and feeds as well as for their derived products. The information in Table IX is not designed to be exhaustive, but rather illustrative of the type of directives in member states of the European Union and in North America. Van Egmond and Dekker[119] indicate that 90 countries are now known to have regulations relating to maximum permissible levels of mycotoxins in various commodities. However, 13 countries are known to have no regulations and for some 50 countries, mostly in Africa, no data are available. Virtually all developed countries have regulations. Most of the existing mycotoxin regulations relate to the aflatoxins. For example, in the UK, port authorities have applied a 10 µg/kg total aflatoxin limit to imported nuts and dried figs[36]. Consignments exceeding this value have been rejected since implementation of regulations in 1988. The statutory limits were amended and extended in 1992 to reflect recommendations that aflatoxin concentrations in susceptible commodities be reduced to the lowest level 'that is technologically achievable', and to take account of improvements in analytical methodology. The regulations were extended to dried fig products which were also considered to be susceptible to aflatoxin contamination. The limits were reduced to 4 µg/kg for nuts, dried figs and their products for sale or for incorporation in any compound food or for import for direct human consumption; for such imports intended for further processing before sale or incorporation in any compound food for human consumption, the limit for total aflatoxin was set at

Table IX Examples of worldwide regulations for mycotoxins *

Countries	Mycotoxins	Commodities	Maximum levels	Status
European Union	Aflatoxin B_1 (µg/kg)	*Feeding stuffs*		Statutory
		Straight feedingstuffs except:	50	
		groundnut, copra, palm kernel, cottonseed, babassu, maize and products derived from the processing thereof	20	
		Complete feedingstuffs for cattle, sheep and goats (with the exception of complete feedingstuffs for calves, lambs and kids)	50	
		Complete feedingstuffs for pigs and poultry (except those for young animals)	20	
		Other complete feedingstuffs	10	
		Human Foods		
		Cereals, nuts, nut products and dried fruits	2–4	
		Babyfood and food for infants	1–2	
	Aflatoxin M_1 (mg/kg)	Milk/milk products, baby foods and food for infants	0.5	Statutory
	Ochratoxin A (µg/kg)	Cereals and cereal products	4	Advisory
		Raw coffee	4	
	Patulin (µg/kg)	Apple products: juice, cider, puree	50	Advisory
USA/ Canada	Deoxynivalenol (mg/kg)	Wheat products e.g. flour, bran	<1	Advisory

* Based on data presented by Food and Agriculture Organization of the United Nations[119] and Trucksess *et al.*[120].

10 μg/kg. In instances where aflatoxin levels between 4 and 10 μg/kg were found, the importer was required to give a written undertaking to process the batch so that it complied with the 4 μg/kg limit. Alternatively, the consignment could be returned to the consignor, or used for a purpose other than human consumption or destroyed. Schedules for food sampling and analysis of aflatoxins were also provided. The latter included performance parameters for the aflatoxin tests. For example, a detection limit of less than or equal to 2 μg/kg was set for foods intended for direct human consumption[36]. European Union regulations have been set for AFB_1[119]. Limits for selected animal feeds and human foods are presented in Table IX. It is clear that the AFB_1 regulations are more stringent for human foods than for animal feeds. Even with the aflatoxins, however, the tolerated levels vary widely between countries, making harmonisation of regulations an important priority. For a few mycotoxins such as patulin and DON, only advisory directives exist, unsupported by legislative measures. A recent study[120] to determine the extent to which manufacturers in the USA were complying with the advisory level for DON indicated that for a substantial number (11%) of wheat products such as bran, white flour and whole wheat flour, concentrations of DON exceeded the 1 mg/kg limit. For the vast majority of other mycotoxins no regulations exist. Of particular concern, here, is the lack of statutory control of the fumonisins.

It should be recognized that no account is taken of co-occurrence of mycotoxins in the formulation of regulatory and advisory directives. Thus animal feeds based on peanut and maize meals may contain aflatoxins, cyclopiazonic acid and fumonisins, while children in Africa and elsewhere are known to be exposed to combinations of aflatoxins and ochratoxin A[114]. Consequently, major difficulties may be anticipated in future attempts to establish tolerance limits and regulatory guidelines for co-occurring mycotoxins[30].

8. *Preventive and Remedial Measures*

It is axiomatic that preventive measures are of paramount importance in reducing the risk of mycotoxin contamination of different commodities intended for human or animal consumption. When fungicides are used effectively to control fungal diseases of crop plants, then the risk may be minimised. However, under certain conditions certain fungicides may enhance mycotoxin production. In the case of *Fusarium* head blight of cereals, it is generally accepted that fungicide control is only partially effective and the potential exists for mycotoxin contamination of grain. There is growing optimism that, in terms of an environmentally acceptable solution, plant selection and breeding offers considerable potential (Table X). Experimental studies show that breeding maize plants that are resistant to colonisation and ear rot caused by *Aspergillus flavus* generally results in lower contamination of grain with AFB_1[62]. Similarly, exploitation of genetic resistance to *Fusarium* head blight in wheat has been used successfully to reduce DON levels in the grain[63]. Selection of Chinese cultivars of wheat which are resistant to *Fusarium* head blight can also result in lower levels of DON in kernels compared with those of grain from susceptible Canadian cultivars[64]. A similar strategy is being advocated in the control of lupinosis, with the development of *Phomopsis*-resistant cultivars of lupins[33].

In attempts to introduce endophyte-free cultivars of tall fescue or perennial ryegrass to prevent alkaloid toxicity in ruminants, however, agronomic disadvantages might accrue, since both productivity and stand persistence of these grasses may be compromised[121]. Under such conditions, management strategies might be more effective in reducing toxicity of endophyte-infected grasses. For example, managing

Table X Control measures: use of resistant cereal genotypes to reduce mycotoxin contamination of grain *

Cereal type	Genotype	Mycotoxins	Concentrations
Maize	Susceptible	Aflatoxin B_1, μg/kg grain	42–5796
	Resistant		18–198
Wheat	Susceptible	Deoxynivalenol (DON), mg/kg grain	up to 37
	Resistant		3.4–4.6
Wheat	Canadian cultivars	DON, mg/kg grain	17–121
	Chinese cultivars		nd#–9.7

* Based on data of Brown *et al.*[62], Snijders and Perkowski[63] and Wong *et al.*[64] respectively. # nd = not detectable.

grazing to maintain the grass in a vegetative state reduces toxicity by preventing consumption of the seedheads which contain higher levels of the alkaloids. Rotational grazing involving pastures free of tall fescue, particularly during hot weather to offset heat stress, has been advocated as an additional measure[121].

When mycotoxin contamination of different matrices has occurred, a number of remedial options are available. For example, sorting of susceptible foods such as nuts is being attempted. Thus, it has been shown that when pistachio nuts were sorted on the basis of quality, a set of process streams with differing aflatoxin levels were obtained[37]. These levels were correlated with pre-harvest physical damage, such as that caused by hull splitting and insect invasion. Hull discoloration was also linked with high aflatoxin content. Methods to remove DON from contaminated cereal grains primarily depend upon physical separation from the more heavily contaminated outer layers of the kernels[120]. The efficacy of de-contamination varies with the procedures used, but none of these has been shown to be completely effective. In the case of aflatoxin-contaminated oilseeds destined for animal feed, specific detoxification procedures are commercially available in a number of countries. Ammoniation of contaminated oilseed meals appears to be the method of choice[6], involving treatment with either ammonium hydroxide or gaseous ammonia at high temperatures and pressure as in commercial feed mills or at ambient temperature and low pressure for small-scale operations. If the ammoniation reactions are allowed to proceed to completion, the detoxification process is irreversible and aflatoxin contamination is virtually eliminated. Providing that the residual ammonia is dissipated, diets containing the de-contaminated meals are readily consumed by animals without ill-effects. Depending upon the efficacy of de-contamination, residues of AFM_1 in the milk of dairy cows are substantially reduced or absent altogether. The use of phyllosilicate clays (hydrated sodium calcium aluminosilicate) to complex with aflatoxins has also been attempted. Such complexes reduce the gut uptake of aflatoxins and residues are, therefore, lower in body tissues and in milk[6].

Aflatoxin contamination of foods is unavoidable even when good manufacturing practices have been applied[6]. In parts of Africa and Asia, basal levels of aflatoxin in human foods often exceed current norms for *animal* feed in Western Europe and North America. Prevention of aflatoxin-induced cancers is one strategy which may be advocated for individuals at risk in Africa and Asia. Experimentally, it has been shown that anti-oxidants, when administered during aflatoxin exposure,

31. J.-C. Kim, Y.-W. Lee and S.-E. Yu, "Sambutoxin-producing isolates of *Fusarium* species and the occurrence of sambutoxin in rotten potato tubers" *Appl. Environ. Microbiol.* **61**, 3750–3751 (1995).
32. J. K. Porter, "Endophyte alkaloids" **In**: (J. P. F. D'Mello, ed.) *Handbook of Plant and Fungal Toxicants* (CRC Press Inc., Boca Raton, 1997).
33. P. F. Gallagher, B. Walker and J. A. Edgar, "Quantitative analysis of phomopsins in *Phomopsis*-resistant and *Phomopsis*-susceptible lupin stubbles" **In**: (S. M. Colegate and P. R. Dorling, eds.) *Plant-Associated Toxins* (CAB International, Wallingford, 1994).
34. A. Flaoyen and A. Froslie, "Photosensitization disorders" **In**: (J. P. F. D'Mello, ed.) *Handbook of Plant and Fungal Toxicants* (CRC Press Inc., Boca Raton, 1997).
35. E. Johanning, R. Biagini, D. Hull, P. Morey, B. Jarvis and P. Landsbergis, "Health and immunology study following the exposure to toxigenic fungi (*Stachybotrys chartarum*) in a water-damaged office environment" *Int. Arch. Occup. Environ. Health* **68**, 207–218 (1996).
36. Ministry of Agriculture Fisheries and Food, "Mycotoxins: third report" *Food Surveillance Paper No.* **36** (HMSO, London, 1993).
37. T. F. Schatzki and J. Pan, "Distribution of aflatoxin in pistachios. 4. Distribution in small pistachios" *J. Agric. Food Chem.* **45**, 205–207 (1997).
38. J. M. Scholten and M. C. Spanjer, "Determination of aflatoxin B_1 in pistachio kernels and shells" *J. AOAC Int.* **79**, 1360–1364 (1996).
39. T. F. Schatzki, "Distribution of aflatoxin in almonds" *J. Agric. Food Chem.* **44**, 3595–3597 (1996).
40. M. A. Doster, T. J. Michailides and D. P. Morgan, "*Aspergillus* species and mycotoxins in figs from California orchards" *Plant Dis.* **80**, 484–489 (1996).
41. M. I. Selim, W. Popendorf, M. S. Ibrahim, S. E. Sharkawy and E. S. E. Kashory, "Aflatoxin B_1 in common Egyptian foods" *J. AOAC Int.* **79**, 1124–1129 (1996).
42. S. I. Phillips, P. W. Wareing, A. Dutta, S. Panigrahi and V. Medlock, "The mycoflora and incidence of aflatoxin, zearalenone and sterigmatocystin in dairy feed and forage samples from Eastern India and Bangladesh" *Mycopathologia* **133**, 15–21 (1996).
43. J. Prieta, M. A. Moreno, S. Diaz, G. Suarez and L. Dominguez, "Survey of patulin in apple juice and children's apple food by the diphasic dialysis membrane procedure" *J. Agric. Food Chem.* **42**, 1701–1703 (1994).
44. S. Krivobok, F. Seigle-Murandi, R. Steiman and E. E. Creppy, "Fungal flora and ochratoxin A production in various food and feed in France" *Appl. Microbiol.* **18**, 455–459 (1995).
45. P. M Scott, "The natural occurrence of trichothecenes" **In**: (V. R. Beasley, ed.) *Trichothecene Mycotoxicosis: Pathophysiologic Effects* (CRC Press Inc., Boca Raton, 1989).
46. H. M. Muller and K. Schwadorf, "A survey of the natural occurrence of *Fusarium* toxins in wheat grown in southwestern area of Germany" *Mycopathologia* **121**, 115–121 (1993).
47. J. Perkowski, R. D. Plattner, P. Golinski, R. F. Vesonder and J. Chelkowski, "Natural occurrence of deoxynivalenol, 15-acetyl-deoxynivalenol, nivalenol, 4,7-dideoxynivalenol and zearalenone in Polish wheat" *Mycotoxin Res.* **6**, 7–12 (1990).
48. V. Hietaniemi and J. Kumpulainen, "Contents of *Fusarium* toxins in Finnish and imported grains in feeds" *Food Add. Contam.* **8**, 171–182 (1991).
49. T. Tanaka, S. Yamamoto, A. Hasegawa, N. Aoki, J. R. Besling, Y. Sugiura and Y. Ueno, "A survey of the natural occurrence of *Fusarium* mycotoxins, deoxynivalenol, nivalenol and zearalenone in cereals harvested in The Netherlands" *Mycopathologia* **110**, 19–22 (1990).
50. Y. Sugiura, K. Fukasaku, T. Tanaka, Y. Matsui and Y. Ueno, "*Fusarium poae* and *Fusarium crookwellense*, fungi responsible for the natural occurrence of nivalenol in Hokkaido" *Appl. Environ. Microbiol.* **59**, 3334–3338 (1993).
51. T. Tanaka, R. Teshima, H. Ikebuchi, J. Sawada and M. Ichinoe, "Sensitive enzyme-linked immunosorbent assay for the mycotoxin zearalenone in barley and Job's-tears" *J. Agric. Food Chem.* **43**, 946–950 (1995).
52. C. Fernandez, M. E. Stack and S. M. Musser, "Determination of deoxynivalenol in 1991 U.S. winter and spring wheat by high-performance thin-layer chromatography" *J. AOAC Int.* **77**, 628–630 (1994).
53. M. W. Trucksess, F. Thomas, K. Young, M. E. Stack, W. J. Fulgueras and S. W. Page, "Survey of deoxynivalenol in U.S. 1993 wheat and barley crops by enzyme-linked immunosorbent assay" *J. AOAC Int.* **78**, 631–636 (1995).
54. G. W. Stratton, A. R. Robinson, H. C. Smith, L. Kittilsen and M. Barbour, "Levels of five mycotoxins in grains harvested in Atlantic Canada as measured by high performance chromatography" *Arch. Environ. Contam. Toxicol.* **24**, 399–409 (1993).
55. M. B. Doko, S. Rapior, A. Visconti and J. E. Schjoth, "Incidence and levels of fumonisin contamination in maize genotypes grown in Europe and Africa" *J. Agric. Food Chem.* **43**, 429–434 (1995).
56. M. B. Doko, C. Canet, N. Brown, E. W. Sydenham, S. Mpuchane and B. A. Siame, "Natural co-occurrence of fumonisins and zearalenone in cereals and cereal-based foods from Eastern and Southern Africa" *J. Agric. Food Chem.* **44**, 3240–3243 (1996).

57. S. N. Chulze, M. L. Ramirez, M. C. Farnochi, M. Pascale, A. Visconti and G. March, "*Fusarium* and fumonisin occurrence in Argentinian corn at different ear maturity stages" *J. Agric. Food Chem.* **44**, 2797–2801 (1996).
58. O. M. Viquez, M. E. Castell-Perez and R. A. Shelby, "Occurrence of fumonisin B_1 in maize grown in Costa Rica" *J. Agric. Food Chem.* **44**, 2789–2791 (1996).
59. M. W. Trucksess, M. E. Stack, S. Allen and N. Barrion, "Immunoaffinity column coupled with liquid chromatography for determination of fumonisin B_1 in canned and frozen sweet corn" *J. AOAC Int.* **78**, 705–710 (1995).
60. P. M. Scott, "Mycotoxins transmitted into beer from contaminated grains during brewing" *J. AOAC Int.* **79**, 875–882 (1996).
61. J. K. Porter, "Analysis of endophyte toxins: fescue and other grasses toxic to livestock" *J. Anim. Sci.* **73**, 871–880 (1995).
62. R. L. Brown, T. E. Cleveland, G. A. Payne, C. P. Woloshuk, K. W. Campbell and D. G. White, "Determination of resistance to aflatoxin production in maize kernels and detection of fungal colonization using an *Aspergillus flavus* transformant expressing *Escherichia coli* B-glucuronidase" *Phytopathology* **85**, 983–989 (1995).
63. C. H. A. Snijders and J. Perkowski, "Effects of head blight caused by *Fusarium culmorum* on toxin content and weight of wheat kernels" *Phytopathology* **80**, 566–570 (1990).
64. L. S. L. Wong, D. Abramson, A. Tekauz, D. Leisle and R. I. H. McKenzie, "Pathogenicity and mycotoxin production of *Fusarium* species causing head blight in wheat cultivars varying in resistance" *Can. J. Plant Sci.* **75**, 261–267 (1995).
65. J. P. F. D'Mello, A. M. C. Macdonald and M. P. Cochrane, "A preliminary study of the potential for mycotoxin production in barley grain" *Aspects Appl. Biol.* **36**, 375–382 (1993).
66. B. Cahagnier, D. Melcion and D. Richard-Molard, "Growth of *Fusarium moniliforme* and its biosynthesis of fumonisin B_1 on maize grain as a function of different water activities" *Lett. Appl. Microbiol.* **20**, 247–251 (1995).
67. C. M. Placinta, A. M. C. Macdonald, J. P. F. D'Mello and R. Harling, "The influence of carbendazim on mycotoxin production in *Fusarium sporotrichioides*" *Proceedings of Brighton Crop Protection Conference* (1996) **Vol. 1**, pp. 415–416.
68. N. F. Sommer, J. R. Buchanan and R. J. Fortlage, "Relation of early splitting and tattering of pistachio nuts to aflatoxin in the orchard" *Phytopathology* **76**, 692–694 (1986).
69. T. Fernando and G. Bean, "Effects of tricyclazole on growth, release of aflatoxin and sterol and fatty acid content, by *Aspergillus* isolates" *Trans. Br. Mycol. Soc.* **87**, 445–449 (1986).
70. R. L. Buchanan, S. B. Jones and H. G. Stahl, "Effect of miconazole on growth and aflatoxin production by *Aspergillus parasiticus*" *Mycopathologia* **100**, 135–144 (1987).
71. J. P. F. D'Mello and A. M. C. Macdonald, "Production and control of mycotoxins from *Fusarium* species pathogenic on cereals" *Proceedings of Brighton Crop Protection Conference* (1996) **Vol. 2**, pp. 517–522.
72. D. Boyacioglu, N. S. Hettiarachchy and R. W. Stack, "Effect of three systemic fungicides on deoxynivalenol (vomitoxin) production by *Fusarium graminearum* in wheat" *Can. J. Plant Sci.* **72**, 93–101 (1992).
73. M. Gareis and J. Ceynowa, "Influence of the fungicide Matador (tebuconazole/triadimenol) on mycotoxin production by *Fusarium culmorum*" *Z. Lebensm. Unters. Forsch.* **198**, 244–248 (1994).
74. F. A. Draughon and J. C. Ayres, "Inhibition of aflatoxin production by selected insecticides" *Appl. Environ. Microbiol.* **41**, 972–976 (1981).
75. F. A. Draughon, M. E. Elahi and D. R. West, "Insecticide inhibition of aflatoxin production in corn" *J. Agric. Food Chem.* **31**, 692–695 (1983).
76. M. T. Wu and J. C. Ayres, "Effects of dichlorvos on ochratoxin production" *J. Agric. Food Chem.* **22**, 536–537 (1974).
77. F. A. Draughon and J. C. Ayres, "Insecticide inhibition of growth and patulin production by *Penicillium urticae*" *J. Food Sci.* **44**, 1232–1234 (1979).
78. Y. C. Berisford and J. C. Ayres, "Use of the insecticide naled to control zearalenone production" *J. Agric. Food Chem.* **24**, 973–975 (1976).
79. R. C. Braunberg, C. N. Barton, O. O. Gantt and L. Friedman, "Interaction of citrinin and ochratoxin A" *Nat. Toxins* **2**, 124–131 (1994).
80. D. Abramson, E. Usleber and E. Martlbauer, "Determination of citrinin in barley by indirect and direct enzyme immunoassay" *J. AOAC Int.* **79**, 1325–1329 (1996).
81. E. Charmley, H. L. Trenholm, B. K. Thompson, D. Vudathala, J. W. G. Nicholson, D. B. Prelusky and L. L. Charmley, "Influence of level of deoxynivalenol in the diet of dairy cows on feed intake, milk production and its composition" *J. Dairy Sci.* **76**, 3580–3587 (1993).
82. B. A. Rotter, B. K. Thompson and M. Lessard, "Effects of deoxynivalenol-contaminated diet on performance and blood parameters in growing swine" *Can. J. Anim. Sci.* **75**, 297–302 (1995).

83. H. L. Trenholm, B. C. Foster, L. L. Charmley, B. K. Thompson, K. E. Hartin, R. W. Coppock and M. A. Albassam, "Effects of feeding diets containing *Fusarium* (naturally) contaminated wheat or pure deoxynivalenol (DON) in growing pigs" *Can. J. Anim. Sci.* **74**, 361–369 (1994).
84. B. Kollarczik, M. Gareis and M. Hanelt, "*In vitro* transformation of the *Fusarium* mycotoxins deoxynivalenol and zearalenone by the gut microflora of pigs" *Nat. Toxins* **2**, 105–110 (1994).
85. M. K. Gibson, S. J. Bursian and R. J. Aulerich, "Effects of deoxynivalenol on feed consumption and body weight gains in mink (*Mustela vison*)" *Bull. Environ. Contam. Toxicol.* **51**, 6–11 (1993).
86. D. B. Prelusky, H. L. Trenholm, R. M. G. Hamilton and J. D. Miller, "Transmission of $[^{14}C]$ deoxynivalenol to eggs following oral administration to laying hens" *J. Agric. Food Chem.* **35**, 182–186 (1987).
87. D. B. Prelusky, B. A. Rotter and R. G. Rotter, "Toxicology of mycotoxins" **In**: (J. D. Miller and H. L. Trenholm, eds.) *Mycotoxins in Grain. Compounds Other Than Aflatoxin* (Eagen Press, St. Paul, 1994).
88. S. Fekete, G. Huszenicza, M. Szilagyi, M. Albert and A. Szilagyi, "Effect of long-term feeding of sublethal quantities of T-2 toxin upon the ovarian activity of the rabbit" *J. Appl. Rabbit Res.* **15**, 583–593 (1992).
89. A. A. Ademoyero and P. B. Hamilton, "Mouth lesions in broiler chickens caused by scirpenol mycotoxins" *Poult. Sci.* **70**, 2082–2089 (1991).
90. A. A. Ademoyero and P. B. Hamilton, "High dietary fat increases toxicity of diacetoxyscirpenol in chickens" *Poult. Sci.* **70**, 2271–2274 (1991).
91. D. W. Friend, B. K. Thompson, H. L. Trenholm, H. J. Boermans, K. E. Hartin and P. L. Panich, "Toxicity of T-2 toxin and its interaction with deoxynivalenol when fed to young pigs" *Can. J. Anim. Sci.* **72**, 703–711 (1992).
92. G. A. Weaver, H. J. Kurtz, C. J. Mirocha, F. Y. Bates, J. C. Behrens, T. S. Robison and W. F. Gipp, "Mycotoxin-induced abortions in swine" *Can. Vet. J.* **19**, 72–74 (1978).
93. M. Etienne and J.-Y. Dourmad, "Effects of zearalenone or glucosinolates in the diet on reproduction in sows: a review" *Livest. Prod. Sci.* **40**, 99–113 (1994).
94. K. C. Williams and B. J. Blaney, "Effect of the mycotoxins, nivalenol and zearalenone in maize naturally infected with *Fusarium graminearum* on the performance of growing and pregnant pigs" *Aust. J. Agric. Res.* **45**, 1265–1279 (1994).
95. M. W. Trucksess, "Mycotoxins" *J. AOAC Int.* **80**, 119–126 (1997).
96. M. A. Diekman and M. L. Green, "Mycotoxins and reproduction in domestic livestock" *J. Anim. Sci.* **70**, 1615–1627 (1992).
97. G. A. Weaver, H. J. Kurtz, J. C. Behrens, T. S. Robison, B. E. Seguin, F. Y. Bates and C. J. Mirocha, "Effect of zearalenone on the fertility of virgin dairy heifers" *Am. J. Vet. Res.* **47**, 1395–1397 (1986).
98. N. R. Towers and J. M. Sprosen, "Zearalenone-induced infertility in sheep and cattle in New Zealand" *N. Z. Vet. J.* **41**, 223–224 (1993).
99. C. O. Miles, A. F. Erasmuson, A. L. Wilkins, N. R. Towers, B. L. Smith, I. Garthwaite, B. G. Scahill and R. P. Hansen, "Ovine metabolism of zearalenone to α-zearalanol (zeranol)" *J. Agric. Food Chem.* **44**, 3244–3250 (1996).
100. S. W. Casteel, J. R. Turk, R. P. Cowart and G. E. Rottinghaus, "Chronic toxicity of fumonisin in weanling pigs" *J. Vet. Diagn. Invest.* **5**, 413–417 (1993).
101. C. Zacharias, G. V. Echten-Deckert, E. Wang, A. H. Merrill and K. Sandhoff, "The effect of fumonisin B_1 on developing chick embryos: correlation between de novo sphingolipid biosynthesis and gross morphological changes" *Glycoconjugate J.* **13**, 167–175 (1996).
102. T. Javed, J. L. Richard, G. A. Bennett, M. A. Dombrink-Kurtzman, R. M. Bunte, K. W. Koelkebeck, L. M. Cote, R. W. Leeper and W. B. Buck, "Embryopathic and embryocidal effects of purified fumonisin B_1 or *Fusarium proliferatum* culture material extract on chicken embryos" *Mycopathologia* **123**, 185–193 (1993).
103. T. Javed, G. A. Bennett, J. L. Richard, M. A. Dombrink-Kurtzman, L. M. Cote and W. B. Buck, "Mortality in broiler chicks on feed amended with *Fusarium proliferatum* culture material or with purified fumonisin B_1 and moniliformin" *Mycopathologia* **123**, 171–184 (1993).
104. G. D. Osweiler, M. E. Kehrli, J. R. Stabel, J. R. Thurston, P. F. Ross and T. M. Wilson, "Effects of fumonisin-contaminated corn screenings on growth and health of feeder calves" *J. Anim. Sci.* **71**, 459–466 (1993).
105. A. L. Barnes, K. P. Croker, J. G. Allen and N. D. Costa, "Lupinosis of ewes around the time of mating reduces reproductive performance" *Aust. J. Agric. Res.* **47**, 1305–1314 (1996).
106. K. Maaroufi, A. Achour, M. Hammami, M. El-May, A. M. Betbeder, F. Ellouz, E. E. Creppy and H. Bacha, "Ochratoxin A in human blood in relation to nephropathy in Tunisia" *Human Exp. Toxicol.* **14**, 609–614 (1995).
107. J. E. Smith and M. O. Moss, *Mycotoxins: Formation, Analysis and Significance* (John Wiley & Sons Ltd., Chichester, 1985).

108. G. S. Bailey, "Role of aflatoxin-DNA adducts in the cancer process" **In**: (D. L. Eaton and J. D. Groopman, eds.) *The Toxicology of Aflatoxins. Human Health, Veterinary, and Agricultural Significance* (Academic Press, Inc., San Diego, 1994).
109. H. D. Tandon and B. N. Tandon, "Pathology of liver in an outbreak of aflatoxicosis in man with a report on the follow up" **In**: (S. Natori, K. Hashimoto and Y. Ueno, eds.) *Mycotoxins and Phycotoxins* (Elsevier Science Publishers, Amsterdam, 1989).
110. O. A. Oyelami, S. M. Maxwell and E. Adeoba, "Aflatoxins and ochratoxin A in the weaning food of Nigerian children" *Ann. Trop. Paediat.* **16**, 137–140 (1996).
111. G. K. Muriuki and G. M. Siboe, "Maize flour contaminated with toxigenic fungi and mycotoxins in Kenya" *Afric. J. Health Sci.* **2**, 236–241 (1995).
112. S. Vasanthi, R. V. Bhat and G. Subbulakshmi, "Aflatoxin intake from maize-based diets in a rural population in Southern India" *J. Sci. Food Agric.* **73**, 226–230 (1997).
113. A. Breitholtz, M. Olsen, A. Dahlback and K. Hult, "Plasma ochratoxin A levels in three Swedish populations surveyed using an ion-pair HPLC technique" *Food Addit. Contam.* **8**, 183–187 (1991).
114. F. E. Jonsyn, S. M. Maxwell and R. G. Hendrickse, "Ochratoxin A and aflatoxins in breast milk samples from Sierra Leone" *Mycopathologia* **131**, 121–126 (1995).
115. M. Miraglia, A. Dominicis, C. Brera, S. Corneli, E. Cava, E. Menghetti, E. Miraglia and A. De-Dominicis, "Ochratoxin A levels in human milk and related food food samples: an exposure assessment" *Nat. Toxins* **3**, 436–444 (1995).
116. J. D. Groopman, "Molecular dosimetry for assessing human aflatoxin exposures" **In**: (D. L. Eaton and J. D. Groopman, eds.) *The Toxicology of Aflatoxins. Human Health, Veterinary, and Agricultural Significance* (Academic Press, Inc., San Diego, 1994).
117. S. Lappalainen, M. Nikulin, S. Berg, P. Parikka, E.-L. Hintikka and A.-L. Pasanen, "*Fusarium* toxins and fungi associated with handling of grain on eight Finnish farms" *Atmos. Environ.* **30**, 3059–3065 (1996).
118. K. M. Hendry and E. C. Cole, "A review of mycotoxins in indoor air" *J. Toxicol. Environ. Health* **38**, 183–198 (1993).
119. Food and Agriculture Organization of the United Nations (FAO), "Worldwide regulations for mycotoxins 1995" *FAO Food and Nutrition Paper* **64** (FAO, Rome, 1997).
120. M. W. Trucksess, D. E. Ready, M. K. Pender, C. A. Ligmond, G. E. Wood and S. W. Page, "Determination and survey of deoxynivalenol in white flour, whole wheat flour and bran" *J. AOAC Int.* **79**, 883–887 (1996).
121. R. E. Joost, "*Acremonium* in fescue and ryegrass: boon or bane? A review" *J. Anim. Sci.* **73**, 881–888 (1995).
122. T. W. Kensler, P. A. Egner, M. A. Trush, E. Bueding and J. D. Groopman, "Modification of aflatoxin B_1 binding to DNA *in vivo* in rats fed phenolic antioxidants, ethoxyquin and a dithiolthione" *Carcinogenesis* **6**, 759–763 (1985).
123. A. F. Rizzo, F. Atroshi, M. Ahotupa, S. Sankari and E. Elovaara, "Protective effect of antioxidants against free radical-mediated lipid peroxidation induced by DON or T-2 toxin" *J. Vet. Med.* **A41**, 81–90 (1994).
124. G. S. Stoewsand, J. B. Babish and H. C. Wimberly, "Inhibition of hepatic toxicities from polybrominated biphenyls and aflatoxin B_1 in rats fed cauliflower" *J. Environ. Pathol. Toxicol.* **2**, 399–406 (1978).
125. J. P. F. D'Mello, "Toxic compounds from fruit and vegetables" **In**: (F. A. Tomas-Barberan and R. J. Robins, eds.) *Phytochemistry of Fruit and Vegetables* (Clarendon Press, Oxford, 1997).

14. NATURAL PLANT TOXINS—BENEFITS AND RISKS

A. B. HANLEY * and S. J. MACDONALD

INTRODUCTION

The term 'natural toxin' contains an internal inconsistency. 'Natural' superficially denotes health and fitness while 'toxin' implies harm. This simplistic approach is, however, inappropriate to describe the complex interplay between living organisms and their environment. Studies on apoptosis (programmed cell death), for example, illustrate that it is essential to the health of the whole organism that is able to remove damaged cells in a controlled manner. Cell death can, therefore, protect the organism against proliferative and uncontrolled cell growth. A toxin may exhibit a range of effects depending on a number of factors including dose, the presence of other bioactive constituents, and the susceptibility of the individual. These effects, while overtly toxic, may be beneficial therefore "toxin" could cover the range of effects.

The use of 'natural' to denote health does not survive even a cursory examination. Some of the most acute toxins known are natural substances for example, ricin, strychnine, heroin, etc.

Plant toxins are consumed on a day to day basis, however, foods do not generally contain acutely toxic components. The toxins present in foods are therefore chronically rather than acutely bioactive. In addition, food is a complex system and the effects of an individual compound (or group of compounds) can be greatly affected by dose, the presence of other components (synergistic effects), bioavailability and individual susceptibility. The general rules which apply to toxins noted above also apply to plant toxins consumed as a part of the diet. The definition of a toxin and the assessment of potential benefits and risks is, therefore, of great importance.

A natural plant toxin is a distinct compound (or group of compounds) which elicit a measurable, abnormal biological effect when ingested. Compounds which are essential to primary metabolism (e.g. carbohydrates, proteins, most lipids and essential vitamins and minerals) can be excluded. The biological effect is most often either beneficial or deleterious, depending upon circumstances (dose, individuals susceptibility etc). The bioactive principle should be identifiable and, where possible, be both chemically and biologically measurable and quantifiable. This brief review will make use of specific examples to illustrate the types of components which constitute natural plant toxins. Various aspects of the chosen toxins will be examined in order to define the most important characteristics. Their biosynthesis must be examined since this helps to define their utility to plants, the similarities and links between different toxins and their occurrence. Quantification of plant toxins is important and methods which have been used for their analysis and structural characterisation will be reviewed. Finally, the biological activity of the toxins is important and must be considered to determine the risks or benefits associated with their presence as dietary components. Before beginning these sections, it is useful to

* CSL Food Science Laboratory, Norwich Research Park, Colney Norwich NR4 7UQ (UK).

consider the natural function of plant toxins and to briefly describe their general occurrence.

Plant toxins appear to have several functions. The most obvious is as a part of the plant defence system to discourage herbivorous attack. Other potential uses of such compounds are, as a means to prevent other plants from growing nearby, as attractants of pollinating insects and as a means to maintain important biosynthetic pathways (even if they are not immediately required). Compounds can be used or classified in ways which are very different from their primary biological purpose to the plant. For example, the indigo plant (*Indigofera spicata*), as the name implies, produces the precursor of indigo, indican. The principal utility of the compound to the plant is presumably not as a dyestuff. The natural functions of toxins may, therefore be substantially different from that employed by man.

After definition and function, a third area of importance in the study of natural plant toxins is their occurrence and relative importance. Most plants contain toxins at varying levels. One possible definition of a food plant is that it is a plant whose toxic load can be tolerated upon consumption and hence for which benefits of ingestion outweigh disadvantages. Clearly some compounds have more immediate effects than others and there is variability in the relative significance from a health standpoint. In addition, the characterisation of some is more complete than others. The examples chosen serve to illustrate these points. It is only by considering a range of representative food toxins that a clear picture of the factors which must be taken into account can be determined, assessed, prioritised and quantified.

The examples of plant toxins have been chosen to illustrate the general principles outlined above. The glucosinolates are a group of amino acid-derived compounds present in brassicaceous plants; phytoestrogens—a structurally diverse group of compounds linked by a common biological activity—occur widely in food plants such as soya; pyrrolizidine alkaloids are a group of acutely toxic compounds which occur in certain herbal remedies and as contaminants in some cereal crops, and, finally, the neurotoxin 3-nitropropanoic acid, which occurs both as the free acid and as its glycoside in a range of plants including indigo and has a very specific mode of biological activity. Each of these groups will serve to illustrate different aspects of the area of plant toxins—their risks and benefits.

BIOSYNTHESIS

The biosynthesis of secondary metabolites has been an area of considerable interest to organic chemists for a number of years[1]. Once the structures of compounds had been established, it became clear that certain of them were related in structure and biogenesis[2]. This promoted a reductionist approach to biosynthesis where a given molecule was defined in terms of its skeleton or backbone[3]. Additional structural features (e.g. methylation, acylation, acetylation) were largely ignored in the establishment of biosynthetic theories. Such approaches are the intellectual precursors to some more recent synthetic strategies. Once a theory concerning the biosynthesis of a given compound or group of compounds has been developed, the next stage is to test that theory. Adopting a reductionist approach once again, the most convincing proof is to isolate every enzyme involved in the biosynthetic pathway and to demonstrate that, when they perform their reactions sequentially starting with the putative precursor, the final product is identical to that found in the plant. This rigorous approach requires a huge level of effort and, in some cases, will not succeed because of the instability either of the intermediates or the enzymes or both[4].

A simpler strategy involves finding natural mutants in which the pathway is incomplete and isolating those stable intermediates which are produced. This approach can be successful in plants however it has generally been of greater value in the study of biosynthesis in prokaryotes (e.g. bacteria) where natural and induced mutants are easier to find or create. An alternative approach is to determine the precursorial relationship of series of putative compounds as intermediates in the proposed pathway by administering the compounds individually to the plant and looking for incorporation into the final product. Clearly the precursor must be 'tagged' in some way and this is generally carried out by radiolabelling (incorporation of ^{14}C or ^{3}H) and looking for radiolabelled product. This is an immensely powerful technique, however these experiments must be meticulously planned and the results interpreted with care. In addition, it will only provide limited information on mechanisms of formation of products. A more powerful and useful strategy is to use multiply labelled precursors containing either stable or radiolabelled atoms. This can provide significant information on precursorial relationships, likely pathways, mechanisms of formation and the stereochemistry of reactions involved in the steps in a pathway. The examples chosen will illustrate some of these principles and provide a framework for the study of the biosynthesis of plant toxins generally. Finally, one potential result from an understanding of the biosynthesis of plant toxins is the ability to manipulate a pathway so as to produce either more or less of a given product depending upon its relative value as a dietary or agronomic constituent. This is an area of great interest and examples of improved plant derived foods are becoming more and more common place.

3-NITROPROPANOIC ACID

The neurotoxin, 3-nitropropanoic acid occurs as its glycoside, hiptagin in a range of plants of the Fabaceae family[5]. It was first isolated from *Hiptage mandblata* in 1920[6] and has subsequently been found in *Carynocarpus laevigata* and *Indigofera spicata*[7]. In the case of the latter species, it was isolated as the free acid rather than the glycoside. The major studies on 3-nitropropanoic acid have been carried out in fungi (especially in *Penicillium* and *Aspergillus*) genera[8]. It is an inhibitor of succinate dehydrogenase with a clearly defined mode of action[9] and this has encouraged both biosynthetic and other biological investigations. Its biosynthesis has been extensively studied in fungi where it has been demonstrated to be derived from aspartate via (S)-nitrosuccinate with retention of all hydrogen atoms of the precursor[10]. The final step in the biosynthesis has been demonstrated to be a spontaneous decarboxylation[11]. The biosynthesis of the glycoside and the free acid in plants follows a completely different pathway although it has been comparatively little studied. Radiolabelling studies suggested that both malonic acid and malonylmonohydroxamate were precursors, however, incorporation of C-2 labelled malenic acid occurred at both C-2 and C-3[12]. A plausible mechanism of direct formation of 3-nitropropanoic acid is shown in Fig. 1; however, this would not lead to the incorporation of ^{14}C at C-3. One alternative potential mechanism is also shown in Fig. 1. Conversion of malonic acid to oxaloacetic acid (perhaps induced by feeding labelled malonic acid thereby reversing the normal direction of the reaction catalysed by the enzyme oxaloacetate decarboxylase) could occur and subsequent addition of a nitro group with concomitant elimination of the carboxyl would give 3-nitropropanoic acid with a label at C-3. Direct nitration is, biologically, a rare reaction however the pathway could proceed by initial addition of nitrogen at a

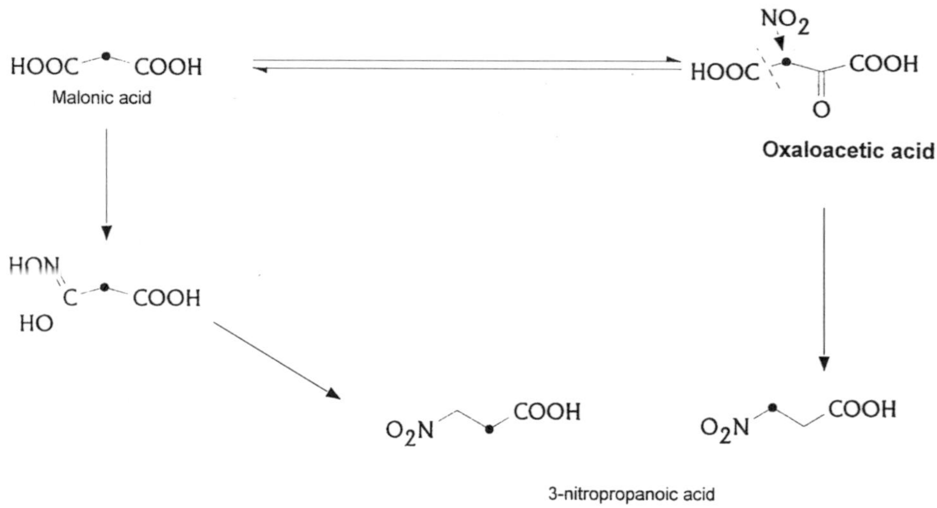

Figure 1 The biosynthesis of 3-nitropropanoic acid in indigo.

lower oxidation state followed by oxidation to give the nitro substituent. The combination of the direct and the indirect routes would result in the production of 3-nitropropanoic acid with the observed (C-2 and C-3) pattern.

One major lesson to be learnt from this example is that biosynthetic pathways can be significantly different in different organisms. The discovery of the same toxin in two different organisms therefore may not be of any great significance from a biological or biosynthetic standpoint.

GLUCOSINOLATES

The biosynthesis of glucosinolates has been extensively studied and many details of their biosynthesis have been worked out (Fig. 2)[13]. They are derived from amino acids which are chain lengthened by addition of acetate units. The pathway proceeds via an aldoxime which is converted into a thiohydroxamate and thence to a desulpho glucosinolate and finally to the glucosinolate (Fig. 2). A putative pathway involving the intermediacy of a nitro compound[14] remains unproven however, by analogy with 3-nitropropanoic acid, spontaneous decarboxylation of such a compound could result in formation of the aldoxime. The major interest in glucosinolates arises from their breakdown by a co-occurring thioglucosidase enzyme, myrosinase which leads to the formation of a range of products including organic nitriles, isothiocyanates and thiocyanates (see below)[15]. One of the most potentially toxic of the glucosinolates is progoitrin (2S-but-3-enyl glucosinolate) which, gives an isothiocyanate, which, in turn, spontaneously cyclises to give goitrin (vinyl oxazolidimethione) (Fig. 3). There is some interest in preventing the formation of progoitrin by inhibiting the hydroxylation of the precursor, but-3-enyl glucosinolate in the plant. If this could be carried out then the resulting plant would be likely to have enhanced agronomic characteristics as well as improved nutritional value since the enzymic product from but-2-enyl glucosinolate (but-3-enyl isothiocyanate) is comparatively non-toxic to humans but is an insect antifeedant.

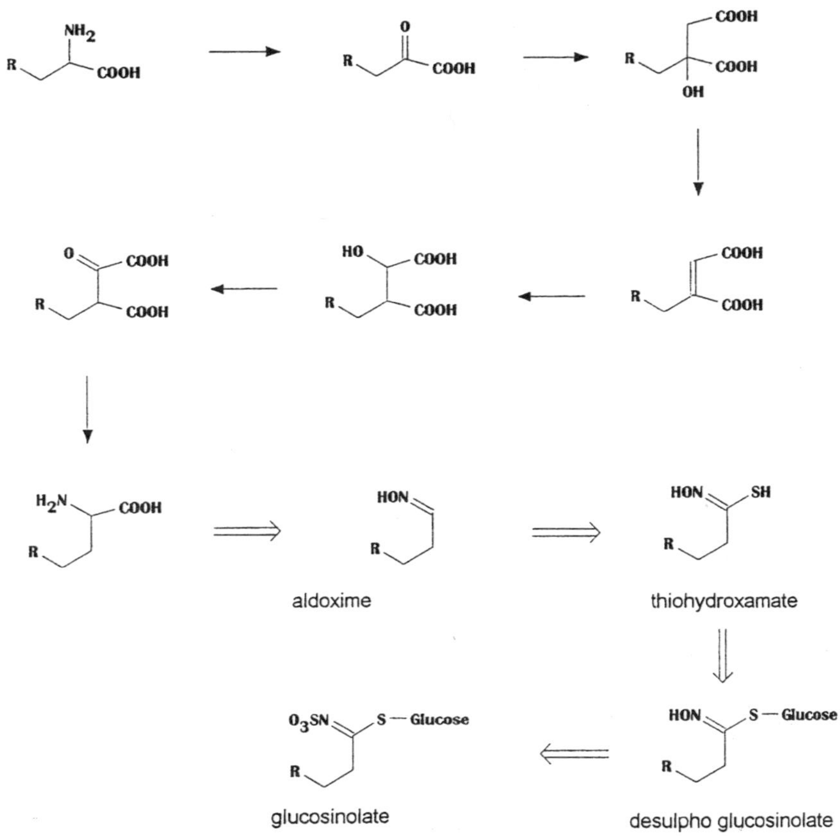

Figure 2 Biosynthesis of glucosinolates {⇒} from chain extended amino acids {→}.

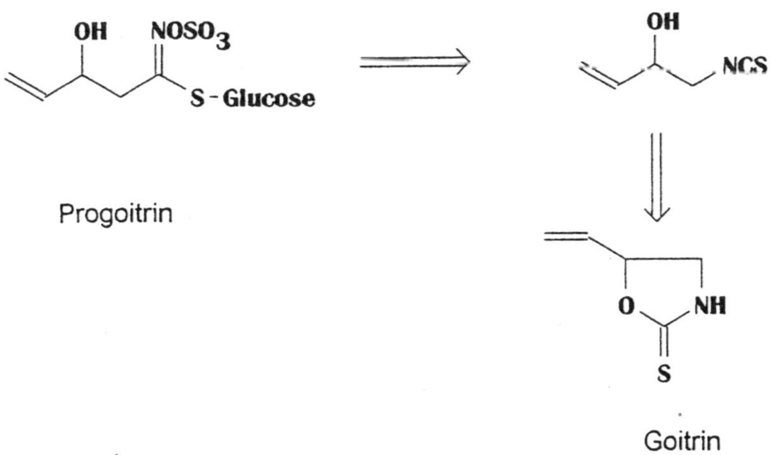

Figure 3 Formation of goitrin.

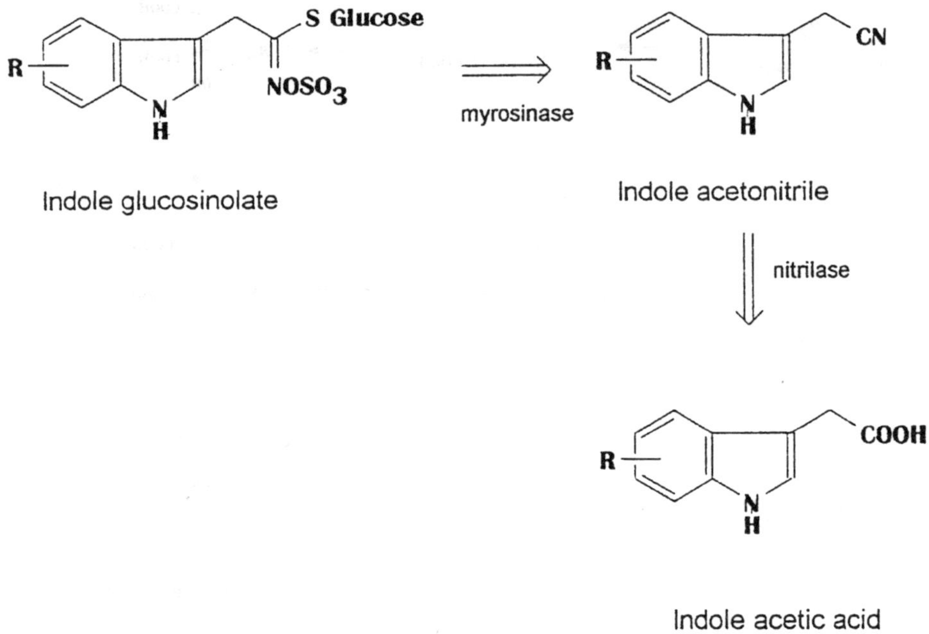

Figure 4 The formation of indole acetic acid from indole glucosinolate.

One other aspect of the biosynthesis of glucosinolates is the production of indole glucosinolates from tryptophan[10]. There are five members of the class (Fig. 4) and, while their biosynthesis appears to follow the same general pathway as other glucosinolates, the genetic control of the levels of indole glucosinolates appears to be different. Interestingly, it has been noted that the major auxin of higher plants, indole-3-acetic acid, is synthesised by a non-tryptophan pathway in the cruciferous plant *Arabadopsis thaliana*[17]. The discovery of the genes for two nitrilase enzymes which convert indole-3-acetonitrile (an indole glucosinolate breakdown product) into indole-3-acetic acid (Fig. 4) suggests that indole glucosinolates may be important precursors of the auxin[17]. The role of indole glucosinolates may be as a means of promoting growth rather than being feeding inhibitors for insects and preventing insect damage. It would appear, therefore, that different members of the same class of compounds have different roles to play in the growth of the host plant.

ALKALOIDS

The most extensively studied group of plant toxins are probably the alkaloids. They comprise a vast number of structural types with the common characteristic of containing a nitrogen atom. Alkaloids which occur in plants can be divided into true alkaloids, protoalkaloids and pseudoalkaloids[1]. The first of these categories are defined as those compounds in which the nitrogen forms part of a heterocyclic ring and whose precursors are amino acids. The presence of a nitrogen in a ring system made early studies into the isolation of alkaloids relatively easy to carry out. By contrast, glucosinolates were difficult to isolate and were rapidly broken down in macerated plant tissue yielding volatile organic molecules[15]. This probably explains why a relatively large number of alkaloids were isolated prior to 1900 while the total number of identified glucosinolates remained at two until well into the 20th century. Studies into the biosynthesis of alkaloids were at the forefront of chemistry research for much of the 19th and 20th centuries. The following description of the unravelling of the biosynthesis of the group of pyrrolizidine alkaloids can be considered to be typical of the sort of study which has been carried out on alkaloids over a number of years.

Pyrrolizidine alkaloids comprise two distinct parts—a base portion containing a nitrogen which is esterified to one or two acids (Fig. 5). The postulated biosynthesis of the base portion of the molecule is from two molecules of ornithine or putrescine (Fig. 6) and the incorporation of 14-C ornithine was first demonstrated in 1962[18]. One problem with radiochemical experiments is the determination of the site of incorporation. This is generally carried out by degradation of the molecule in a

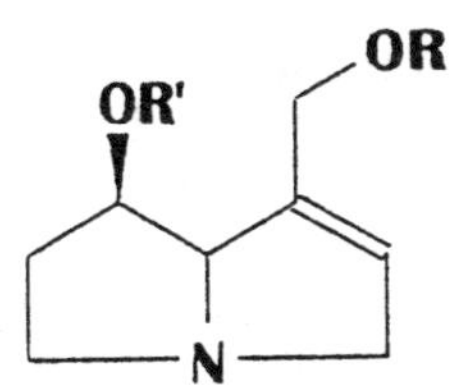

Figure 5 The structure of necine pyrrolizidine alkaloids.

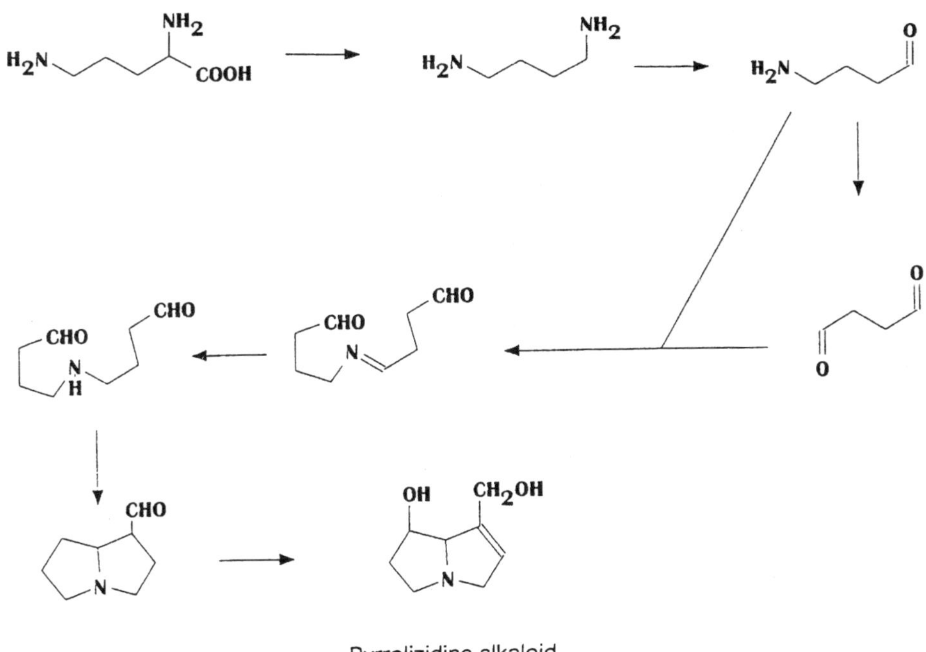

Figure 6 Biosynthesis of pyrrolizidine alkaloids.

controlled way to yield fragments which are able to be characterised. With the base portion of the pyrrolizidine alkaloid, retronecine, treatment with osmium tetroxide and sodium periodate gave formaldehyde derived from C-9 (Fig. 7)[19], however, it proved difficult to devise alternative fragmentation strategies which could provide evidence for incorporation at other sites. Multiple labelling experiments where more than one labelled precursor is fed have been of great value in the determination of competing pathways, however, the difficulties in devising suitable fragmentation reactions for the products remained a major obstacle. The use of ^{13}C labelled precursors enables the direct assay of the incorporation into a biosynthetic product by nmr spectroscopy[20]. In an elegant series of experiments, Robins and co-workers were able to demonstrate the advantages of using multiply labelled precursors to investigate not only the site of incorporation, but also the maintenance of the integrity of C–C bonds in the biosynthesis of retronecine from putrescine (Fig. 8)[21,22]. There is insufficient space here to describe in detail many of the other experiments which have been carried out on pyrrolizidine alkaloids and the interested reader is referred to the excellent review by Robins[23].

While 13-C labelling provides considerable information on the precise site of incorporation, the next stage in the construction of a mechanism of formation of a compound is the determination of the stereochemistry of the reaction. The use of ^{2}H labelling is a very powerful tool in this area. For example, a study using [1,4-$^{2}H_4$]-putrescine as a precursor was able to suggest a mechanism based upon the reaction of labelled putrescine with endogenous, unlabelled material to give the intermediate shown (Fig. 9)[23]. This compound will preferentially cyclise as shown because the unlabelled part of the molecule will be more readily converted into an aldehyde leading to the product as shown. One of the deuterium atoms is lost preferentially from C-9 suggesting that the reduction of the aldehyde to give the alcohol is a stereospecific reaction. This experiment illustrates both a strength and a weakness of ^{2}H labelling experiments. For the former, it is possible to determine the movement

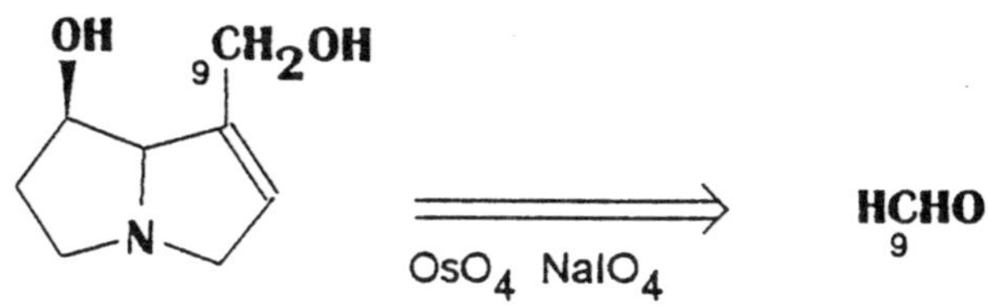

Figure 7 Fragmentation of labelled pyrrolizidine alkaloid.

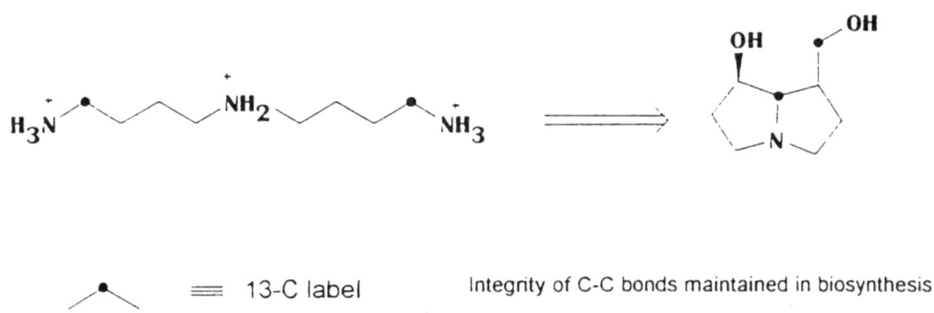

Figure 8 Double labelling experiment with pyrrolizidine alkaloid.

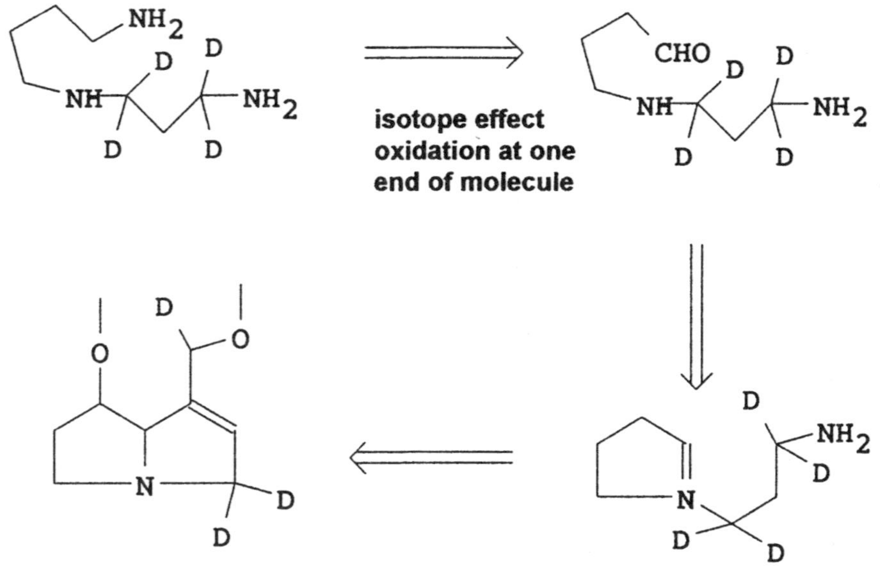

Figure 9 Deuterium isotope effect in biosynthesis of pyrrolizidine alkaloids.

of hydrogen atoms and hence to find out, in much more precise detail, the mechanism of a reaction. The major drawback is that the different strengths of the ^{2}H–C and ^{1}H–C bonds mean that a precursor containing ^{2}H atoms may be treated differently by an enzyme than the same, unlabelled compound.

FUTURE STUDIES

The study of the biosynthesis of plant toxins is crucial to an understanding of why they are present and how their levels might be manipulated[24]. The use of genetic manipulation to alter food plants has become an area of particular interest however, it is not possible to carry out meaningful experiments if the basic knowledge of the pathways which are to be manipulated is not available. The examples given above demonstrate some approaches to biosynthesis from the classical approach to work on the manipulation of pathways. Other examples are present in the literature however a single further approach will serve to delineate the basis of some future studies.

The xanthine alkaloid, caffeine is present in a range of foods including tea and coffee[25]. There is a growing market for caffeine-free products and, while this is achievable using post harvest extraction techniques, a more effective method would be to have naturally low caffeine plants. Unfortunately, the production of caffeine is probably a part of the plant defence system against predators hence a low caffeine variety is likely to be extensively consumed by insects and other pests. One approach is to increase the breakdown of caffeine at the end of the plant's life cycle just prior to harvest. The breakdown pathway of caffeine involves sequential oxidative demethylation to eventually give xanthine which is decomposed further by the normal purine metabolic system (Fig. 10). If the genes which carry out the demethylation reactions could be identified and isolated then they could be cloned back into the plant and placed under the control of a suitable, inducible promoter. This means

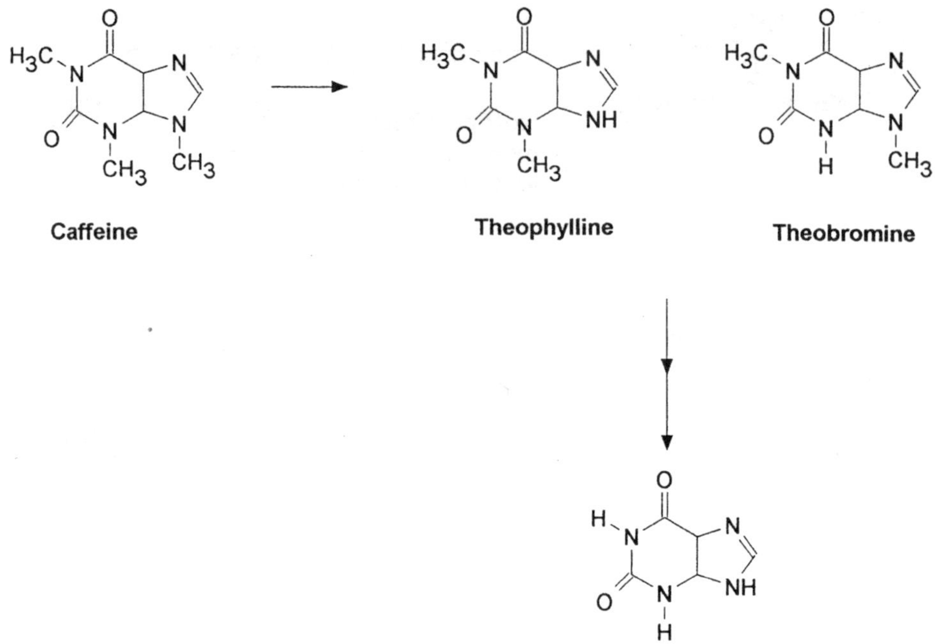

Figure 10 Sequential demethylation of caffeine.

that, in principle, the decomposition of caffeine in the plant could be regulated by the grower and hence could be induced in response to, say brief UV irradiation or spraying with a compound which switches the promoter on.

This type of manipulation of natural plant toxins is likely to increase in the future—particularly if it can be demonstrated that the technique involves simply altering of endogenous genes rather than introducing of foreign genetic material.

ANALYSIS

Considerable efforts have been made to develop methods for the analysis of natural plant toxins. While some limited efforts have been made to develop bioassays, the major methods adopted have tended to use standard chemical analytical procedures. Two specific examples will be used to illustrate the general approaches which have been adopted for the analysis of natural plant toxins.

1. *Glucosinolates*

The analysis of glucosinolates has been carried out in major feed crops for a number of years[26]. Early attempts at quantification were relatively unsuccessful due to the presence of the co-occurring thioglucosidase, myrosinase. When plant tissue is disrupted (e.g. by grinding prior to extraction) the myrosinase hydrolyses any glucosinolates present to give volatile products (isothiocyanates and nitroles). These were used as analytical targets. Methods for the analysis of the parent glucosinolates were less common although hot aqueous methanol extraction followed by examination by thin layer or paper chromatography offered some options for quantification. From an analytical standpoint, the major problem with glucosinolates is the presence of a sulphonate moiety which renders them insoluble to organic solvents and involatile and hence not amenable to gas chromatography (GC). A major breakthrough occurred with the discovery that a sulphatase enzyme from the edible snail, *Helix pomatia*, would hydrolyse the sulphonate grouping to give the desulpho glucosinolate which could then be derivatised and analysed by GC[27]. This basic approach involving adsorption of intact glucosinolates onto a suitable ion exchange resin (e.g. DEAE Sephadex), washing to remove uncharged contaminants, hydrolysis with sulphatase and elution of the uncharged desulphoglucosinolates for chromatographic analysis, formed the basis of a number of analytical approaches[28,29]. More recently, intact glucosinolates have been analysed by paired ion chromatography[30], on a porous graphitised carbon column[31], and reverse phase chromatography[32] has been carried out. Novel approaches using micellar electrokinetic capillary chromatography[33] and capillary electrophoresis[34] have been carried out. The major problem with chromatographic methods is the time taken for each analysis and a screening method based upon near infrared analysis has been developed for glucosinolates[35].

2. *Phytoestrogens*

The analysis of phytoestrogens mirrors the situation for glucosinolates in many ways. Most phytoestrogens are either isoflavones or lignans and, of these, the isoflavones are by far the most significant. They generally occur as glycosides in plants and their analysis can be either as the whole molecule or as the aglucone. Typically, HPLC analysis on a reversed phase support using aqueous/organic solvent systems have been carried out[36,37]. A range of detectors have been used for quantification including UV, electro chemical and thermospray mass spectrometric systems[38]. The use of GC has been particularly useful for measuring low levels of isoflavones[39] and for the analysis of metabolites in urine and blood[40,41]. Capillary electrophoresis has also been carried out coupled both to UV and to electrospray ionisation MS[42]. Bioassays have been developed for phytoestrogens[43]. This involves the use of a human breast cancer cell line (MCF-7) whose growth is stimulated in the presence of estrogenic compounds[44]. In a variation of the simple assay, levels of ^{35}S labelled exoproteins from MCF-7 cells exposed to phytoestrogens in the presence of ^{35}S-methionine were measured[45]. The use of a genetically modified cell line which produces chloramphenicol acetyltransferase (CAT) when exposed to estrogenic compounds has been carried out[46] and this type of specific assay would appear to offer some advantages when rapid screening has to be carried out.

BIOLOGICAL EFFECTS

1. *Overview*

Natural plant toxins have a range of biological effects. These can vary from the acute to the chronic. In the case of the former, it is relatively easy to determine if a given plant is giving rise to a toxic effect since the consequence of ingestion is generally short term or immediate. In some cases, the presence of an acute effect can be of use since the bioactive principle may be useful in the treatment of a disease (e.g. quinine, taxol, vinblastine), however, for most individuals, exposure occurs as a result of eating the plant in a normal diet. The resulting chronic effects are both more difficult to detect and to quantify. In addition, the interplay of different biologically active components can give rise to a range of effects which may not be simply either additive or predictable. The major biological effects which will be considered here will fall into a number of categories. We will focus first of all on effects on general nutritional status and consider the overall effect of plant toxins in this context. Plants are a major source of endocrine disrupting chemicals and the effects of goitrogenic and estrogenic chemicals will be reviewed. There are a number of synthetic, environmental endocrine modifying chemicals present and these will be compared and contrasted with the plant toxins. Neurotoxins comprise an important class of plant toxins and these will be briefly considered. Food contains a variety of electrophilic species which can bind to DNA and cause mutations. The putative role of diet in cancer formation and susceptibility is a huge area and we will consider two examples of putative pro- and anti-carcinogens and their likely mechanisms of action. Finally, future areas of interest will be reviewed with emphasis on potential areas of increasing importance such as allergenicity and the use of biomarkers to determine long term biological effects.

2. *Nutritional Effects*

The effects of diet on nutrition are well documented[47]. The traditional approach suggests that many nutrients are also, under some circumstances, toxic. In addition, the diet contains a number of constituents which can adversely affect nutrient availability, uptake or effect. It is relevant, therefore, to also consider plant toxins in terms of availability, uptake and effect.

Most nutrients have an optimal concentration compatible with good health. This may not necessarily correspond to the recommended daily intake since the latter may only be the level required to prevent an adverse health impact. In addition, the range of concentrations over which optimal intake spreads can be wide or narrow. From the standpoint of nutritional effects, it would appear to make sense to prioritise plant toxins on the basis of their action i.e. do they affect a nutritional aspect for which there is a narrow range of tolerance or one for which the levels of intake can vary widely without any significant discernible effect. For example, the tolerance for water soluble vitamins such as ascorbic acid is significantly greater than that of fat soluble vitamins (e.g. vitamin A or D) therefore plant compounds which inhibit the action of vitamins A or D would appear to have the potential to cause a more significant effect. Biotin toxicity has never been reported in humans while, on the other hand, excess folic acid can mask the symptoms of vitamin B_{12} deficiency and dietary inadequacy of folic acid is particularly dangerous to women who intend becoming pregnant. The extensive study of potential risk factors and the range of tolerance levels lies outwith the scope of this chapter, however, it should be recognised as a significant factor in deciding the relative importance of plant toxins.

In terms of generic nutritional effects, plants contain a plethora of substances which have varying activities. For example, a large range of plant components can adversely affect monoamine oxidase activity, similarly, legumes contain trypsin inhibitors[48] and thiamine is inhibited by the enzyme thiaminase[49], by alcohol and by bisulphite[50] (a common preservative).

The effects and importance of plant toxins, from a nutritional standpoint, must take into account both the importance of the nutritional marker and the extent and type of biological effect.

3. *Endocrine Disrupters*

Plants contain a range of compounds which adversely affect the endocrine system. Even in this case, however, there are both benefits and deleterious effects. The endocrine system is controlled by the hypothalamus and includes the adrenal gland, the thyroid, the pancreas and the sex tissues. The effects of compounds which interact directly with the endocrine system are not, however, limited to these areas and can affect a range of secondary targets.

Two of the major plant-derived effects are on the thyroid and on the sex hormone system and these will be dealt with here.

3.1. Thyroid. Goitrogenic substances occur particularly in brassicaceous crops and in plants which contain cyanogenic glycosides. In the case of humans, the greatest risk is due to the former rather than the latter. Brassica vegetables contain two components which are goitrogenic. The first of these arises from the spontaneous hydrolysis of the glucosinolate breakdown product, indolyl-3-methyl isothiocyanate, which produces thiocyanate ion and indole-3-carbinol[51] (Fig. 11). Thiocyanate ion is goitrogenic by direct competition with iodine and its effects can be prevented by iodine supplementation. A second goitrogenic substance is also produced from the further chemical transformation of a glucosinolate breakdown product. In this case, the glucosinolate precursor is 2S-but-3-enyl glucosinolate which is hydrolysed, as described previously, to give the corresponding isothiocyanate under the action of the enzyme, myrosinase (thioglucoside glucohydrolase E.C.4.3.4.1). In this case, the isothiocyanate readily cyclises to give vinyl-oxazolidine thione (Fig. 12) which inhibits the formation of thyroxine[52]. Supplementation with iodine has no effect and the simplest approach is to remove the source of the parent glucosinolate from the diet. While endemic goitre is comparatively rare, there remains a risk both for susceptible individuals and as a result of chronic, longterm exposure. Some glucosinolates are thought to be protective against cancer-inducing chemicals in the diet (see below) hence the complete removal of Brassica vegetables because of the potential risk of goitre may not be, on balance, a good idea. This type of conflict is commonly encountered with food toxins. In some cases, the same compound

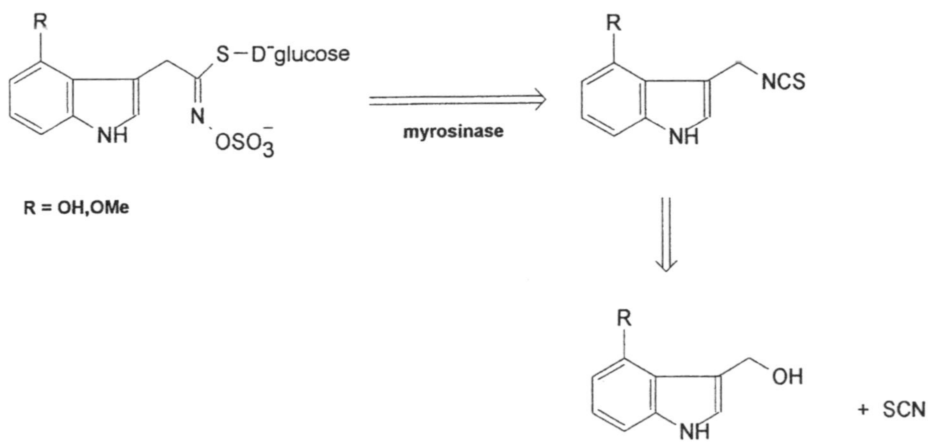

Figure 11 Breakdown of indole glucosinolates.

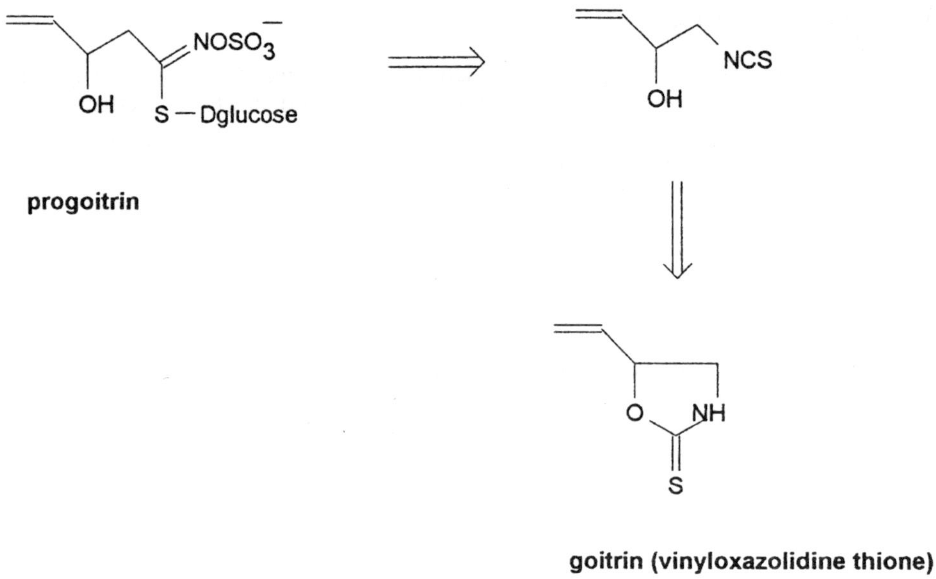

Figure 12 Formation of vinyloxazolidine thione.

(or group of compounds) can have both beneficial and deleterious effects and this is the situation for dietary oestrogens.

3.2. Oestrogens. Oestrogenic compounds occur widely in the environment and have been the subject of much recent speculation[53]. In addition to those derived from plants (phytoestrogens), there are also several environmental oestrogens derived from, the chemical industry (Fig. 13). Dietary phytoestrogens occur in plants such as soya and consumption has been linked with a range of biological effects[54]. In addition to being directly associated with fertility, the effects of phytoestrogens are also manifested through the regulation of gene function by activated steroid hormone receptors[55]. This means that a range of biological outcomes could arise from the ingestion of oestrogenic substances. A further complication arises from the known differential metabolism of phytoestrogens. In soya the major phytoestrogens occur principally as the corresponding glycosides which are largely inactive[56]. The biological effect resulting from ingestion is largely determined by the hydrolysis of the parent glycoside by gut bacteria. This means that the response of different individuals can vary greatly and is dependent, not only on the level of exposure, but also on the level and type of gut bacteria present[57]. Nonetheless, in assessing the biological response to phytoestrogens, or indeed any other plant toxins and determining how important they are in the context of overall exposure, the most important factors are as follows. The first requirement is to develop a way of measuring the biological effect being elicited—preferably in a suitable model system. The second stage is to compare a range of sources of exposure of different compounds to try to decide upon a likely prioritisation. The third part of the investigation should focus upon finding if the effects observed are significant in the context of known or likely human exposure. Finally, the overall risk can be assessed and the need, if any, to take appropriate action decided upon. These steps can be applied to phytoestrogens

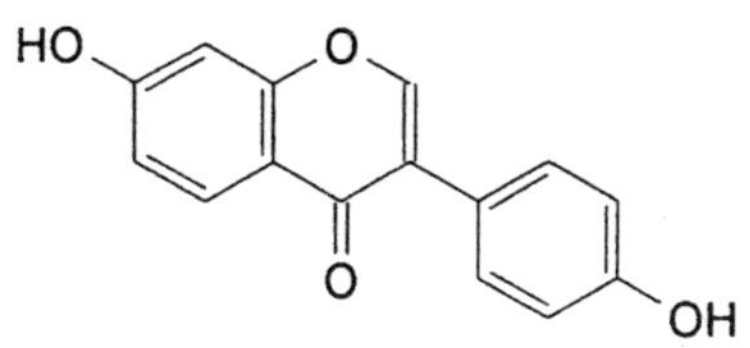

Daidzein

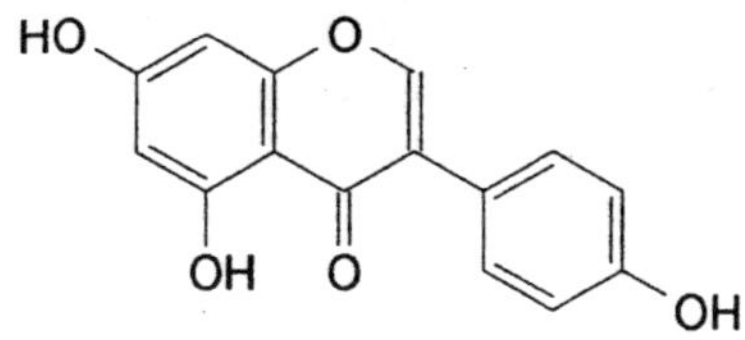

Genistein

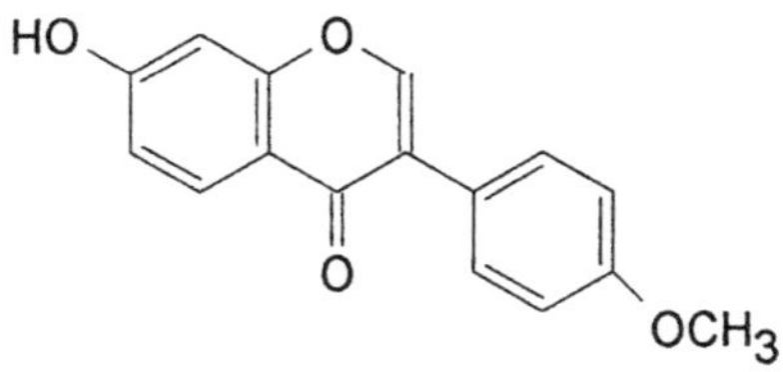

Formononetin

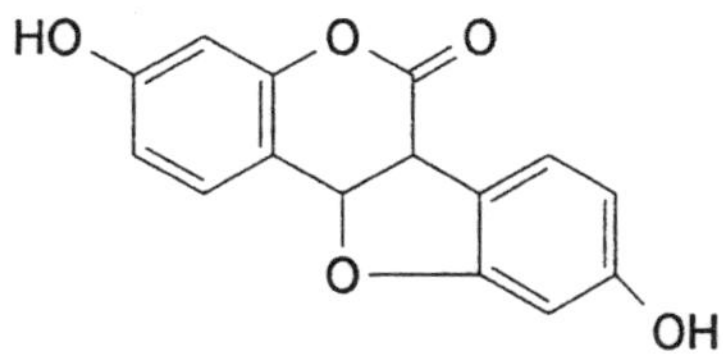

Coumestrol

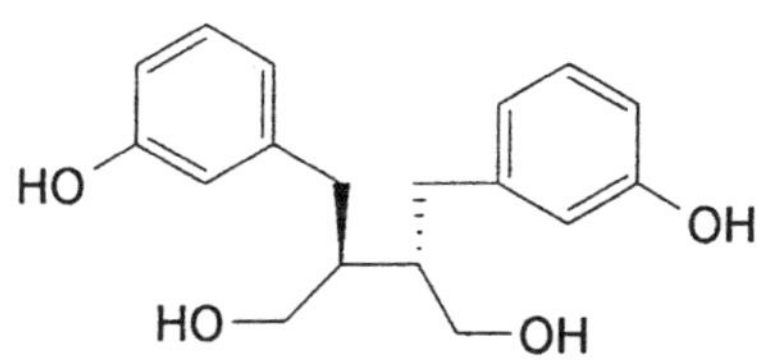

Enterodiol

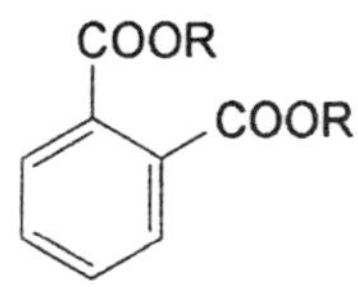

Phthalates

Figure 13 Some representative natural and synthetic estrogens.

as an example of how to assess the benefits and risks inherent in exposure to a plant toxin.

Several investigators have developed model systems for measuring oestrogenic substances. The most frequently used assay has traditionally been the mouse uterotropic assay[56] however, as with all animal based models the relevance to humans is questionable and the speed and cost can impose restrictions on use. In order to overcome some of these problems, a number of workers have attempted to develop *in vitro* assays based upon cells in culture. The simplest systems used have probably been oestrogen-sensitive cell lines (frequently breast cancer cells). When grown in the presence of oestrogens, the growth of the cells is stimulated. This means that a measure of cell proliferation can serve as a marker for oestrogen exposure[58]. One problem with this assay is that compounds like the soya phytoestrogen, genistein, while being an oestrogen, can inhibit the growth of human breast cancer cells[59]. A simple cell growth assay may, therefore be misleading. A related assay system uses the breast cancer cell line, MCF-7. In this case, the induction of an oestrogen-specific exoprotein was a measure of oestrogenic activity[40]. A further refinement of this latter procedure makes use of an oestrogen-responsive element controlling a chloramphenicol acetyltransferase (CAT) gene. In this case a suitable target cell (e.g. mouse L-cells) are transfected with the gene construct and the induction of CAT measured in response to exposure to oestrogenic substances[60]. It has previously been demonstrated that antioestrogens can promote binding of the oestrogen receptor to DNA in an analogous way to oestrogens[61] hence it may not be possible to distinguish between them in a reporter gene based assay. Despite these difficulties, the use of cell based assay methods offers significant advantages in the screening of plant and other dietary toxins. The development of more powerful methods to measure gene transcription offers the opportunity to assess compounds such as oestrogens and antioestrogens with similar but differing activities. Receptor based assays are already commonly used in the pharmaceutical industry and the extension of these to plant toxins is likely to occur with increasing frequency.

Determination of the relative oestrogenic potencies of phytoestrogens remains a major problem. In general, the major dietary sources are soya (isoflavonoids) and certain vegetables such as carrot and asparagus (lignans). There has been a report that the indole glucosinolate-derived breakdown product, indole-3-carbinol, exhibits oestrogenic activity[62], however, further confirmation is necessary before brassica vegetables can be added to the list of phytoestrogen-containing vegetables. In terms of relative potency, all assays suggest that phytoestrogens have significantly less activity than the natural hormones (e.g. β-estradiol) by factors of 1000 or greater. Nonetheless, the levels of exposure can be significant for certain sub-groups of the population (vegetarians, infants fed on soya formula) and these represent groups who should be targeted in future research. It is difficult to establish a priority list based upon exposure and biological potency when there is so much individual variation in response. This poses a significant problem in the study of many natural plant toxins. In the case of phytoestrogens, there is a clear requirement to develop a clear picture of exposure and effect in a human population. In order for this to be carried out, it is essential to be able to carry out population studies and to correlate exposure with biological effects.

The risk posed by phytoestrogens is difficult to quantify. It would seem unlikely that exposure would cause major health problems since diets have contained compounds of this type for hundreds of years. Nonetheless, changes in diet may be significant for certain sub-groups of the population. For example, the increased prevalence of soya-based infant formula over the past twenty years or so may have a long term effect in the West. The major problem in assessing the potential risk (or benefit) is that the effects of compounds such as phytoestrogens may be varied and unpredictable. Under such circumstances, studies which lead to an increased understanding of the correlation between dietary exposure, short term measures of the effect and long term health risks are of particular importance. A further complication is the apparent increased dietary load of synthetic oestrogens including estradiol, PCBs, dioxins and plasticisers. Some of these compounds will affect the oestrogen receptor directly by binding as agonists or antagonists while others will have more indirect effects. Under these circumstances, the best option is probably to compare specific compounds against an index of known biological activity. In this way it is possible to obtain an indicative list of compounds and their relative effects. This type of approach can be used with plant and other toxins of defined, principal biological activity however it relies upon this type of information being available.

4. *Neurotoxins*

Plant-derived neurotoxins comprise a wide variety of compounds. Apart from those substances used in a social context (e.g. nicotine, cannabis, etc.), a number of plant toxins have significant toxicological activity. One of the most common neurotoxins is N^3-oxalyldiaminopropanoic acid which occurs in the Indian pea (*lathyrus sativus*) and in some *Crotalaria* species[1]. The major effects of poisoning (*neurolathrism*) are muscular rigidity, paralysis, convulsions and, in extreme cases, death. Several other amino acid-related neurotoxins are produced by plants including 3-cyanoalanine from vetch and lupin. One of the best defined neurotoxins is 3-nitropropanoic acid which is produced from the corresponding glycoside by acid hydrolysis (see above). This compound acts as a suicide inhibitor of succinate dehydrogenase (Fig. 14) and hence affects energy metabolism[63]. In addition to having a known mechanism of action, 3-nitropropanoic acid produces basal ganglia degeneration and extra pyramidal symptoms in humans[64]. This suggests that, while it is a general inhibitor of succinate dehydrogenase, its effect is particularly severe on GABAergic neurones. Clearly ingestion of 3-nitropropanoic acid is not recommended however, the pathology of its action has been of interest to workers attempting to understand Huntington's disease and it has been suggested as a model for this type of neurological dysfunction[65]. The importance or value of a toxin may not simply reside in its biological activity but it may also shed light on functional aspects of toxicity.

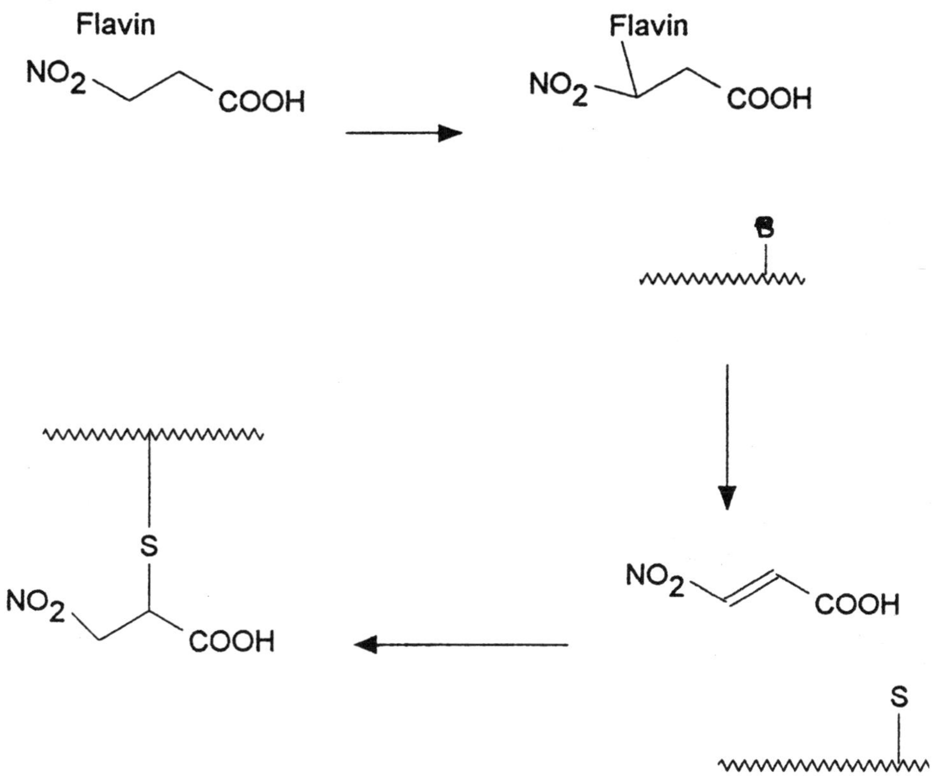

Figure 14 Mode of action of 3-nitropropanoic acid.

5. *Mutagens and Carcinogens*

Plants contain a range of compounds which cause mutations in DNA and others which appear to be protective against mutagenicity and cancer. The former are, on the whole, easier to detect than the latter since electrophilic species will be likely to bind to DNA and can be identified by consideration of chemical structure. This simplistic approach does not take account of metabolic activation of potential mutagens and this may considerably compound matters. In the case of protective agents, the situation is far more complex. The potential mechanisms by which a compound may exert a protective effect against mutagenicity or cancer formation are many. Antioxidants may prevent the initial activation of compounds. Enzyme inhibitors could have the same effect by preventing the activating enzymes (generally cytochromes P450's) from carrying out oxidative activation. Other groups of compounds may induce enzymes involved in the conjugation and subsequent removal of activated compounds from the cell. Finally, some plant constituents are differentially toxic to tumour cells and may induce apoptosis (programmed cell death) or other forms of cell destruction. It is not feasible to consider even a small proportion of the wide range of plant toxins which induce these effects therefore the following discussion is limited to two groups of compounds—pyrrolizidine alkaloids, a group of mutagenic compounds which occur in certain herbs and glucosinolates, a group of putative protective agents which occur in Brassica vegetables (see above).

Pyrrolizidine alkaloids are found in plants growing in most environments and in all parts of the world[1]. On the whole, they are not common contaminants of food plants and ingestion generally occurs as a result of eating certain herbs (in particular, comfrey (*Symphytum officinale*) or because of contamination of cereal crops with, for example, *Crotalaria* species. The major symptom of acute poisoning with pyrrolizidine alkaloids is veno occlusive disease of the liver, however because of the limited level of exposure this does not generally occur in humans. They have also been implicated in primary liver cancer however, they are generally considered to be fairly weak carcinogens and evidence of human cancers resulting from ingestion is lacking[66]. One possible reason is that the alkaloids generally occur in the plant principally as the corresponding N-oxide and this compound is considerably less active than the parent alkaloid (Fig. 15). Activation of pyrrolizidine alkaloids is thought to occur *via* the 7-hydroxy intermediate which then undergoes dehydration to give the corresponding electrophilic pyrrole[67] which is the proposed mutagen[68]. Activation is carried out by a cytochrome P450 (CYP 3A4) which also deactivates the parent alkaloid via the N-oxide[69] (Fig. 15). While the deactivation pathway was previously thought to be irreversible, recent studies suggest that conversion of the N-oxide to the parent alkaloid (and hence to the active pyrrole) can occur catalysed by liver microsomal preparations[70]. The presence of a mutation does not prove carcinogenic potential however recent studies have implicated loss of function of specific cell cycle-related proteins with proliferative cell growth and loss of cell cycle control. One such gene is the p53 tumour suppresser gene which is known to be involved in the induction of apoptopic pathways in cells subjected to DNA damage. Several recent studies have suggested that mutations in the gene, at specific points, could give rise to a functionally inactive protein and that mutations at some of these 'hot spots' were indicative of specific types of cancer[71]. For example, mutations at codon 249 has been found in liver tumours[72]. The possibility of organ specificity in mutational occurrence, combined with the known association between certain carcinogens and site of activity, offers the thought that certain carcinogens may induce mutations at the 'hot spots' in the p53 gene. Recent studies have suggested

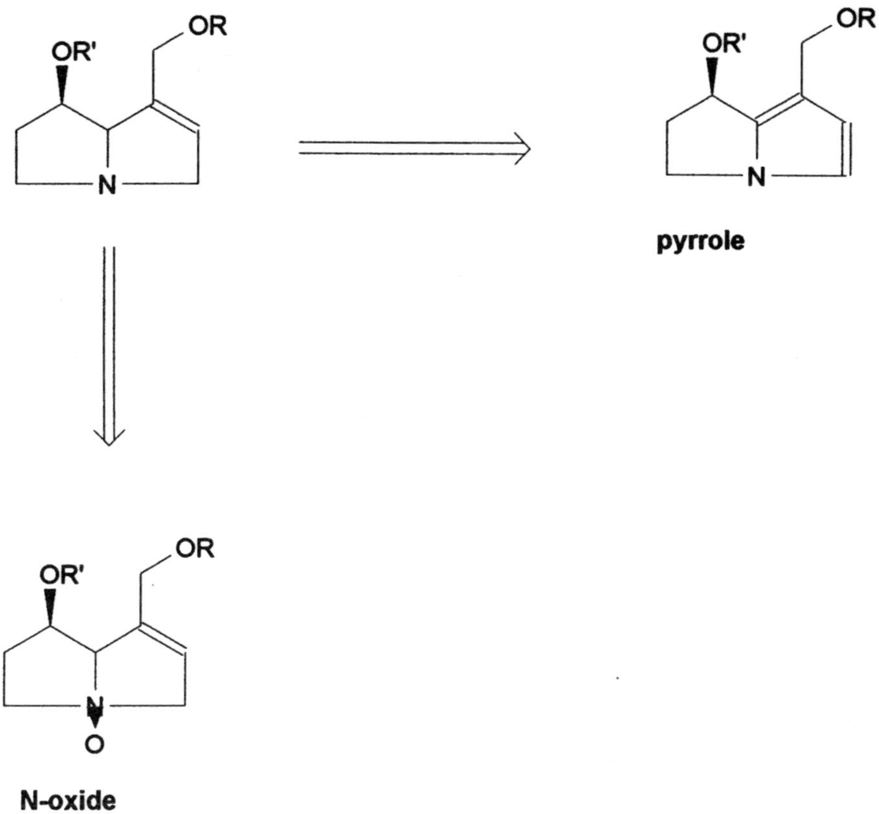

Figure 15 Activation and deactivation of pyrrolizidine alkaloids.

this may be the case for benz-pyrene (a lung carcinogen)[73] and an investigation showed that the pyrrolizidine alkaloids present in comfrey (*Symphytum officinale*) could give rise to a specific mutation at codon 249 in the p53 gene in human liver cells in culture[74]. If such a phenomenon were shown to be generally applicable, it could allow predictions to be made concerning the likely site of action and, perhaps, potency of a range of known and suspected carcinogens.

The importance of plant toxins as potential carcinogens is of less significance than their role as protective factors. There is considerable epidemiological evidence to suggest that a diet rich in vegetables and fruit can help to protect against some specific forms of cancer. The mode of protection varies and many compounds have been isolated from plants which block various stages in the carcinogenesis process. These include chemicals which prevent formation of activated electrophiles, compounds which stimulate the conjugation and excretion of activated compounds, and substances which prevent neoplasia in cells which are precancerous. In many cases, the mechanism of action of the protective effect is unclear.

In the instance of brassica vegetables, the occurrence of glucosinolates and their breakdown products (principally isothiocyanates) has been suggested to be a major factor in the apparent protection offered by these vegetables. One major problem is correlating observed effects *in vitro*, with apparent effects *in vivo*. Organic isothiocyanates have been reported to block tumour production in rodents[75] and this was suggested to be due to suppression of carcinogen activation by cytochromes P450 enzymes and the induction of Phase 2 enzymes such as glutathione transferases and NAD (P)H quinone reductase. More recently, several studies have examined this effect more carefully and found isothiocyanates to induce quinone reductase activity in human cell lines in culture[76]. In addition to this apparent cancer-blocking activity, isothiocyanates have also been found to be cytotoxic[77] and it has been suggested that glucosinolates and their breakdown products exhibit a differential toxic effect on tumour cells. When attempting to correlate this type of activity with protective effects in humans, the situation becomes more complex. Verhagen *et al.*[78] have demonstrated that a diet rich in Brussels sprouts appears to reduce the levels of oxidised DNA by 28%, albeit in a small sample size (5 test subjects and 5 controls). In attempting to link the *in vivo* with the *in vitro* studies, the same investigators found that consumption of Brussels sprouts led to an increase in plasma levels of class alpha glutathione-5-transferase but only in males, not in females. No effect was observed on glutathione levels. Antiproliferative effects were observed for mercapturic acid; pathway metabolites of phenylethylisothiocyanate. For example, S-(N-phenylethylthiocarbamoyl) cysteine was a potent inhibitor of human leukaemia 60 cells *in vitro* but had a low toxicity to corresponding differentiated cells in culture[79]. The possible mechanism of the effect remains unclear however both N-phenylethylthiocarbomylation of nucleophilic groups in the active sites of enzymes and glutamine antagonist activity have been suggested. It was observed that mechanisms in addition to inhibition of DNA synthesis were operative.

It is important to be able to link the various protective effects observed in different systems into a complete pattern and this has clearly been attempted with glucosinolates. While there are still significant gaps in knowledge, the progress of understanding from epidemiological evidence and *in vitro* mechanistic theories to a predictable hypothesis in humans is clear[80]. For glucosinolates and their breakdown products—and, indeed for most dietary components studied hitherto—there appear to be several mechanisms and levels of protection from inhibition of initial DNA damage to selective cytotoxicity for proliferative cells. Further work is necessary to clarify the situation for individual compounds and groups of compounds and to then deal with diets of increasing complexity.

CONCLUSION

The assessment of the benefits and risks which could result from the ingestion of natural plant toxins is difficult to carry out—except in a very qualitative manner. Determination of no effect levels for isolated compounds in animal models provides, at best, only a partial answer. Most toxins which occur in food plants are chronically rather than acutely biologically active and exposure to them occurs over a long time period. The biological effects and the balance between risk and benefit must be determined over a similarly lengthy time scale. It is only when the balance of risk and benefit has been established that steps can be taken to enhance or reduce the levels of given components using processing or, with greater specificity, manipulation of biosynthetic pathways. The sequence of events is, therefore, fairly clear. The biological effects of components in the food matrix must be determined. The levels present in the food and the levels of exposure must be measured and, finally, the balance of risk and benefit must be used to assess if the amounts present in the diet need to be manipulated. In each of these areas, there is considerable information already available and considerable opportunity for future work.

1. *Biological Effects*

The use of traditional toxicological methods to assess natural plant toxins poses a number of problems. In the case of mutagens or carcinogens there is some merit in Ames testing and the use of animal models however there is considerable variability in both sensitivity and response and the extrapolation of any data to humans is fraught with difficulties. One approach which is becoming increasingly important is the use of biomarkers. Briefly, a biomarker is a measure made from a biological system which is indicative of a clinical effect or an exposure to a bioactive principle. One of the most important aspects of biomarker research is that the measurement should provide information about long term clinical outcomes—i.e. a biomarker is a short term measure indicative of long term health status. The simplest (conceptually, at least) biomarkers are those which relate directly to exposure. For example, the discovery of lead in blood is indicative of exposure to lead. Similarly, it is possible to monitor metabolites of plant toxins in blood, urine, faeces or any other, accessible, body product. While biomarkers of exposure can be generally indicative of health or disease effects, a direct link to clinical conditions is rare. A second approach is to determine biomarkers of effect. These are necessarily much more complex nonetheless some examples have been investigated including measurement of DNA damage caused by mutagens, prevention of oxidative damage to cellular proteins, lipids and DNA, induction of specific gene expression in response to exposure to bioactive constituents and immuno suppression. It is important to validate biomarkers of effect and one way of doing this is to develop the approaches in cell culture before investigating humans directly. Unfortunately, most cell culture involves repeated seeding and passaging of cells on a weekly basis and this makes long term exposure experiments difficult to carry out. A recent study in this laboratory suggests that a flow cell bioreactor may provide a suitable method for long term exposure (*ca.* 6 months) experiments.

The application of biomarker approaches to human diet and risk assessment is currently being considered by an EU Concerted Action under the FAIR program. Reassessment of existing knowledge in the light of novel approaches and the application of these to new problems are priority areas.

2. *Analytical Approaches*

The importance of more sensitive analytical approaches is of particular importance in the measurement of exposure to plant toxins. Since, as discussed above, most plant toxins have a chronic rather than an acute effect, the determination of extremely low levels of compounds in the plant is not of particular importance. Nonetheless, there is a clear requirement to increase analytical efficiency and throughput and the increased use of automated techniques is likely. There is a clear trend of increasing analytical trial specificity which has employed both chemical (e.g. phase partitioning, volatility, solubility) and biological (e.g. binding to antibodies,) techniques. Developments in analytical methodology are likely in both these areas.

3. *Manipulation of Plant Toxins*

The use of processing to modify deleterious components in plants has been carried out for a considerable period of time. For example, kidney beans contain an acutely toxic haemaglutanin which is destroyed by cooking and cassava contains toxic cyanogenic glycosides which are hydrolysed when the plant is crushed prior to cooking. There are numerous other examples. Dietary supplementation has been carried out for a number of years with the addition of vitamins and minerals to a range of foods. The concept of "functional foods" is becoming increasingly adopted with foods being advertised as having added health benefits (e.g. high fibre, low fat, high in polyunsaturates). The use of genetic manipulation to alter biosynthetic pathways and hence increase or decrease the levels of certain compounds has already been carried out and is likely to increase. Such approaches will complement and supersede traditional breeding programmes and the concept of "designer food" is likely to become a reality.

Natural plant toxins comprise a vast number of compounds. Their toxicity is chronic rather than acute and, for some members of the class, there are both beneficial and deleterious aspects to their occurrence and ingestion in the diet. While food is, almost by definition of benefit, diet related disease is a clear indication that monitoring and, where necessary, altering exposure to natural plant toxins could have a significant effect on health and fitness.

References

1. M. L. Vickery and B. Vickery, *Secondary Plant Metabolism* (MacMillan Press, London, 1981).
2. R. Robinson, *Structural Relations of Natural Products* (Clarendon Press, Oxford, 1955).
3. J. March, *Advanced Organic Chemistry*, Fourth Edition (John Wiley & Sons, 1991).
4. J. N. Abelson and M. I. Simon, *Methods in Enzymology* **Vol. 267**, *Combinatorial Chemistry* (Academic Press, London, 1996).
5. C. L. Carter and W. J. McChesney, *Nature* **164**, 575 (1949). "Hiptagenic Acid Identified As β-Nitropropionic Acid."
6. K. Gorter, *Bull Jard. Bot. Buitenz.* **Ser. III**(3), 187 (1920).
7. M. P. Morris, C. Pagen and H. E. Warmke, *Science* **119**, 322 (1954). "Hiptagenic Acid, A Toxic Component of *Indigofera endecaphylla*."
8. W. B. Turner, *Fungal Metabolites* (Academic Press, London, 1971) pp. 303–305.
9. R. B. Silverman, *Mechanism Based Enzyme Inactivation: Chemistry and Enzymology* **Vol. 1** (CRC Press Inc., Boca Raton) pp. 79–80.
10. R. L. Baxter, A. B. Hanley, H. W.-S. Chan, S. L. Greenwood, E. M. Abbott and I. J. McFarlane "Fungal biosynthesis of 3-nitro propanoic acid" *J. Chem. Soc. Perkin Trans.* **I**, 2495 (1992).
11. R. L. Baxter, S. L. Smith, J. R. Martin and A. B. Hanley, "The fungal biosynthesis of 3-nitropropanoic acid: is the dicarboxylation of L-nitrosuccinate an enzymic reaction" *J. Chem. Soc. Perkin Trans.* **I**, 2297 (1994).
12. E. Candlish, L. J. La Crox and A. M. Unran, "The biosynthesis of 3-nitropropionic acid in creeping indigo (*Indigogera spicata*)" *Biochemistry* **8**, 182 (1969).
13. M. G. Ettlinger and A. Kjaer, "Glucosinolates" *Recent Adv. in Phytochem.* **1**, 59–144 (1968).
14. G. W. Dawson, A. J. Hick, R. N. Bennett, A. Donald, J. A. Pickett and R. M. Wallsgrove, "Synthesis of glucosinolate precursors and investigations into the biosynthesis of phenyalkyl and methylthioalkyl glucosinolates" *J. Biol. Chem.* **268**, 27154–27159 (1993).
15. G. R. Fenwick, R. K. Heaney and W. J. Mullin, "Glucosinolates and their breakdown products in food and food plants" *CRC Crit. Rev. in Food Sci. Nutr.* **18**, 123–201 (1983).
16. M. Kutachek, Z. Prochazka and K. Veres, *Nature* **104**, 393–394 (1963).
17. D. Bartling, M. Seedorf, R. C. Schmidt and E. W. Weiler, "Molecular characterisation of two cloned nitrilases from Arabadopsis thaliana: key enzymes in biosynthesis of the plant hormone indole-3-acetic acid" *Proc. Natl. Acad. Sci. USA* **91**, 6021–6025 (1994).
18. E. Nowacki and R. V. Byerrum, *Life Sci.* **1**, 157 (1962).
19. W. Bottomley and T. A. Geissman, *Phytochem.* **3**, 357 (1964).
20. R. L. Baxter, A. B. Hanley, H. W.-S. Chan, S. L. Greenwood, E. M. Abbot, I. J. MacFarlane and K. Milne, "Fungal biosynthesis of 3-nitropropanoic acid" *J. Chem. Soc. Perkin. Trans.* **I**, 2495–2502 (1992).
21. D. J. Robins, "Pyrrolizidine alkaloid biosynthesis. Synthesis of putrescine-1, 2-13C2 and its incorporation into retronecine" *J. Chem. Res.* **12**, 326–327 (1983).
22. H. A. Kahn and D. J. Robins, "Pyrrolizidine alkaloid biosynthesis–incorporation of C-13-labelled putrescines into retronecine" *J. Chem. Soc. Chem. Comm.* 554 (1981).

23. D. J. Robins, "Biosynthesis of pyrrolizidine alkaloids" *Chem. Soc. Rev.* **18**, 375–408 (1989).
24. G. R. Pettit, "Progression the discovery of biosynthetic anticancer drugs" *J. Nat. Prod.* **59**, 812–821 (1996).
25. E. Looser, T. W. Bauman and H. Warner, "The biosynthesis of caffeine in the coffee plant" *Phytochem.* **13**, 252–317 (1974).
26. D. I. McGregor, W. J. Mullin and G. R. Fenwick, "Analytical methodology for determining glucosinolate composition and content" *J. Assoc. Off. Anal. Chem.* **66**, 825–849 (1983).
27. R. K. Heaney and G. R. Fenwick, "The analysis is of glucosinolates in *Brassica* species using gas chromatography" *J. Sci. Food Agric.* **31**, 593–599 (1980).
28. R. K. Heaney and G. R. Fenwick, "A microcolumn method for the rapid determination of total glucosinolate content of cruciferous material" *Z. Pflanzensuchtg* **87**, 89–95 (1981).
29. E. A. Spinks, K. Sones and G. R. Fenwick, "The quantitative analysis of glucosinolates in cruciferous vegetables, oilseeds and forages using high performance liquid chromatography" *Fette, Seifen, Anstrichmittel* **86**, 228–231 (1984).
30. T. Prestera, J. W. Fahey, W. D. Holtzclaw, C. Abeygunawardana, J. L. Kachinski and P. Talalay, "Comprehensive chromatographic and spectroscopic methods for the separation and identification of intact glucosinolate" *Analytical Biochemistry* **239**, 168–179 (1996).
31. C. Elfakir and M. Dreux, "Simultaneous analysis of intact and desulfated glucosinolates with a porous graphitized carbon column" *J. Chromatogr.* **727**, 71–82 (1996).
32. J. M. Bez and W. D. Fox, "High performance liquid chromatographic determination of glucosinolates in Brassica vegetables" *AC Symposium Series* **546**, 181–196 (1994).
33. C. Bjergegaard, S. Michaelson, P. Moller and H. Sorensen, "Separation of desulphoglucosinolates by micellar electrokinetic capillary chromatography based on bile salt" *J. Chromatogr. A* **717**, 325–333 (1995).
34. F. A. Tomasbarberan, "Capillary electrophoresis—A new technique in the analysis of plant secondary metabolites" *Phytochemical Analysis* **6**, 177–192 (1995).
35. J. K. Daun, K. M. Clear and P. Williams, "Comparison of 3 whole seed near-infrared analysers for measuring quality components of Canola seed" *J. American Oil Chemists Society* **71**, 1063–1068 (1994).
36. A. E. Jones, K. R. Price and G. R. Fenwick, "Development of an HPLC method for the analysis of phytoestrogens" *J. Sci. Food Agric.* **46**, 357–364 (1989).
37. Y. Kitada, Y. Ueda, M. Yamamoto, M. Ishikawa, H. Nakazawa and M. Fujita, "Determination of isoflavones in soy bean by high-performance liquid chromatography with amperometric detection" *J. Chromatogr.* **366**, 403–406 (1986).
38. K. D. R. Setchell, M. B. Welsh and C. K. Lim, "High-performance liquid chromatographic analysis of phytoestrogens in soy protein preparations with ultraviolet, electrochemical and thermospray mass spectrometric detection" *J. Chromatogr.* **386**, 315–323 (1987).
39. W. Mazur, T. Fotsis, K. Wahala, S. Ojala, A. Salakka and H. Adlercreutz, "Isotope dilution gas chromatographic mass spectrometric method for the determination of isoflavonoids, coumestrol, and lignans in food samples" *Anal. Biochem.* **233**, 169–180 (1996).
40. G. E. Kelly, C. Nelson, M. A. Waring, G. E. Joannou and A. Y. Reeder, "Metabolites of dietary (soya) isoflavones in human urine" *Clin. Chim. Acta.* **223**, 9–22 (1993).
41. L. Coward, M. Kirk, N. Albin and S. Barnes, "Analysis of plasma isoflavones by reversed-phase HPLC-multiple reaction ion monitoring-mass spectrometry" *Clin. Chim. Acta.* **247**, 121–142 (1996).
42. Z. K. Shihabi, T. Kute, L. L. Garcia and M. Hinsdale, "Analysis of isoflavones by capillary electrophoresis" *J. Chromatogr. A* **680**, 181–185 (1994).
43. L. Markiewicz, J. Garey, H. Adlercreutz and E. Gurpide, "*In vitro* bioassays of non-steroidal phytoestrogens" *J. Steroid Biochem.* **45**, 399–405 (1993).
44. W. V. Welshons, G. E. Rottinghaus, D. J. Nonneman, M. Dolan-Timpe and P. F. Ross, "A sensitive bioassay for detection of dietary estrogens in animal feeds" *J. Vet. Diag. Invest.* **2**, 268–273 (1990).
45. U. E. Mayr, A. Butsch and S. Schneider, "Validation of two *in vitro* test systems for estrogenic activities and zearalenone, phytoestrogens and cereal extracts" *Toxicology* **74**, 135–149 (1992).
46. U. E. Mayr, "Estrogen-controlled gene expression in tissue culture cells by zearalenone" *FEBS, Lett.* **239**, 223–226 (1988).
47. D. V. Parke, C. Ioannides and R. Walker, *Food, Nutrition and Chemical Toxicity* (eds.) (*Smith Gordon Nishimura*, 1993).
48. D. Gallaher, B. O. Schneeman, "Nutritional and metabolic response to plant inhibitors of digestive enzymes" *Adv. Exp. Med. and Biol.* **177**, 299–320 (1984).
49. A. Fujita, *Advances in Enzymology* **15**, 389 (1954).
50. J. A. Zoltowicz and G. M. Kaifinan, *J. Amer. Chem. Soc.* **99**, 3134–3136 (1977).
51. A. B. Hanley, K. R. Parsley, J. A. Lewis and G. R. Fenwick, "Chemistry of indole glucosinolates: intermediacy of indol-3-yl methyl isothiocyanates in the enzymic hydrolysis of indole glucosinolates" *J. Chem. Soc. Perkin Trans.* **1**, 2273–2296 (1990).

52. E. B. Astwood, M. A. Greer and M. G. Ettlinger, "L-5-Vinyl-2-thiooxazolidinone, an antithyroid compound from yellow turnip and from *Brassica* seeds" *J. Biol. Chem.* **181**, 121–125 (1949).
53. S. Safe, B. Astroff, M. Harris, T. Zacharewski and R. Dickerson, 2,3,7,8-tetra chlorodibenzo-p-dioxins (TCDD) and related compounds as antiestrogen" *Pharmacol. Toxicol.* **69**, 400–409 (1991).
54. K. Reinli and G. Block, "Phytoestrogen content of foods—a compendium of literature values" *Nutrition and Cancer* **26**, 123–148 (1996).
55. A. O. Brinkmann, "Steroid hormone receptors; activators of gene transcription" *J. Pediatric. Endocrin.* **7**, 275–281 (1994).
56. E. M. Bickoff, A. L. Livingston, A. P. Hendrickson and A. N. Booth, "Relative potencies of several estrogen-like compounds found in forages" *J. Agric. Food Chem.* **10**, 410–412 (1962).
57. X. Xu, K. S. Harris, H. J. Wang, P. A. Murphy and S. Hendrich, "Bioavailability of soyabean isoflavones depends upon gut microflora in women" *J. Nutr.* **125**, 2307–2315 (1995).
58. W. V. Welshons, G. E. Rottinghans, D. J. Nonneman, M. Dolan-Timpe and P. F. Ross, "A sensitive bioassay for the detection of dietary estrogens in animal feeds" *J. Vetn. Diagn. Invest.* **2**, 268–273 (1990).
59. M. C. Pagliacci, M. Smacchia, G. Miglioratri, F. Grinani, C. Riccardi and I. Nicoletti, "Growth inhibitory effects of the natural phytoestrogen genistein in MCF-7 human breast cancer cells" *Europ. J. Cancer* **30A**, 1675–1682 (1994).
60. U. E. Mayr, A. Butsch and S. Schneider, "Validation of two *in vitro* test system for estrogenic activity with zearalenone, phytoestrogens and cereal extracts" *Toxicology* **74**, 135–149 (1992).
61. T. A. Pham, J. F. Elliston, Z. Nawaz, D. P. McDonnell, M. J. Tsai and B. W. O'Malley, "Antiestrogen can establish non productive receptor complexes and alter chromatin structure at target enhancers" *Proc. Natl. Acad. Sci. USA* **68**, 3125–3129 (1991).
62. H. L. Bradlow, J. J. Michnovicz, N. T. Telang and M. P. Osborne, "Effects of dietary indole-3-carbinol on estradial metabolism and spontaneous mammary tumours in mice" *Carcinogenesis* **12**, 1576–1584 (1991).
63. C. J. Coles, D. E. Edmonson and T. P. Singer, "Inactivation of succinate dehydrogenase by 3-nitropropionate" *J. Biol. Chem.* **254**, 5161–5167 (1979).
64. B. Hassel and U. Sonnewald, "Selective inhibition of the tricarboxylic acid cycle of GABAergic neurons with 3-nitropropionic acid *in vivo*" *J. Neurochem.* **65**, 1184–1191 (1995).
65. U. Wullner, A. B. Young, J. B. Penney and M. F. Beal, "3-Nitropropionic acid toxicity in the striatum" *J. Neurochem.* **63**, 1772–1781 (1994).
66. IARC, *Monographs on the Evaluation of the Carcinogenic Risk of Chemicals to Humans* **31**, 231–246 (1983).
67. E. Roader, "Medicinal plants in Europe containing pyrrolizidine alkaloids" *pharmazia* **50**, 83–98 (1995).
68. H. Niwa, T. Ogawa, O. Okamoto and K. Yamada, "Alkylation of nucleosides by dehydromoncrotaline, the putative toxic metabolite of the carcinogenic pyrrolizidine alkaloid, monocrotaline" *Tetrahedron Lett.* **32**, 927–930 (1991).
69. C. L. Miranda, R. L. Reed, F. P. Guengerich and D. R. Buhler, "Role of cytochrome P450 II A4 in the metabolism of the pyrrolizidine alkaloid senecronine in human liver" *Carcinogenesis* **12**, 515–519 (1991).
70. C. E. Couet, J. Hopley and A. B. Hanley, "Metabolic activation of pyrrolizidine alkaloids by human, rat and avocado microsomes" *Toxicon.* **34**, 1058–1061 (1996).
71. A. Puisceaux, S. Lim, J. Groopinan and M. Ozturk, "Selective targeting of p53 gene mutational hotspots in human cancers by ecologically defined carcinogens" *Cancer Res.* **51**, 6185–6189 (1991).
72. E. P. H. Yap, K. Cooper, B. Mahway and J. O. McGee, "p53 codon 249 hotspot mutation in HBV-negative hepatocellular carcinoma" *Lancet* **341**, 251 (1993).
73. M. F. Denissenko, A. Pao, M.-S. Tang and G. P. Pfeifer, "Preferential formation of benzo(a)pyrene adducts at lung cancer mutational hotspots in p53" *Science* **274**, 430–432 (1996).
74. C. E. Couet, C. Crews and A. B. Hanley, "Analysis, separation and bioassay of pyrrolizidine alkaloids from comfrey" *Nat. Toxins* **4**, 163–167 (1966).
75. Y. Zhang and P. Talalay, "Anticarcinogenic activities of organic isothiocyanates: chemistry and mechanisms" *Cancer Res.* **54**, 1976S–1981S (1994).
76. N. Tawfig, R. K. Heaney, J. A. Plumb, G. R. Fenwick, S. R. Musk and G. Williamson, "Dietary glucosinolates as blocking agents against carcinogenesis: glucosinolate breakdown products assessed by induction of quinone reductase activity in murine hepalclc7 cells" *Carcinogenesis* **16**, 1191–1194 (1995).
77. S. R. Musk, T. K. Smith and I. T. Johnson, "On the cytotoxicity and genotoxicity of allyl and phenethyl isothiocyanates and their parent glucosinolates sinigrin and gluconasturtiin" *Mutat. Res.* **348**, 19–23 (1995).

78. H. Verhagen, H. E. Poulsen, S. Loft, G. van Poppel, M. I. Willems and P. J. van Bladeren, "Reduction of oxidative DNA-damage in humans by Brussels sprouts" *Carcinogenesis* **16**, 969–970 (1995).
79. A. Adesida, L. G. Edwards and P. J. Thornalley, "Inhibition of human leukaemia 60 cell growth by mercapturic acid metabolites of phenylethyl isothiocyanate" *Food Chem. Toxicol.* **34**, 385–392 (1996).
80. Biomarkers of Exposure and Effect in Relation to Quality of Life and Human Risk Assessment, *Concerted Action ERB FAIR CT 96/178* (1996).

15. PLANT-ASSOCIATED TOXINS IN THE HUMAN FOOD SUPPLY

STEVEN M. COLEGATE *, JOHN A. EDGAR * and
BRYAN L. STEGELMEIER **

I. INTRODUCTION

Naturally occurring toxins (natural toxicants) associated with plants are those synthesized or accumulated by plants or by the many fungi and microorganisms that invade or grow saprophytically on, or endophytically within the plants. Natural toxicants, which include low molecular weight secondary metabolites through to high molecular weight proteins, are amongst the most damaging and deadly chemicals known. They are capable of eliciting a spectrum of biological effects from sudden death to prolonged chronic disease in livestock and humans. The harmful effects on humans resulting from the ingestion of common household or garden plants have been reviewed[1] and will not be considered in this present review. The use of herbs and other plants by humans for health purposes can also result in poisoning episodes. Apart from simply collecting the wrong plant, the dangers of toxins associated with herbal and medicinal plant preparations are similar to those of foods, i.e. natural toxicants can be intrinsic to the medicinal plants or can be present as contaminants. The influence that developing regulations concerning human exposure to natural toxicants in herbal preparations may have on regulations concerning human exposure to natural toxicants in food is considered in Section V.

The effects on livestock and on farming enterprises resulting from poisonous plants have been well researched and documented[2–5]. Exposure of food-producing animals to plant-associated toxins can result from grazing natural range land on which poisonous plants occur, through to contamination of otherwise safe, improved pasture and feed with opportunistic toxic weeds. Such exposure occurs worldwide[4,5]. For example, livestock deaths occurred when stock were first introduced onto rangelands in the western United States of America. In fact, larkspurs (*Delphinium* spp.) and locoweeds (*Astragalus* and *Oxytropis* spp.) continue to poison thousands of animals each year in those areas. Cheeke and Shull describe examples of feed contamination with poisonous plants such as the infestation of grain fields with the pyrrolizidine alkaloid (PA)-containing plant *Amsinkia intermedia*, and the piperidine alkaloid-containing poison hemlocks (*Conium* spp.)[6]. Some poisonous plants such as PA-containing *Senecio* or *Cynoglossum* spp. are seldom eaten in open range or pasture conditions, however these plants are readily consumed by livestock when they are included in hay or silage. A study has shown that whilst the PA content of hay contaminated with *Senecio alpinus* remained constant over months, the PAs in silage contaminated with *Senecio alpinus* were destroyed to a great extent,

* Plant Toxins Unit, Commonwealth Scientific and Industrial Research Organisation, Division of Animal Health, Australian Animal Health Laboratory, Private Bag 24, Geelong, Victoria 3220, (Australia).
** Poisonous Plant Research Laboratory, Agriculture Research Service, US Department of Agriculture, Logan, Utah (USA).

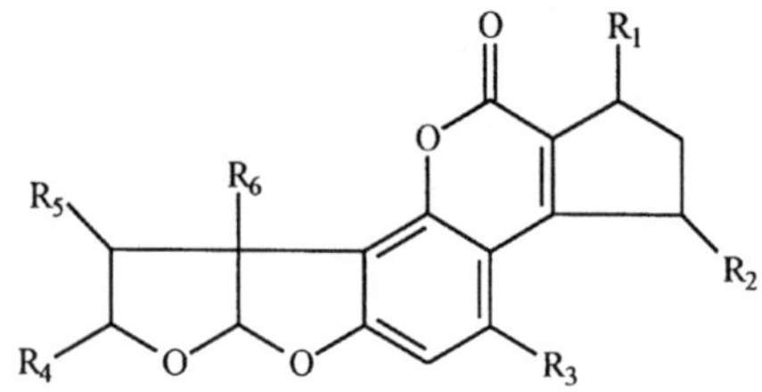

1 - 4

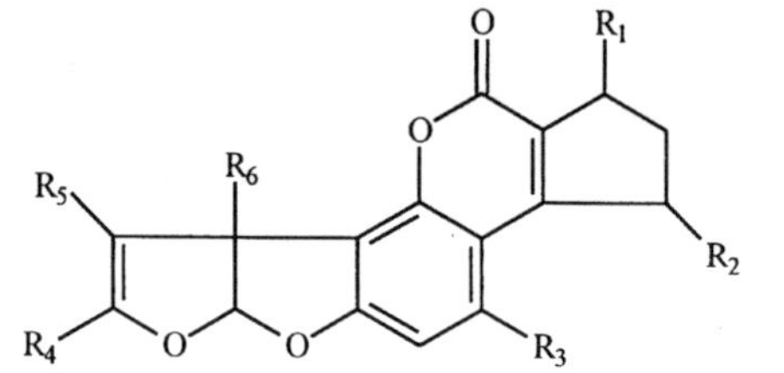

5 - 10

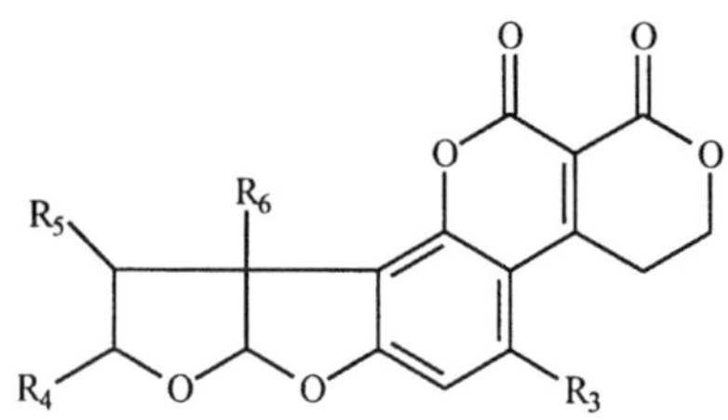

11

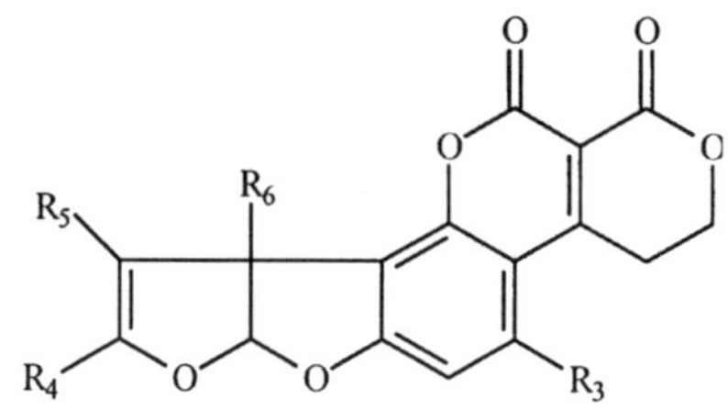

12

Structure	Name	R_1	R_2	R_3	R_4	R_5	R_6
1	aflatoxin B_2	=O	H	OCH_3	H	H	H
2	aflatoxin M_2	=O	H	OCH_3	H	H	OH
3	aflatoxin B_{2a}	=O	H	OCH_3	OH	H	H
4	aflatoxin B_1 2,3-oxide	=O	H	OCH_3	- O -		H
5	aflatoxin B_1	=O	H	OCH_3	H	H	H
6	aflatoxin M_1	=O	H	OCH_3	H	H	OH
7	aflatoxicol	OH	H	OCH_3	H	H	H
8	aflatoxin H_1	OH	OH	OCH_3	H	H	H
9	aflatoxin P_1	=O	H	OH	H	H	H
10	aflatoxin Q_1	=O	OH	OCH_3	H	H	H
11	aflatoxin G_2	-	-	OCH_3	H	H	H
12	aflatoxin G_1	-	-	OCH_3	H	H	H

Figure 1 Structures of aflatoxins.

but the degradation of PAs was much less complete in the lower concentration range. Indeed, silage comprising 3.5–23% *S. alpinus* still contained hepatotoxic, macrocyclic PAs in a concentration of about $20\,mg\,kg^{-1}$ wet weight. Despite the low concentrations, such silage cannot be recommended as safe for cattle due to the cumulative effects of PAs[7].

In addition to endogenous poisonous chemicals produced by plants, some plant species can be infected with toxin-producing microorganisms. For example, food crops such as peanuts, corn, rice, fruits, wheat and cassava can be infected with the fungi *Aspergillus flavus* or *A. parasiticus* which can produce the aflatoxins, a family of mutagenic, teratogenic and carcinogenic bisfuranocoumarins (**1**–**12**, Fig. 1). This review will briefly consider the poisonous chemicals intrinsic to plants which are ingested by humans for nutritional purposes, before considering in more detail the environmental toxicology issue of contamination of human food by plant-associated poisons. This review distinguishes between chemical **contaminants** of food and **residues** in food (e.g. from deliberately applied pesticides and veterinary chemicals). This distinction becomes important in the assessment of risk associated with chemicals in food intended for human consumption[8].

II. HUMAN EXPOSURE TO NATURAL TOXICANTS

The human and food plant interaction has evolved over millennia and plants continue to be the most important food source for man, providing us with essential energy, nutrients and vitamins. Plant-derived antioxidants have also been shown to be essential in minimising inflammatory effects and avoiding many diseases including cancer[9,10].

The interaction between humans and plants has evolved to the extent that various cultures have acquired empirical knowledge about which foods to avoid, or how to treat plants in order to prevent intoxication. Cultures are also generally aware of circumstances which might lead to increased toxicity, such as the increase in steroidal glycoalkaloids in greening potato tubers[11]. A concern arises however with the worldwide trend to experience foods of other cultures or to utilise plants as novel foods to ostensibly provide nutrition, a novel flavour or flavour that we currently obtain from a mainstream plant. Sometimes there can be an associated lack of awareness about how to render the plant safe for consumption. In addition, "new" plant foods may also contain chemicals for which long term effects of chronic, low level exposure may be unknown. Of major concern however, is the contamination of known safe and nutritious or flavoursome food plants with toxic plants that are normally avoided, and the contamination of food products from food-producing animals that have consumed toxic plants.

Human exposure to hazardous, naturally-occurring, plant-associated chemicals in the diet can vary:

1. The plants can form part of the usual diet. Incomplete detoxification of those that need to be treated in a specific way to render the food safe can lead to poisoning episodes, or a long term exposure to low levels of natural toxicants may lead to unattributable, or misattributed, toxic effects.
2. The plants are purposefully sought for use as alternative "natural" foods, medicines or for their consciousness-altering properties.
3. The toxic plants are mistakenly collected when other plants are sought.
4. An otherwise safe food has been contaminated with a naturally occurring toxic chemical *via* contamination of food plants with toxic plant species, or sub-clinical ingestion of toxic plants by food-producing animals resulting in contamination of the carcass or milk with toxic chemicals.
5. Ingestion of normally safe foods, but which occasionally contain poisonous chemicals, e.g. honey produced by bees collecting the nectar of toxic plants.

Whilst the acute toxic effects of many naturally-occurring chemicals may be well known, the possible effects of long term, low level exposure are not always fully appreciated. In addition, the vagaries of growing, collecting/harvesting, processing and storing of plants means that toxin content can vary considerably and unpredictably. For example, the PA concentrations of *Senecio ridellii* vary from 0.03% to 17.9%. The concentrations varied with the season, plant phenotype, location, and other unidentified factors that make it nearly impossible to predict what the toxicity may be from year to year[12,13].

III. TOXINS INTRINSIC TO THE FOODS WE EAT

The relevance of, and the extent to which toxic chemicals occur naturally in the foods we eat have been reviewed by many authors[14,15]. Examples of chemicals with known toxicities that occur naturally in our food are shown in Table I.

A culture evolving in the presence of plants, soon learns to avoid eating those plants, or parts of plants, that produce acute toxic effects. However, long term effects of chronic, low level exposure to a toxic chemical are more difficult to recognise and assess.

Some cultures also learn how to treat plants that produce ill effect in order to render them edible. For example, the Australian Aborigine had discovered that the fruiting bodies (sporocarps) of Nardoo fern (*Marsilea* spp.) could be eaten safely after heat treatment (which destroys the plant's thiaminase activity)[6,16]. Robert Burke and William Wills, explorers of the Australian continent in the mid-19th Century are suspected of meeting their death as a result of ingesting untreated Nardoo on their last ill-fated expedition in 1861. They may have observed Aborigines collecting the plant but were unaware of the need for adequate cooking. Combined with a postulated thiamine deficiency resulting in clinical symptoms of beri-beri, the thiaminase activity of the nardoo would have been fatal[17,18]. In a similar way, the seeds of *Canavalia ensiformis* are toxic to man if eaten raw but, are apparently eaten safely by people of those cultures which have acquired the empirical knowledge to

Table I Some examples of naturally occurring toxins intrinsic to food

Food plant	Toxic chemical constituent	Effect
Potatoes	Steroidal glycoalkaloids	Gastrointestinal (vomiting, diarrhoea, and abdominal pain), neurological (restlessness, drowsiness, confusion, delerium)
Canola, cabbage	Glucosinolates	Goitragenic
Mushrooms	Hydrazines	Protoplasmic
Celery	Psoralen	Photosensitivity
Alfalfa sprouts	Phorbol esters	Co-carcinogenic
	Canavanine	Arginine analogue (antinutritional, renal toxicity, immunocompromising)
Black pepper	Safroles	Carcinogenic, hepatotoxicity
Red Kidney beans (*Phaseolus vulgaris*)	Lectins	Abdominal pain, nausea, vomiting, diarrhoea
Chick peas (*Lathyrus sativa*)	Non-protein amino acids	Neurotoxic, osteolathyrogenic
Fava beans (*Vicia faba*)	Aminopyrimidinyl-glucosides	Favism: headache, dizziness, nausea, yawning, vomiting, abdominal pain, acute haemolytic anaemia

treat the seeds by cooking. It has however been shown that the toxic constituent, canavanine (an arginine anti-metabolite, **13**, Fig. 2), is not completely destroyed by cooking and therefore could still cause a health problem if *Canavalia* seeds form a major part of the diet, as might occur in times of other food shortages, or when consuming diets with marginal arginine concentrations[19,20].

The difficulties associated with investigating and attributing ill effects in humans to plant-associated toxins are exemplified by the attempts to find a cause for a high incidence of neurodegenerative disorders amongst the Chamorros inhabitants of the island of Guam in the Pacific Ocean. Many of these Chamorros people developed central neurodegenerative disorders characterised by tremors, paralysis or dementia. All these changes are common to other motor neurone-related diseases such as Alzheimer's and Parkinson's diseases. The epidemiological work towards elucidating possible causal factors has been reviewed by Duncan[21]. A likely hypothesis was that the consumption of seeds of the cycad *Cycas circinalis*, a common food of the Chamorros people, was a factor in development of the disease. To avoid the acute emetic effect of cycad seeds, the Chamorros treated them by soaking the fleshy, white inner tissue of the seeds in water before drying and grinding the product to produce a flour. A suspicion that incomplete treatment of the cycad seeds could result in a neurotoxin-tainted flour resulted from the observation that untreated cycads have

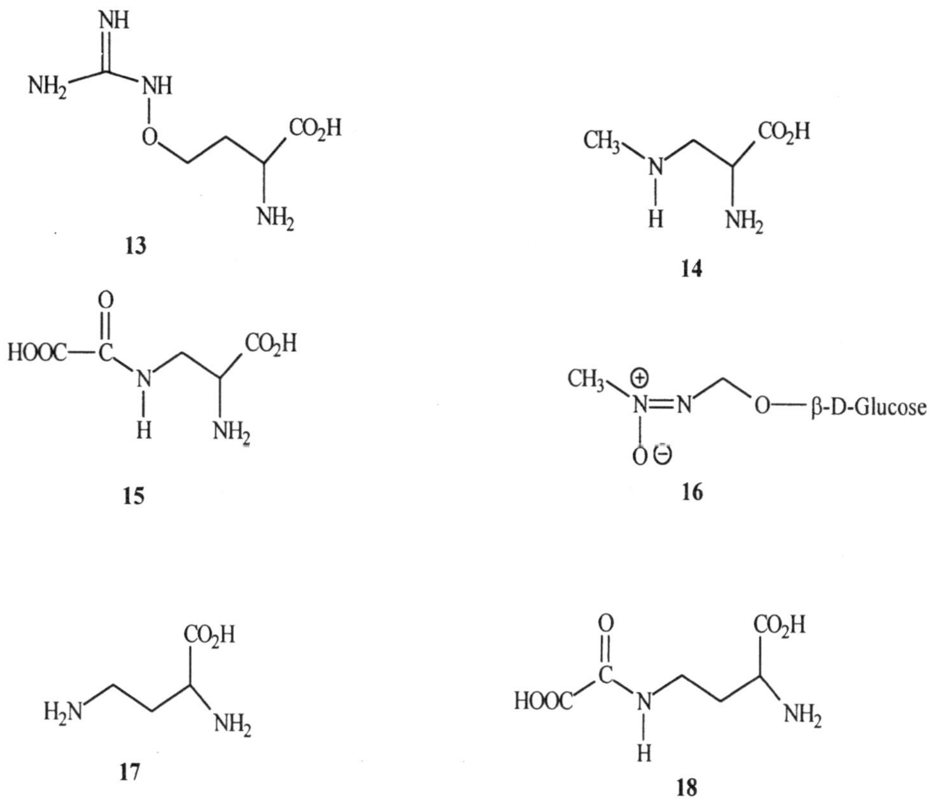

Figure 2 Non-protein amino acids and a methylazoxyglycoside from *Canavalia* spp., *Cycas* spp. and *Lathyrus* spp.

been shown to induce locomotor dysfunction in grazing animals[22]. Focus initially centred on the potential neurotoxicity of an amino acid, β-methylamino-L-alanine (BMAA, **14**, Fig. 2), present in the cycad and similar to β-N-oxalylamino-L-alanine (BOAA, **15**, Fig. 2), an unusual amino acid isolated from *Lathyris sativa* and which produced lathyrism type signs in primates[23]. However, the best evidence which indicated that BMAA is indeed neurotoxic to primates still required a dose orders of magnitude greater than the dietary exposure of the Chamorros people to BMAA. The focus of ongoing investigations to determine an environmental trigger to the Guam neurodegenerative disease has changed to include the azoxyglycosides, such as cycasin (**16**, Fig. 2), also found in *Cycas* spp[6]., and the possible involvement of zinc contamination of the cycad flour[24]. Although much of the epidemiological evidence is suggestive of an environmental or food-borne aetiology, the problem remains an enigma since the causative factor has not been conclusively identified.

Toxic chemicals intrinsic to the food we eat could become a problem if overindulgence in a particular food plant, to the detriment of a balanced diet, were to occur. For example, sporadic incidents of lathyrism occur in India when diets include a higher than usual proportion of seeds from *Lathyrus* spp., usually *L. sativa*, as a consequence of a scarcity of other foods. Young men are most severely affected with degeneration of motor neuron tracts resulting in paralysis that can be permanent even when the diet is corrected. Lathyrogenic chemicals present in species of *Lathyrus* include the unusual amino acids BOAA (**15**, Fig. 2)[25], L-α,γ-diaminobutyric acid (**17**, Fig. 2)[26] and the γ-oxalyl derivative of L-α,γ-diaminobutyric acid (**18**, Fig. 2)[27]. Another example is that involving cassava (*Manihot esculenta*, also referred to as tapioca or manioc) which is a tropical tuber that contains cyanogenic glycosides. Treatment of the grated tuber with water hydrolyses the glycoside and releases the cyanide. Alternatively, cooking destroys the enzymes that usually convert the heat stable glycoside to hydrogen cyanide thereby rendering the product safer. However, cultivation of cassava has increased in drought affected countries to the extent that epidemics of spastic paraparesis have occurred as a result of a greater consumption of incompletely detoxified cassava[28].

Also of concern is the introduction of novel or altered foods into our diet. Foods which are novel to a particular culture may cause problems if the transmission of empirical, cultural food-safety knowledge was inefficient and the food is not treated correctly. For example, incomplete detoxification of bitter lupin seeds led to intoxication of a woman who ingested two handfuls of bitter lupin beans at her evening meal and a smaller quantity the following morning. Anticholinergic symptoms of intoxication (dilated pupils, blurred vision, dry mouth, difficulty in swallowing, tachycardia, etc.) by the lupin alkaloids began immediately after the morning helping of lupin seeds[29]. The report cites other incidences of lupin bean intoxication and made recommendations for appropriate labelling of bitter lupin beans including comprehensive instructions for an acceptable method for removing the toxic alkaloids.

The use of a novel food source may cause problems if there are unrecognised antimetabolites or toxins in the plant that are capable of causing clinical effects following long term, low level exposure to the toxin, or causing an irreversible, acute sub-clinical effect that may only manifest itself clinically after a period of time. For example people in the Shaanxi province of China have supplemented their diet with rapeseed oil resulting in high levels of erucic acid in children of the region. Erucic acid has been shown to cause cardiac lipidosis and necrosis in laboratory animals and also exacerbates selenium deficiency that is associated with endemic cardiomyopathy (Keshan disease) in some areas[30]. Whether erucic acid does in fact contribute to endemic disease in areas of high rapeseed intake remains to be proven and highlights the uncertainty and difficulties of attributing clinical effect to dietary causes.

Some foods have been genetically altered. This includes those that have been selected by breeding programs and those manipulated at a molecular level to acquire more desirable characteristics. Such alterations are potentially dangerous since relative levels of endogenous chemicals may change, giving rise to concentrations of toxic chemicals higher than the maximum tolerated dose and leading to clinical effect. This latter consequence has been observed with efforts to cross breed commercially cultivated potatoes with wild varieties in efforts to improve insect resistance. A resultant cultivar, named Lenape, showed much promise for its genetic resistance to certain potato beetles but was withdrawn from the U.S. market due to high levels of toxic solanidine glycosides in the tubers[31]. It was a cogent lesson to potato breeders, and indeed to all agronomists involved in cross breeding experiments, to be aware of the possibility of inadvertently introducing undesired levels or classes of toxic chemicals.

IV. CONTAMINATION OF THE FOODS WE EAT BY EXTRINSIC PLANT-ASSOCIATED TOXINS

Co-evolution of human cultures and animals with plants has led to procedures for, and mechanisms of dealing with potentially toxic plants. Thus plants are avoided, detoxified before ingestion or only ingested as a minor part of the overall diet as is the case with browsing animals. In addition, metabolic pathways may evolve to detoxify the plant chemicals *in vivo* (for example, see the following paragraph describing fluoroacetate tolerance). Some animals can safely store the toxins and have come to depend upon the toxins for their survival and procreation. For example, the Monarch butterfly has evolved such that its larvae consume cardioactive glycoside-containing milkweed (*Asclepias* spp.) thereby acquiring an exclusive food source. Further evolution has led to the larvae and butterflies accumulating cardioactive glycosides from the plant since the emetic effect of the accumulated chemicals provides a competitive advantage by deterring birds from feeding on the larvae and adults[32]. Other butterflies of the Danainae and Ithomiinae families have become similarly dependent upon toxic pyrrolizidine alkaloids. In this case they use the plant-derived chemicals not only for defence but also in their courtship and territorial behaviour[33].

Problems arise when people, animals and/or plants are relocated to areas where poisonous plants occur to which they have not adapted or where natural biological controls are absent. For example, Australian marsupials that evolved in the presence of fluoroacetate-containing species of *Gastrolobium* and *Oxylobium* have developed a tolerance for the poison which affects cellular respiration by disrupting the tricarboxylic acid (TCA) cycle. Similar, island-bound marsupials, and exotic animals that did not evolve in the presence of the plants are extremely susceptible to this potent poison[34]. This relative tolerance of native fauna in Western Australia to fluoroacetate has enabled the use of fluoroacetate-contaminated baits to control populations of feral carnivores such as foxes and cats.

An additional problem is the potential lack of natural, biological controls for toxic plants that have been introduced into a new environment. Given suitable growing conditions, and unchecked by natural disease and co-evolved herbivores, the newly established plant can rapidly extend its range to pest proportions and thereby expose food-producing animals to dangerous levels of undesirable natural chemicals. Even indigenous plants can become a pest and a danger to animal health when agricultural practices disrupt the finely balanced checks that have taken eons to establish. For example in the western United States of America, halogeton (*Halogeton glomeratus*), an oxalate-containing plant, often grows and replaces native grasses and shrubs alongside roads, trails and in other disturbed areas. Livestock that are driven through these areas can ingest lethal amounts of oxalate and deaths of hundreds of animals have been reported[35].

Toxic plants/weeds pervade human agricultural endeavours. Effects of poisonous plants on animal health and the economics of farming include:

- animal ill health and death
- reproductive effects
- decreased productivity e.g., reduced wool, meat, milk or egg production, decreased productive life span
- increased costs e.g., fencing off infested pasture, herbicidal management of weeds, veterinary charges, removal of toxic plants, reduced real estate values, special animal management procedures.

Movement of these plant-associated toxic chemicals into the human food supply can occur *via* contamination of our usual, otherwise safe food material with toxic plants or by toxin-producing microbes. The consequent effects have the potential to be acute and with obvious clinical symptoms in cases of heavy contamination, through to less recognisable or attributable symptoms resulting from an insidious, long term, low level exposure. The long term, low level exposure is of concern in that little is known of the consequences of such exposure.

IV.1. *Contamination of Grain*

Grain for human consumption can become contaminated with the seeds of toxic plants. This is common when crops are infested with the toxic weeds and the harvesting procedures fail to discriminate between them. Such contaminated grain can be cleaned up considerably by post-harvest processing (sieving, winnowing, etc). However, the level of scrutiny or regulation to which the final product can be subjected is affected by, among other considerations, recognition of potential health problems. Thus, levels of contamination could become acceptable which may prevent acute intoxication, with obvious clinical signs, but may in fact lead to low level, long term exposure to the toxic chemicals with consequent clinical effects not easily attributed to a particular cause. The cleaning up process will also generate a product, in which the contaminating material has been concentrated, rejected for human consumption but often presumed to be useful for animal nutrition. Such a use of the reject grain can cause overt toxicity in animals, but can also cause less obvious, deleterious effects upon productivity, and result in transfer of the toxic chemicals to human food products such as meat, milk and eggs (see Sections IV.2 and IV.3).

Pyrrolizidine alkaloid-containing plants have featured prominently in animal toxicoses and have been extensively researched and well reviewed[36–38]. The toxic PAs all possess structural requirements[39] that enable hepatic, primary metabolism to their corresponding pyrrolic (didehydro-PAs) entities which are potent *in vivo* alkylating agents. The short-lived (highly reactive) pyrrolic metabolites can react with tissue components such as liver enzymes and DNA causing effects such as megalocytosis, centrilobular necrosis, hepatic fibrosis leading to cirrhosis and hepatic

Table II Human poisoning episodes resulting from contaminated grain (data and references cited by Huxtable[42])

Plant genera	Representative alkaloids	Poisoning episodes
Crotalaria	Monocrotaline, fulvine	India, 1973, 1975
Heliotropium	Heliotrine, lasiocarpine	Afghanistan, 1974
Senecio	Senecionine, seneciphylline	South Africa, 1920s, 1950s; Iraq, 1970s
Trichodesma	Trichodesmine, incanine	Central Asia, 1950

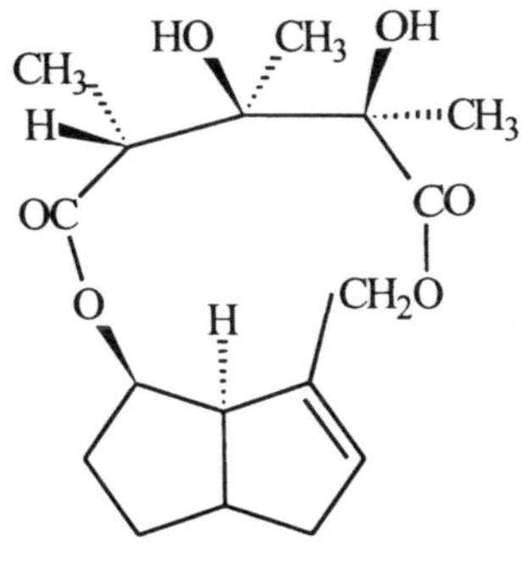

19

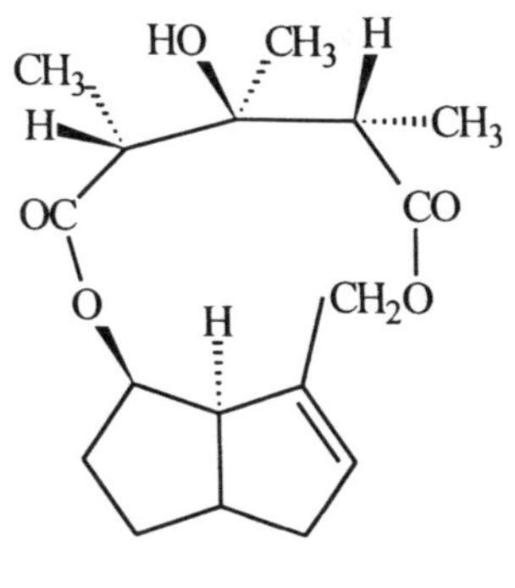

20

Figure 3 Pyrrolizidine alkaloids from contaminated millet.

failure, pulmonary hypertension and cancers[40–42]. In a comprehensive review of human health-related effects of PAs, Huxtable describes the toxic actions of PAs, discusses human exposure to PAs and details case studies of PA intoxication of humans[42]. Ingestion of PAs by humans often results in a characteristic veno-occlusive syndrome which is characterised by hepatic vascular fibrosis, portal hypertension and subsequent ascites[42].

Huxtable reviewed reports that indicate species of the plant genera *Senecio*, *Trichodesma*, *Heliotropium* and *Crotalaria*, within the plant families Compositae, Boraginaceae and Leguminosae, are prominent in being implicated in human toxicoses following contamination of food grain (Table II)[42].

In a number of villages in India, the ingestion of locally-grown millet grain contaminated with seeds of *Crotalaria nana* has resulted in veno-occlusive disease and death[43,44]. Two PAs, monocrotaline (**19**, Fig. 3) and fulvine (**20**, Fig. 3), were indeed isolated from the contaminated grain.

A massive poisoning of humans by PAs occurred in 1992/93 in Tadjikistan. Milling flour was contaminated with seeds of *Heliotropium* (see Section IV.3 and Fig. 10 for structures of some heliotrope alkaloids) resulting in more than 4000 people hospitalised with acute liver injury[45].

Jimson weed (*Datura stramonium*, Solanaceae), otherwise known as common thornapple, stramonium or false castor oil, whilst probably of North American origin, is now commonly found in temperate and sub-tropical regions of the world. Containing the parasympatholytic tropane alkaloids scopolamine (**21**, Fig. 4) and hyoscyamine (**22**, Fig. 4), the plant has a long history of human and animal poisonings, and use as a narcotic and hallucinogenic drug[6,16,46].

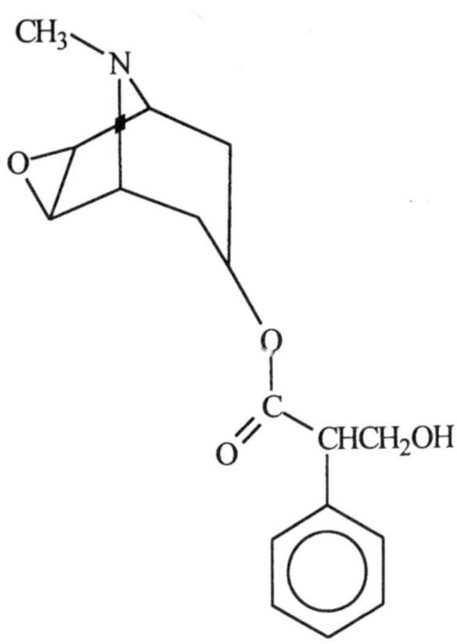

21

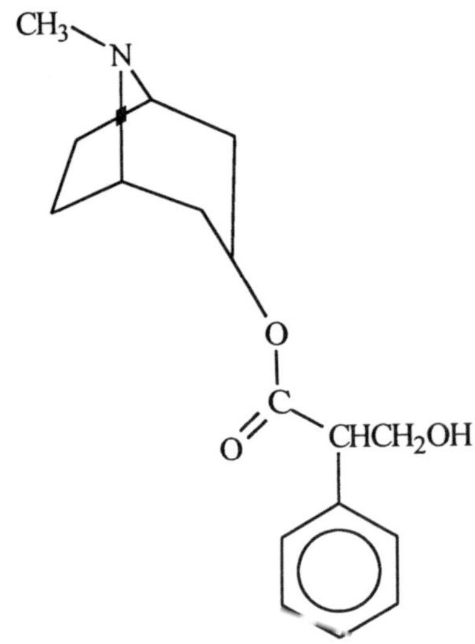

22

Figure 4 Tropane alkaloids from *Datura* spp.

Since *Datura* spp. can grow as a weed in many cereal crops there is the potential for contamination of grain with this plant. As with other cereal crop contaminants, regulations can limit levels of *Datura* seed impurities. For example, in Australia the amount of *Datura* seed in sorghum grain for export is limited to 1.5 seeds kg^{-1}, whereas for stock feed the limit is 6 seeds kg^{-1} [6].

Intoxication of humans from ingesting *Datura*-contaminated grain has been reported. For example, contamination of millet with *D. stramonium* seeds resulted in ten people, suffering from symptoms of atropine-like intoxication, to be hospitalized in Usangi, Africa after ingesting a porridge made from the millet. The patients presented with psychosis and hallucinations, rapid heart rate, fixed, dilated pupils and a dry mouth. Examination of official records indicated that a similar type of food poisoning had previously occurred in at least eight other regions[47]. A similar case involving contamination of grain with *D. stramonium* was reported from Tanzania[48]. In this instance, a clinical syndrome typical of atropine poisoning was reported to be frequently encountered in an area where *D. stramonium* grows as a weed in wheat fields. The weeds were being harvested along with the wheat and there was often no sorting at the local mills. As a result the flour was contaminated with the atropine-related alkaloids from *D. stramonium* leading to poisoning of people when bread or chapattis made from the flour were ingested.

Incorporation of *Aspergillus* spp.-infected grain into the diet can result in poisonings. For example an aflatoxin-contaminated corn crop in Western India resulted in the death of 100 out of 400 affected people. In this instance, exposure to aflatoxins (specifically AFB_1, Fig. 1) was estimated to be 2–6 mg kg^{-1} day^{-1} for one month. Clinical symptoms included massive gastro-intestinal bleeding, jaundice, ascites and portal hypertension[49].

The previous examples represent documented cases of overt clinical illness as a result of attributable contamination of staple foods. There are also examples of where contamination of a staple with a toxic plant can be demonstrated but where bulking and screening of the staple lowers the level of contamination in the human food product quite substantially. However, in some cases the levels of toxins, derived from the toxic plants, to which humans are exposed is still unknown. Further, it is also generally unknown what level of exposure can be considered safe for humans. Thus, long term, low level ingestion of some toxins may result from using contaminated, albeit low level contamination, grain. The difficult problem that this poses relates to the recognition of expected, or predicted symptoms, and eventual attribution of the clinical effect to the plant toxin. As an example of this predicament, cereal crops in some countries are often contaminated with annual ryegrass (*Lolium rigidum* Gaudin, Gramineae) which is harvested along with the crop. The seedheads of annual ryegrass (ARG) can become toxic to animals, causing annual ryegrass toxicity (ARGT) which has been reported in Australia[50], South Africa[51] and possibly in the United States of America[6]. Farming practices in Oregon, including the burning of crop stubble, may have diminished or eradicated the incidence of ARGT in North America (P. R. Cheeke, personal communication)[6]. The ARGT toxins are produced in the ARG seedheads by a bacterium (*Clavibacter toxicus*) which is carried to the seedhead by a nematode (*Anguina funesta*). The toxins, referred to as corynetoxins (**23**, Fig. 5), have been identified as a group of glycolipids which are characterised by a common core made up of uracil, a C_{11} amino sugar (tunicamine) and N-acetylglucosamine, and differentiated by the fatty acid side chain attached to the tunicamine entity by an amide linkage[52,53]. The fatty acid side chain, varying in length from C_{15} to C_{19}, can be saturated, αβ-unsaturated or β-hydroxylated, and terminate in a normal (CH_3CH_2–), iso ($CH_3(CH_3)CH$–) or anteiso ($CH_3CH_2(CH_3)CH$–) methyl branching[54].

In this structural aspect, the corynetoxins are related to the tunicaminyluracil antibiotics such as the tunicamycins, streptovirudin and antibiotic MM19290[55]. The spread of herbicide-resistant ARG, the occurrence of outbreaks of ARGT in grazing animals, the biochemical and pathological aspects of intoxication, and the isolation

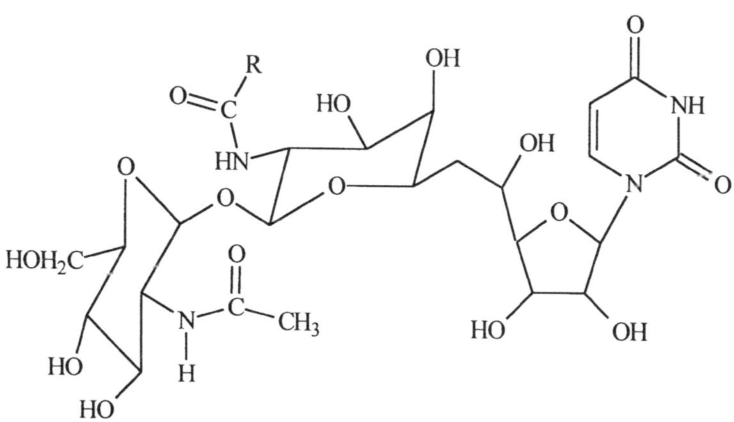

23

R = C_{15} - C_{19} aliphatic chain (saturated, αβ-unsaturated, β-hydroxylated)

Figure 5 General structure of the corynetoxins.

and identification of the corynetoxins have been reviewed[56,57]. From an environmental toxicology aspect, any risk analysis of potential problems associated with corynetoxin-contamination of grain destined for human consumption must include the following considerations:

- corynetoxins are very poisonous, inhibiting the tissue-ubiquitous enzyme, uridine diphospho-N-acetylglucosamine:dolichol-phosphate N-acetylglucosamine-1-phosphate transferase[58,59]
- the corynetoxins are cumulative in their effect[60,61]
- observation of clinical signs in animals is delayed some 60 hours after initial dosing with toxins[62]
- there are no data relating to the long term, low level exposure of humans to corynetoxins in the food chain
- what might a "No Observable Effect Limit" (NOEL) be for humans. This will form the basis of setting Provisional Tolerable Weekly Intake and Provisional Maximum Tolerable Daily Intake standards[8]
- little is known about the metabolism, pharmacodynamics or pharmacokinetics of the corynetoxins
- what are the levels of the corynetoxins in food products, i.e. what is the actual exposure of humans to corynetoxins?
- assay procedures will have to be developed in order to monitor food products for contamination by corynetoxins
- what might be the expected clinical effects in humans exposed to low levels of corynetoxins for a long time?

The corynetoxins also have the potential to contaminate animal-derived food products such as meat, milk and eggs (see Sections IV.2 and IV.3). It is important to realise that evidence, available in the scientific literature, only indicates the potential for contamination of the human food supply. More research is needed to elucidate the reality of exposure and associated risks.

Considerations similar to those described for the corynetoxins will apply to all other natural toxicants that have the potential to contaminate human food. An example of where potential contamination has been recognised and acted upon involves the use of lupins as animal feed and for production of food for humans. Ingestion of lupins by animals has been associated with teratogenic effects as a result of quinolizidine-type alkaloids[63]. The development of low alkaloid lupins has led to an increase in the use of this leguminous grain crop as a source for animal feed (grain and post-harvest stubble) and flour for human food products. However, some lupins can be colonised by the fungus *Diaporthe toxica* (anamorph *Phomopsis* sp.) which will persist on the stubble after the grain has been harvested with dire implications for grazing animals (see Section IV.2). The fungus produces hepatotoxic, modified hexapeptides, the phomopsins (**24**, Fig. 6)[64,65], which elicit their effect by binding to tubulin and thereby affecting microtubule assembly within cells[66]. Since the phomopsins specifically target the liver[67], can cause cancer[68] and chromosomal aberrations[69], and since assembly of microtubules is fundamental to cellular processes such as formation of the mitotic spindle and intracellular transport[70,71], and therefore have human health implications, the Australian Food Standards Code limits phomopsin A content in any food for human consumption to 5 ppb[72]. In this regard it is important to note that a trial fermentation process aimed at detoxifying phomopsins and thereby rendering affected seed useable, did not achieve the desired detoxification[65].

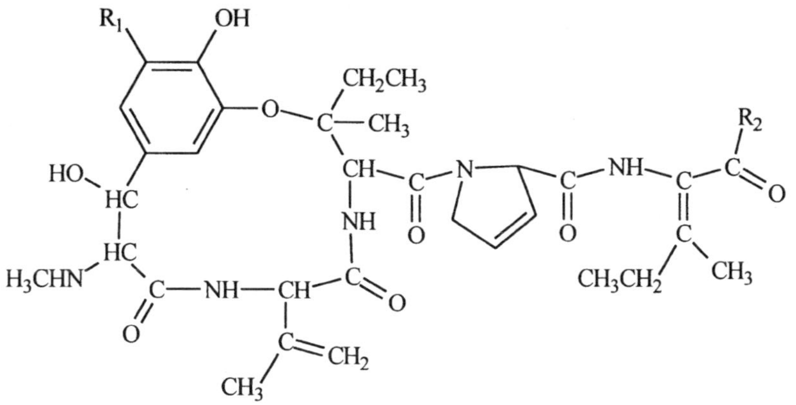

24

	R_1	R_2
Phomopsin A	Cl	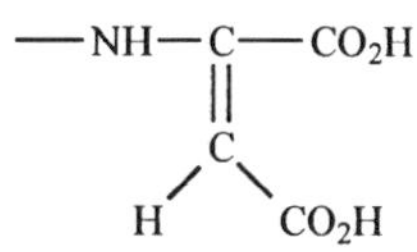
Phomopsin B	H	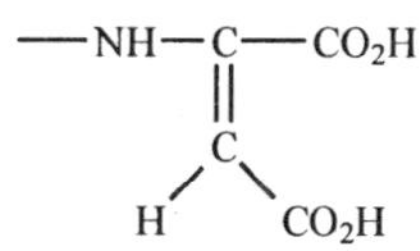
Phomopsin D	Cl	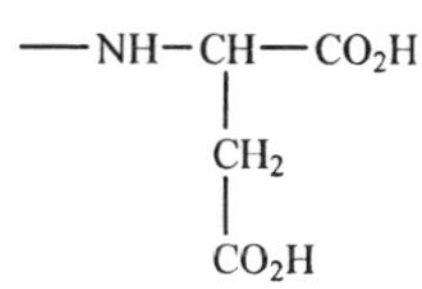

Figure 6 Phomopsins A, B and D.

A complicating factor in assessing the potential toxicity of a grain product is that the total removal of contaminating seed may not alleviate the toxicity of the grain. Dust material generated from the toxic plant parts during the harvest and grain processing procedures can still contain the chemicals responsible for the toxicosis. It has been shown, for example, that following exhaustive clean-up of a stock feed contaminated with seeds of *Heliotropium europaeum*, the pyrrolizidine alkaloids responsible for the poisoning of poultry and pigs[73] were still to be found in washings of the "clean" grain (Edgar, J.A., unpublished data). Similarly, aflatoxin B_1 (**5**, Fig. 1) has been detected in dust samples obtained from grain silos in central Illinois in the USA[74]. These facts have enormous implications from an environmental toxicology aspect in that visual inspections for contaminating seed, or detection of toxin-producing organisms, may not be sufficient to ensure against low level exposure of human populations to the toxic chemicals derived from contaminating plants or microorganisms. Both cases, which can be extrapolated to include all contaminating chemicals in grain, also have occupational health and safety implications for workers exposed to the dust from grain harvesting, transport, storage and processing.

IV.2. *Contamination of Meat Products*

It is a common observation that animal meat can be tainted by flavours derived from the food eaten by the animals[75]. Such contamination is readily recognised by the consumer and represents practical demonstrations that chemicals intrinsic to livestock feed can be translocated to food for humans. This observation has particular implications for plant-derived toxins that may be tasteless to humans. The potential for plant-associated chemicals to cause ill-health in humans consuming animal tissue will depend upon the metabolism of the chemical, and the tissue distribution and pharmacokinetics of the metabolites or parent compounds within the animal, i.e. what are the actual concentrations of parent compound or metabolites in the animal tissue that is ingested by humans?

The potential contamination of meat products for human consumption is intricately associated with contamination of natural and cultivated pastures, and animal feed stock derived from cultivated grains. An illustrative example is a problem which occurred several years ago in Australia when 100,000–200,000 chickens and 1000–4000 pigs died from PA poisoning as a result of heliotrope-contaminated stock feed[73]. Whilst in this case it was evident from the mortality rates that a problem existed, the possible consequences of low level exposure of the animals to the pyrrolizidine alkaloids are harder to define, especially when the PAs can induce a slow, progressing liver failure rather than sudden death[76]. This has important implications for human health when asymptomatic, intoxicated animals are sent to slaughter and resultant meat products contaminated with PAs or their metabolites are ingested by humans.

It has been shown that PA-related mortalities of livestock can occur many months after the last exposure to the plants. For example, cattle have died up to 17 months after being exposed to PA-containing *Senecio* plants as grazing calves[77]. If this were a delayed death due to a slow clinical manifestation of an initial insult then there would be little to be concerned about since the toxin and/or metabolites would presumably have been cleared from the edible tissue. However, evidence suggests that PAs, or more specifically the didehydro-PA ("pyrrolic") metabolites, form potentially labile adducts with endogenous thiol groups[78] and nitrogen atoms (enabling the cross-linking of DNA strands)[79]. If these adducts are labile as suggested[80] the possibility arises that an alkylating pyrrolic entity can be released from ingested food and cause damage leading to human health problems. Although possible, such human exposure to pyrrolizidine alkaloid adducts and consequent intoxication has never been documented. Additional research is needed to better define what concentrations of pyrrolic adducts may be present in animal products and whether those concentrations represent a significant risk to human health.

There are however, some well documented examples of secondary poisoning of animals eating contaminated meat from an intoxicated animal. *Indigofera linnaei* S. Ali (Leguminosae), known in Australia as Birdsville Indigo, has been reported to be only toxic to horses which run on large, open areas and have little supplementary food offered to them[16]. This neurologic disease of horses is referred to as "Birdsville

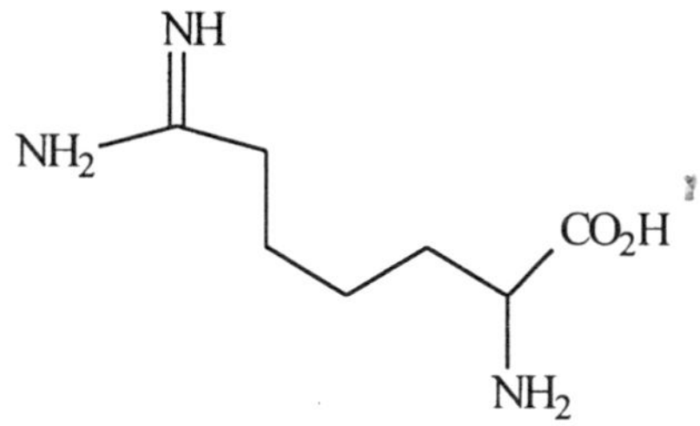

25

Figure 7 Indospicine, an arginine analogue from *Indigofera* spp.

horse disease" and does not produce liver damage. However, horse meat derived from animals that had fed upon the plant was fatally hepatotoxic to dogs[81]. It was later shown that the non-protein amino acid, indospicine (**25**, Fig. 7), that is present in *Indigofera* spp., is hepatotoxic to dogs whether administered as naturally contaminated horse meat or as the pure compound[82]. This is a clear demonstration of secondary poisoning following ingestion of meat from an apparently healthy herbivore.

This study also demonstrated a progressive accumulation of indospicine in the blood and tissues of dogs which were fed small amounts of contaminated horse meat over a long term. An implication of this work is that human ingestion of meat from other species that might graze on *Indigofera* spp. might result in a similar exposure to indospicine with resultant chronic liver damage. In this instance, and despite an Australian Department of Health report that consumption of contaminated horse meat was unlikely to pose a human health risk[83], a voluntary code not to source animals for human or pet food from *Indigofera* areas was established between the horse meat industry and the Australian Quarantine and Inspection Service[84].

As with the PA metabolites, which have a long tissue residency due to electrophilic bonding to endogenous thiol groups and DNA, lipophilic toxins also pose a special problem when considering the application of withholding periods before slaughter. The situation is further complicated when the bioactivity of the lipophilic chemical is cumulative, or is suspected of being cumulative within humans. This is the case with the effects of ciguatera toxins in man[85]. The lipophilic ciguatoxins are produced in some tropical species of fish by the marine dinoflagellate *Gambierdiscus toxicus*. The toxins can gradually accumulate in humans ingesting the fish until a threshold is reached and overt illness results such as reported by Geller and Olson[86]. Subsequent exposure to even small amounts of ciguatoxins can precipitate further illness because of the cumulative nature of the poisons. A similar scenario can be postulated for ingestion of corynetoxins (**23**, Fig. 5), the glycolipids that cause ARGT in stock grazing on pasture infested with toxic ryegrass, or in stock that are given feed contaminated with infected ryegrass seed (see Section IV.1). In addition to ARGT resulting from ingestion of infected ryegrass, the neurological effects of the corynetoxins have now also been recorded in animals grazing upon the grasses *Agrostis avenacea* C. Gemelin, (blown grass) and *Polypogon monspeliensis* (L) Desf. (annual beardgrass) in New South Wales and South Australia respectively[87]. Whilst it has been clearly demonstrated that the effects of corynetoxins are cumulative in sheep[60,61], the effects in humans are unknown and, until proven otherwise, should be treated with caution. Searching for secondary poisoning, Bourke has fed meat from tunicamycins-intoxicated cows to pigs. Although the pigs consumed more than twice their own weight of the meat, no clinical signs of intoxication were observed[61]. However, since it has been shown that inhibition of the liver N-acetylglucosamine transferase activity (see Section IV.1) is a sensitive biochemical indicator of intoxication[58,59], this assay should be employed to assess exposure *via* the meat of intoxicated animals. Therefore, as is the case with the ciguatoxins, the possibility remains that low level exposure of humans to corynetoxins *via* a contaminated meat diet (derived from asymptomatic animals sourced from ARGT-endemic areas) may result in a gradual accumulation of corynetoxins within human tissues until a threshold concentration is reached resulting in subtle clinical symptoms that may not be attributable to ingestion of corynetoxins[88].

Many plants, such as the foxglove (*Digitalis purpurea*), oleanders (*Nerium oleander* and *Thevetia peruviana*) and the milkweeds (*Asclepias* spp.), contain cardioactive glycosides (cardenolides, **26**, Fig. 8, and bufadienolides, **27**, Fig. 8) that can cause acute toxic effects in grazing animals and humans who choose to ingest the plant[6]. However, it is the chronic effects of cumulative cardioactive glycosides that can pose a potential health problem for humans ingesting the meat from intoxicated animals.

In South Africa, chronic intoxication of sheep and goats with cumulative bufadienolides derived from ingested plants can result in an ovine and caprine paretic syndrome[51]. This is colloquially referred to as krimpsiekte after the

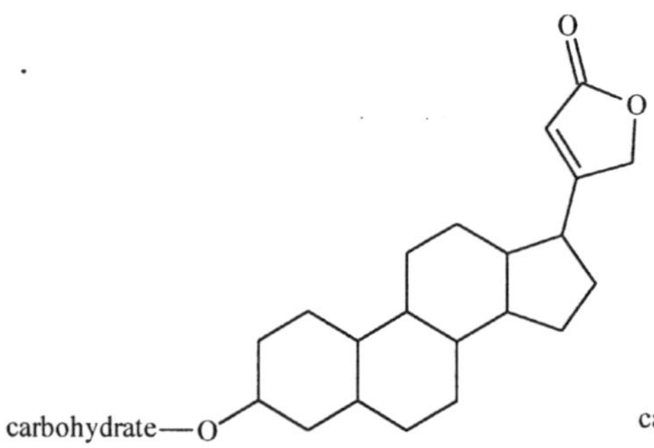

26

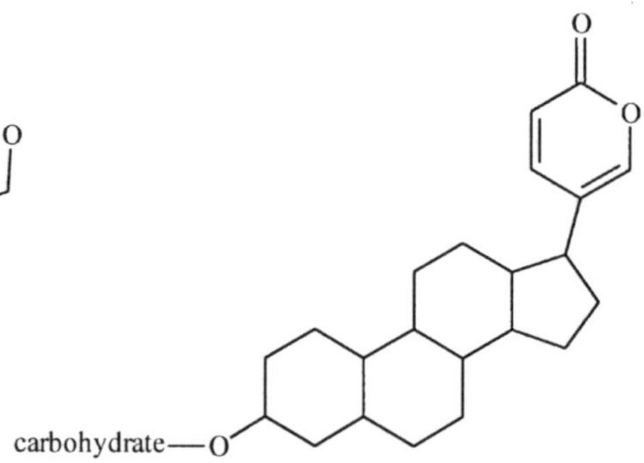

27

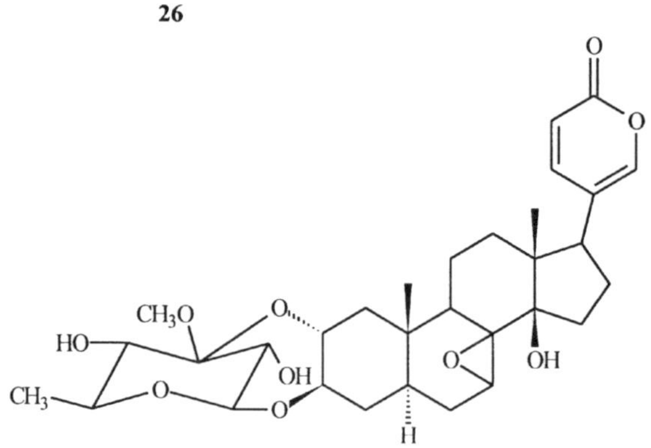

28

Figure 8 Cardioactive glycosides.

characteristic posture assumed by intoxicated animals. Krimpsiekte has reportedly been caused by some species of *Tylecondon* and *Kalanchoe*, and by *Cotyledon orbiculata* as well as by the bufadienolides isolated from some of these plants. For example, Kellerman cites the cumulative activity of cotyledoside (**28** Fig. 8), a bufadienolide isolated from *T. wallichii*, as causing acute cardiac glycoside intoxication in sheep after an intravenous injection of 0.05 mg kg^{-1}, mainly respiratory effects with 0.025 mg kg^{-1}, and signs of krimpsiekte following repeated doses of 0.01 mg kg^{-1} [51]. Secondary intoxication of meat eaters i.e., dogs and humans, can follow ingestion of animal-derived meat contaminated with the cumulative bufadienolides and, since the compounds are heat stable, cooking the meat will not necessarily detoxify the food[51].

As previously mentioned in Section IV.1, lupin seed and post-harvest stubble can be infected with the phomopsins-producing fungus *Diaporthe toxica* (anamorph *Phomopsis* sp.). The hepatogenic disease, lupinosis, resulting from ingestion of infected lupins has been induced experimentally in a wide variety of species and has been observed as a natural intoxication in sheep, cattle, horse, goats, donkeys and pigs[64]. Because the phomopsins (**24**, Fig. 6) bind to tubulin and consequently prevent the formation of microtubules, a cross-species ubiquitous process, it is not unreasonable to expect a similar effect in humans. Therefore the chronic intoxication of food animal species by phomopsins provides a source of concern for human health until it can be rigorously demonstrated that phomopsins do not accumulate under such circumstances and therefore are not present in meat ingested by humans.

IV.3. *Contamination of Milk, Milk Products and Eggs*

The transfer of plant components to the milk of grazing animals is highlighted when the contaminant taints the flavour of the milk making it unpalatable and thus unsaleable. This is demonstrated by the transfer of the sesquiterpene lactone tenulin (**29**, Fig. 9) from bitterweed (*Helenium amarum*) to the milk of grazing cows. The resultant bitter taste with as little as 1 ppm of tenulin in the milk makes it unacceptable for human consumption[89]. It has also been well documented that plant-associated mycotoxins, such as aflatoxins[90], fumonisins[91] and cyclopiazonic acid (**30**, Fig. 9)[92] in animal feed can be translocated to milk and eggs for human consumption[93].

The potential for transfer of plant-associated toxins (see Table III) to the milk of lactating animals that have grazed upon toxic plants or have been given contaminated feed is dependent upon several factors. These influencing factors include the *in vivo* distribution and metabolism, the lipophilicity, and physico-chemical properties such

Table III Examples of plant-associated toxins transferred milk

Plant name		Toxins	Refs.
Eupatorium rugosum	White snakeroot	Tremetone (**31**), dehydrotremetone, hydroxytremetone	[89]
Senecio jacobea	Tansy ragwort	Jacoline (**32**)	[96,97]
Astragalus and *Oxytropis* sp.	Locoweeds	Swainsonine (**33**)	[98]
Pteridium aquilinum	Bracken fern	Ptaquiloside (**34**)	[99,100]
Lupinus latifolius	Broadleaved lupine	Anagyrine (**35**)	[101]

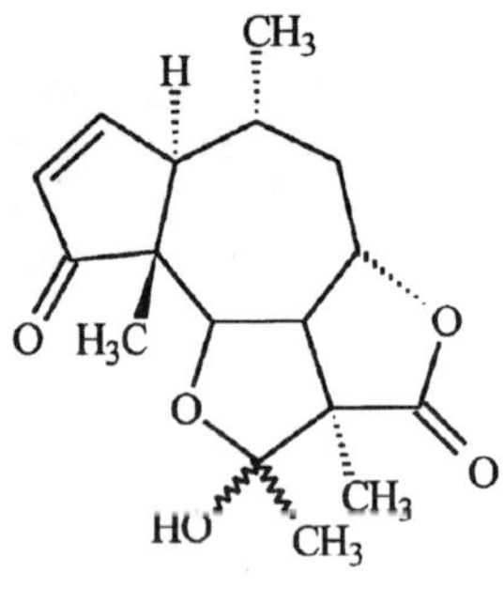

29

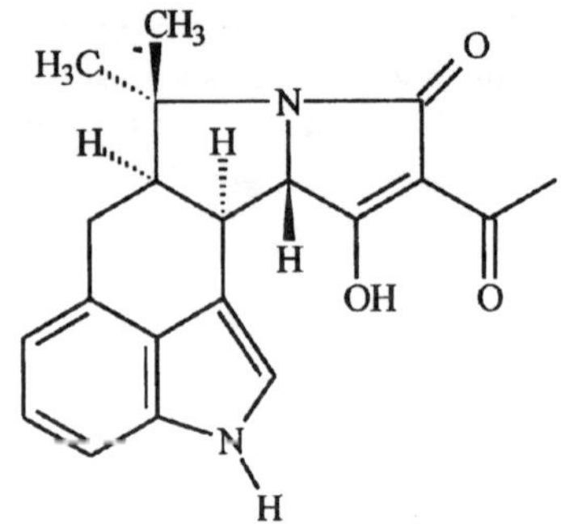

30

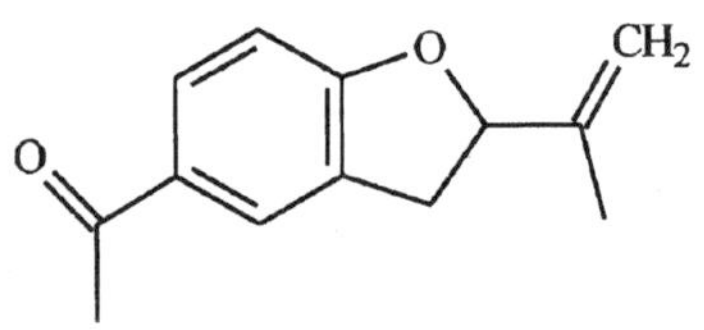

31

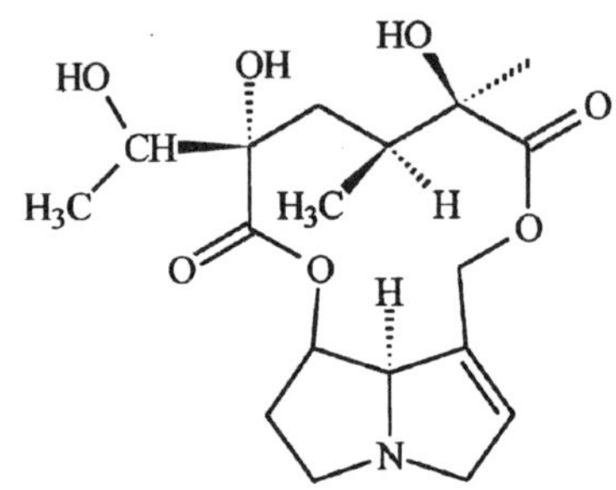

32

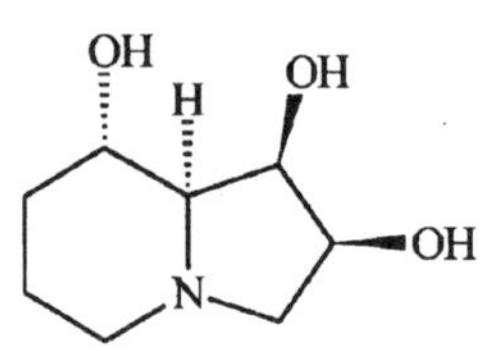

33

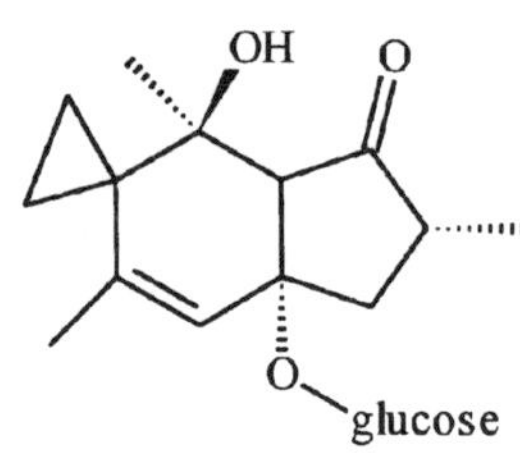

34

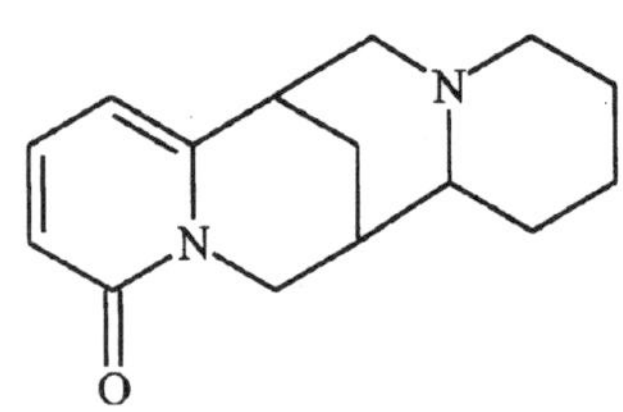

35

Figure 9 Structures of some natural toxicants transferred to milk.

as pH of the toxic chemicals. Panter and James *et al*. have reviewed literature reports of the transfer of plant-derived toxins to the milk of lactating grazing animals[94,95]. Some of the important issues that they raise include:

- contaminated milk, sufficient to cause overt toxicosis in suckling young or humans, can be obtained from an asymptomatic animal
- physico-chemical properties of the toxins may lead to favoured distribution and concentration in milk
- the complexity of milk (emulsified fats in an aqueous solution of protein and minerals) makes it a suitable sink for virtually any toxin that is bound to plasma proteins, freely circulating in the plasma or dissolved in blood lipids
- chronic, low level, repetitive exposure of animals to toxins may lead to accumulation in the milk, and may result in a chronic, low level, repetitive exposure of humans to the toxins
- young animals and young children may be at more risk to milk-borne, plant-associated toxins since they may experience a greater exposure and may not be able to detoxify or eliminate the toxins as efficiently as adults
- some toxins are preferentially eliminated *via* the mammary gland and may be bound to milk protein or occur in the aqueous phase or milk fat
- modern methods of pooling and processing milk will dilute toxin concentrations but increased risk exists when the milk comes from a few animals such as in subsistence farming or on a family farm
- chronic damage to organs, such as the kidney or liver, as a result of ingestion of toxic plants may affect the ability of the lactating animal to detoxify the xenobiotics and thereby increase transfer *via* the milk

An incident that occurred in South Australia in 1987 demonstrated the transfer of natural toxicants to eggs. An outbreak of reduced egg production on a poultry farm was traced back to feed, obtained from a neighbouring farm, that was contaminated with *Heliotropium europeum*. The heliotrope alkaloids, heliotrine (**36**, Fig. 10), europine (**37**, Fig. 10) and lasiocarpine (**38**, Fig.10), were isolated from the feed and identified by gas chromatography-mass spectrometry. The eggs produced during, and just after the poisoning episode also contained the heliotrope alkaloids. Additionally, the eggs were shown to contain PAs indicative of feed contamination by *Echium plantagineum* (Paterson's Curse), another common weed in the area. The investigation revealed that the eggs from chickens fed with PA-contaminated grain contained up to 9.7 μg of PAs (J. Edgar, unpublished data), a real concern from a public health aspect in light of emerging regulations on PA contamination (see Section V).

VI.4. *Contamination of Honey*

Depending upon the species and the relative variety of flowering plants available to foraging bees, honey has the potential to include poisonous chemicals derived from the nectar of plants toxic to animals. There have been many reports on the investigation of pesticide and other synthetic chemical residues in honey because it is known where such chemicals are applied. Less well known is the contact that honey bees have with plants that express chemicals toxic to humans in the nectar. As with other food sources, the risk associated with ingestion of contaminated honey is lessened if bulking/diluting practises are utilised. Increased risk accompanies the

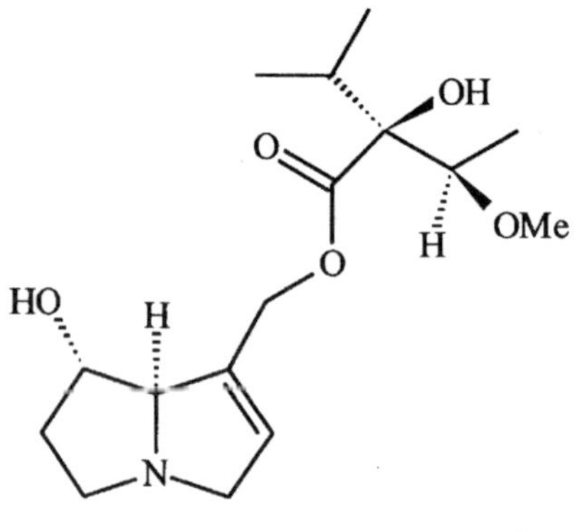

36

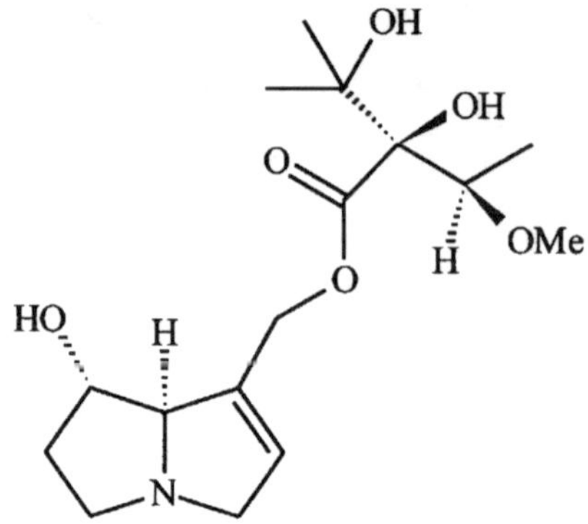

37

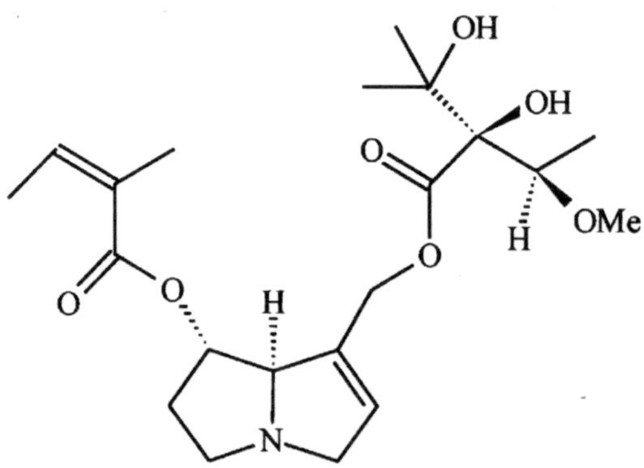

38

Figure 10 Heliotrope-derived pyrrolizidine alkaloids isolated from eggs.

ingestion of honey obtained from a producer that may only have a few hives in an area where the toxic plant is predominant or plentiful.

Grayanotoxins (**39**, Fig. 11) are diterpenes that have been isolated from rhododendrons and other members of the Ericaceae Family such as *Pieris* spp., the mountain laurel (*Kalmia latifolia*) and sheep laurel (*K. angustifolia*). Intoxication of humans by grayanotoxins usually occurs via ingestion of contaminated honey but can also result from ingestion of the nectar, leaves and flowers of the plants. The effects of grayanotoxins are similar to those of the veratrum and aconite alkaloids in that excitable cells (muscle and nerve) are maintained in a state of depolarisation as a result of the grayanotoxins binding to the sodium channels in cell membranes. Symptoms such as salivation, vomiting, low blood pressure, sinus bradycardia, progressive muscular weakness and incoordination can result from intoxication[16]. Cardiac arrhythmia has been reported in a patient intoxicated by grayanotoxin-contaminated honey[102]. According to the US Food and Drug Administration Center for Food Safety and Applied Nutrition all people are believed to be susceptible to such honey intoxication and many cases of grayanotoxin intoxication

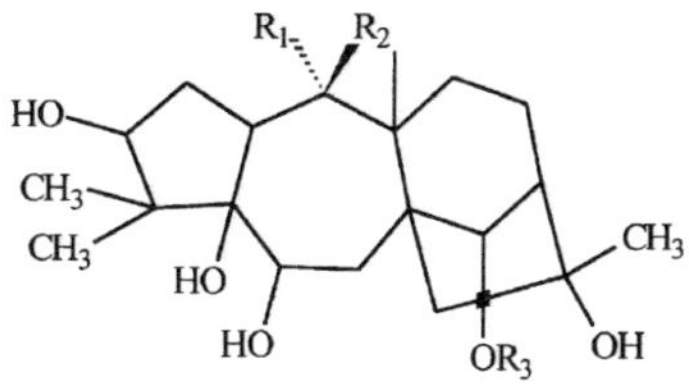

39

	R_1	R_2	R_3
Grayanotoxin I	OH	CH_3	COCH
Grayanotoxin II	CH_2=		H
Grayanotoxin III	OH	CH_3	H

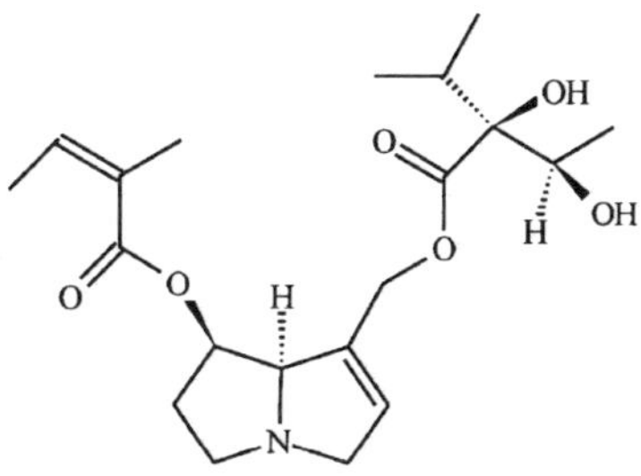

40

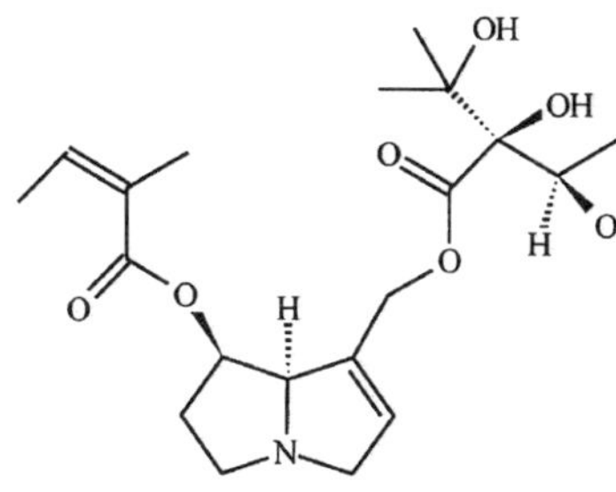

41

Figure 11 Natural toxicants isolated from honey.

have been recently documented involving local use of honey and the export of honey to other countries[103].

Another documented example of the potential for honey to be contaminated with toxic chemicals derived from the plants that honey bees visit in their foraging concerns the pyrrolizidine alkaloid-containing plant *Echium plantagineum*. Known colloquially in Australia as Salvation Jane or Paterson's Curse, the invasive weed has a history of poisoning grazing animals[16] but it is prized by apiarists as a good source of nectar for foraging bees in late winter/early spring. Belonging to the same Family (Boraginaceae) as heliotrope, associated with contamination of grain (Section IV.1), the major concern with honey derived from the nectar of this plant is the potential for contamination of the honey with pyrrolizidine alkaloids. Major hepatotoxic PAs isolated from *E. plantagineum* are the diesters echiumine (**40**, Fig. 11) and echimidine (**41**, Fig. 11)[104]. *Echium*-derived honey has been reported to contain between 0.2 and 0.9 μg of PAs per g of honey[105]. Not only does this have health implications for people for whom *Echium*-derived honey comprises their major source of honey, it is also a clear indication that PAs from other flowering plant sources could contaminate honey. Indeed, bees foraging on the PA-containing plant *Senecio jacobaea* (tansy ragwort) have produced honey containing from 0.3 to 4 μg of PAs per gram[106]. With increasing awareness of food contamination, and the likelihood of regulations concerning the levels of tolerable PA contamination in the wake of the German regulations controlling herbal products (see Section V), food-safety implications of PA-contamination of honey need to be addressed.

V. RISK ASSESSMENT, REGULATORY CONCERNS AND FUTURE DIRECTIONS

In the area of human food safety, natural toxicants have been the subject of less review and regulation than the synthetic food additives, pesticides and growth enhancers etc. As the awareness, especially consumer awareness, of the potent toxicity of natural toxicants with respect to the synthetic veterinary and agri-chemicals increases, and as it becomes accepted that little is known about long term, low level exposure of humans to natural toxicants contaminating food, then there will be increasing official concern for the regulation of contamination by natural toxicants.

The potential for chronic illness, or long term, irreversible effects, or simply a lower than optimum health status (e.g., chronic fatigue, immunoincompetence) cannot be overlooked when humans are exposed to low levels of toxic plant-associated chemicals *via* the diet. Epidemiological studies can indicate possible associations between ill-health and the intake of environmental toxins. For example, several such studies have indicated a possible association of idiopathic Parkinson's Disease with rural living, where one might presume that affected people may have had access to animal-derived food products from limited sources e.g., the family farm[107].

Phytochemical analysis of plants in suspect areas, and toxicological evaluation of the chemicals isolated from the plants are necessary to support the epidemiological studies. Thus, the observation that equine nigropallidial encephalomalacia (ENE), which has some features of human parkinsonism, occurs in areas where *Centaurea solstitialis* (yellow starthistle) is abundant[108], has led to an investigation of the potential neurotoxicity of sesquiterpene lactones[107]. In this regard it should be noted that sesquiterpene lactones have been demonstrated to transfer into milk of cows feeding on *Helenium amarum* (Section IV.3).

The use of herbs and other plants for medicines has a long history and is so steeped in folk lore that it is not uncommon for a particular plant to be prescribed for a litany of ailments. The general public unawareness of risks associated with the use of herbal products is reflected in the increasing use and trust in herbal remedies. However, with increasing scientific knowledge of the chemical constituents of some medicinal plants and herbs comes a slowly increasing concern for the safety of these products. Consequently, official regulatory bodies are, or should be, required to assess the risks associated with these herbs and medicinal plants, and to set limits on exposure to any of their known toxic constituents. However, very few of the therapeutic claims have received scientific attention such that it is difficult to make any risk/benefit assessment for regulatory purposes. Faced with an absence of any validated therapeutic benefit, and with the scientifically demonstrated presence of toxic compounds, regulatory bodies may have no choice but to refuse registration for the sale of such products to the public. It is of note that the bioactive chemical can, indeed, be restricted but can still inadvertently find its way onto the market as a component in a herbal preparation. For example, the neurotoxic alkaloid aconitine[46] is a Schedule 4 poison restricted in Australia to availability *via* a medical prescription (Dr. P. Di Marco, Western Australian Health Department, personal communication). However, a man was fatally poisoned in Perth, Western Australia after ingesting the aconitine-containing monkshood (*Aconitum* spp.) supplied by a herbalist.

A major current concern of regulatory bodies is the pyrrolizidine alkaloid content of herbal and medicinal plant preparations. For example, German regulations (which came into effect on the 1st January 1993) affecting the sale of herbal and medicinal plant preparations have restricted the daily oral intake of PAs to 0.1 μg, and the daily topical exposure to 10 μg. In the case of pregnant women, the daily oral intake is set at zero whilst topical exposure is allowed only on the advice of a medical doctor. If use is limited to 6 weeks each year, then the daily oral dose and topical exposure are limited to 1 μg and 100 μg respectively[109]. The potential exists for these regulations to be extended to include PA-contamination of food products, and then to provide the impetus for addressing other natural toxicants that have not yet been assessed or regulated.

There are however, many difficulties associated with the safety assessment of natural, plant-associated toxins for human exposure[110]. One of the major obstacles facing such research is the generally held conception that "natural" is healthy or best. This will be overcome as the "chemical" nature of food components is accepted and that a chemical's bioactivity, rather than its source, should form the basis of concern and study. It can be claimed that food is safe because no ill-effects are observed within the population of consumers. However, such claims often only refer to acute effects and give no attention to the potential, and possibly subtle, health effects (including mutagenic, carcinogenic, embryotoxic, and chronic ill-health) of long term, low level exposure to these natural toxicants. To begin to assess safety issues for natural toxicants, the chemicals and the degree of human exposure have to be well understood. Therefore plants toxic to animals have to be fully investigated and the causative chemicals isolated and identified. For effective accumulation of the toxicological data needed[111] for safety assessment and exposure regulation the putative toxins have to be available in sufficient quantities and of sufficient purity. This will require chemical synthesis, microbial synthesis by modified organisms or large scale isolation of toxins from the plant source. In addition, highly sensitive, specific and unambiguous assay procedures for the toxins in the various food matrices in which they might be found need to be developed in order to quantitate the human exposure to these chemicals. Examples of this include the development of enzyme linked immunosorbent assays (ELISAs) for pyrrolizidine alkaloids[112].

When considering the potential for secondary poisoning, a knowledge of the *in vivo* behaviour and tolerable levels of intake of the toxic chemicals is necessary in order to enable the recommendation of suitable withholding periods. This would ensure that all plant-associated toxic chemicals, and their metabolites, are effectively reduced to safe levels within the animal before slaughter. This is exemplified by the determination of the *in vivo* half-life of swainsonine (**33**, Fig. 9), the causative toxin in *Swainsona* spp. poisoning in Australia[113] and locoweed poisoning in the USA[114], as approximately 20 hours in sheep and cattle[115,116]. This would suggest a 6–7 day withholding period to ensure clearance of swainsonine from serum and tissues of intoxicated animals.

VI. CONCLUSIONS

Contamination of food for human consumption by poisonous, plant-associated chemicals is a documented reality. In many cases the acute poisonous effects of these toxins are known and high levels of contamination will be readily recognised. From an environmental toxicology aspect, the long term, low level exposure of humans to poisonous, plant-associated chemicals in their diet is of major concern. The levels of actual exposure and the possible health effects are usually unknown. Since risk assessment and management cannot proceed without such data, it is imperative that exposure levels be determined and research on bioavailability, pharmacology and toxic bioactivities of low levels of plant-associated poisonous chemicals be pursued.

References

1. F. W. Oehme, "The hazard of plant toxicities to the human population" **In**: *Effects of Poisonous Plants on Livestock*, (R. F. Keeler, K. R. Van Kampen and L. F. James, eds.) (Academic Press, New York, 1978) pp. 67 80.
2. R. F. Keeler, K. R. Van Kampen and L. F. James (eds.), *Effects of Poisonous Plants on Livestock* (Academic Press, New York, 1978).
3. A. A. Seawright, M. P. Hegarty, L. F. James and R. F. Keeler (eds.), *Plant Toxicology* (Queensland Poisonous Plants Committee, Yeerongpilly, Australia, 1985).
4. L. F. James, R. F. Keeler, E. M. Bailey, Jr., P. R. Cheeke and M. P. Hegarty, (eds.), *Poisonous Plants: proceedings of the third international symposium* (Iowa State University Press, Ames, Iowa, USA, 1992).
5. S. M. Colegate and P. R. Dorling (eds.), *Plant-associated Toxins: Agricultural, Phytochemical and Ecological Aspects* (CAB International, UK, 1994).
6. P. R. Cheeke, and L. R. Shull, *Natural Toxicants in Feeds and Poisonous Plants* (AVI Publishing, Westport, Connecticut, USA, 1985).
7. U. Candrian, J. Luthy, P. Schmid, C. Schlatter and E. Gallasz, "Stability of pyrrolizidine alkaloids in hay and silage" *J. Agric. Food Chem.* **32**(4), 935–937 (1984).
8. Food and Agriculture Organisation of the United Nations and the World Health Organisation, "Application of risk analysis to food standards issues" *Report of the Joint FAO/WHO Expert Consultation*, Geneva, 13–17 March (1995).
9. B. N. Ames, L. S. Gold and W. C. Willett, "The causes and prevention of cancer" *Proc. Natl. Acad. Sci. USA* **92**(12), 5258–5265 (1995).
10. R. K. Heaney and G. R. Fenwick, "Natural toxins and protective factors in *Brassica* species, including rapeseed" *Nat. Toxins* **3**(4), 233–237 (1995).
11. W. M. J. Van Gelder, "Chemistry, toxicology, and occurrence of steroidal glycoalkaloids: potential contaminants of the potato (*Solanum tuberosum* L.)" **In**: *Poisonous Plant Contamination of Edible Plants* (A.-F. M. Rizk, ed.) (CRC Press, Boca Raton, 1991) pp. 117–156.
12. E. A. Johnson, R. J. Molyneux and B. G. Merrill, "Chemistry of toxic range plants. Variation in pyrrolizidine alkaloid content of Senecio, *Amsinckia* and *Crotolaria* species" *J. Agr. Food. Chem.* **33**(1), 50–55 (1985).
13. E. A. Johnson, R. J. Molyneux and M. H. Ralphs, "*Senecio*: A dangerous plant for man and beast" *Rangelands* **11**(6), 261–264 (1989).
14. Committee on Food Protection, *Toxicants Occurring Naturally in Foods* (National Academy of Sciences, Washington, D.C., 1973).
15. J. P. F. D'Mello, C. M. Duffus and J. H. Duffus (eds.), *Toxic Substances in Crop Plants* (The Royal Chemical Society, Cambridge, UK, 1991).
16. S. L. Everist, *Poisonous Plants of Australia*, Revised Edition (Angus & Robertson Publishers, Melbourne, 1981).
17. T. Bergin, *In the Steps of Burke and Wills* (Australian Broadcasting Commission, Sydney, NSW, 1981).
18. A. Moorehead, *Coopers Creek* (Harper and Row, NY, 1963).
19. E. A. Bell, "Aminonitriles and amino acids not derived from proteins" **In**: *Toxicants Occurring Naturally in Foods* (Committee on Food Protection, National Academy of Sciences, Washington, D.C., 1973) pp. 163–169.
20. J. P. F. D'Mello, "Toxic amino acids" **In**: *Toxic Substances in Crop Plants* (J. P. F. D'Mello, C. M. Duffus and J. H. Duffus, eds.) (The Royal Chemical Society, Cambridge, UK, 1991) pp. 21–48.
21. M. W. Duncan, "Environment and neurodegenerative disorders" *Today's Life Science* (February, 18–24, 1992).
22. M. G. Whiting, "Toxicity of Cycads" *Econ. Bot.* **17**, 270 (1963).
23. P. S. Spencer, D. N. Roy, A. Ludolph, J. Hugon and M. P. Dwived, "Lathyrism: evidence for the role of the neuroexcitatory amino acid BOAA" *Lancet* **2**(8515), 1066–1067 (1986).
24. P. S. Spencer, P. B. Nunn, J. Hugon, A. Ludolph, S. M. Ross, D. N. Roy and R. C. Robertson, "Guam amyotrophic lateral sclerosis–parkinsonism–dementia linked to a plant excitant neurotoxin" *Science* **237**, 517–522 (1987).

25. S. L. N. Rao, L. K. Adiga and P. S. Sarma, "The isolation and characterisation of β-N-oxalyl-L-α, β-diaminopropionic acid: A neurotoxin from seeds of *Lathyrus sativus*" *Biochemistry* **3**, 432–436 (1964).
26. V. V. S. Murti, T. R. Seshadri and T. A. Venkitasubramanian, "Neurotoxic compounds of the seeds of *Lathyrus sativus*" *Phytochemistry* **3**, 73 (1964).
27. E. A. Bell and J. P. O'Donovan, "The isolation of α and γ-oxalyl derivatives of α,γ-diaminobutyric acid from seeds of *Lathyrus latifolius* and the detection of the α-oxalyl isomer of the neurotoxin α-amino-β-oxalylaminopropionic acid which occurs together with the neurotoxin in this and other species" *Phytochemistry* **5**, 1211 (1966).
28. E. Casadei, J. Cliff and J. Neves, "Surveillance of urinary thiocyanate concentration after epidemic spastic paraparesis in Mozambique" *J. Trop. Med. Hyg.* **93**, 257–261 (1990).
29. R. J. Lowen, F. K. A. Alam and J. A. Edgar, "Lupin bean toxicity" *Med. J. Aust.* **162**, 256–257 (1995).
30. M. D. Laryea, Y. F. Jiang, G. L. Xu and I. Lombeck, "Fatty acid composition of blood lipids in Chinese children consuming high erucic acid rapeseed oil" *Ann. Nutr. Metab.* **36**, 273–278 (1992).
31. A. Zitnak and G. R. Johnston, "Glycoalkaloid content of B5141-6 potatoes" *Am. Potato J.* **47**, 256–260 (1970).
32. L. P. Brower and C. M. Moffit, "Palatability dynamics of cardenolides in the monarch butterfly" *Nature* London **249**, 280–283 (1974).
33. J. A. Edgar, "Parsonsieae: Ancestral larval foodplants of the Danainae and Ithomiinae" **In**: *The Biology of Butterflies* (R. I. Vane-Wright and P. R. Ackery, eds.) (Academic Press, London, 1984) pp. 91–93.
34. L. E. Twigg and D. R. King, "The impact of fluoroacetate-bearing vegetation on native Australian fauna: A review" *OIKOS 61*, 412–430 (1991).
35. S. D. Lincoln and B. Black, "Halogeton poisoning in range cattle" *J. Am. Vet. Med. Assoc.* **176**, 717–718 (1980).
36. Various authors, *Effects of Poisoning Plants on Livestock*" (R. F. Keeler, K. R. Van Kampen and L. F. James, eds.) (Academic Press, New York, 1978).
37. Various authors, *Plant Toxicology* (A. A. Seawright, M. P. Hegarty, L. F. James and R. F. Keeler, eds). Queensland Poisonous Plants Committee, Yeerongpilly, Australia, 1985).
38. H. J. Segall, D. W. Wilson, M. W. Lamé and D. Morin, "Metabolism of pyrrolizidine alkaloids" **In**: *Handbook of Natural Toxins,* **Vol. 6**: *Toxicology of Plant and Fungal Compounds* (R. F. Keeler and A. T. Tu, eds.) (Marcel Dekker, New York, 1991) pp. 3–26.
39. C. C. J. Culvenor, J. A. Edgar, M. V. Jago, A. Outleridge, J. E. Peterson and L. W. Smith, "Hepato- and pneumotoxicity of pyrrolizidine alkaloids and derivatives in relation to molecular structure" *Chem.-Biol. Interact.* **12**, 299–324 (1976).
40. H. J. Segall, D. W. Wilson, M. W. Lamé, D. Morin, L. C. Pan and A. D. Jones, "Monocrotaline pulmonary toxicity: Mechanisms and pathology" **In**: *Plant-associated Toxins: Agricultural, Phytochemical and Ecological Aspects*" (S. M. Colegate and P. R. Dorling, eds.) (CAB International, UK, 1994) pp. 131–136.
41. A. R. Mattocks, "Pyrrolizidine alkaloids: What metabolites are responsible for the extrahepatic tissue damage in animals?" **In**: *Poisonous Plants: proceedings of the third international symposium* (L. F. James, R. F. Keeler, E. M. Bailey, Jr., P. R. Cheeke and M. P. Hegarty, eds.) (Iowa State University Press, Ames, Iowa, USA, 1992) pp. 192–197.
42. R. J. Huxtable, "Human health implications of pyrrolizidine alkaloids and the herbs containing them" **In**: *Toxicants of Plant Origin*, **Volume 1**, *Alkaloids* (P. R. Cheeke, ed.) (CRC Press, Boca Raton, 1989) pp. 41–86.
43. B. N. Tandon, H. D. Tandon, R. K. Tandon, M. Narendranathan and Y. K. Joshi, "Epidemic of veno-occlusive disease in central India" *Lancet* **2**(7980), 271–272 (1976).
44. M. A. Siddiqui, K. A. Suri, M. P. Suri and C. K. Atal, "Novel pyrrolizidine alkaloid from *Crotalaria nana*" *Phytochemistry* **17**, 2143 (1978).
45. F. Mayer and J. Lüthy, "Heliotrope poisoning in Tadjikstan" *Lancet* **342**, 246–247 (1993).
46. P. R. Dorling, S. M. Colegate and C. R. Huxtable, "Plant neurotoxins" **In**: *Handbook of Neurotoxicology* (L. W. Chang and R. S. Dyer, eds.) (Marcel Dekker, Inc., 1995) pp. 611–636.
47. H. T. Rwiza, "Jimson weed food poisoning. An epidemic at Usangi rural government hospital" *Trop. Geogr. Med.* **43**, 85–89 (1991).
48. A. van-Meurs, A. Cohen and P. Edelbroek, "Atropine poisoning after eating chapattis contaminated with *Datura stramonium*" *Trans. R. Soc. Trop. Med. Hyg.* **86**, 221 (1992).
49. K. A. V. R. Krishnamachari, R. V. Bhat, V. Nagarajon and T. B. G. Tilak, "Hepatitis due to aflatoxins. An outbreak in Western India" *Lancet* **1**(7915), 1061–1063 (1975).
50. C. C. J. Culvenor, "Economic loss due to poisonous plants in Australia" **In**: *Plant Toxicology* (A. A. Seawright, M. P. Hegarty, L. F. James and R. F. Keeler, eds.) (Queensland Poisonous Plants Committee, Yeerongpilly, Australia, 1985) pp. 3–13

101. W. W. Kilgore, D. G. Crosby, A. L. Craigmill and N. K. Poppen, "Toxic plants as possible human teratogens: circumstantial evidence points to lupine toxin in goats' milk as cause of human birth defects" *California Agriculture* **35**, 6 (1981).
102. H. Gossinger, K. Hruby, A. Haubenstock, A. Pohl and S. Davogg, "Cardiac arrhythmias in a patient with grayanotoxin-honey poisoning" *Vet. Hum. Toxicol.* **25**, 328–329 (1983).
103. US Food and Drug Administration Center for Food Safety and Applied Nutrition, *Foodborne Pathogenic Microorganisms and Natural Toxins* (1992. Internet access *via* http://vm.cfsan.fda.gov/~mow/intro.html.
104. C. C. J. Culvenor, "The alkaloids of *Echium plantagineum* L. I. Echiumine and echimidine" *Aust. J. Chem.* **9**, 512–520 (1956).
105. C. C. J. Culvenor, J. A. Edgar and L. W. Smith, "Pyrrolizidine alkaloids in honey from *Echium plantagineum*" *J. Agric. Food Chem.* **29**, 958–960 (1981).
106. M. L. Deinzer, P. A. Thomson, D. M. Burgett and D. L. Isaacson, "Pyrrolizidine alkaloids: their occurrence in honey from tansy ragwort (*Senecio jacobea* L.)" *Science* **195**, 497–499 (1977).
107. R. J. Riopelle and K. L. Stevens, "*In vitro* neurotoxicity bioassay: neurotoxicity of sesquiterpene lactones" **In**: *Bioactive Natural Products: Detection, Isolation and Structural Determination* (S. M. Colegate and R. J. Molyneux, eds.) (CRC Press, Boca Raton, 1993) pp. 457–463.
108. D. R. Cordy, "Nigropallidial encephalomalacia in horses associated with ingestion of yellow starthistle" *J. Neuropathol. Exp. Neurol.* **13**, 330–342 (1954).
109. Federal Health Bureau, *Notification of the Permission and Registration of Drugs* Berlin (June, 1992).
110. G. E. Diggle, "Risk assessment and natural toxins" *Nat. Toxins* **1**, 71–72 (1992).
111. G. J. A. Speijers, "Toxicological data needed for safety evaluation and regulation on inherent plant toxins" *Nat. Toxins* **3**, 222–226 (1995).
112. T. Langer, E. Möstl, R. Chizzola and R. Gutleb, "A competitive enzyme immunoassay for the pyrrolizidine alkaloids of the senecionine type" *Planta Med.* **62**, 267–271 (1996).
113. S. M. Colegate, P. R. Dorling and C. R. Huxtable, "A spectroscopic investigation of swainsonine: an α-mannosidase inhibitor isolated from *Swainsona canescens*" *Aust. J. Chem.* **32**, 2257–2264 (1979).
114. R. J. Molyneux and L. F. James, "Loco intoxication: indolizidine alkaloids of spotted locoweed (*Astragalus lentigenosus*)" *Science* **216**, 190–191 (1982).
115. B. L. Stegelmeier, L. F. James, K. E. Panter and R. J. Molyneux," Serum swainsonine concentration and alpha-mannosidase activity in cattle and sheep ingesting *Oxytropis sericea* and *Astragalus lentiginosus* (locoweeds)" *Am. J. Vet. Res.* **56**, 149–154 (1995).
116. B. L. Stegelmeier, L. F. James, K. E. Panter and R. J. Molyneux, "Tissue and serum swainsonine concentrations in sheep ingesting *Astragalus lentiginosus* (locoweed)" *Vet. Hum. Toxicol.* **37**, 336–339 (1995).

16. TOXIC CYANOBACTERIAL BLOOMS

ALAN HOWARD*

INTRODUCTION

Cyanobacteria are amongst the most ancient groups of organisms on earth with calcareous accretions formed by cyanobacterial mats in the estuarine waters of Western Australia being more than a billion (1×10^9) years old[1]. Over 2000 species of cyanobacteria have been identified and can be found across all climatic zones in aquatic environments ranging in size from garden ponds to oceans. This chapter is concerned with the toxic bloom-forming cyanobacteria found in freshwater impoundments with relatively long retention times such as lakes and reservoirs and also in some rivers, particularly within dead zones. The most common bloom forming toxic cyanobacteria include *Microcystis*, *Anabaena*, *Oscillatoria* and *Aphanizomenon* and like all cyanobacteria these genera are excellent at surviving and often flourishing in marginal environments.

In recent years there has been widespread media coverage of the cyanobacterial bloom problem prompted by the rapid deaths of many animals after immersion in water containing toxic cyanobacteria and, in 1996, the apparent fatal poisoning of 43 dialysis patients in Brazil after receiving water supplied by a bloom-affected river. Public concern about the potential threat of toxic cyanobacteria is usually justified. The apparent increase in the frequency of problem blooms may reflect increased monitoring but is also a symptom of the global eutrophication problem. Indeed, toxic cyanobacterial blooms have probably done more to give eutrophication its bad name than any other consequence of lake enrichment[2].

The principal problem associated with cyanobacterial blooms is the potential release of toxins into a water-body used for drinking water supply or for recreational activities. The three groups of cyanobacterial toxin are: neurotoxins, hepatotoxins which, with one exception, are all cyclic heptapeptides collectively known as microcystins[3], and endotoxins which include lipopolysaccharides. Clinical symptoms associated with hepatoxicity include liver and circulatory damage (often involving necrosis of tissue); with neurotoxicity, muscle tremors, hyper-salivation, respiratory distress and paralysis may be observed; and less seriously, endotoxicity may give rise to contact irritation and minor gastrointestinal illness. *Microcystis* is more often associated with hepatoxicity; *Oscillatoria* and *Anabaena* with neurotoxicity and endotoxicity with *Oscillatoria* and *Aphanizomenon*[4] although such toxin-genera specificity is not always observed.

In addition to often being toxic, cyanobacterial blooms release odorous substances. For example, waterbodies affected by *Oscillatoria* can release geosmin and 2-methylisoborneol which give rise to earthy and musty odours[5] whilst *Microcystis* can produce large quantities of β-cyclocitral, a moderately odorous substance with a smoky tobacco odour[6]. The presence of odour agents is not however a sensitive, consistent or reliant indicator of the presence of toxin[5] and the seriousness of the

* Cyanobacterial Modelling Research Group, Aquatic Environments Research Centre, The University of Reading, Reading (UK).

cyanobacterial bloom problem can best be highlighted by reference to examples of animal and human poisonings caused by ingestion of cyanobacterial toxin.

EXAMPLES

A male black Labrador that became ill within four days of swimming in Rutland Water was one of 15 dogs thought to have been killed after ingestion of the toxic *Microcystis* scum that extended around the perimeter of the waterbody in late September 1989. Initial symptoms of vomiting and diarrhoea developed into kidney problems and subsequent death. The post-mortem examination[7] showed massive necrosis in the liver, further necrosis in the kidney, gastrointestinal mucosal haemorrhage and necrotising arteritis. Although it cannot be proved toxicologically, these findings provide reliable evidence of poisoning by hepatotoxins released by the Rutland Water bloom.

Rutland Water is a major impoundment supplying 450,000 people with drinking water; it is also a popular water-sports centre and general recreational site. The sudden deaths of 15 dogs and 20 sheep understandably created some public anxiety reinforced by the popular press seeking to sensationalise the problem. 'Bloom hysteria' reached such a level that questions were being asked at Governmental level. These concerns were amplified when shortly afterwards two soldiers became seriously ill with atypical pneumonia after canoeing and swimming in Rudyard Lake, Staffordshire which also contained a large *Microcystis* bloom. Consequently the newly formed National Rivers Authority (now the Environment Agency) set-up a task group of experts to investigate the causes, consequences and possible management of this apparently serious water quality and public health problem. The group's conclusions were published in 1990[8]. The report identified that of 594 waterbodies in England and Wales considered to have cyanobacterial problems in 1989, 68% were shown to be toxic. This is comparable to findings in Portuguese fresh waters in 1991–92 where 60% of samples taken tested positive for toxicity[9]. The fact that not all blooms are toxic and that the level of toxicity is variable is an important characteristic of cyanobacterial blooms.

Problem cyanobacterial blooms are not restricted to lowland, nutrient rich impoundments. For example, several cases of cyanobacterial toxicoses resulting from the formation of a large bloom of benthic *Oscillatoria* at Loch Insh in Scotland's Highland Region have been reported[10]. Over a ten-day period four dogs (two Scottish terriers, an adult springer spaniel and a collie bitch) died shortly after walking along the loch shore. Symptoms included convulsions, limb twitching and hypersalivation with death occurring 10–30 minutes after onset of illness. The collie bitch, however, displayed no clinical symptoms and died within 15 minutes of exposure to the water. Additionally one of a pair of springer spaniels began hypersalivating after retrieving sticks from the loch; this dog made a full recovery after both spaniels were hospitalised overnight and treated with emetics and intravenous fluids. Microscopic examination of the stomach contents of the deceased dogs revealed the presence of diatom and cyanobacterial filaments. Neurotoxin contained in these filaments would have caused the toxicoses observed in the five dogs.

Further research[11] suggests that dogs may drink from shoreline water containing cyanobacteria even though clean water is located nearby. At Loch's Awe and Avich dogs were seen to actually eat shoreline deposits of cyanobacteria resulting in the two dogs suffering symptoms of neurotoxin poisoning. In this case both dogs made a full recovery but the observations suggest that dogs may be attracted to the cyanobacteria by the emission of some substance. Whether this substance which may cause a possible 'fatal attraction' to cyanobacteria is one of the odour agents previously described is unknown.

Governmental task forces have been set up in several countries. For example, in Australia the Blue-Green Algae Task Force studied the occurrence of the 'world's biggest' bloom which occurred in the Murray–Darling River in 1991 and extended for 1000 km. This massive *Anabaena circinalis* bloom was triggered by a combination of nutrient enrichment, low flows, lowering of turbidity, anoxic bottom waters, bacterial action and resolubilization of sediment band phosphorus[12,13]. Australia has a major problem with cyanobacteria and several studies have attempted to examine the risk to human health arising from toxic blooms.

One review[14] of eight published studies investigated the evidence of causality between contact with cyanobacteria and subsequent illness using a subjective method of assessing different criteria and the weight of evidence. This work concluded that most studies demonstrate a weak but consistent association between exposure to cyanobacteria and an adverse health outcome. Investigation of specific and detailed epidemiological studies also seem to strongly confirm an adverse human health risk. For example, Falconer and co-workers[15] reported the retrospective study of human liver damage using data from blood tests carried out on patients living in a region of Australia in which drinking water supply was known to contain microcystin. The results showed a highly significant increase in ϒ-glutamyl transferase (an indicator of toxic liver injury) in the blood of patients who had obtained water from the supply during a bloom. Additionally a severe attack of hepato-enteritis was reported[16] after 150 people, mainly children on Palm Island (an island off the Australian coast) received drinking water supplied from a water body immediately following the treatment of a cyanobacterial bloom with copper sulphate. This treatment would have caused the release of a large pulse of cyanobacterial toxin resulting in the intoxication of any animal or human ingesting affected water.

Further concern about the human health risk posed by cyanobacteria arises from experimental work[17] which has found that skin tumours on mice given microcystins were heavier than those on mice not given the toxin indicating that microcystins may act as tumour promoters. The mechanism by which tumour formation or enlargement may occur is the inhibition of protein phosphatases caused by the microcystin toxin[18,19].

Cyanobacteria may also be responsible for maintaining endemicity of cholera in some developing countries. In endemic areas of Bangladesh, cholera epidemics occur twice a year during which time *V. cholerae* can be isolated from patients and surface waters; during the inter-epidemic seasons it disappears from the environment. Laboratory findings suggest that *V. cholerae* is able to enter the mucilaginous sheath of *Anabaena variablis* and continue to replicate by binary fission and remain detectable for up to 15 months[20]. This would be long enough to maintain a population during the inter-epidemic periods. These experimental findings are supported by samples taken from a pond in Dhaka city during the 1988–89 inter-epidemic period which showed the presence of *V. cholerae* in 16 out of 24 samples of *Anabaena* sp[20].

The seriousness of the problem is therefore well known. Cyanobacterial toxins can quickly kill dogs and livestock and may cause serious illness in humans. In the case of cholera, for example, cyanobacteria may even act as hosts for other pathogens. It was/is inevitable that ingestion of cyanobacterial toxins will result in human fatalities. In April 1996 it was reported that 126 patients receiving haemodialysis treatment in a hospital in Caruara, NE Brazil, became intoxicated with what preliminary investigations showed to be microcystin LR toxin. Within a few weeks 43 patients died from toxic hepatitis as a result of, it is claimed, ingesting this cyanobacterial toxin. Water containing the toxin had been obtained from a river containing a large bloom of *Microcystis* and subsequently transported to the hospital in tankers. Treatment of the water with large amounts of chlorine took place in-transit. The water was then given to the patients intra-venously. Professor Wayne Carmichael, involved in investigating the incident, reported that analysis of liver and serum from patients who had died and/or had been hospitalised confirmed microcystins at levels associated with acute or lethal toxicites [Cyano Tox Home Page, World Wide Web]. Carmichael remarks however that further investigations were required to ensure that the microcystin was the primary toxin involved since the source water could have contained other toxic organics including pesticides. The current indication is however that the Brazilian haemodialysis tragedy will be the first confirmed case of human fatalities directly attributed to the ingestion of cyanobacterial toxin.

Under normal circumstances, microcystin toxin should not enter the water supply in significant amounts as the toxin can be removed with activated charcoal and microcystin will not cross perfectly working reverse osmosis filters. In the Brazilian case the water bypassed normal treatment and relied entirely upon chlorination. Addition of chlorine would have caused the rupture of the cyanobacterial cells resulting in the release of toxin. Although in sufficient concentrations the chlorine can deactivate the microcystin toxin, in this case sufficient active toxin was available to cause acute damage once in the liver. Transmission to the liver was hastened by the water entering the patients intra-venously. Although an exceptional case, the incident raises worrying implications for areas of the world lacking modern water treatment facilities where cyanobacterial blooms in water sources are a common problem.

MANAGEMENT

There have been several recently published reviews of the management options available to prevent or to eliminate problem cyanobacterial blooms[8,21,22]. The NRA Toxic Algae Task Group identified management strategies which included short-term and long-term action plans (see Fig. 1). Most of the short-term actions were related to the practicalities of dealing with incidents including informing landowners and recreational users about the nature of the problem and precautions to be taken[22]. In the longer term, a variety of potential solutions are possible including nutrient removal and controlling light by artificial mixing. The three main strategies can be identified as[21]:

1. sink the cyanobacteria;
2. kill the resident population before/after blooming; and
3. prevent large growth from occurring.

The first strategy realises that toxic cyanobacteria are only likely to become a serious problem when they are accessible to animals and humans. This usually happens when the population of cyanobacteria migrates to the lake surface and is blown to the shoreline by light winds. The ability to regulate buoyancy and therefore change position in the water column is an important characteristic of bloom-forming cyanobacteria which has allowed them to gain a competitive advantage over

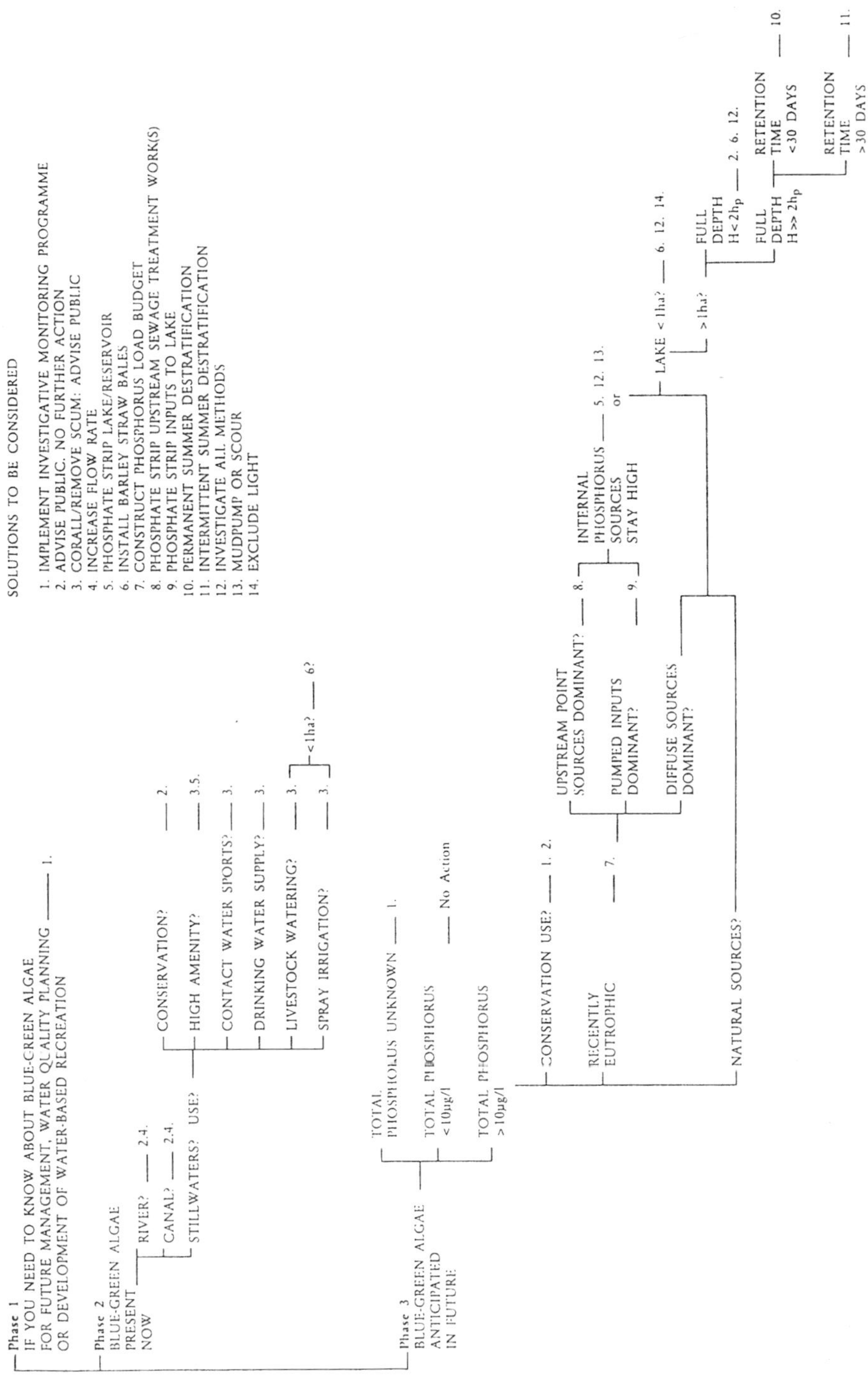

Figure 1 Suggested toxic algae management plans, reproduced from NRA (1990)[8] with kind permission of the Environment Agency.

deterministic model called PROTECH (Phytoplankton Response to Environmental Change) which requires additional data on lake nutrient balance and weather conditions.

These two approaches provide general information on phytoplankton productivity which can assist operational managers in devising and executing action plans. At present, these models remain the unpublished and exclusive property of the Environment Agency. There are however many other models published in the scientific literature which may have a role in bloom management.

One of the most promising new techniques is inductive neural-network modelling which has already been successfully applied to predicting algal populations in lakes and rivers[30,31]. The technique is a logical extension of statistical and time-series techniques[32] and has the potential to work on-line to provide operational managers with early warning of a bloom event. A neural network consists of many processing elements organised into groups called layers. A typical network consists of a sequence of layers with full or random connections to the outside world; an input buffer where data is presented to the network, and an output buffer which holds the response of the network to a given input. Layers distinct from the input and output buffers are called hidden layers. Neural computing differs from artificial intelligence and traditional computing in several important ways. Unlike traditional expert systems where the knowledge is made explicit in the form of rules, neural networks generate their own rules by learning from the examples shown to them[31]. The technique does however require further evaluation and testing since there is a tendency for neural network models to under predict the magnitude of a bloom although predictions of the timing are usually accurate.

The SCUM series of models simulate the growth and movement of cyanobacteria and exclude all other competing algal species. The most recent version, SCUM'96[24], provides a realistic simulation of buoyancy change in *Microcystis*. SCUM'96 attempts to model how buoyancy change occurs in *Microcystis* and how this will be affected by lake mixing. The model calculates the photosynthetic rate at 600 s time intervals using P/I curves. The glycogen photosynthate is then apportioned between growth (K), ballast (B) and respiration (R) using the model's principal algorithm[24]:

(i) If $P_{qi} - R \leqslant C_{gmax}$ THEN $K = P_{qi} - R$
with no addition to carbon 'in store'

(ii) If $P_{qi} - R > C_{gmax}$ THEN $K = C_{gmax}$
and ballast is generated at the rate
$B = P_{qi} - R - K$

(iii) Where $P_{qi} - R < 0$ THEN $K = 0$
and
$B = P_{qi} - R$ (and will be negative)

where P_{qi} is the photosynthetic in mol C (mol C)$^{-1}$ s^{-1}, R is the respiration rate in the dark at 20°C (about 0.55×10^{-6} mol C (mol C)$^{-1}$ s^{-1}, K represents growth and C_{gmax} is the maximum amount of carbon that can be allocated to growth at the given temperature. Where conditions (ii) and (iii) apply, a change in cell density will occur. Under (ii) density will increase with each gram of carbon assimilated as B producing 2.38 g of carbohydrate (glycogen) ballast (B_g); under (iii) carbohydrate ballast is used to meet demand from respiration and B_g will be a negative value equal to $2.38 \times R$.

The change in cell density ($\Delta\rho_{cel}$) is calculated as:

$$\Delta\rho_{cel} = (B_g \times \text{Cell}_{carbon})/67,$$

where the *Microcystis* cell volume is taken to be 67 μm³, and Cell_{carbon} is the amount of carbon contained in each cell. For a *Microcystis* colony of radius 100 μm, the resultant change in colony density ($\Delta\rho_{col}$) will be:

$$= \text{CellsPerColony} \times 67 \times \Delta\rho_{cel}/(4/3) \times \Pi \times (100\,\mu\text{m})^3,$$

where CellsPerColony is the number of cells in each colony (for *Microcystis*, about 12032[33]). During a 24-hour period it is usual for all three conditions (i)–(iii) to operate at some point depending on the surface light intensity and the depth of the colony. SCUM'96 predicts that the density of *Microcystis* will vary from about 992 to 1000 kg m^{-3} under average environmental conditions (and without lake mixing) resulting in upwards and downwards vertical movements determined using Stokes's equation. Additionally, lake mixing is simulated by calculating the depth of mixing as a function of wind speed and profile width (fetch) using the Wedderburn number[34]. The velocity of water (and colony) movement within the mixed layer can then be calculated as a function of wind velocity (U), drag (c), density of water (ρ_{wat}) and density of air (ρ_{air})[35]:

$$v_*^2 = \rho_{air} \cdot c \cdot U^2/\rho_{wat}.$$

The direction of movement of colonies within the mixed layer is determined by a random walk routine[24].

Sample model output (Fig. 2) demonstrates a clear inverse relationship between mean colony density and mixing depth. When the mixed depth is >5 m, the mean

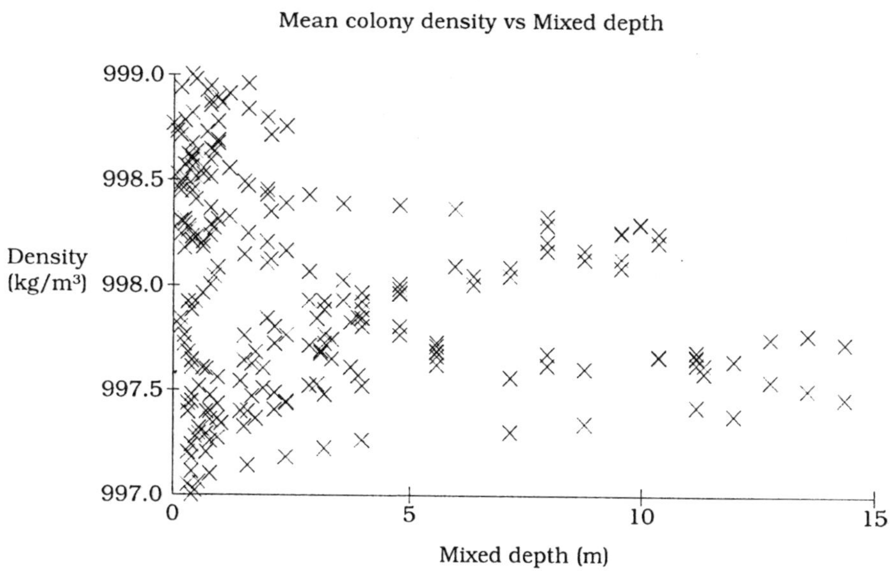

Figure 2 Graph produced by SCUM'96[24] showing mean colony density of *Microcystis* vs. mixed depth (Maximum surface sunlight intensity = 600 W m^{-2}; Water extinction coefficient = 2; colony form resistance = 1).

colony density rarely exceeds the density of water ($998.2\,kg\,m^{-3}$). For much of the time, the population will therefore be positively buoyant and will migrate towards the lake surface during periods of calm and may form problem blooms. Under nutrient-limited conditions, growth will be limited, ultimately resulting in more photosynthate being assimilated as ballast, and the colony sinking further. When the system is carbon limited, then there is lowered photosynthesis, less ballast and more prospect of a bloom[2]. This behaviour is well anticipated by the SCUM'96 model and when combined with realistic real-time reservoir simulations may have success in predicting bloom formation allowing management strategies to be made operational.

Future modelling objectives should include the realisation that the distribution and frequency of major blooms is likely to increase as a result of climate change. Reduced flow in rivers will create optimum conditions for cyanobacterial growth. In recent years, cyanobacterial problems in Europe have largely been restricted to impoundments but with changes in the environment, river blooms may become as common as those experienced in Australian rivers most notably the River Darling. New models therefore need to predict the likely future distribution of problem cyanobacterial blooms.

CONCLUSIONS

Cyanobacterial blooms have been clearly demonstrated to present a worrying water quality and public health problem. Already diverse in their distribution, the occurrence of problem blooms is likely to increase as agricultural land is farmed more intensively with fertilisers resulting in an increased eutrophication problem; additionally, climate change will result in reduced flow in rivers and elevated water temperatures making conditions ideal for large algal growths. Amelioration of the bloom problem is unlikely to occur through legislation such as the Urban Wastewater Treatment Directive and in any case would have little relevance to developing countries where cyanobacterial toxins may become life threatening if drinking water has not been properly treated. In such areas short-term action plans should be devised to manage and minimise the risk that toxic blooms may pose to drinking water supplies especially where the normal treatment is inadequate. Incidents such as the Brazilian haemodialysis tragedy make such strategies urgent and essential.

References

1. R. J. Shiel and J. D. Green, "Cyanobacteria: a problem in perspective?" *The Victorian Naturalist* **109**, 225–232 (1992).
2. C. S. Reynolds, "Cyanobacterial Water Blooms" **In**: (J. A. Callow, ed.) *Advances in Botanical Research* (Academic Press, London, 1987) pp. 68–143.
3. H. James, "Blue-green algal toxins in water" *Lab. Practice* **41**, 11–13 (1992).
4. W. M. Repavich, W. C. Donzogni, J. H. Standridge, R. E. Wedepohl and L. F. Meisner, "Cyanobacteria (blue-green algae) in Wisconsin waters: acute and chronic toxicity" *Water Res.* **24**, 225–231 (1990).
5. S. L. Kenefick, S. E. Hrudey, E. E. Prapas, N. Notkovsky and H. G. Peterson, "Odorous substances and cyanobacterial toxins in Prairie drinking water sources" *Wat. Sci. Tech.* **25**, 147–154 (1992).
6. F. Juttner, "Biochemistry of biogenic off-flavour compounds in surface waters" *Wat. Sci. Tech.* **20**, 107–116 (1988).
7. D. F. Kelly and R. Pontefract, "Hepatorenal toxicity in a dog after immersion in Rutland Water" *Veterinary Record* **127**, 453–454 (1990).

8. NRA, "Toxic Blue-green Algae" *Water Quality Series No. 2* (National Rivers Authority, London, 1990).
9. V. M. Vasconcelos, "Toxic cyanobacteria (blue-green algae) in Portuguese fresh waters" *Arch. Hydrobiol.* **130**, 439–451 (1994).
10. G. J. Gunn, G. C. Rafferty, N. Cockburn, N. C. Edwards, K. A. Beattie and G. A. Codd, "Fatal canine neurotoxicosis attributed to blue-green algae (cyanobacteria)" *Veterinary Record* **130**, 301–302 (1992).
11. G. A. Codd, C. Edwards, K. A. Beattie, W. M. Barr and G. J. Gunn, "Fatal attraction to cyanobacteria?" *Nature* **359**, 110–111 (1992).
12. Blue-Green Algae Task Force, "Blue-green Algae" *Final Report of the New South Wales Blue-Green Algae Task Force* (Dept. Water Resources, Parramatta, 1992).
13. J. Gutteridge, G. F. Haskins and H. Davey, "Investigation of nutrient pollution in the Murray–Darling River system" (Murray–Darling Basin Commission, 1992).
14. B. Jalaludin and W. Smith, "Blue-green algae (cyanobacteria)" *Med. Jour. Austral.* **156**, 744 (1992).
15. I. R. Falconer, A. M. Beresford and M. T. C. Runnegar, "Evidence of liver damage by toxin from a bloom of blue-green algÍ, *Microcystis aeruginosa*" *Med. Jour. Austral.* **1**, 511–514 (1983).
16. A. T. C. Bourke, R. B. Hawes, A. Neilson and N. D. Stallman, "An outbreak of hepatoenteritis (the Palm Island mystery disease) possibly caused by algal intoxication [abstract]" *Toxicon Suppl.* **3**, 45–48 (1983).
17. I. R. Falconer, "Tumor promotion and liver injury caused by oral consumption of cyanobacteria" *Env. Toxicol. & Water Quality* **6**, 177–184 (1991).
18. J. E. Erikkson, D. Toivala, J. A. O. Merilvoto, H. Karaki, Y. G. Han and D. Hartshorne, "Hepatocyte deformation induced by cyanobacterial toxins reflects inhibition of protein phosphatases" *Biochem. Biophys. Res. Communications* **173**, 1347–1353 (1990).
19. C. Mackintosh, K. A. Beattie, S. Klumpp, P. Cohen and G. A. Codd, "Cyanobacterial microcysin-LR is a potent and specific inhibitor of protein phosphatases 1 and 2A from both mammals and higher plants" *FEBS Letters* **264**, 189–192 (1990).
20. M. S. Islam, B. S. Drasar and R. B. Sack, "Probable role of blue-green algae in maintaining endemicity and seasonality of cholera in Bangladesh: a hypothesis" *J. Duarrhoeal, Dis. Res.* **12**, 245–256 (1994).
21. A. Howard, A. T. McDonald, P. E. Kneale and P. G. Whitehead, "Cyanobacterial (blue-green algal) blooms in the UK: a review of the current situation and potential management options" *Prog. Phys. Geog.* **20**, 53–61 (1996).
22. A. J. D. Ferguson, "The role of modelling in the control of toxic blue-green algae" *Hydrobiologia* **349**, 1–4 (1997).
23. A. Howard, "SCUM: simulation of cyanobacterial underwater movement" *Comput. Applic. Biosciences* **9**, 413–419 (1993).
24. A. Howard, A. E. Irish and C. S. Reynolds, "A new simulation of cyanobacterial underwater movement (SCUM'96)" *J. Plankton Res.* **18**, 1375–1385 (1996).
25. A. Braband, B. A. Faafeng, T. Kallqvist and J. P. Nilssen, "Biological control of undesirable cyanobacteria in culturally eutrophic lakes" *Oecologia* **60**, 1–5 (1983).
26. M. Haddon, "Broadlan brouhaha" *Water Bulletin* **645**, 7–8 (1995).
27. EC, "Council Directive concerning urban waste water treatment" *Official Journal of the European Communities* **L 135**, 40–47 (1991).
28. J. W. G. Lund and B. Moss, "Eutrophication in the United Kingdom" *Report to the UK Soap and Detergent Industry Association* (University of Liverpool, Liverpool, 1990).
29. Anon., "The UK nitrates 'timebomb'—fact or fiction?" *World Water* 21–25 (2 December, 1980).
30. F. Recknagel, M. French, P. Harkonen and K.-L. Yabunaka, "Artificial neural network approach for modelling and prediction of algal blooms" *Ecol. Modelling* **96**, 11–28 (1997).
31. P. G. Whitehead and G. Hornberger, "Modelling algal behaviour in the River Thames" *Wat. Res.* **18**, 945–953 (1984).
32. P. G. Whitehead, A. Howard and C. Arulmani, "Modelling algal populations in the River Thames using artificial neural networks" *Hydrobiologia* **349**, 39–46 (1997).
33. C. S. Reynolds and G. H. M. Jaworski, "Enumeration of natural *Microcystis* populations" *Br. Phycol. J.* **13**, 269–277 (1978).
34. J. Imberger and P. F. Hamblin, "Dynamics of lakes, reservoirs and cooling ponds" *Ann. Rev. Fluid Mech.* **14**, 153–187 (1982).
35. C. S. Reynolds, R. L. Oliver and A. E. Walsby "A review of mixing processes relevant to phytoplankton dynamics in lakes" *N.Z. J. Mar. Freshwater Res.* **21**, 392–405 (1987).

17. USE OF STABLE ISOTOPE RATIOS IN FRESHWATER AND MARINE BIOMAGNIFICATION STUDIES

KAREN A. KIDD *

INTRODUCTION

Ecologists are using measurements of naturally-occurring stable carbon, nitrogen and sulfur isotopes to characterize energy flow and trophic interrelationships in marine, freshwater and terrestrial systems[1]. These stable isotope ratios have helped determine the contribution of terrestrial organic matter to aquatic food webs[2], an organism's reliance on benthic versus pelagic[3] or freshwater versus marine[4] carbon, the relative trophic positioning of biota within a food web[5], and, more recently, to improve our understanding of the biomagnification (increasing concentrations of contaminants with increasing trophic level) of persistent contaminants through improved food web characterization[6]. Previous studies of contaminant accumulation in aquatic organisms have defined food webs using conventional stomach content analyses or estimated interspecies relationships[7]. Now the combination of contaminant and stable isotope analyses has enabled researchers to accurately characterize food web interactions, and, both qualify and quantify the trophic transfer of contaminants up the food chain. This review examines the recent use of stable isotope ratios in understanding contaminant movement through freshwater and marine food webs, and discusses potential applications of these isotopic measurements to future research on contaminant biomagnification.

The main route of contaminant uptake for algae and small-bodied organisms, such as zooplankton, is through adsorption or absorption directly from water[8] because these organisms have high surface area to volume ratios. Organochlorine contaminants are lipophilic, and, therefore, are present at higher concentrations in fatter organisms[7,9] and tissues[10]. These compounds biomagnify (increase in concentration from prey to predator) up the food chain, and are generally found at the highest concentrations in top predators[11–13]. Although uptake from water is the main route of contaminant exposure for lower-trophic-level biota, the main source of organochlorines for top predators is diet. For example, lake trout from Lake Ontario had concentrations of PCBs that were an order of magnitude higher than expected from the concentrations of PCBs in the water[7]. The authors concluded that this discrepancy between predicted (based on a compound's lipophilicity) and observed concentrations in these fish was due to significant accumulation of organochlorines from food. Similarly, a model developed by Thomann and Connolly[14] indicated that dietary uptake accounted for 99% of the PCBs in lake trout from Lake Michigan. However, within one species significant variability in contaminant concentrations exists among sites[15], and it remained important to ascertain that this variability was due to differences in dietary habits, and not other factors such as contaminant inputs.

* Freshwater Institute, 501 University Crescent, Winnipeg, Manitoba R3T 2N6 (Canada).

Photosynthetic enzymes in C_4 plants are less discriminating against $^{13}CO_2$ than C_3 plants, resulting in a difference in $\delta^{13}C$ of at least 8‰ between these plant types[40,41]. Variability in the $\delta^{13}C$ signals between terrestrial and aquatic plants[42,43, but see 2], and benthic and planktonic algae[3,43,44], can be exploited to determine the importance of these carbon sources to upper-trophic-level organisms[e.g. 3].

Sulfur isotopes are also useful in characterizing food web structure because this ratio is conserved from prey to predator, with a slight average enrichment of 0.2‰[1] with each trophic level. In regions where the geological or anthropogenic sources of sulfur are isotopically distinct[45–48], $\delta^{34}S$ could be used in the same manner as $\delta^{13}C$ to measure the relative importance of different food sources in the diets of consumers. Sulfur isotopes enabled Hesslein *et al.*[47] to distinguish migratory from non-migratory fishes in the Mackenzie Delta, Northwest Territories, Canada because of regional geological differences in $\delta^{34}S$ signals. As in this study, $\delta^{34}S$ can provide trophic information that may not be ascertainable from $\delta^{13}C$ or $\delta^{15}N$ analyses alone.

STABLE NITROGEN ISOTOPE RATIOS IN BIOMAGNIFICATION STUDIES

Studies examining the transfer of persistent contaminants through the food web have been limited by an inability to accurately quantify the trophic positioning of individuals. For this reason, most of the current contaminant studies using stable isotope ratios have focused on $\delta^{15}N$ to quantify long-term feeding habits. The potential for $\delta^{15}N$ to be used in biomagnification studies was demonstrated by Rolff *et al.*[49] and Broman *et al.*[50]. They examined the relationships between biotic concentrations of polychlorinated dibenzo-*p*-dioxins (PCDDs) and dibenzofurans (PCDFs) and $\delta^{15}N$ in the pelagic (phytoplankton → seston → zooplankton → mysids → herring → cod) and littoral (phytoplankton → seston → mussels → eider duck) food chains in the Baltic. Their studies indicated that some of the dioxin and furan congeners were significantly related to $\delta^{15}N$ through these food chains, while total concentrations of PCDDs and PCDFs in these organisms decreased with increasing trophic level. These relationships were modeled using the exponential equation

$$[\text{contaminant}] = A * e^{[B * \delta^{15}N]}$$

and a significant positive relationship was indicative of contaminant biomagnification within the system. A negative exponent ($B < 0$) indicated that the compound was not taken up, or was metabolized and excreted, while a positive exponent ($B > 0$) reflected increasing concentrations of the contaminant, and the magnitude of this increase, through the food chain. For example, Rolff *et al.*[49] found that concentrations of OCDD ($pg\,g^{-1}$ dry weight) decreased with $\delta^{15}N$ through the food web ($B = -0.39$), which the authors believed was a function of high adsorption of OCDD to phytoplankton, and some metabolism of this compound in upper-trophic-level organisms. In contrast, the concentration of three PCDD and PCDF congeners (2, 3, 7, 8-TCDD, 2, 3, 4, 7, 8-PnCDF and 1, 2, 3, 7, 8-PnCDD) increased significantly ($B = 0.26$) with $\delta^{15}N$, and explained 69% of the variability in the contaminant data. For those contaminants that are assimilated by an organism and do not undergo significant metabolism and excretion, their results demonstrated that $\delta^{15}N$ is a significant predictor of contaminant concentrations through marine food chains.

Studies of other marine food webs, off of the coast of California[51,52] and in the Canadian Arctic[13], have found similar relationships to those of Rolff *et al.*[49] and Broman *et al.*[50]. Jarman *et al.*[52] analysed whole invertebrates (euphasiids) and fishes, bird eggs, and blubber and muscle from sea lions, from the Gulf of the Farallones, and found that concentrations of all organochlorines and Hg (ng or $\mu g\, g^{-1}$ dry weight) in these organisms were significantly related to their tissue $\delta^{15}N$ (using the ln-transformed concentrations, B for ΣDDT and ΣPCB was 0.79 and 0.88, respectively). In the same location, Jarman *et al.*[53] found no relationship between lipid-normalized concentrations of PCDDs and PCDFs and $\delta^{15}N$ through an abbreviated food web consisting of eggs from four species of birds and sea lion tissues. This lack of a relationship may be due to an ability of these organisms to metabolize dioxins and furans, to the small sample size examined, or to the choice of tissues analysed in the study (see discussion below). In the eastern Canadian Arctic, Muir *et al.*[13] found significant correlations between the log-transformed wet weight concentrations of ΣPCB, ΣDDT and Σchlordane, and $\delta^{15}N$ among two populations of walrus, and through two trophic levels (mussel → walrus; $r^2 = 0.25$ to 0.73). However, the latter correlations were based on the results from one pooled mussel sample, and need to be corroborated by further examination of organochlorine-$\delta^{15}N$ relationships in arctic marine food webs. Within two Pacific Ocean food webs (invertebrates → fishes), one offshore of Los Angeles, California that was strongly influenced by sewage input, and one from a more pristine location near Santa Barbara, ΣDDT and PCBs (Aroclor 1254) were significantly related to $\delta^{15}N$[51]. These results demonstrate that $\delta^{15}N$ is an important predictor of contaminant con centrations in a variety of systems, differing in species composition.

The use of $\delta^{15}N$ in marine bioaccumulation studies has been successfully extended to freshwater systems. Several studies have demonstrated that $\delta^{15}N$ is a significant predictor of both chlorinated contaminants[6,11,25,54], and mercury[28,55,56] in biota. In subarctic Lake Laberge, Yukon Territory, Canada, Kidd *et al.*[6] found a sign i-ficant relationship between concentrations of hexachlorocyclohexane (ΣHCH), ΣDDT and toxaphene (chlorinated bornanes, ΣCHB), and $\delta^{15}N$ through the food web (zooplankton, benthic invertebrates → broad and lake whitefish, longnose sucker, cisco → lake trout and burbot). The relationship between ΣDDT and $\delta^{15}N$ is shown in Fig. 1, and is described by the linear regression equation

$$\log \Sigma\mathrm{DDT} = 0.32\ (\pm 0.03)\delta^{15}\mathrm{N} - 0.72(\pm 0.24), \quad n = 27,\ P < 0.01,\ r^2 = 0.81.$$

When two other subarctic lakes were added to the study, it was found that $\delta^{15}N$ was a significant predictor of toxaphene concentrations through all three food webs, and that the rates of toxaphene accumulation, based on the Σtoxaphene–$\delta^{15}N$ slopes, were the same in each lake[54]. Contaminant and $\delta^{15}N$ analyses of food-web organisms from Lake Baikal, Russia indicated that ln-transformed ΣPCB concentrations ($ng\, g^{-1}$ wet weight) and $\delta^{15}N$ were significantly related[11] ($r^2 = 0.52$ including all data, $r^2 = 0.88$ excluding high-lipid 5 to 10 year-old sculpin; relationships for other organochlorines were not examined in this study). In the Lake Ontario pelagic food web[25] (plankton → mysids, amphipods → alewife, rainbow smelt, sculpin → lake trout), log-transformed *p,p'*-DDE (the main metabolite of DDT), mirex and ΣPCB concentrations ($ng\, g^{-1}$ wet weight) were related to $\delta^{15}N$, and trophic positioning explained between 54 and 64 % of the variability in the contaminant data. $\delta^{15}N$ is also a significant predictor of Hg levels in freshwater biota; significant relationships have been found between Hg ($\mu g\, g^{-1}$ wet weight) and $\delta^{15}N$ in lake trout from seven Class 1 to 3 lakes in Ontario and Quebec[28] ($r = 0.89$), and within

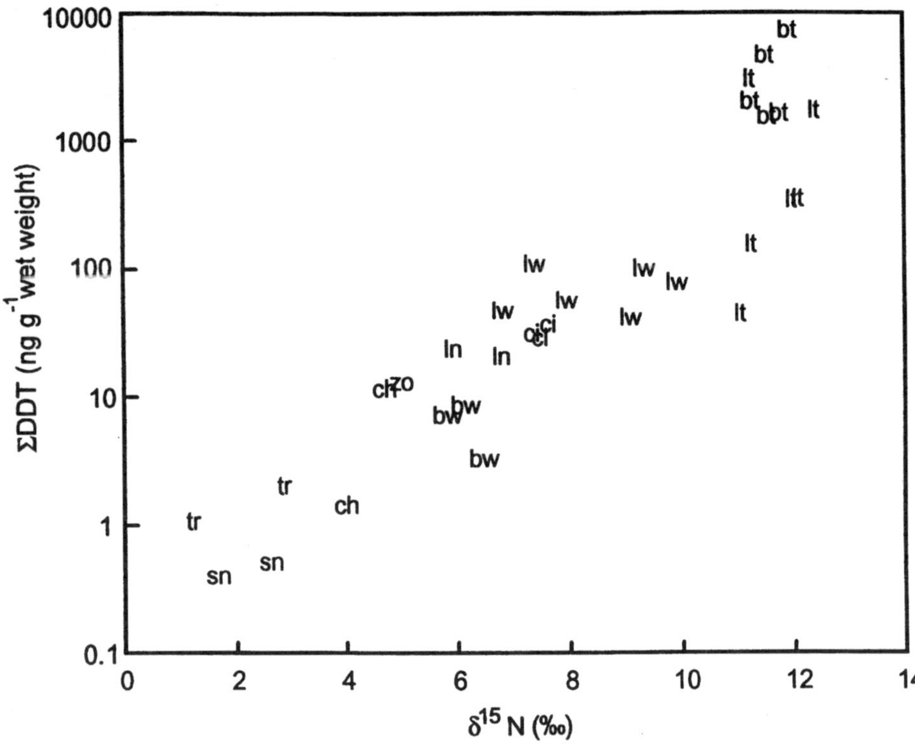

Figure 1 Relationship between log-transformed concentrations of ΣDDT (ng g^{-1} wet weight) and $\delta^{15}N$ (‰) for fishes and invertebrates from Lake Laberge, Yukon Territory [tr—trichopteran, sn—snail (Family Lymnaeidae), ch—chironomid, zo—zooplankton, (whole body); bw—broad whitefish, lw—lake whitefish, ln—longnose sucker, ci—cisco and lt—lake trout (muscle); and bt—burbot ($\delta^{15}N$ in muscle, ΣDDT in liver) taken from reference [6]. Reprinted from *The Science of the Total Environment* **Vol. 160/161**; K. A. Kidd, D. W. Schindler, R. H. Hesslein and D. C. G. Muir; "Correlation between stable nitrogen isotope ratios and concentrations of organochlorines in biota from a freshwater food web"; pp. 381–390, 1995; with kind permission of Elsevier Science—NL, Sara Burgerhart street 25, 1055 K. V. Amsterdam, The Netherland.]

six lakes in northwestern Ontario[56] ($r^2 = 0.47$ to 0.91). Yoshinaga *et al.*[55] expanded upon the previous studies by measuring concentrations of several elements (Na, Mg, Al, P, K, Ca, Cr, Fe, Mn, Cu, Zn, Sr, total Hg, organic and inorganic Hg, and Pb) and $\delta^{15}N$ in terrestrial, freshwater and marine organisms from Papua New Guinea. They found that $\delta^{15}N$ was positively correlated to concentrations of K, Ca, Sr, total Hg, organic and inorganic Hg, and negatively correlated to concentrations of Fe, Cu, and Zn. Results from these studies demonstrate that $\delta^{15}N$ can be used to examine the food-chain transfer of a large number of organic and inorganic contaminants.

The slope of the contaminant–$\delta^{15}N$ relationship is a quantitative measure of the biomagnification of these contaminants from primary producers through to the top predators, and can be compared for several chemicals within one food web or for one chemical across several food webs. If trophic positioning is the most important factor influencing biomagnification, one would expect to find similar relationships between a contaminant and $\delta^{15}N$ for food webs differing in species composition. The relationships between organochlorine concentrations and $\delta^{15}N$ were compiled from the published literature in order to examine among- and within-site variability in the slopes. These relationships were quantified using least-squares regression analyses of the wet weight log-transformed organochlorine concentrations versus $\delta^{15}N$. These analyses are included herein only for the purposes of general comparisons because data were limited for some studies. Relationships between a commonly measured contaminant, DDT (ng g^{-1} wet weight), and $\delta^{15}N$ (‰) for marine and freshwater food webs are shown in Figs. 2 and 3, respectively, and the slopes of all organochlorine-$\delta^{15}N$ regressions are given in Table I. Mean ΣDDT[57] and $\delta^{15}N$[35] data for cod, seal and polar bear from Barrow Strait in the Canadian Arctic were included in Fig. 2 although the two variables were not significantly related. ΣDDT accumulated at a similar per-trophic-level rate (slopes of 0.15 and 0.16) in the sewage-distorted and reference systems studied by Spies *et al.*[51]. It is interesting to note

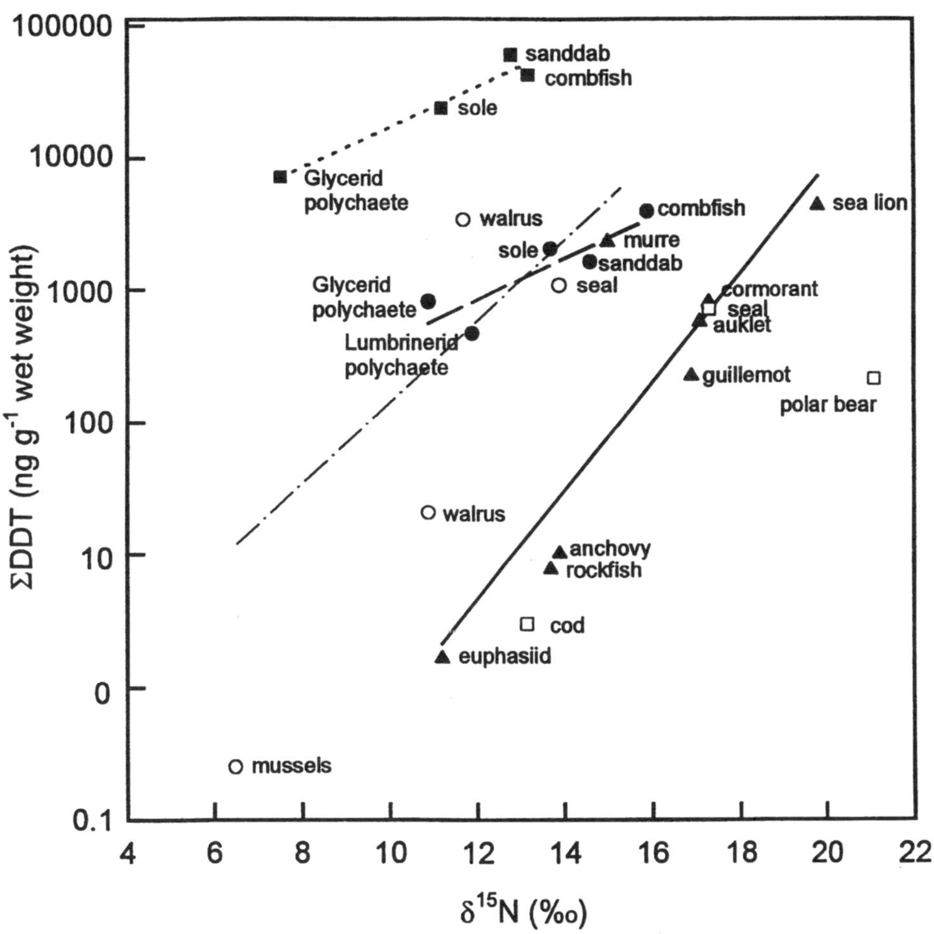

Figure 2 Relationship between mean log-transformed concentrations of ΣDDT (ng g^{-1} wet weight) and mean $\delta^{15}N$ (‰) in marine biota from Gulf of the Farallones, California coast ([52], ▲—), Los Angeles, California sewage-influenced coastal food web ([51]; ■------), Santa Barbara, California pristine coastal food web ([51]; ●– – –), Hudson Bay ([13], O–·–·–·–·–), and Barrow Straight ([57,35]; □). See Table I for slopes, *n* of relationships and DDT congeners quantified.

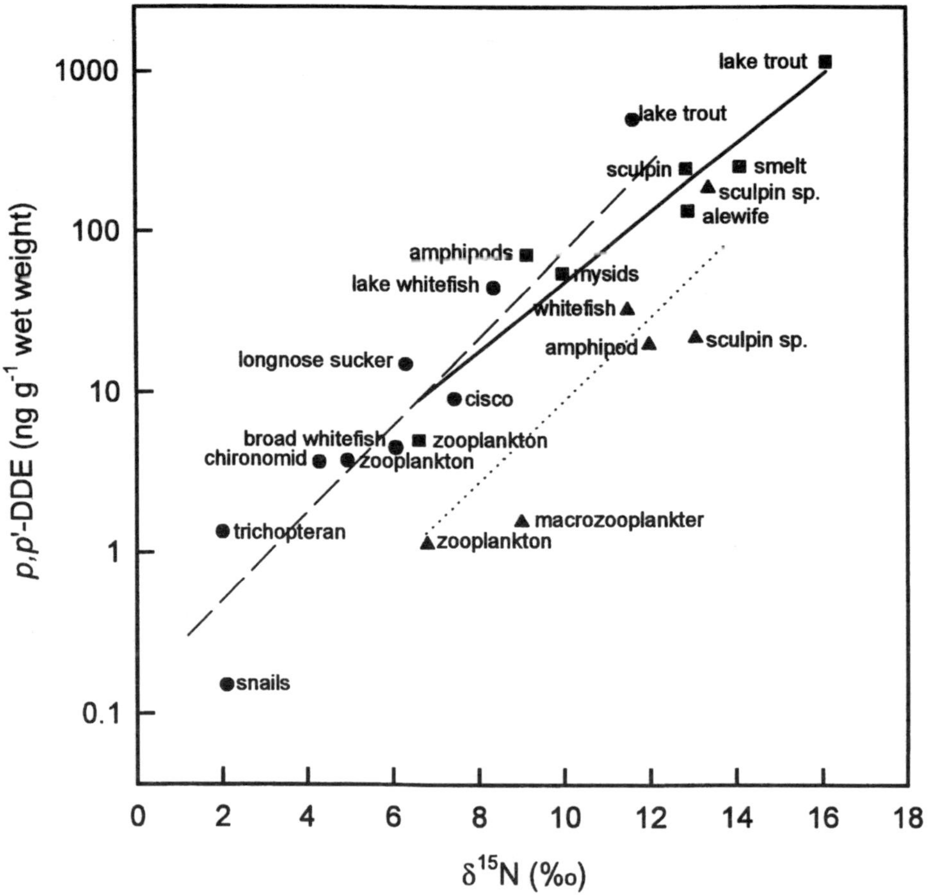

Figure 3 Relationship between mean log-transformed concentrations of p,p'-DDE ($ng\,g^{-1}$ wet weight) and mean $\delta^{15}N$ (‰) in freshwater biota from Lake Laberge, Yukon Territory (K. Kidd, unpublished data, ●– – –), Lake Ontario ([25]; ■—), and Lake Baikal, Russia ([11]; ▲·········). For Lakes Laberge and Baikal, regressions were plotted using individual or year-class data while symbols represent means of all samples for each major taxon. See Table I for slopes and *n* of relationships.

that the ΣDDT–$\delta^{15}N$ intercept for the food web receiving contaminated effluent from Los Angeles was higher than at the reference site. This implies that the intercept of the organochlorine–$\delta^{15}N$ regression may be used as a measure of actual contaminant inputs to the base of the food web, as was suggested in Broman *et al.*[50]. The slopes from Spies *et al.*'s study[51] were considerably lower than those found for the Gulf of the Farallones (0.41) or Hudson Bay (0.30) biota, indicating greater rates of ΣDDT accumulation through the latter two food webs. For the Barrow Strait, the ΣDDT–$\delta^{15}N$ relationship from cod to their main predator seal appears to be higher than when all three species (cod, seal and polar bear) are considered, likely because polar bears are able to metabolize ΣDDT[58]. Within the freshwater food webs (Fig. 3 and Table I), the slopes of p,p'-DDE versus $\delta^{15}N$ were 0.15, 0.25 and 0.27 in Lakes Ontario, Baikal and Laberge (excluding burbot liver data), respectively, also indicating that the per-trophic-level biomagnification of this contaminant differs among freshwater systems. Relationships between

Table I Least-squares regressions ($P < 0.05$) of log-transformed concentrations of hexachlorocyclohexane (ΣHCH), ΣDDT (or metabolite DDE), Σtoxaphene (chlorinated bornanes), polychlorinated biphenyls (ΣPCB), *cis*- and *trans*-chlordane, *cis*- and *trans*-nonachlor, Σchlordane, dieldrin, hexachlorobenzene (HCB), and mirex (ng g^{-1} wet weight), and $\delta^{15}N$ (‰) in organisms from freshwater and marine food webs

		Lowest trophic level analysed	Top predator(s)	Contaminant	*n*	Slope	Ref.
Freshwater	Lake Laberge, Yukon Territory, Canada[1]	zooplankton/benthic invertebrates	trout/burbot	ΣHCH	23	0.16	[6]
				ΣDDT[2]	27	0.32	[6]
				ΣDDT[2*]	23	0.28	[6]
				p,p'-DDE[3]	23	0.27	
				Σtoxaphene	40	0.23	[54]
	Fox Lake, Yukon Territory, Canada	zooplankton/benthic invertebrates	trout/burbot	Σtoxaphene	18	0.19	[54]
	Kusawa Lake, Yukon Territory, Canada	zooplankton/benthic invertebrates	trout	Σtoxaphene	24	0.25	[54]
	Lake Baikal, Russia[4]	zooplankton	pelagic sculpin	ΣPCB	20	0.23	[11]
				ΣDDT[5]		0.26	
				p,p'-DDE		0.25	
				Σtoxaphene		0.09[6]	
				dieldrin		0.11[7]	
				trans-chlordane		0.17	
				cis-chlordane		0.18	
				trans-nonachlor		0.17	
				cis-nonachlor		0.15	
				HCB		0.22	
	Lake Ontario	zooplankton	trout	*p,p'*-DDE	100	0.15	[25]
				ΣPCB		0.14	
				mirex		0.15	

Table I (*Continued*)

		Lowest trophic level analysed	Top predator(s)	Contaminant	n	Slope	Ref.
Marine	Gulf of Farallones[8,9], California, U.S.A.	euphausiids (invertebrates)	sea lion	ΣHCH	8	0.23	[52]
				ΣDDT[10]		0.41	
				Σchlordane		0.34	
				HCB[11]		0.36	
				ΣPCB		0.42	
	Hudson Bay, Canada[3]	mussel	seal	ΣDDT[2]	30	0.30[12]	[13]
	Barrow Strait, Canada	cod	polar bear	ΣDDT[2]	3	NS	[35,57]
	Los Angeles sewage outfall, California coast, U.S.A.[8]	polychaetes	combfish	ΣDDT[2]	4	0.15	[51]
	Santa Barbara,California coast, U.S.A.[8]	polychaetes	combfish	ΣDDT[2]	5	0.16	[51]

[1] for ΣDDT* and p,p'-DDE regressions exclude burbot liver data from Lake Laberge. [2] sum of o,p'-DDD, -DDE and -DDT, and p,p'-DDD, -DDE and -DDT. [3] K. Kidd, unpublished data. [4] $\delta^{15}N$ and organochlorine (Hudson Bay only) data for individuals and year classes obtained from authors. [5] Sum of p,p'-DDE and p,p'-DDT, samples below detection limits were based on 1/2 of the detection limit. [6] $P=0.13$. [7] $P=0.09$. [8] Organochlorine-$\delta^{15}N$ slopes were calculated using mean organochlorine concentrations and $\delta^{15}N$ for each major taxon. [9] Organochlorine concentrations were converted to a wet weight basis using the mean % H_2O given for each species, bird eggs were analysed for organochlorines and $\delta^{15}N$, and sea lion blubber and liver data were used for organochlorine and metal concentrations respectively. [10] sum of p,p'-DDD, -DDE and -DDT. [11] HCB not measured in sea lion blubber. [12] $P=0.1$.

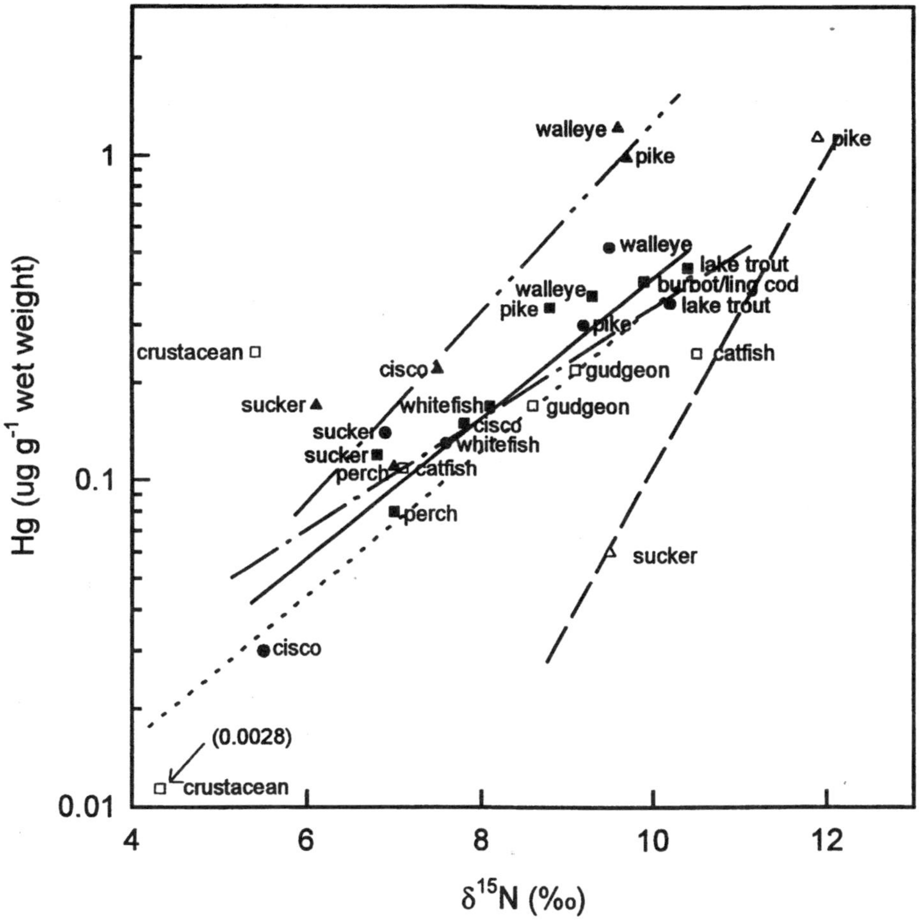

Figure 4 Relationship between mean log-transformed concentrations of Hg (μg g^{-1} wet weight) and mean δ^{15}N (‰) in freshwater biota from Papua New Guinea ([55]; □·········), Green Lake (△-------), Trout Lake (●—), Sydney Lake (■–·—·–), and Orange Lake, northwestern Ontario ([56]; ▲–··–··–). For the lakes, regressions were plotted using individual-fish data while the mean Hg and mean δ^{15}N for each taxon were indicated by the symbols. See Table II for slopes and *n* of relationships.

log-transformed concentrations of Hg or Pb and δ^{15}N for freshwater and marine food webs are shown in Fig. 4 and Table II. Hg–δ^{15}N relationships was similar (slopes of 0.17 to 0.21) across most freshwater lakes, with the exception of Green Lake in northwestern Ontario (slope = 0.48), and generally lower than was found in marine systems (slopes of 0.36 and 0.32). Negative Pb–δ^{15}N slopes (−0.07 and −0.27) were found for the Gulf of the Farallones and Papua New Guinea marine systems. Differences in the magnitude of the contaminant–δ^{15}N slopes for marine and freshwater systems indicate that trophic positioning is not the sole factor determining the accumulation of contaminants through these food webs (see below).

It should be noted that among-site comparisons in the slopes of these relationships may be confounded by the species or tissues analysed, or by intraspecies variability in age, growth rate, size or sex. This slope within one food web (e.g. cod → seal →

Table II Least-squares regressions of log-transformed Hg or Pb concentrations (μg g^{-1} wet weight) versus $\delta^{15}N$ (‰) in organisms from freshwater and marine food webs

	Site	Lowest trophic level analysed	Top predator(s)	Metal	n	Slope	Ref.
Freshwater	Trout Lake	least cisco	trout/pike/walleye	Hg	30	0.21	[56]
	Sydney Lake	white sucker	trout/burbot/pike/walleye	Hg	39	0.17	
	Musclow Lake	white sucker	walleye/pike	Hg	25	0.21	
	Linge Lake	white sucker	trout/walleye/pike	Hg	31	0.23	
	Orange Lake	white sucker	walleye/pike	Hg	24	0.29	
	Green Lake, NW Ontario, Cananda	white sucker	pike	Hg	12	0.48	
	Papua New Guinea	clams	catfish	Hg	6	0.22^{1}	[55]
Marine	Gulf of Farallones[2]	euphausiids (invertebrates)	sea lion	Hg	8	0.36	[52]
				Pb		-0.07^{3}	
	Papua New Guinea	clams	shark	Hg	5	0.32	[55]
				Pb		−0.27	

[1] $P = 0.12$. [2] Slope calculated from mean Hg, Pb and $\delta^{15}N$ for each species. [3] $P = 0.17$.

polar bear) may be different than another (cod → seal; see Fig. 2) simply because the top predator in the former system can metabolize and excrete the contaminant of interest. Use of more than one tissue type for contaminant and stable isotope analyses may skew the pollutant–isotope relationship if these tissues represent different time periods[21,59]. For example, Jarman *et al.*[52,53] used bird egg albumen to determine the dietary habits of the laying female, and whole eggs to measure persistent contaminant concentrations. They found that common murres were outliers in the food web organochlorine–$\delta^{15}N$ relationship because these birds fed upon isotopically lighter amphipods just prior to egg protein formation, and mobilized contaminants into the eggs from tissues stores formed during the remainder of the year when the birds were feeding upon more contaminated fishes. Interpretation of contaminant–isotope relationships may also be confounded by the composition of the tissues analysed. Lipids are depleted in ^{13}C when compared to whole tissues[19] because of the loss of heavier CO_2 during lipid synthesis[60]. Measurements of tissues or organisms varying considerably in fat content may skew relationships, and, for this reason, lipids are sometimes extracted prior to ^{13}C analyses[26]. In addition, contaminant concentrations in biota may vary between males and females, or with weight, length, age, and growth rates of individuals[10,15,61–65]. Food webs with older, slow-growing biota may have higher contaminant–$\delta^{15}N$ slopes than ones containing younger, fast-growing organisms.

The accumulation of organochlorine contaminants in biota is a function of lipophilicty (determined by the *n*-octanol to water partition coefficients or K_{ow}), with the more lipophilic contaminants accumulating to higher levels in organisms than the less lipophilic contaminants e.g.[7,52,66]. Kidd *et al.*[6] noted that the slopes of the organochlorine–$\delta^{15}N$ regressions were higher for the more lipophilic ΣDDT (log K_{ow}s of 5.69[67] to 6.95[68] for the main metabolite of DDT, *p,p′*-DDE) and Σtoxaphene (6.44[69] for Σtoxaphene) than the less lipophilic ΣHCH (log K_{ow} of 3.8[70]) through the Lake Laberge food web. A within-site comparison of the organochlorine–$\delta^{15}N$ relationships in Table I generally reveals that compounds of similar lipophilicity have comparable slopes, and that the more lipophilic contaminants have greater slopes than the less lipophilic contaminants. With the exception of toxaphene in the Lake Baikal food web, the slopes increase [dieldrin (log K_{ow} 3.69[67]) < *trans*- and *cis*-chlordane, *trans*- and *cis*-nonachlor (log K_{ow} 5.54 for Σchlordane[70]) < HCB (log K_{ow} 5.5[71]) < ΣDDT] with the lipophilicity of the compounds. In Lake Ontario the slopes are the same (0.15) for compounds of comparable lipophilicity, mirex (log K_{ow} 6.89[67]) and *p,p′*-DDE. Similarly, the slopes through the Gulf of the Farallones food web increase as follows: ΣHCH < Σchlordane < HCB < ΣDDT. Some contaminants, such as PCBs and toxaphene, have a large number of congeners which vary considerably in their lipophilicity, and which are differentially accumulated through the food web. For example, log K_{ow}s increase from 4.3 to 8.3 for mono- through to deca-chlorinated PCBs[71], and lower-trophic-level organisms have higher proportions of the less chlorinated PCBs when compared to upper-trophic-level organisms[7,52]. Technical toxaphene is composed of a large number of congeners (up to 32,768 different compounds[72]), and the differences between the original pesticide mixture and what is found in the biotic and abiotic environment[73,74] have made it difficult to quantify. A more precise comparison of the relationship between organochlorine–$\delta^{15}N$ slopes and lipophilicity could be made if individual chemical compounds were examined. However, the trend of increasing slope with increasing lipophilicity found herein supports the results of other studies; the more lipophilic contaminants biomagnify to a greater degree than the less lipophilic contaminants.

The trophic transfer of persistent organochlorines through the food web has typically been quantified by calculating a biomagnification factor (BMF) as follows:

$$\text{BMF} = \frac{\text{Predator (mass contaminant. g}^{-1}\text{ lipid)}}{\text{Main prey (mass contaminant. g}^{-1}\text{ lipid)}}$$

BMFs are calculated on a lipid weight basis to remove the effects of differences in lipid content among individuals. However, this method of quantifying biomagnification is limited because it does not account for any variability in the feeding habits of individuals within a population. Kidd *et al.*[6], Kucklick *et al.*[11] and Kiriluk *et al.*[25] have found significant relationships between lipid-normalized organochlorine concentrations and δ^{15}N in biota. The slope of this relationship can be converted into a biomagnification factor as follows:

$$\text{BMF} = \text{antilog}(B) * 3.4;$$

B is equal to the slope of log-transformed lipid weight concentrations of organochlorines versus δ^{15}N, and 3.4 is the average per-trophic-level increase in δ^{15}N. This BMF removes the uncertainty associated with assigning trophic positions, and quantifies the average accumulation of these contaminants through the food chain. For Lake Laberge[6], BMFs of 5.4, 5.1 and 3.8 were calculated for ΣDDT ($B = 0.2$), Σtoxaphene ($B = 0.18$), and ΣHCH ($B = 0.05$; Table III) respectively. From studies on Lake Ontario in 1981–84[7] and 1992[25], BMFs were calculated using both the predator–prey pairs outlined in Oliver and Niimi[7], and the slopes of the log-transformed organochlorine concentrations (lipid weight) versus δ^{15}N[25]. When compared to BMFs calculated using defined predator and prey interactions, the δ^{15}N-based BMFs were slightly higher for ΣPCB, and comparable for mirex and *p,p′*-DDE (Table II). The δ^{15}N-based BMF for ΣPCB in Lake Ontario was similar to that calculated for Lake Baikal, indicating that PCB accumulation is comparable through these two food webs. Total, organic and inorganic Hg (log-transformed, wet weight) increased by a factor of 5 with each trophic level (3.5‰) for biota from Papua New Guinea[55] ($B = 0.21$, 0.21 and 0.22, respectively). The per-trophic-level (3.5‰) accumulation of Hg was comparable (BMF of 5) in biota from northwestern Ontario lakes, with the exception of Green Lake which had a BMF of 10.6. Recently, Vander Zanden and Rasmussen[18] found that BMFs which incorporate omnivory into the calculations are higher than those BMFs using discrete trophic level assignments, demonstrating the need for a more precise measure of BMFs. Using the contaminant—δ^{15}N slopes to calculate a food-web BMF eliminates the uncertainty associated with traditional BMF calculations, although this approach does not provide species-specific information.

STABLE CARBON ISOTOPE RATIOS IN BIOMAGNIFICATION STUDIES

To date, the relationship between contaminant concentrations and δ^{13}C in food webs has rarely been examined. Organochlorine concentrations were not significantly correlated with δ^{13}C in walrus tissues from eastern Hudson Bay[13], whereas Spies *et al.*[51] found a significant positive relationship between ΣDDT and δ^{13}C through both the sewage-distorted and reference food webs off of the coast of California. In the latter study, the authors attributed this relationship to an enrichment of δ^{13}C and contaminant concentrations in biota with increasing trophic level. Despite little evidence of its utility thus far, measurements of δ^{13}C may prove beneficial in

Table III Biomagnification factors [calculated as (ng contaminant g^{-1} lipid in predator/ng contaminant g^{-1} lipid in prey) for specific predator–prey relationships outlined below, or as (antilog of the slope of log- (or ln-) transformed organochlorine concentrations (ng g^{-1} lipid) versus $\delta^{15}N$ (‰) $*$ 3.4[1]) for the food web] of *p,p'*-DDE, ΣDDT, ΣPCB, ΣHCH (hexachlorocyclohexane), ΣTox (toxaphene, chlorinated bornanes) and mirex through the Lake Ontario, Baikal, and Laberge food webs

Lake		*p,p'*-DDE	ΣDDT[2]	ΣPCB	Mirex	ΣHCH	ΣTox	Ref.
Ontario	lake trout : smelt[3]	1.7–1.2	1.8–1.2	2.5–1.1	2.5–1.2			[7]
		1.9		2.1	2.8			[25]
Ontario	smelt : mysids[3]	2.4–1.7	2.5–1.7	3.2–1.4	5.0–2.4			[7]
		4.1		3.3	6.2			[25]
Ontario	mysids : plankton	3.1	2.1	1.1	1.0			[7]
		4.3		2.4	2.7			[25]
Ontario	food web	4.3		4.2	3.7			[25]
Baikal	food web			4.4				[11]
Laberge	food web		5.4			3.8	5.1	[6]

[1] Average increase in $\delta^{15}N$ from prey to predator[16]. [2] Sum of *p,p'*-DDE, *p,p'*-DDT, *p, p'*-DDD. [3] BMFs were calculated separately for large and small smelt and presented as range.

locations where the sources of carbon at the base of the food web are isotopically distinct, and can be used to determine the relative importance of different prey items to the contaminant burden of the predator.

POTENTIAL APPLICATIONS OF STABLE ISOTOPES TO BIOMAGNIFICATION STUDIES

Stable isotope ratios may be useful in understanding atypical concentrations of contaminants in organisms. For example, Muir *et al.*[13] found concentrations of PCBs and DDT in walrus from Inukjuak, Quebec that were orders of magnitude higher than the same species from other locations in the Canadian Arctic. These high concentrations appear to be the result of the unusual dietary habits of these walrus; measurements of muscle $\delta^{15}N$ and $\delta^{13}C$ indicated that Inukjuak walrus fed upon seals and mussels, rather than a typical diet consisting only of mussels. Likewise, cannibalism[26] or ontogenic shifts in diet[44] may also result in differences in contaminant concentrations among individuals within a population. Tissue stable isotope ratios could be used to ascertain whether such individuals had atypical dietary habits.

Distinct $\delta^{13}C$, $\delta^{15}N$ or $\delta^{34}S$ signals of plants or biota could be valuable in interpreting the importance of various food sources to a consumer's contaminant burdens. The contribution or proportion of species A (P_A) versus species B in a predator's diet may be calculated as in Hobson and Sealey[75]

$$P_A = (\delta R_{measured} - \delta R_{expected\ B})/(\delta R_{expected\ B} - \delta R_{expected\ A})$$

where the predator's measured isotopic ratios ($\delta R_{measured}$) and expected tissue values for a diet consisting only of species A ($\delta R_{expected\ A}$) or species B ($\delta R_{expected\ B}$) are incorporated. This model can be used for both conservative and fractionating isotopes, and in Hobson and Sealey's study was used to determine the proportion of two trophically-distinct species (cod and amphipods) in the diets of seabirds. Other such applications include calculating the relative importance of allochthonous versus autochthonous[2,42,43,76], benthic versus pelagic[3], and freshwater versus marine[4,75,77] production in an organism's diet. The following are examples of contaminant studies in which biotic stable isotope analyses may have provided significant ecological information that could have been used to further interpret the results. Evans *et al.*[12] found that the benthic amphipod *Pontoporeia hoyi* had considerably higher concentrations of persistent organochlorines than its more pelagic counterpart the mysid (*Mysis relicta*) in Lake Michigan. The authors speculated that upper-trophic-level fishes feeding on the amphipods would accumulate higher concentrations of organochlorines than those fish feeding upon mysids. Similarly, in a study of four species of forage fishes with similar trophic positioning, Hebert and Haffner[78] found that habitat usage was an important factor in their contaminant uptake: benthivorous species were much higher in organochlorine concentrations than pelagic species because of a greater exposure to more highly contaminated sediments. Mathers and Johansen[79] found that the top predators walleye and northern pike differed considerably in their Hg concentrations. The authors believed that walleye were highest in Hg because they consumed more smelt, a forage fish that is occasionally piscivorous and high in Hg, than the northern pike. In such studies, stable isotope ratios may be used to prove or disprove such hypotheses by qualifying important routes of contaminant exposure for individuals at comparable trophic levels but with different food sources.

For decades, bird eggs from the Laurentian Great Lakes have been collected and analysed for persistent organochlorines[80], and temporal trends in these data were interpreted as real changes in contaminant inputs to these systems. Some of these measurements were done on eggs from migratory birds assuming that the contaminants and nutrients present in these eggs were obtained locally and not from the birds' overwintering grounds. Recently, sulfur isotope measurements have shown that these eggs were formed from local foodstuffs and can therefore be used to interpret contaminant trends in the Great Lakes[81]. As in this study, tissue stable carbon, nitrogen or sulfur isotope ratios may enable researchers to choose an appropriate species for biomonitoring studies.

In conclusion, there appears to be significant impetus to incorporate measurements of stable isotope ratios in future biomagnification studies. Stable carbon, nitrogen and sulfur isotope ratios provide a method to accurately quantify the trophic positioning of organisms both within a system, and among food webs differing in species composition. As such, the magnitude of contaminant transfer from one trophic level to another, and the relative contribution of different food sources to a predator's contaminant burdens may be ascertained. Concentrations of Hg and most persistent organochlorines in biota were significantly related to their trophic position, as determined by $\delta^{15}N$, through both marine and freshwater food webs. The magnitude of the slope of this relationship appears to be related to the nature of the compound, with more lipophilic contaminants having greater slopes than less lipophilic contaminants. Although measurements of $\delta^{13}C$ and $\delta^{34}S$ have not been widely applied in contaminant research, they may be used to quantify differences in resource use and, as with $\delta^{15}N$, provide significant insights into the processes underlying the biomagnification of contaminants through the food web.

References

1. B. J. Peterson and B. Fry, "Stable isotopes in ecosystem studies" *Ann. Rev. Ecol. Syst.* **18**, 293–320 (1987).
2. R. France, "Critical examination of stable isotope analysis as a means for tracing carbon pathways in stream ecosystems" *Can. J. Fish. Aquat. Sci.* **52**, 651–656 (1995).
3. R. E. Hecky and R. H. Hesslein, "Contributions of benthic algae to lake food webs as revealed by stable isotope analysis" *J. N. Am. Benthol. Soc.* **14**(4), 631–653 (1995).
4. M. A. Ramsay and K. A. Hobson, "Polar bears make little use of terrestrial food webs: evidence from stable-carbon isotope analysis" *Oecologia* **86**, 598–600 (1991).
5. M. Minagawa and E. Wada, "Stepwise enrichment of ^{15}N along food chains. Further evidence and the relation between $\delta^{15}N$ and animal age" *Geochim. Cosmochim. Acta* **48**, 1135–1140 (1984).
6. K. A. Kidd, D. W. Schindler, R. H. Hesslein and D. C. G. Muir, "Correlation between stable nitrogen isotope ratios and concentrations of organochlorines in biota from a freshwater food web" *Sci. Total Environ.* **160/161**, 381–390 (1995).
7. B. G. Oliver and A. J. Nimi, "Trophodynamic analysis of polychlorinated biphenyl congeners and other chlorinated hydrocarbons on the Lake Ontario ecosystem" *Environ. Sci. Technol.* **22**(4), 388–397 (1988).
8. D. L. Swackhamer and R. S. Skoglund, "Bioaccumulation of PCBs by algae: kinetics versus equilibrium" *Environ. Toxicol. Chem.* **12**(5), 831–838 (1993).
9. D. Mackay, "Correlation of bioconcentration factors" *Environ. Sci. Technol.* **16**, 274–276 (1982).
10. D. C. G. Muir, C. A. Ford, N. P. Grift, D. A. Metner and W. L. Lockhart, "Geographical variation of chlorinated hydrocarbons in burbot (*Lota lota*) from remote lakes and rivers in Canada" *Arch. Environ. Contam. Toxicol.* **19**, 530–542 (1990).
11. J. R. Kucklick, H. R. Harvey, P. H. Ostrom, N. E. Ostrom and J. E. Baker, "Organochlorines in the water and biota of Lake Baikal, Siberia" *Environ. Sci. Technol.* **28**(1), 31–37 (1994).
12. M. S. Evans, G. E. Noguchi and C. P. Rice, "The biomagnification of polychlorinated biphenyls, toxaphene and DDT compounds in a Lake Michigan offshore food web" *Arch. Environ. Contam. Toxicol.* **20**, 87–93 (1991).

13. D. C. G. Muir, M. D. Segstro, K. A. Hobson, C. A. Ford, R. E. A. Stewart and S. Olpinski, "Can seal eating explain elevated levels of PCBs and organochlorine pesticides in walrus blubber from eastern Hudson Bay (Canada)?" *Environ. Pollut.* **90**(3), 335–348 (1995).
14. R. V. Thomann and J. P. Connolly, "Model of PCB in the Lake Michigan lake trout food chain" *Environ. Sci. Technol.* **18**(2), 65–71 (1984).
15. J. B. Rasmussen, D. J. Rowan, Dave R. S. Lean and J. H. Carey, "Food chain structure in Ontario lakes determines PCB levels in lake trout (*Salvelinus namaycush*) and other pelagic fish" *Can. J. Fish. Aquat. Sci.* **47**, 2030–2038 (1990).
16. G. Cabana, A. Tremblay, J. Kalff and J. B. Rasmussen, "Pelagic food chain structure in Ontario lakes: A determinant of mercury levels in lake trout" *Can. J. Fish. Aquat. Sci.* **51**, 381–389 (1994).
17. D. J. R. Rowan and J. B. Rasmussen, "Why don't Great Lakes Fish reflect environmental concentrations of organic contaminants?—An analysis of between-lake variability in the ecological partitioning of PCBs and DDT" *J. Great Lakes Res.* **18**(4), 724–741 (1992).
18. M. J. Vander Zanden and J. B. Rasmussen, "A trophic position model of pelagic food webs: Impact on contaminant bioaccumulation in lake trout" *Ecological Monographs* **66**(4), 451–477 (1996).
19. G. W. Kling, B. Fry and W. J. O'Brien, "Stable isotope and planktonic trophic structure in arctic lakes" *Ecology* **73**(2), 561–566 (1992).
20. R. H. Hesslein, K. A. Hallard and P. Ramlal, "Replacement of sulfur, carbon and nitrogen in tissue of growing broad whitefish (*Coregonus nasus*) in response to a change in diet traced by δ^{34}S, δ^{13}C and δ^{15}N" *Can. J. Fish. Aquat. Sci.* **50**, 2071–2076 (1993).
21. L. L. Tieszen, T. W. Boutton, K. G. Tesdahl and N. A. Slade, "Fractionation and turnover of stable carbon isotopes in animal tissues: Implications for δ^{13}C analysis of diet" *Oecologia* (Berlin) **57**, 32–37 (1983).
22. K. W. Steele and R. M. Daniel, "Fractionation of nitrogen isotopes by animals: a further complication to the use of variations in the natural abundance of ^{15}N for tracer studies" *J. Agric. Sci.* **90**, 7–9 (1978).
23. J. L. Bada, M. J. Schoeninger and A. Schimmelmann, "Isotopic fractionation during peptide bond hydrolysis" *Geochim. Cosmochim. Acta* **53**, 3337–3341 (1989).
24. O. H. Gaebler, T. G. Vitti and R. Vukmirovich, "Isotope effects in metabolism of ^{14}N and ^{15}N from unlabeled dietary proteins" *Can. J. Biochem.* **44**, 1249–1257 (1966).
25. R. M. Kiriluk, M. R. Servos, D. M. Whittle, G. Cabana and J. B. Rasmussen, "Using ratios of stable nitrogen and carbon isotopes to characterize the biomagnification of DDE, mirex, and PCB in a Lake Ontario pelagic food web" *Can. J. Fish. Aquat. Sci.* **52**, 2660–2674 (1995).
26. K. A. Hobson and H. E. Welch, "Cannibalism and trophic structure in a high Arctic lake: insights from stable-isotope analysis" *Can. J. Fish. Aquat. Sci.* **52**, 1195–1201 (1995).
27. B. Fry, "Stable isotope diagrams of freshwater food webs" *Ecology* **72**(6), 2293–2297 (1991).
28. G. Cabana and J. B. Rasmussen, "Modelling food chain structure and contaminant bioaccumulation using stable nitrogen isotopes" *Nature* **372**(17), 255–257 (1994).
29. B. Gu, D. M. Schell and V. Alexander, "Stable carbon and nitrogen isotopic analysis of the plankton food web in a subarctic lake" *Can. J. Fish. Aquat. Sci.* **51**, 1338–1344 (1994).
30. G. Cabana and J. B. Rasmussen, "Comparison of aquatic food chains using nitrogen isotopes" *Proc. Nat. Acad. Sci.* (*USA*) **93**, 10844–10847 (1996).
31. B. Gu, C. L. Schelske and M. Brenner, "Relationship between sediment and plankton isotope ratios (δ^{13}C and δ^{15}N) and primary productivity in Florida lakes" *Can. J. Fish. Aquat. Sci.* **53**, 875–883 (1996).
32. M. L. Estep and S. Vigg, "Stable carbon and nitrogen isotope tracers of trophic dynamics in natural populations and fisheries of the Lahontan Lake system, Nevada" *Can. J. Fish. Aquat. Sci.* **42**, 1712–1719 (1985).
33. B. Peterson, B. Fry and L. Deegan, "The trophic significance of epilithic algal production in a fertilized tundra river ecosystem" *Limnol. Oceanog.* **38**, 872–878 (1993).
34. B. Fry and E. B. Sherr, "Critical examination of stable isotope analysis as a means for tracing carbon pathways in stream ecosystems" *Can. J. Fish. Aquat. Sci.* **52**, 651–656 (1995).
35. K. A. Hobson and H. E. Welch, "Determination of trophic relationships within a high Arctic marine food web using δ^{13}C and δ^{15}N analysis" *Mar. Ecol. Prog. Ser.* **84**, 9–18 (1992).
36. B. Fry and C. Armold, "Rapid ^{13}C/^{12}C turnover during growth of brown shrimp (*Penaeus aztecus*)" *Oecologia* (*Berl*) **54**, 200–204 (1982).
37. G. H. Rau, A. J. Mearns, D. R. Young, R. J. Olson, H. A. Schafer and I. R. Kaplan, "Animal ^{13}C/^{12}C correlates with trophic level in pelagic food webs" *Ecology* **64**(5), 1314–1318 (1983).
38. G. D. Farquhar, M. H. O'Leary and J. A. Berry, "On the relationship between carbon isotope discrimination and the intercellular carbon dioxide concentration in leaves" *Aust. J. Plant Physiol.* **9**, 121–137 (1982).

39. C. B. Osmond, N. Valaane, S. M. Haslam, P. Uotila and Z. Roksandic, "Comparisons of $\delta^{13}C$ values in leaves of aquatic macrophytes from different habitats in Britain and Finland: some implications for photosynthetic processes in aquatic plants" *Oecologia* (*Berl*) **50**, 117–124 (1981).
40. B. N. Smith, J. Oliver and C. McMillan, "Influence of carbon source, oxygen concentration, light intensity and temperature on $^{13}C/^{12}C$ ratios in plant tissues" *Bot. Gaz.* **137**, 99–104 (1976).
41. B. N. Smith and S. Epstein, "Two categories of $^{13}C/^{12}C$ ratios for higher plants" *Plant Physiol.* **47**, 380–384 (1971).
42. M. Junger and D. Planas, "Quantitative use of stable carbon isotope analysis to determine the trophic base of invertebrate communities in a boreal forest lotic system" *Can. J. Fish. Aquat. Sci.* **51**, 52–61 (1994).
43. G. H. Rau, "Carbon-13/carbon-12 variation in subalpine lake aquatic insects: food source implications" *Can. J. Fish. Aquat. Sci.* **37**, 742–746 (1980).
44. J. R. Keough, M. E. Sierszen and C. A. Hagley, "Analysis of Lake Superior coastal food web with stable isotope techniques" *Limnol. Oceanogr.* **41**(1), 136–146 (1996).
45. J. O. Nriagu and H. H. Harvey, "Isotopic variation as an index of sulphur pollution in lakes around Sudbury, Ontario" *Nature* **274**, 883–885 (1978).
46. J. O. Nriagu and R. D. Coker, "Sulphur in sediments chronicles past changes in lake acidification" *Nature* **303**, 692–694 (1983).
47. R. H. Hesslein, M. J. Capel, D. E. Fox and K. Hallard, "Stable isotopes of sulfur, carbon and nitrogen as indicators of trophic level and fish migration in the lower Mackenzie River basin, Canada" *Can. J. Fish. Aquat. Sci.* **48**, 2258–2265 (1991).
48. B. J. Peterson, R. W. Howarth and R. H. Garritt, "Sulfur and carbon isotopes as tracers of salt-marsh organic matter flow" *Ecology* **67**(4), 865–874 (1986).
49. C. Rolff, D. Broman, C. Naf and Y. Zebuhr, "Potential biomagnification of PCDD/Fs—New possibilities for quantitative assessment using stable isotope trophic position" *Chemosphere* **27**(1–3), 461–468 (1993).
50. D. Broman, C. Naf, C. Rolff, Y. Zebuhr, B. Fry and J. Hobbie, "Using ratios of stable nitrogen isotopes to estimate bioaccumulation and flux of PCDDs and dibenzofurans (PCDFs) in two food chains from the northern Baltic" *Environ. Toxicol. Chem.* **11**, 331–345 (1992).
51. R. B. Spies, H. Kruger, R. Ireland and D. W. Rice, "Stable isotope ratios and contaminant concentrations in a sewage-distorted food web" *Mar. Ecol. Prog. Ser.* **54**, 157–170 (1989).
52. W. M. Jarman, K. A. Hobson, W. J. Sydeman, C. E. Bacon and E. B. McLaren, "Influence of trophic position and feeding location on contaminant levels in the Gulf of the Farallones food web revealed by stable isotope analysis" *Environ. Sci. Technol.* **30**(2), 654–660 (1996).
53. W. M. Jarman, W. J. Sydeman, K. A. Hobson and P. Bergqvist, "Levels of polychlorinated dibenzo-*p*-dioxins (PCDDs) and polychlorinated dibenzofurans (PCDFs) and their relationship to stable-nitrogen isotope abundance in marine birds and mammals from the Gulf of the Farallones National Marine Sanctuary" *Environ. Toxicol. Chem.* **16**, 1010–1013 (1997).
54. K. A. Kidd, D. W. Schindler, D. C. G. Muir, W. L. Lockhart and R. H. Hesslein, "High concentrations of toxaphene in fishes from a subarctic lake" *Science* **269**, 240–242 (1995).
55. J. Yoshinaga, T. Suzuki, T. Hongo, M. Minagawa, R. Kawabe, T. Inaoka and T. Akimichi, "Mercury concentration correlates with the nitrogen stable isotope ratio in the animal food of Papuans" *Ecotoxicol. Environ. Safety* **24**, 37–45 (1992).
56. K. A. Kidd, R. H. Hesslein, R. J. P. Fudge and K. A. Hallard, "The influence of trophic level as measured by $\delta^{15}N$ on mercury concentrations in freshwater organisms" *Water, Air Soil Pollut.* **80**, 1011–1015 (1995).
57. D. C. G. Muir, R. J. Norstrom and M. Simon, "Organochlorine contaminants in arctic marine food chains: Accumulation of specific polychlorinated biphenyls and chlordane-related compounds" *Environ. Sci. Technol.* **22**(9), 1071–1079 (1988).
58. R. J. Norstrom, M. Simon, D. C. G. Muir and R. E. Schweingburg, "Organochlorine contamination in arctic marine food chains: Identification, geographical distribution and temporal trends in polar bears" *Environ. Sci. Technol.* **22**, 1063–1071 (1988).
59. K. A. Hobson, D. M. Schell, D. Renouf and E. Noseworthy, "Stable carbon and nitrogen isotopic fractionation between diet and tissues of captive seals: implications for dietary reconstructions involving marine mammals" *Can. J. Fish. Aquat. Sci.* **53**, 528–533.
60. M. J. DeNiro and S. Epstein,"Mechanism of carbon isotope fractionation associated with lipid synthesis" *Science* **197**, 261–263 (1977).
61. C. A. Stow, "Factors associated with PCB concentrations in Lake Michigan salmonids" *Environ. Sci. Technol.* **29**, 522–527 (1995).
62. U. Borgmann and D. M. Whittle, "Contaminant concentration trends in Lake Ontario lake trout (*Salvelinus namaycush*): 1977 to 1988" *J. Great Lakes Res.* **17**(3), 368–381 (1991).
63. C. P. Madenjian, S. R. Carpenter and P. S. Rand, "Why are the PCB concentrations of Salmonine individuals from the same lake so highly variable?" *Can. J. Fish. Aquat. Sci.* **51**, 800–807 (1994).

64. J. Hammar, P. Larsson and M. Klavins, "Accumulation of persistent pollutants in normal and dwarfed arctic char (*Salvelinus alpinus* sp. complex)" *Can. J. Fish. Aquat. Sci.* **50,** 2574–2580 (1993).
65. P. Larsson, L. Olka and L. Collvin, "Reproductive status and lipid content as factors in PCB, DDT and HCH contamination of a population of pike (*Esox lucius* L.)" *Environ. Toxicol. Chem.* **12**, 855–861 (1993).
66. R. V. Thomann, "Bioaccumulation model of organic chemical distribution in aquatic food chains" *Environ. Sci. Technol.* **23**, 699–707 (1989).
67. L. R. Suntio, W. Y. Shiu, D. Mackay, J. N. Seiber and D. Glotfelty, "Critical review of Henry's Law Constants for Pesticides" *Rev. Environ. Contam. Toxicol.* **103**, 1–60 (1988).
68. J. F. de Bruijn, Bussers, W. Seinen and J. L. M. Hermens, "Determination of octanol/water partition coefficients for hydrophobic organic chemicals with the 'slow-stirring' method" *Environ. Toxicol. Chem.* **8**, 499–512 (1989).
69. M. A. Saleh, "Toxaphene: Chemistry, biochemistry, toxicity and environmental fate" *Rev. Environ. Contam. Toxicol.* **118**, 1–85 (1991).
70. P. H. Howard, *Handbook of Environmental Fate and Exposure Data for Organic Chemicals*, **Volume III** (Lewis Publishers, Michigan, 1991).
71. D. Mackay, W. Ying Shiu and K. Ching Ma, *Illustrated Handbook of Physical-chemical Properties and Environmental Fate for Organic Chemicals*, **Volume I** (Lewis Publishers, Inc., Michigan, 1992).
72. W. Vetter and B. Luckas, "Theoretical aspects of polychlorinated bornanes and the composition of toxaphene in technical mixtures and environmental samples" *Sci. Total Environ.* **160/161**, 505–510 (1995).
73. G. A. Stern, D. C. G. Muir, C. A. Ford, N. P. Grift, E. Dewailly, T. F. Bidleman and M. D. Walla, "Isolation and identification of two major recalcitrant toxaphene congeners in aquatic biota" *Environ. Sci. Technol.* **26**, 1838–1840 (1992).
74. T. F. Bidleman, R. L. Falconer and M. D. Walla, "Toxaphene and other organochlorine compounds in air and water at Resolute Bay, NWT, Canada" *Sci. Total Environ.* **160/161**, 55–63 (1995).
75. K. A. Hobson and S. G. Sealey, "Marine protein contributions to the diet of Northern Saw-whet owls on the Queen Charlotte Islands: A stable-isotope approach" *The Auk* **108**(2), 437–440 (1991).
76. R. France, "Source variability in $\delta^{15}N$ of autotrophs as a potential aid in measuring allochthony in freshwaters" *Ecography* **18**(3), 318–320 (1995).
77. R. J. Smith, K. A. Hobson, H. N. Koopman and D. M. Lavigne, "Distinguishing between populations of fresh- and salt-water harbour seals (*Phoca vitulina*) using stable-isotope ratios and fatty acid profiles" *Can. J. Fish. Aquat. Sci.* **53**, 272–279 (1996).
78. C. E. Hebert and G. D. Haffner, "Habitat partitioning and contaminant exposure in Cyprinids" *Can. J. Fish. Aquatic. Sci.* **48**, 261–266 (1991).
79. R. A. Mathers and P. H. Johansen, "The effects of feeding ecology on mercury accumulation in walleye (*Stizostedion vitreum*) and pike (*Esox lucius*) in Lake Simcoe" *Can. J. Zool.* **63**, 2006–2012 (1985).
80. K. E. Pettit, C. A. Bishop, D. V. Weseloh and R. J. Norstrom, "An atlas of contaminants in the eggs of fish-eating colonial birds of the Great Lakes (1989–1992)" *Can. Wild. Ser. Tech. Rep. Ser.* No. 193 (Canadian Wildlife Service, Ontario Region, 1994).
81. K. A. Hobson, K. D. Hughes and P. J. Ewins, "Using stable-isotope analyses to identify endogenous and exogenous sources of nutrients in eggs of migratory birds: Applications to Great Lakes contaminants research" *The Auk* **114**, 467–478 (1997).

18. ENVIRONMENTAL RISK ASSESSMENT FOR THE INTERACTION BETWEEN AGRICULTURAL LAND AND SURFACE WATERS

T. M. ADDISCOTT and P. SMITH *

INTRODUCTION

There is much talk about risk but a poor understanding of it by the public. Politicians tend to be swayed by forceful lobbying as much as by scientific evidence, and as a result, environmental decisions can easily be taken on the wrong basis[1]. In the UK, for example, we have a stringent and very costly limit on nitrate in potable water, but no random breath-testing for drunk drivers, implying a set of priorities very much at variance with the mortality statistics. No-one in the UK has been killed by nitrate in the mains water supply, and there is now evidence[2] that nitrate intake is beneficial. Far too many people have been killed by drunken drivers. There is a far more stringent limit on pesticides in potable water. This is obviously prudent, but it seems to have been imposed without any consideration of the fact that practically all our pesticide intake comes from naturally-produced pesticides in plants. In fact, 99.99% of all known pesticides are produced quite naturally by plants wishing to repel insects. Cabbage, for example, contains 49 natural pesticides, and farmers would not be allowed to use some plant-produced pesticides because they can cause cancer in laboratory rats[3]. Coffee is particularly rich in rodent carcinogens, and there is certainly no point in worrying about synthetic pesticides in the water used to brew it. However, the reality is that many people worry, and even campaign, about pesticides in water, while few even think about the natural chemicals in coffee. Risk assessment might help the worriers to direct their energies to greater effect.

Risks to humans are not the only consideration. Most species in the environment do not have any choice about the substances to which they are exposed, particularly those thrust upon them by human activity. There is therefore a moral imperative that demands that we should think very carefully about the risks to which we are subjecting other species, and there is also an aesthetic imperative that requires that we leave the natural world in good condition for our own enjoyment and that of posterity.

These essentially altruistic reasons for risk assessment are backed by political and economic considerations. The resources available for keeping the environment wholesome are limited and must be used where the risks are most serious. This can only be achieved if risks can be assessed realistically. Technological advance is widely perceived as a source of risk, despite the fact that new technologies are almost always safer than old ones. The synthetic pyrethroid insecticides are, for example, surely preferable to the arsenic compounds used earlier in the century. Perhaps the real problem technological advance has brought in the present context is that chemicals and other substances can now be measured, and therefore worried about, at far smaller concentrations than before. The EC limit for any one pesticide in potable water was set at 0.01 ppm, which was the limit of detection at the time. This

* IACR-Rothamsted, Harpenden, Herts AL5 2JQ (UK).

concentration thus seems to have been set on the basis of analytical chemistry rather than biology or risk assessment. If too many decisions of this nature are made without proper risk assessment, we could find ourselves in an over-rigid regulatory system, that inhibits activity without contributing to safety. The reality is that there is no such thing as absolute safety, and that all we can do is to try to assess risks on as logical a basis as possible and make sure that we optimize the balance between productive or enjoyable activity and the risks that it entails.

DEFINITIONS

Certain terms need to be defined clearly for the purposes of risk assessment, notably 'Hazard' and 'Risk', and we sought definitions first in two dictionaries.

The definitions of 'hazard' as a noun given by the Shorter Oxford Dictionary seemed to owe much to the gaming and billiards tables and to the tennis court and golf course. The most relevant offering was 'Risk of loss or harm', but this included the other word to be defined. The Dictionary of Environmental Science and Technology[4] did not list 'hazard' *per se*, but it gave the United States Environmental Protection Agency's definition of a hazardous pollutant.

Hazardous Pollutant—One to which even slight exposures may cause serious illness or death.

With respect to the definition of 'risk' as a noun, the Shorter Oxford Dictionary gave 'hazard' as the first option, which was again not helpful because it was the other word to be defined. This was followed by 'danger' and 'exposure to mischance or peril'. The Dictionary of Environmental Science and Technology[4] gave a more useful definition:

Risk—The chance that some undesired event or effect will occur.

We do not feel that, for the purposes of risk assessment, the words 'hazard' and 'risk' are as synonymous as the Shorter Oxford Dictionary seems to imply. We also feel that a more specific definition of 'hazard' is needed, and that the definition of 'risk' must include the word 'probability', because while there are some substances to which exposure carries the certainty of ill effect or death, there are many others for which the risk can only be assessed in terms of statistical probability. Another good reason for assessing risk in this way is that using a statistical basis allows risks to be compared. It can be instructive, when considering a risk, to see how it compares with a risk such as travelling in a car on the motorway that most people would take as a matter of course. Some comparisons of this nature presented by Fumento[1] show that several of the risks that cause most public concern are less than risks we take every day. The Organization for Economic Co-operation and Development (OECD) has taken an interest in risk assessment, and a workshop organized under its auspices[5] also saw value in a probabilistic approach to risk assessment. Such an approach has already been used in the USES system (Uniform System for the Evaluation of Substances) developed in the Netherlands and reviewed recently by Smith and Hart[6]. With these considerations in mind, Addiscott and Smith[7] opted for the following definitions:

Hazard—Something exposure to which may cause undesirable effects,

Risk—The probability that an undesirable effect will occur.

Politicians and the public like certainties, and expressing risks as probabilities may not appeal to them, but as Benjamin Franklin observed, in this world nothing can be said to be certain except death and taxes.

Another form of definition is that needed to establish the rules for environmental risk assessment in specific cases. This involves defining:

- what exactly is to be protected,
- whether the centre of concern is a particular species or an ecological community whose individual species may respond to pollutants in different ways,
- whether pollution needs to be considered in terms of concentration or load,
- what the standard is (e.g. an ecotoxicological "no effect" level) and how it should be determined.

TIERING

Tiering is a very useful way of making risk assessment more efficient[5,7,8]. The procedure involves making a series of appraisals of increasing intensity and detail until a definitive assessment can be made. A tiered system might start, for example, with simple look-up tables but extend if necessary to large-scale monitoring or modelling. The key point is that if an appraisal at any level beneath the highest one shows that the risk is negligible, higher-intensity appraisals can then be omitted and attention turned to the next risk. One reason for tiering can be seen in the ever-increasing number of substances entering the environment. There will never be enough money or appropriately-skilled people to assess the risks posed by all substances to all species in all environments, so it is important not to waste time and money where even a quick check can show that the risk is negligible or non-existent, particularly since this may delay or prevent the identification of a more serious risk. It is absolutely essential to establish as early as possible in the assessment whether the substance is actually toxic. This point may seem to be self-evident, but recent events suggest that it still needs to be made.

The US Environmental Protection Agency (USEPA) routinely uses a tiered risk assessment approach, and a draft scheme for hazardous air pollutants has been produced. The basic elements of this scheme, which starts with look-up tables and then proceeds to models of various levels of complexity, are shown in simplified form in Table I. The tiering includes two main elements, the emission and transport of the pollutant and the risk posed by the pollutant to the organism that is the source of concern (man), and similar concerns arise in many environmental systems. Our particular interest lies in the losses of pollutants from agricultural land, their transport to a surface water and the resulting risk to organisms in the surface water, and we shall discuss approaches to risk assessment in the context of this problem.

MODELS

The USEPA system outlined in Table I included computer models, as do several other risk assessment systems. Indeed, a working group at the OECD workshop cited above[5] decided that models were "essential tools" with several important roles in risk assessments on potential pollutants. These included the following:

- understanding the spread of the pollutant in the environment and its ultimate fate,

Table I A summary of the USEPA draft tiered risk assessment approach for emissions of potentially hazardous air pollutants. (Derived from Committee on Risk Assessment of Hazardous Air Pollutants[8])

Tier 1: Look-up tables	
Short- and long-term tables based on a model	
Inputs:	Emission rates, release height, fenceline distance
Outputs:	Maximum offsite concentrations and cancer risk (chronic and acute)
Tier 2: Screening dispersion	
Based on an EPA model	
Inputs:	Tier 1 + stack diameter, exit velocity and temperature and rural/urban classification and building dimensions
Outputs:	Maximum offsite concentration and downwind distance. Cancer risk and/or non-cancer hazard index
Tier 3: Site-specific dispersion model	
Based on two EPA models	
Inputs:	Tier 1 + Tier 2 + local meteorology, release point and fence layout, terrain features, release frequency and duration
Outputs:	Short- and long-term receptor-specific risk and hazard indices. Ambient monitoring used to enhance modelling or as alternative in difficult cases
Tier 4: Site-specific dispersion and exposure model	
Based on more advanced model	
Inputs:	Tier 3 + population model
Outputs:	Maximum offsite concentration, exposure distribution and population risk with optional characterisation of uncertainties
Personal monitoring used as an alternative on a case-by-case basis for difficult modelling applications	

- providing insights into the influence of external factors on its spread,
- aiding decisions about exposure or risk, e.g. in screening and in planning research,
- predicting environmental concentrations in the absence of monitoring data as may be necessary with new substances.

The group added that in certain circumstances "predicted concentrations from models may be more cost-effective than implementation of monitoring programmes." These four roles seem strongly relevant, particularly the fourth one; risk assessment will be practically impossible without extrapolating from the behaviour of known to unknown substances and from monitored to unmonitored sites. However, a cautionary note must be sounded about the concept of models as a cost-effective alternative to monitoring. Models can certainly be used to give added value to measurements and to extrapolate from them, but we need to be cautious about viewing them as an alternative to measurements. This is possible and often necessary, but the following three conditions must be met for it to be valid.

- The model must have been subjected to a proper, statistically-based evaluation process.
- Its parameters must have been obtained in a way that minimises the possibility that the values obtained have been influenced by any deficiencies in the model or any fitting procedures used.
- The model must be used with a clear understanding of the implications of any variability in its parameters. This usually implies spatial variability, but could mean temporal variability.

These conditions are discussed below.

1. *Evaluation of Models*

Modellers sometimes need to be reminded that 'justification by faith,' though central to St Paul's theology, has no place in modelling. Any discussion of model evaluation needs to begin with the clear understanding that no model can be unequivocally justified, or 'validated' as the conventional jargon puts it. All that can be done is to assess how small the probability is that the model has been refuted. Whether this probability is acceptable is an essentially subjective decision, but one that should be based on previous decisions about acceptable levels of probability taken *before* the evaluation. This problem was discussed elsewhere in more detail by Addiscott *et al.*[9] The probability of refutal is assessed statistically, and the approach used depends on the nature and quality of the measured data available for evaluating the model. One key question is whether or not the measurements are replicated.

If the measurements are replicated, the model can be assessed by comparing the extents to which (a) the simulations differ from the measurements and (b) the measurements differ within themselves. The model is obviously not refuted if the differences between the simulations and the measurements are no more serious than those between the replicates of the measurements. A Lack of Fit procedure developed by Whitmore[10] partitions the sum of squares of the deviations between measurement and simulation into those attributable to lack of fit and those due to pure error, and uses the F-test to assess the significance of the ratio of mean square lack of fit to mean square error, thereby enabling the probability that the model is refuted to be assessed.

There is sometimes little or no replication of the measurements, making a different approach necessary. The basis for comparison will usually be sets of measurement/simulation pairs, and we need to ask (a) to what extent the simulations are *associated* with the measurements, and (b) to what extent they *agree* with them. Association and agreement are fairly similar concepts, and it might be asked why both have to be assessed. This can be explained in terms of two hypothetical situations in which the simulation is not satisfactory. In the first, the simulation is found always to be 7.86 times the measurement, showing perfect association but very poor agreement. In the second, the mean of the simulations is exactly the same as the mean of the measurements, but plotting one against the other gives an entirely random distribution of points; Here the agreement is perfect but the association very poor. In the statistical assessment, association is measured by the product moment correlation and agreement by the mean difference between simulation and measurement, which is obtained simply by summing the individual differences and dividing by the number of pairs[11]. The model is refuted if either the correlation coefficient is *not* significant or the mean difference *is* significant according to the predetermined criteria.

There are several other ways of assessing the quality of simulations, and accounts of these are provided by Loague and Green[12] and Smith *et al.*[13]. One problem that affects all the procedures is that evaluation of models becomes increasingly difficult as the area over which they are applied increases[9], and we cannot, unfortunately, assume that an evaluation of a model made at one scale will necessarily be valid at a larger or smaller one. An additional problem discussed below and by Addiscott *et al.* [9] is that failure to take account of the interaction between any non-linearity in the model and the intrinsic variability of its parameters could invalidate the evaluation.

2. *Obtaining Reliable Parameter Values*

The reliability of parameters is just as important as the validity of models, but obtaining reliable parameter values may be a considerable problem in many of the contexts in which we wish to make risk assessments. There are several ways of obtaining parameter values, but they are not all of equal status[9] or equally practicable. They also differ in the extent to which the model has to be involved in the evaluation of the parameter through a fitting procedure.

(i) The best option is direct measurement without any fitting procedure. The value is obtained without any risk that it has been influenced by deficiencies in the model or the fitting procedure. However, for soil parameters in particular, the variance of the value obtained may lead to complications that are discussed below. Also, if parameters have to be measured on a large scale, modelling pollutant behaviour may not be any easier than monitoring it in the field.

(ii) If the parameter cannot be measured directly, the next best option is to obtain it by fitting to data, but not data of the type to be simulated. This will often involve *direct fitting*, in which a parameter used in an expression within the model is obtained by fitting to measurements that relate directly to that expression. The resulting value will be influenced by any inadequacies in the expression or the fitting procedure but not by the shortcomings of other parts of the model.

(iii) Where direct fitting is not possible, we may need to resort to *indirect fitting* procedures that involve the whole model. This sometimes involves fitting all the parameters at the same time, which may yield several sets of values, none of which gives a clearly better fit than the rest. Even if only one parameter is obtained by fitting, the rest being measured, the value obtained will be influenced by any deficiency anywhere in the model.

(iv) Some modellers obtain parameter values by *fitting to the data* to be *simulated*. This practice is not helpful in the present context.

In general, there is a decrease in real information in the model as we proceed from Option (i) to Option (iv). There is an additional problem with all fitting procedures that non-linearity in the model can lead to errors in the parameter values obtained[9]. Non-linearity is explained below. The criteria for evaluating models for risk assessment clearly need to include the number of parameters the model needs and the ease and reliability with which they can be obtained.

3. *The Implications of Parameter Variability*

Many risk assessments will have to be made in contexts in which variability in the soil or vegetation will lead to variability in the parameters of models. This needs to be taken into account, because even the simplest mathematical functions may have some surprises for us when we examine the effects of variability in their parameters. Some very simple functions are evaluated with replicated parameter values in Table II in two ways,

- by evaluating the function for each parameter value and then averaging,
- by averaging the parameter values and then evaluating the function.

For only one of the functions do the two procedures give the same result, and this function is also the only one that gives a straight line when plotted. Also, the discrepancy vanishes for the functions that can be linearized by taking logarithms when this is done. This introduces the important point that the discrepancy between the ‘evaluate first’ and ‘average first’ procedures occurs with *non-linear functions or models* and *parameters that vary*[9]. What happens in the evaluation of simple relations such as those in Table II can also happen when computer models are run for the purpose of risk assessment. This implies that there is a risk of a misleading result from using a non-linear model with a single-valued parameter based on a property of the soil or any other component of the environment that varies from point to point. The root of the problem is an interaction between the non-linearity and the variance of its parameters[14], which can be put as a simple equation:

$$S(\text{evaluate first}) = S(\text{average first}) + I,$$

where S represents the simulated value and I the interaction. This shows why there can be a problem in evaluating and parameterizing models and why the interaction is something we need to consider when choosing models for use in risk assessment systems. The problem for evaluating models can be seen directly from the equation. The true value of S is that obtained by evaluating first, but the evaluation is more likely to be made with the 'average first' value. If this is so, the term I will lead to a faulty evaluation. Essentially the same problem arises if the model is run in a risk assessment without taking account of the variance of the parameters.

The problem for parameterization can be seen by rearranging the equation:

$$S(\text{average first}) = S(\text{evaluate first}) - I.$$

Parameterization usually involves putting in a series of trial values for the parameter and selecting the one that gives the best fit of the model to the data used. These trial values are usually put in with no allowance for any inherent variance they may have, so S(average first) corresponds to the use of such values. However, S(evaluate first), the true value, is likely to be closer to the data used for the parameterization, implying that the term I will lead to a faulty evaluation of the parameter if no account is taken of the variance.

Risk assessments involving substantial areas of land will almost always need more data than are available, necessitating the use of special procedures for spatial

Table II Evaluation of some simple functions of x that contain a parameter A that is replicated, using two procedures: (1) evaluate function for all A, then average (evaluate first); (2) average the values of A, then evaluate function (average first). Values of A are 1, 2 ,3. Mean of A is 2. Functions evaluated for $x=4$

Function	Procedure	Result
$f(x)=3Ax$	Evaluate first	24
	Average first	24
$f(x)=x^{2A}$	Evaluate first	1456*
	Average first	256
$f(x)=\ln(x/A)$	Evaluate first	0.789
	Average first	0.693
$f(x)=e^{Ax}$	Evaluate first	55263*
	Average first	2981

* Discrepancy between 'evaluate first' and 'average first' vanishes when logarithm is taken.

averaging and interpolation that take account of whether the variability is random or whether it increases with distance. The discrepancy between 'evaluate first' and 'average first' extends to these procedures, and needs to be taken into account when models are used with parameters obtained with them. Models are quite often used with "effective" parameter values obtained by fitting the model over large areas. Unless the model is linear with respect to the parameter, the results of such an exercise are likely to be misleading, and may not be very useful for risk assessment.

It seems clear that for risk assessment purposes it will be advisable to aim to use models that are either fairly linear with respect to their parameters or whose parameters are not too variable.

PROBABILITY

We defined risk as the probability that an undesirable effect will occur. To compute this probability, we need to consider all the uncertainties acting in the ecosystem in which we are interested. In terrestrial ecosystems, these may arise from the weather, the soil, the way the land is used and the nature and behaviour of the pollutants identified as potential hazards. They cannot all be treated the same way, because weather varies in time while soil properties vary mainly in space. Land use may vary in both, depending on the scales involved. The nature of a particular pollutant is fixed, but its behaviour may well depend on its precise location within the ecosystem. Its behaviour will obviously differ between air, water and soil, but in soil, for example, may well be influenced by variations in the acidity of the soil.

The simplest way of taking account of the year-to-year variability in weather is simply to assemble weather files for a number of site/year combinations within a given region and run the model for each of them. This gives a distribution of computed values for the relevant output from which the mean and median values can be obtained, together with the 'worst case' value, which might be the upper or lower quintile value. The same effect can be obtained in a more sophisticated way by using a stochastic weather generator.

The soil properties used as parameters in models differ greatly in their degree of variability, particularly the soil water properties that influence the transport of pollutants. The hydraulic conductivity can vary by one to three orders of magnitude over a fairly short range, but the volumetric moisture content is far less variable and has a coefficient of variation of 10–15%. These two properties are used in somewhat different types of transport model. The hydraulic conductivity is a measure of *rate* of movement, and it is used in models that are described as 'rate' or 'mechanistic' models. 'Mechanistic' implies that the model incorporates the most fundamental description of the process possible with the current level of understanding, and this type of model is exemplified by the LEACHM model of Wagenet and Hutson[15].

The volumetric water content is a measure of the soil's *capacity* to hold water and thence to hold solutes back against leaching. There are several models that use capacity parameters derived from the volumetric moisture content to simulate solute transport. These 'capacity' models are usually simpler than the rate models and do not set out to be fully mechanistic. Addiscott and Wagenet[16] categorized them as 'functional' models. Examples include the models of Burns[17] and Addiscott[18].

Both types of model have potential uses in risk assessment. Rate models will be relevant where the mechanism of transport has a bearing on the risk, but their dependance on the hydraulic conductivity is a problem in unsaturated soils, in which it is very difficult to measure. Most soils are unsaturated, and obtaining a distribution of values involves a great deal of work. One way round this problem is to use the rate model with a 'pedo-transfer function'[19], which is a means of summarizing information about a particular soil unit. Comparing rate and capacity models does not usually show the complexity resulting from the mechanistic nature of the former result to give better simulations than those given by the simpler capacity models[20,21]. The latter may be more useful for many risk assessments, because estimates of the volumetric water content are usually available or readily calculable and its coefficient of variation can be assumed to be 10% without undue error.

The choice of model also depends on the relative importance of rainfall variability and soil variability. If the weather variability is dominant, the precise mechanism of transport probably does not matter, and a capacity model will be just as useful as a rate model. This view was taken by Hutson[22]. However, the model chosen must allow for preferential flow in soils in which it occurs. This is a particular problem in heavy clay soils. They crack when dry, forming pathways for preferential flow, and they usually have field drains to carry away excess water to a ditch or stream. A crack that intersects a field drain provides a very rapid transit for a pesticide or other pollutant from the soil surface to the watercourse. Worm and root holes also provide pathways for preferential flow.

Losses of pollutants from land to water obviously depend on the purpose for which the land is used. Nitrate and phosphate are lost during most forms of agricultural production, but pesticides mainly from arable or horticultural land. The importance of variability of use obviously depends on the scale at which the risk assessment is made. A single field usually carries a single crop, although the use may change from year to year. With a catchment or region, however, we may need to consider not only a multiplicity of crops but also several types of land use, arable, grassland and forest, for example. Set-aside is often a complicating factor, not least because it takes more than one form. Gathering cropping statistics seems likely to be one of the largest problems in a large-scale risk assessment.

Pollutants do not change their behaviour *per se* if moved in space, but pesticide sorption depends strongly on the pH and organic carbon content of the soil[23], and that of phosphate on pH[24]. These properties do vary spatially, implying that the behaviour of the pollutants will do so too, and this may be a consideration for risk assessment, particularly where land has not been subjected to regular applications of lime.

ASSESSING THE PROBABILITY OF POLLUTANT LOSSES FROM LAND TO WATER

We need to find a way of pooling the information about the weather, the soil and other contributing factors that allows us to draw from them a useful assessment of risk. The weather information is in the site/year combinations and that for the soil in the probability distributions of the parameters of the transport model. The model can be run with one or more parameters represented by probability distributions to yield outputs in the form of probability distributions. This may be done using a Monte Carlo procedure or the 'sectioning method' of Addiscott and Wagenet[25], which divides each distribution into sections and runs the model with all combinations of the section medians. It is possible in principle to run a transport model for every weather file (real or generated), for each land use within an area, with the soil transport parameters as probability distributions; distributions of pH and soil organic carbon could be included if appropriate. A *very* large number of output values would result, from which median and 'worst case' values could be obtained. The procedure, though quite feasible, is very demanding in terms of resources, and needs to be curtailed if possible. We need to find out whether any particular input dominates the variance of the output from the model to the extent that its probability distribution is the only one that needs to be considered. We are currently investigating two of the models we use regularly, the SLIM leaching model[26] and the SUNDIAL nitrogen turn-over model[27,28] to see which inputs or parameters dominate the variance of the output. Preliminary results suggest that the weather, specifically rain fall variability, is responsible for most of the variance in leaching losses simulated by the leaching model but that soil parameters play a larger role in the output from the turn-over model.

Assuming that one form of variability is dominant is one way of lessening the load of computing. Another seemingly plausible way of cutting the load is to take the 'worst case', that is, upper or lower quintile, parameter values and combine them through the model to obtain the 'worst case' output values. Referring to the earlier section on parameter variability shows that this might work if the model was linear with respect to all its parameters, but could be very misleading if the model was not linear. The same problem would arise with the medians.

Research on risk assessment has been in progress at Rothamsted for some years, albeit under a different heading. In 1990, we developed a simple expert system, CRAFTNEL, that estimated the effects of various combinations of antecedent and current crops on the losses and concentrations of nitrate in water draining from the soil during autumn, winter and spring. The values were generated by running the SLIM leaching model[26] with simple models for mineralisation and crop uptake. CRAFTNEL was demonstrated at the Royal Show (England's premier agricultural event) but not published. Table III shows some output from the system, but it is a very small proportion of the information generated with the models for the expert system. The difference between the 'worst case' and the median varied considerably.

Table III A page from the CRAFTNEL system. Median and 'worst case' nitrate-N losses and nitrate concentrations (average) for autumn, winter and spring and the whole year. The values take account of region, current and antecedent crops, soil type and weather, but allow for variability in weather only. (Unpublished study by T. M. Addiscott, G. Tuck and N. J. Bailey). The example is for a retentive (clay) soil in the Western Region of England

		Autumn	Winter	Spring	Year
Winter wheat after winter wheat					
N loss (kg/ha)	Median	4	3	32	40
	'Worst case'	4	4	38	47
Nitrate conc. (mg/l)	Median	29	22	100	61
	'Worst case'	37	30	138	83
Winter wheat after potatoes					
N loss (kg/ha)	Median	6	3	36	47
	'Worst case'	8	7	47	64
Nitrate conc. (mg/l)	Median	62	23	138	87
	'Worst case'	81	45	177	101

ASSESSING THE RISKS TO ORGANISMS POSED BY POLLUTANT CONCENTRATIONS

Up to now, we have been discussing how best to assess the concentrations of pollutants likely to occur in water that drains from the soil and makes its way into surface waters. We also need to consider the effects of these pollutant concentrations on organisms living in the surface waters. Most procedures seem to involve subjecting the organism to increasing concentrations of the potential pollutant until it shows some kind of adverse effect. The next concentration below that giving the adverse effect is selected. For humans exposed to toxic substances in the workplace, this was described originally as the "threshold limit value" (TLV)[8]. As the tests became more systematized, the TLV became the "no-observable-effect level" (NOEL). Subsequently 'observable' was changed to 'observed', without any change of acronym, but later on the word 'adverse' was added to give "no-observed-adverse-effect level" (NOAEL), a less satisfactory acronym but a more precise definition[8]. The level that gave the adverse effect is also recognized as the "lowest-observed adverse-effect level" (LOAEL). The ultimate adverse effect is death, and this is recognised in the frequently used LD50 statistic, which gives the concentration at which 50% of the organisms die. Much of the testing that has been done seems to have been concerned with human safety, and an additional term, the acceptable daily intake (ADI) was set by dividing the NOEL by a safety factor of 100. This consideration does not usually seem to be extended to other species.

Those concerned with human safety enjoy the great advantage that they are dealing with a single organism, and environmental risk assessment is rather more complicated. We have to consider a range of organisms that may differ appreciably in their sensitivity to pollutants, and which are integrated in food webs. Thus a pollutant accumulated by one organism may be transferred to another, and a decline in the population of one lessens the food supply of another. An increase in the population of one species may also have adverse effects on another, by depriving it of food, oxygen or light, for example. We really need to be able to define NOAELs for ecosystems, but most of the information is for specific organisms. This problem was addressed in the report of a workshop[5] which suggested the use of micro- and mesocosms as a likely approach to this problem, but it seems clear that much still needs to be done.

Current risk assessments seem to rely heavily on certain representative or 'model' organisms. The ubiquitous laboratory rat appears frequently as a 'model' in human toxicity studies, perhaps reflecting political rather than biological realities. For freshwater ecosystems the three main 'model' organisms seem to be algae, *Daphnia* and *fish*[5], and all seem relevent. Algae are part of the food web as well as a major part of the problem of eutrophication. *Daphnia* are important food web organisms, and fish are probably the organism most desired, with trout the most desirable of all and often used in tests.

POLLUTANT LOSSES FROM AGRICULTURAL LAND TO WATER: TACKLING THE RISK ASSESSMENT PROBLEM

1. *Reasons for the Problem*

Nitrate, phosphate and pesticides have been the main sources of the concern about pollution from agricultural land. Public interest has centred mainly on the health risk believed to be posed by nitrate, and this has led to the imposition of the EC's limit on nitrate in potable water, but there are indications that this risk was probably non-existent. Infantile methaemoglobinaemia seems to be associated with water from wells rather than from the mains water supply[29,30], limiting the risk to a very small proportion of the population in the more developed countries (but not in the Third World). Furthermore, the supposed link between nitrate in water and stomach cancer has been shown to be non-existent in several epidemiological studies [e.g. 31,32], and workers in a factory making ammonium nitrate showed exactly the same mortality due to stomach cancer as workers in comparable jobs[33]. To complete the picture, recent medical evidence[2] suggests that dietary nitrate is beneficial and nitrate is probably produced in the human body. It forms the beginning of a chain of microbial and chemical reactions that form nitric oxide in the stomach and thereby destroy *Salmonella* and several other very undesirable bacteria in the gut.

There is also considerable public worry about pesticides in water supplies, but the review by Ames and Gold[3] discussed in the Introduction suggests that they are not a serious threat to human health. They may, however, be a far greater threat to the health of organisms in the aqueous environment and, if a rationale is needed for risk assessments with respect to pesticides the strongest one is probably to be found there.

Although the evidence that nitrate is a risk to health seems negligible, nutrients washed from agricultural land definitely cause problems of eutrophication in surface waters[34]. Nitrogen and phosphate not only encourage the growth of crop plants, they also encourage that of water plants to an undesirable extent[34,35]. Reeds can grow to excess, narrowing waterways and possibly overloading and damaging banks, while underwater plants foul fishing tackle and the propellers of boats. Water supply conduits can become clogged and machinery damaged. The greatest problem, however, is caused by algal blooms, which are now often seen as unsightly scum-like masses on the surface of natural waters. They are as much of a problem dead as alive, because the microbes that decompose them use oxygen needed by other species, which may decline in number or even die out. As a result, the food web and ecological balance of the stream or lake change, with the loss, in some cases, of the more desirable species of fish. The fact that some algal blooms include cyanobacterial species that are toxic to mammals, including humans and dogs[36] has added to public concern. This latter problem is not a new one; it was recorded as long ago as 1853[36].

To control algal blooms we need to know which nutrient, nitrogen or phosphate, most influences their formation. Evidence reviewed recently by Ferguson *et al.*[34] for lakes includes regression models that relate measures of phytoplankton, such as chlorophyll content, to the quantity or concentration of each nutrient in the lake. Relationships can be found for both nutrients, but that for phosphate is clearly the stronger and extends over five orders of magnitude in phosphate availability. These authors and most other limnologists consider that phosphate has the greater influence of the capacity of a lake to sustain phytoplankton.

The overall conclusion about these 'problem nutrients' seems to be that nitrate is a threat to health only when there is no mains water supply, and to natural waters only when phosphate is present. Indeed, it could be argued that were it not for EC legislation there would be no nitrate problem. However, assuming that what is true of lakes is broadly true of rivers and streams, phosphate is a very definite threat to surface waters. This suggests that there may be different considerations for protecting underground aquifers and surface waters. Because of the EC limit for nitrate in potable water, the aquifers need primarily to be protected from nitrate. Phosphate is not a problem, because it is not perceived as a health risk. Also, so far as the UK is concerned, it will probably be sorbed by the chalk and limestone that form the matrix of many of our aquifers for at least as long as it takes us to sort these nutrient problems. On the other hand, phosphate is an immediate threat to surface waters, and for protecting surface waters, restricting phosphate inputs will often be the main objective. There can, however, be no firm distinction, because in many water systems there is continual interaction between surface water and aquifers[37,38] and in some areas most of the water for the mains supply is drawn from rivers.

2. *Towards a Risk Assessment System*

What are the prior assumptions for a risk assessment system for pollutant losses from agricultural land to water? We assume that the water is surface rather than groundwater, so we are concerned with the welfare of fish and other organisms that inhabit it. Taking groundwater would have simplified the problem to a question of keeping the nitrate concentration within the EC limit of 50 mg/l but it would not have been useful for demonstrating the complexities of risk assessment. Surface water provides a rather more complex problem, because we are concerned with two compartments, the soil and the water, with two hazards, eutrophication and pesticides, and with two considerations in the assessment, the emission and transport of the hazard and the threat posed to the organisms in the water. There is also a question as to whether we consider one, or perhaps two, specific organisms or try to take account of the whole ecosystem and the range of vulnerabilities it encompasses. We do not yet have a risk assessment system, but we do have some ideas about what would be needed in such a system.

The system would definitely be tiered, and it would introduce models and probability at the appropriate levels. It would also assume that questions about toxicology need to be asked before questions about probable concentrations; there is no point in assessing likely concentrations of a potential pollutant before you know whether or not it is toxic to anything. The four or five tiers that we anticipate will be needed are outlined in Table IV, and details are given in the text and other tables.

Tier 1. Tier 1 is the level at which we ask whether there is a problem at all. The first question is toxicological: what evidence is there that this substance is toxic, or will lead indirectly to toxic effects, to an organism in the ecosystem? The question is unnecessary for nitrate and phosphate, because we know that both contribute to eutrophication, and that phosphate is the limiting nutrient for the formation of algal blooms in fresh water. The answer can be sought for pesticides and similar chemicals by looking up tables, making comparisons with chemically related substances and referring to expert opinion. If there is a reasonable suspicion that the sustance is toxic to one or more organisms in the ecosystem, the assessment proceeds. The second question relates to the transport of the substance into the critical compartment of the ecosystem: is there any possibility that the no-observed-adverse-effect level (NOAEL) will be reached? At this level a 'back of the envelope' type calculation may be very useful. What happens if the maximum possible amount of the substance become dissolved in the minimum rainfall likely to occur in the period of interest: is the NOAEL exceeded? If there is a reasonable suspicion that the NOAEL may be exceeded in addition to the reasonable suspicion that the substance is toxic, the assessment proceeds to Tier 2.

Tier 2. Index-based systems and expert systems have a role to play at Tier 2. The toxicological question most likely to be relevent to the eutrophication problem is whether the phosphate concentration in the water compartment exceeds the limiting value for algal bloom formation. This becomes a toxicological question in the context of the toxic effects of the decomposing blooms. The limiting concentration is probably of the order of 0.02 mg/l, but more comprehensive data should be available from the sources cited by Ferguson *et al.*[34]. Toxicological information on pesticides can be obtained from the PETE expert system (Physicochemical Evaluation: The Environment[39]), which is summarized in Table VI. The transport assessment for phosphate could be made at this level by the type of index-based system outlined in Table V, while that for pesticides is provided by the PETE expert system (Table VI).

Tier 3. At Tier 3 the toxicological question with respect to phosphate remains unchanged from Tier 2, and that concerning pesticides would consider *Daphnia* and

Table IV Outline of a tiered risk assessment for losses of nutrients and pesticides from the soil compartment to the water compartment. Each tier has toxicological (*Tox*) and transport (*Trans*) aspects

Tier 1.	Basic questions to assess whether there is a problem at all.
Tox.	What evidence is there that the substance will be toxic to any organism? Assess by reference to expert opinion, look-up tables and comparisons with related substances.
Trans.	Is there any possibility that enough of the substance will be transported to the water to cause the adverse effect? Assess through simple 'back of the envelope' calculation.
Tier 2.	Index-based and expert systems.
Tox.	Use limiting phosphate contration for algal blooms to assess eutrophication risk and PETE expert system (Table VI) to assess pesticide risks.
Trans.	Use phosphate transport index similar to that in Table V and pesticide transport assessment from PETE (Table VI).
Tier 3.	More detailed toxicology and modelling/probability approach to transport.
Tox.	Eutrophication as in Tier 2. Use *Daphnia* and trout to assess toxic risk.
Trans.	Assess risk of exceeding toxic concentrations using capacity-type transport models with probability approach based on variability of weather but not soil parameters.
Tier 4.	Ecosystem risk assessed. More detailed transport assessment.
Tox.	Use results from macrocosm studies to essess risk to whole ecosystem rather than individual species.
Trans.	As in Tier 3, but partition water flow between percolation and surface runoff. If appropriate, allow for transport on particulate and colloidal material. Include nitrate leaching model in case nitrate becomes the limiting nutrient.
Tier 5.	Field-scale ecotoxicological assessment. Mechanistic transport models with allowance for variability of soil parameters as well as weather. (Probably not practicable.)
Tox.	Use results from field-scale ecosystem study that involves monitoring and modelling.
Trans.	Mechanistic models that take account of variability of soil parameters, including those that depend on pH and organic matter.

Table V Outline of an index-based system for estimating losses of phosphate from agricultural land to water. It is anticipated that the system be broadly similar in nature to the DRASTIC system for nitrate[43], with differing weights allocated to the scores for the factors listed

- The proximity of the agricultural land to the surface water
- The slope of the land
- The susceptibility of the soil to erosion
- The nature of the land use, particularly whether arable or grassland
- The use of farmyard manure and slurry
- The long-term use of phosphate fertilisers
- The tillage practices used

Table VI Outline of the PETE expert system[39]. This assesses the likely behaviour of organic chemicals in the environment from basic physicochemical properties using a data-base that contains most of the properties of more than 600 compounds

Information provided on the behaviour of compounds in the soil compartment:
• physical chemistry
• sorption, volatilisation, biodegradation and leaching in soils
• uptake and translocation in plants.
Information provided on their behaviour in the surface water compartment:
• sorption by sediments
• LD50 and related data for freshwater organisms ranging from daphnia to trout
• bioconcentration factors for proteins, lipids, daphnia, fish muscle and whole fish, estimated from the partition between octanol and water

fish, probably trout, as in the PETE system. With respect to the transport aspects of the assessment, Tier 3 seems the appropriate level at which to introduce both models and probability principles. The simpler capacity-type transport are appropriate at this level, and the uncertainty accommodated in the probability approach should probably be limited to that arising from the weather variability. We suggest that the movement of pollutants should be considered down to the base of the rooting zone, and losses beyond this level assumed to constitute a risk, which is assessed as the probability that a specified concentration will be exceeded.

Although there are many models for nitrate leaching in the soil profile, there seem to be few for that of phosphate. This is probably a consequence of the fairly widespread perception that phosphate is so strongly sorbed by soil components such as clay and the hydrous oxides of iron and aluminium, that it does not suffer any appreciable leaching. The senior author has recently developed a simple model for phosphate leaching in the soil profile, but this has yet to be evaluated. Progress on the transport aspects of Tier 3 for eutrophication will have to await developments in modelling. There are, however, many more models for leaching of pesticides than there are for that of phosphate, and the problem is not so much to identify a model as to select the most suitable one. Although a capacity-type model will be appropriate, it will need to allow for preferential flow, which can cause pesticide losses by removing them rapidly from the topsoil before they become sorbed. The relatively old model described by Nicholls *et al.*[20] might still prove useful, because it allows for mobile and immobile water in the soil. Hall's[40] leaching model, which has been adapted for pesticides[41,42], makes more explicit allowance for rapid preferential flow but requires an additional parameter. Both these models could easily be run with multiple sets of weather data to provide a distribution of pesticide concentrations leaving the soil profile that could be used in two ways: to assess the probability that a specified concentration would be exceeded or to give median and 'worst case' concentrations.

Tier 4. At Tier 4 we need ideally to extend the ecotoxicological assessment from individual species to small ecosystems, using information from macrocosm studies where they have been made. This would be an important advance for the assessment of both phosphate and pesticide problems. We also envisage that transport processes other than movement in the soil profile will be included. Incoming rainfall will be partitioned between surface runoff and percolation into the soil. This is potentially an important upgrading, because both phosphate and pesticides can be carried into surface waters by surface runoff. This is also the tier at which lateral movement in the subsoil should be considered for both pollutants. When the capacity-type phosphate model of the type discussed for Tier 3 comes to fruition, we intend to extend it to include phosphate transport on particulate and colloidal material. This development could relevantly be included at Tier 4, and it may prove appplicable to pesticides as well as phosphate. It may also prove relevant to include a nitrate leaching model with the phosphate model for use in circumstances in which algal blooms may be limited by either nutrient.

Tier 5. We do not anticipate that a Tier 5 will be practicable. If it were attempted, the toxicological assessment might assess the effects of eutrophication and pesticides in the surface water in a field-scale ecosystem study that integrated modelling and monitoring to examine direct and indirect impacts at various trophic levels in the food chain. The transport assessment might use mechanistic models and consider the effects on phosphate and pesticide movement of variability in the pH and organic matter content of the soil.

CONCLUSIONS

To justify effort on the development of environmental risk assessment systems, we need look no further than the nitrate problem. Nitrate is not only non-toxic, it is probably good for us[2], and there is no evidence that it is directly toxic to any other organism. It is also not the limiting nutrient in our main environmental pollution problem, algal blooms in freshwater[34], although it is the main culprit in seawater. Many millions of pounds could probably have been saved if a proper risk assessment had been made 25 years ago when the problem first began to surface. Fortunately, the 'nitrate story' is not one of total disaster, because it led to the funding of valuable research on nitrogen cycling in ecosystems that has proved relevent to studies of environmentally important gaseous emissions from the soil. It would, however, have been better if the research had been done for the right reason. Assessing the risks to surface waters from agricultural pollutants is a complex matter, and we are well aware that much needs to be done before the scheme we have outlined becomes feasible, but we feel that the work would be well worthwhile. It is vital that we make every effort to deal with real threats to the health of the environment, and we can only do so effectively if we make sure that we do not waste resources on non-existent problems.

We suggested that the 'back of the envelope' calculation has a useful role in Tier 1 of the system. It can also have a certain entertainment value, as illustrated by a calculation made by the senior author. He was interested in the implications of the EC limits for nitrate and pesticides in potable water, and asked himself how lethal simazine (say) would be at a concentration ten times greater than the EC limit for any one pesticide. He took the LD50 for rats and considered himself as an 85 kg mega-rat to estimate what a lethal accumulation of simazine would be. Remembering his mis-spent youth, in which he had once managed to consume 12 pints of beer (6.8 litres) during one evening, he calculated how long he would survive drinking a daily 12 pints of beer, ten times over the EC limit in simazine. He concluded that his liver was unlikely to last the 165,000 years needed to acquire a lethal dose of simazine.

Acknowledgements

Gill Tuck and Nicola Bailey contributed to the risk assessments in Table III. IACR-Rothamsted receives grant-aided support from the Biotechnology and Biological Sciences Research Council of the United Kingdom. We have had helpful discussions with Dutch and Belgian colleagues in EC-EGXII Environment project EVSV-CT 94-0480.

References

1. M. Fumento, *Science under Siege—balancing technology and the environment* (Morrow, New York, 1993).
2. R. S. Dykhuizen, R. Frazer, C. Duncan, C. C. Smith, M. Golden, N. Benjamin and C. Leifert, "Antimicrobial effect of acidified nitrite on gut pathogens: Importance of dietary nitrate in host defence" *Antimicrobial Agents and Chemotherapy* **40**, 1422–1425 (1996).
3. B. N. Ames and L. S. Gold, "Misconceptions on pollution and the causes of cancer" *Angewandte Chemie, International version in English* **29**, 1197–1208 (1990).
4. A. Porteous, *Dictionary of Environmental Science and Technology* (John Wiley, Chichester, 1992).
5. OECD, Report of the OECD Workshop on *Environmental Hazard/Risk Assessment, London*, UK, 24–27 May 1994. (OECD Environment Monograph, 1995).
6. P. Smith and A. D. M. Hart, *Comparison of Ecological Hazard/Risk Assessment Schemes: A summary document*, (OECD Environment Monograph, 1995).
7. T. M. Addiscott and P. Smith, "Using models in risk assessment of pollutant losses from agricultural land to water" **In**: *Ecotoxicology: Responses, Biomarkers and Risk Assessment*, an OECD Workshop (J.T. Zeltikott, ed.) (SOS Publications, Fair Haven, N.J., USA, 1997) pp. 489–506.
8. Committee on Risk Assessment of Hazardous Air Pollutants, *Science and Judgement in Risk Assessment* (National Academy Press, Washington, DC, USA, 1994).
9. T. M. Addiscott, J. U. Smith and N. J. Bradbury, "Critical evaluation of models and their parameters" *J. Environ. Qual.* **24**, 803–807 (1995).
10. A. P. Whitmore, "A method for assessing the goodness of computer simulation of soil processes" *J. Soil Sci.* **42**, 289–299 (1991).
11. T. M. Addiscott and A. P. Whitmore, "Computer simulation of changes in soil mineral nitrogen and crop nitrogen during autumn, winter and spring" *J. Agric. Sci. Camb.* **109,** 141–157 (1987).
12. K. Loague and R. E. Green, "Statistical and graphical methods for evaluating solute transport models: Overview and application" *J. Contaminant Hydrol.* **7**, 51–73 (1991).
13. J. U. Smith, P. Smith and T. M. Addiscott, "Quantitative methods to evaluate and compare soil organic matter (SOM) models" **In**: *Evaluation of Soil Organic Matter models* (D. S. Powlson, P. Smith and J. U. Smith eds.) (NATO ASI Series Vol. I38, Springer Verlag, Heidelberg, 1996) pp. 181–199.
14. T. M. Addiscott, "Simulation modelling and soil behaviour" *Geoderma* **60**, 15–40 (1993).
15. R. J. Wagenet and J. L. Hutson, "LEACHM: Leaching Estimation and Chemistry Model—a process-based model of water and solute movement, transformations, plant uptake and chemical reactions in the unsaturated zone" *Continuum* **Vol. 2**. (Water Resources Institute, Cornell University, Ithaca, NY, USA, 1989).
16. T. M. Addiscott and R. J. Wagenet, "Concepts of leaching in soils: A review of modelling approaches" *J. Soil Sci.* **36**, 411–424 (1985).
17. I. G. Burns, "A model for predicting the redistribution of salts applied to fallow soils after excess rainfall or evaporation" *J. Soil Sci.* **25**, 165–178 (1974).
18. T. M. Addiscott, "A simple computer model for leaching in structured soils", *J. Soil Sci.* **28**, 554–563 (1977).
19. J. Bouma, "Using soil survey data for quantitative land evaluation" **In**: *Advances in Soil Science* **9** (B. A. Stewart ed.) (Springer, New York, USA, 1989) pp. 225–239
20. P. H. Nicholls, R. H. Bromilow and T. M. Addiscott, "Measured and simulated behaviour of fluometuron, aldoxycarb and chloride ion in a structured soil" *Pesticide Sci.* **13**, 475–483 (1982).
21. P. De Willigen, "Nitrogen turn-over in the soil–crop system: Comparison of fourteen simulation models" *Fertilizer Res.* **27**, 141–149 (1991).
22. J. L. Hutson, "Applying one-dimensional deterministic chemical fate models on a regional scale" *Geoderma* **60**, 201–212 (1993).
23. P. H. Nicholls, "Organic contaminants in soils and groundwaters" **In**: *Organic Contaminants in the Environment* (K. C. Jones ed.) (Elsevier Applied Science, London, 1991) pp. 87–132.
24. I. D. Lopez-Hernandez and C. P. Burnham, "The effect of pH on phosphate adsorption in soils" *J. Soil Sci.* **25**, 207–216 (1974).
25. T. M. Addiscott and R. J. Wagenet, "A simple method for combining soil properties that show variability" *Soil Sci. Soc. Amer. J.* **49**, 1365–1369 (1985).
26. T. M. Addiscott and A. P. Whitmore, "Simulation of solute leaching in soils of differing permeabilities" *Soil Use Manage.* **7**, 94–102 (1991).
27. N. J. Bradbury, A. P. Whitmore, P. B. S. Hart and D. S. Jenkinson, "Modelling the fate of nitrogen in crop and soil in the years following application of ^{15}N-labelled nitrogen fertilizer to winter wheat" *J. Agric. Sci. Camb.* **121**, 363–379 (1993).

28. J. U. Smith, N. J. Bradbury and T. M. Addiscott, "SUNDIAL: A PC-based system for simulating nitrogen dynamics in arable land" *Agron. J.* **88**, 38–43 (1996).
29. T. M. Addiscott, A. P. Whitmore and D. S. Powlson, *Farming, Fertilizers and the Nitrate Problem* (CAB International, Wallingford, 1991) pp. 176.
30. T. M. Addiscott, "Fertilizers and nitrate leaching" *Issues Environ. Sci. and Tech.* **5**, 1–26 (1996).
31. D. Forman, A. Al-Dabbagh and R. Doll, "Nitrate, nitrite and gastric cancer in Great Britain" *Nature* **313**, 620–625 (1985).
32. S. A. Beresford, Is nitrate in drinking water associated with gastric cancer in the urban UK? *Int. J. Epidemiology* **14**, 57–63 (1985).
33. S. Al Dabbagh, D. Forman, D. Bryson, I. Stratton and R. Doll, "Mortality of nitrate fertilizer workers" *Brit. J. Ind. Med.* **43**, 507–515 (1986).
34. A. J. D. Ferguson, M. J. Pearson and C. S. Reynolds, "Eutrophication of natural waters and toxic algal blooms" *Issues Env. Sci. Tech.* **5**, 27–41 (1996).
35. The Royal Society, *The Nitrogen Cycle of the United Kingdom: A Study Group Report* (The Royal Society, London, 1983).
36. S. G. Bell and G. A. Codd, "Detection, analysis and risk assessment of cyanobacterial toxins" *Issues Environ. Sci. Tech.* **5**, 109–122 (1996).
37. T. P. Burt, "The hydrological role of floodplains within the drainage basin system" **In**: *Buffer Zones. Their processes and potential in water protection* (N. E. Haycock *et al*., eds.) (Samara Publishing, 1997).
38. K. W. T. Goulding, L. S. Matchett, G. Heckrath, C. P. Webster, P. C. Brookes and T. P. Burt, "Nitrogen and phosphorus flows from agricultural hillslope" **In**: *Advances in Hillslope Processes* (M. G. Anderson and S. M. Brooks, eds.) (John Wiley and Sons, Chichester, 1996) pp. 213–227.
39. P. H. Nicholls, "'Physicochemical evaluation: The environment'. An expert system for pesticide preregistration assessment" *Proceedings of the Brighton Crop Protection Conference—Pests and Diseases—1994, 1994*) pp. 1337–1341.
40. D. G. M. Hall, "An amended functional leaching model applicable to structured soils. I. Model description" *J. Soil Sci.* **44**, 579–588 (1993).
41. D. G. M. Hall, "Simulation of Dichlorprop leaching in three texturally distinct soils using the Pesticide Leaching Model" *J. Environ. Sci. Health* **A29**, 1211–1230 (1994).
42. P. H. Nicholls and D. G. M. Hall, "Use of the Pesticide Leaching Model (PLM) to simulate pesticide movement through macroporous soils" *BCPC Monograph No. 62*, (1995) pp. 187–192.
43. L. Aller, T. Bennet, J. H. Lehr and R. J. Petty, *DRASTIC: A Standardised System for Evaluating Groundwater Pollution Potential using Hydrogeological Settings: USEPA No. 600/2-85/018*, (United States Environmental Protection Agency, Ada, Oklahoma, 1985) pp. 1–163.

INDEX

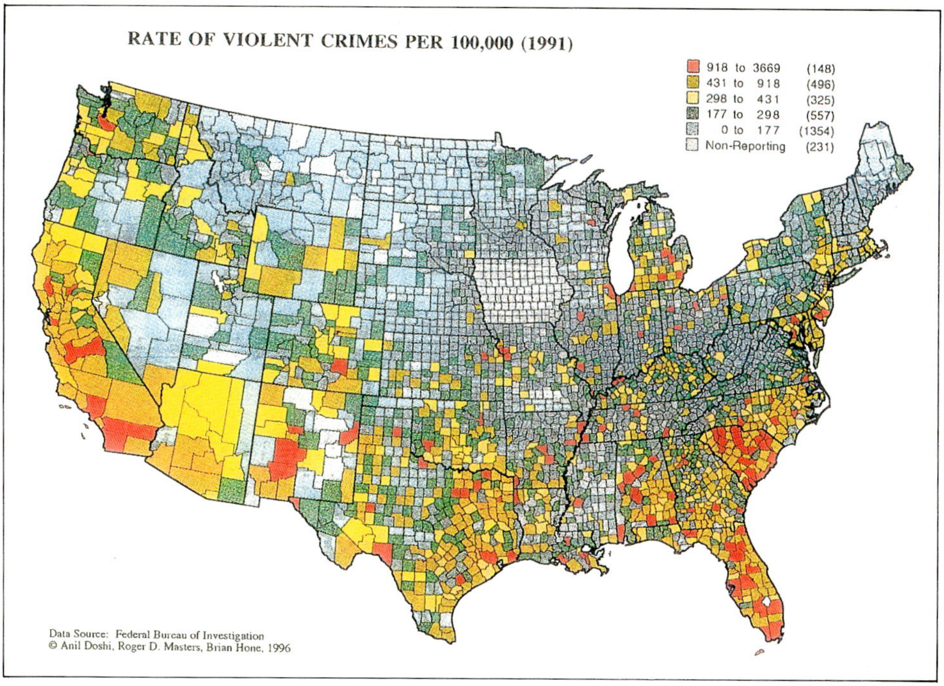

See ROGER D. MASTERS *ET AL.*, Map 1, Page 14.

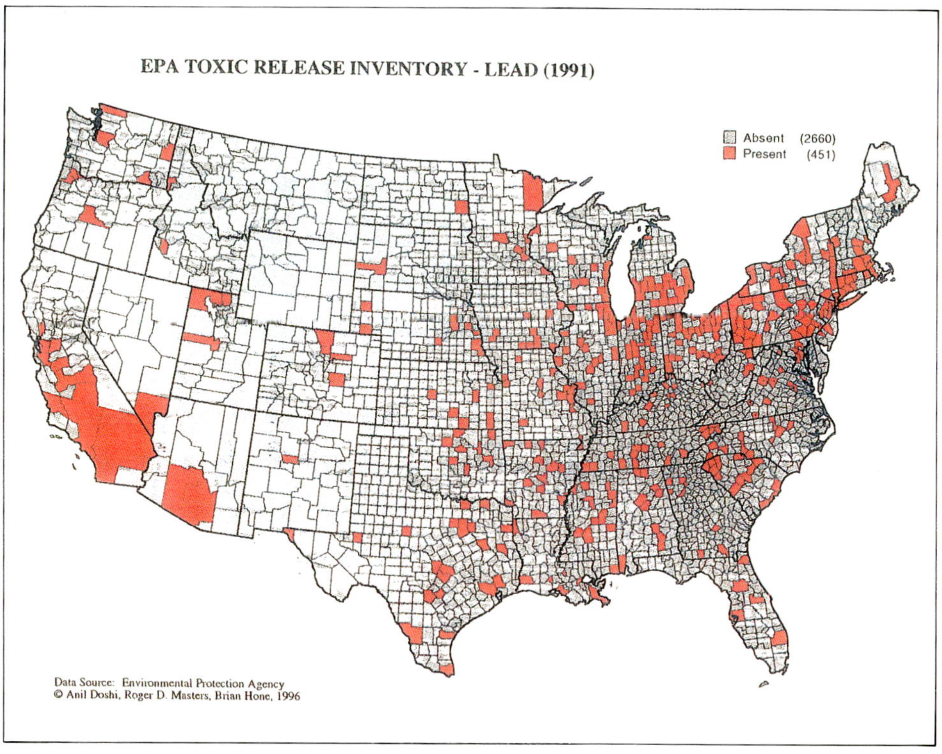

See ROGER D. MASTERS *ET AL.*, Map 2, Page 16.

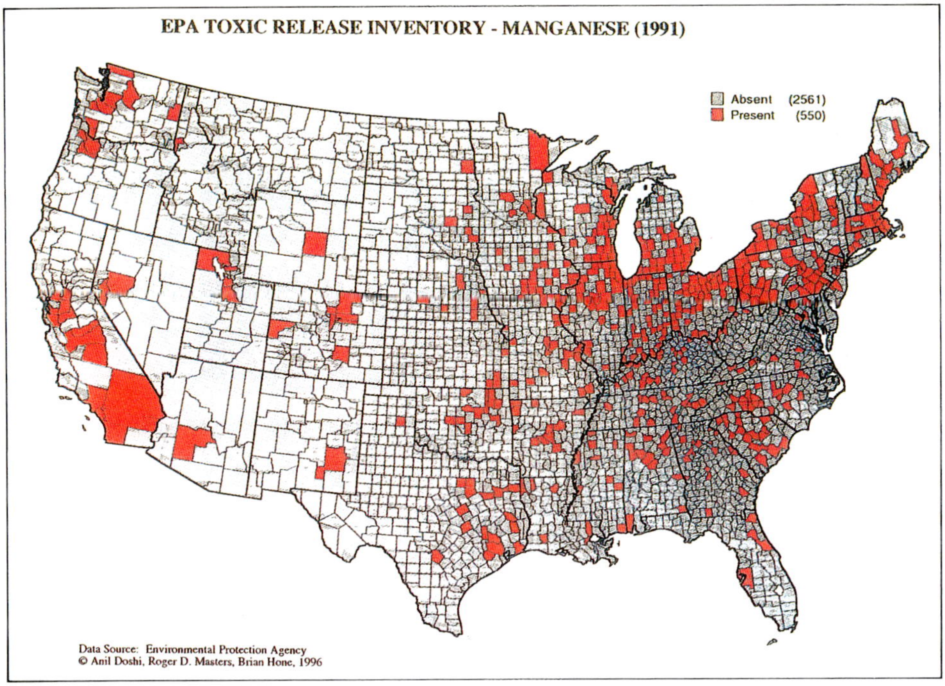

See ROGER D. MASTERS *ET AL.*, Map 3, Page 17.

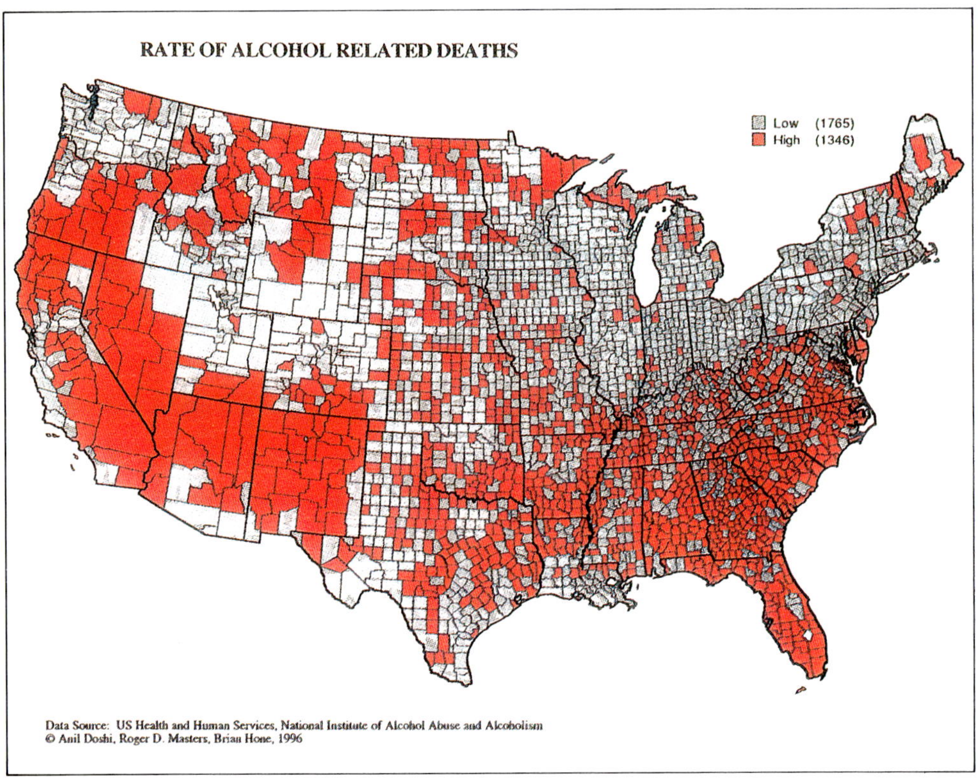

See ROGER D. MASTERS *ET AL.*, Map 4, Page 17.

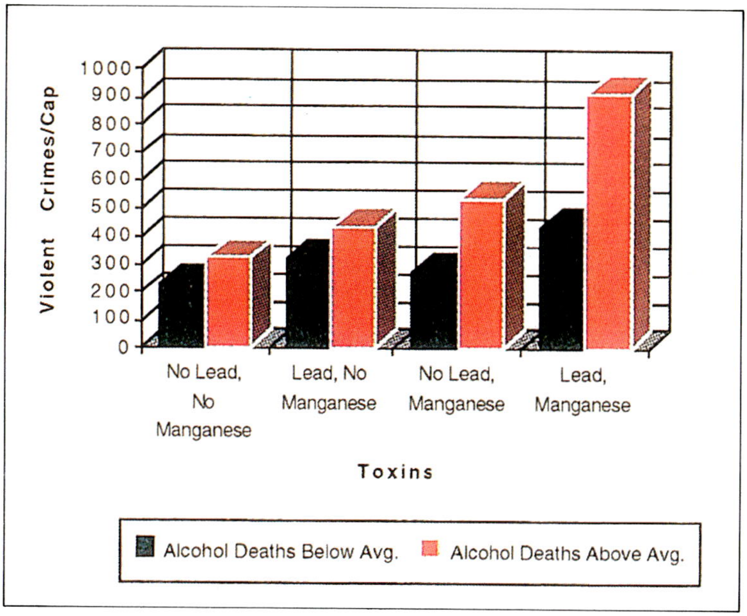

See ROGER D. MASTERS *ET AL.*, Figure 1, Page 18.

See HELENA BJÖRN *ET AL.*, Figure 1, Page 159.